MW01343885

Republic's P-47

THUNDERBOLT

From Seversky to Victory

In Memory of
Lt. Gen. James H. Doolittle
Commander, U.S. Army Eighth Air Force

Republic's P-47
THUNDERBOLT
From Seversky to Victory

WARREN M. BODIE

edited By JEFFREY L. ETHELL

color art work By BOB BOYD

DEDICATION

My dearest, lovely, warm-hearted Catherine Cecelia Bodie will have gone to Heaven almost exactly five long years when this book is ready for distribution. Catherine was very special, for I always found her to be interesting, appealing and loveable for close to a half century. Even with cancer eating away at her life, she looked beautiful at age 61; doctors routinely described her in records as a 50-year-old woman. "Kay" played a significant role in molding my style of non-fiction writing. Therefore, her contribution to the formulation of this book was extraordinary.

I'm certain she would not misunderstand my wish for her to share this Dedication with other deserving souls.

Therefore, it is with the deepest appreciation I also dedicate this work to those daring and talented young men who, a half century or so ago, piloted Thunderbolts into wartime skies to face dedicated enemies or participate in ever-dangerous flight tests.

In that context, every avenue of information has been explored over a quarter century to attain the highest possible level of information accuracy and truthfulness.

*Nihil est veritatis luce dulcius**. – Cicero.

***Nothing is sweeter than the light of truth.**

Published in the United States of America by:
Widewing Publications
Post Office Box 238
Hiawassee, Georgia 30546-0238

Registered Office:
Rt. 1, Box 255 C
Hayesville, NC 28904

ISBN 0-9629359-1-3

Book design by Julie Ethell

Printed by WorldPrint Ltd., Hong Kong

Widewing Publications holds International distribution rights. Motorbooks International/Zenith Books holds North American distribution rights to the Trade.

CONTENTS

FOREWORD

A fading octogenarian memory brings to mind a special event, back yonder, that dovetails with the extensive elaborations of this history of the Thunderbolt. Warren Bodie's book goes to the bone of the previously unresolved historical problem: How did the Seversky Aircraft and Republic Aviation fighters evolve?

In my case, involvement with the P-47 Thunderbolt was lengthy and intensive. It began with graduation from the Army Air Corps Flying School in the dead of that 1937 winter when I was assigned as a Pursuit Pilot in the 36th Fighter Squadron, 8th Pursuit Group, at Langley Field, Virginia. Late in the spring of that year, the squadron commander – Major Ned Schramm – thought it was an opportune time to warm up the squadron's mixture of Curtiss P-6Es and Consolidated PB-2As by taking an orientation cross-country flight with his fledgling aviators. In doing so, he must have reasoned, why not expose them to the gratifications of a night on New York City, landing on a strange, grassy airfield in the remoteness of Long Island and, not least of all, expose his brood to the inner workings of a small but budding aircraft factory. That just happened to be the Seversky Aircraft Company at Farmingdale, L.I., N.Y..

We were briefed before takeoff that the company president, chief aircraft designer and (probably) majority stockholder was a one-legged former Czarist Russian pilot named Alexander P. de Seversky. Somehow or other, Major Schramm and Alex de Seversky were more than just handshaking acquaintances because we made several formation passes at the "cow patch" Seversky Airport before executing pursuit flareout landings. Alighting from our "Peashooter" aircraft, we stood at ease before the props of our individual airplanes. Escorted by Major Schramm, a smiling gent with a slight limp shook our hands and welcomed us to Seversky Aircraft. For de Seversky, himself, to welcome this carefree band of aviators came as something uniquely strange and special. But, why to each of us? That will never be known.

Following those casual introductions, we strode along with a handful of de Seversky's staff into the administrative section of the modest factory. Spread on a temporary sheet-covered table was a catered array of hors d'oeuvres for timely consumption. In due time an escorted tour of the inner sanctum of the factory followed. Work crews busily cut and riveted sheets of aluminum or parts shaped over forming dies. That was something new and different for most of us. Exactly what aircraft were in production at Seversky Aircraft at the time is unclear, but it is probable that the new P-35 pursuit had moved into the manufacturing processes. The facts remained with us that this small aircraft company was a "gung-ho" organization that was destined to go places. We were impressed.

Digressing for a moment to portray the economic realities of the Depression-Recession period, some of us in the Army Air Corps were serving as Flying Cadets on active duty. With a restrictive total authorized force of 1600 officers and 15,000 enlisted men, Congress hadn't gotten around to voting funds to commission a number of us as 2nd Lieutenants in the Reserves. Wearing our blue cadet uniforms, we lived neither as officers nor enlisted men although we served as trained aviation pilots in the first line of national defense. Compensation for such duty brought a basic paycheck of $75.00 per month, or about $17.50 per week. Frugality had been and was continuing as a way of life. By a supreme effort, the inner circles of the Air Corps had eked out funds from a miserly War Department to purchase thirteen Boeing B-17 heavy bombers. At that time they reposed impressively on the flight line of the 2nd Bombardment Group at Langley Field. Rumor had it that the bottom of the barrel was being scraped in an effort to refit the beleaguered "Peashooters" (pilots) with something better than our well-worn P-6s, P-26s and PB-2s.

The passing of five more years would see fantastic strides taken as Seversky Aircraft reorganized as Republic Aviation Corporation and expanded far beyond de Seversky's wildest dreams. In the throes of a grueling World War, this once diminutive aircraft manufacturing company turned out more than 15,300 Thunderbolts that progressed from short-range interceptors to close support and long-range escort fighters. Nearly every theater of operations saw variants of the P-47 series in frontline action, playing a major role in gaining control of the skies. It was the most produced American fighter of World War II.

One of the unassuming young pilots who shook hands with the smiling "Sasha" de Seversky that spring day in 1937 was none other than Lowery "Brab" Brabham, destined to become the Chief Test Pilot and Director of Operations for Republic Aviation Corp. for all of the war years and beyond. Seriously, great doubt exists that Lowery ever visualized that he'd be pushing the throttle forward on anything like a P-47N Thunderbolt that pressed 442 miles per hour at 30,000 feet with a radial engine which generated at least four times the power of his PB-2A's liquid-cooled Curtiss Conqueror engine. Neither could he (or I) have even dreamed of anything like a P-47N fighter that was capable – with external fuel tanks – of transporting its pilot to a destination 2000 miles away without refueling.

Every 36th Fighter Squadron pilot who set down on that Seversky "Cow Patch" that day nearly six long decades ago went on to play far more important roles. Some became Tactical Air Force commanders. Each and every one served at least one tour as a Combat Group Commander. Only the passing of time thinned their ranks as they served their country well.

As they served they gained knowledge and talent via numerous experiences through which their great successes could be properly measured.

With definite enthusiasm, prolific research and extensive investigative coverage of the subject, Warren Bodie provides in this book overviews and insights into the mass of problems to be overcome in developing, producing and utilizing the P-47 Thunderbolt. Showing courage where many an author hesitates to tread, Bodie delves into the various military, governmental and civilian intrigues affecting development and production of a combat fighter. This book overshadows the numerous piecemeal attempts made to tell the Thunderbolt story. With its extensive pictorial coverage and definitive narrative, its readable shelf life will grow as the years pass. These recordings will endure in the passing years.

HUB ZEMKE
Colonel, USAF (Retired)

ACKNOWLEDGEMENTS

Anybody who can write approximately 200,000 words about the history of one airplane type and has been forced to bury a few hundred thousand additional words in the files has done more research than he wants to acknowledge. Those "buried" words are not just space fillers; most are of quality and should be in the book, but I am not seeking bankruptcy. Of course, unlike fiction which rolls out from imaginative brains, non-fiction requires the dedication of untold hours devoted to correspondence, travel, research in often-musty old file rooms and, of course, some real public relations work. The latter is sometimes daunting because no two people will accept your motives in the same light. Some few are found to be iconoclastic, creating suspicion and arguments. Then there are some long-term imbibers of certain beverages; those with dark personalities are a crushing problem. But that all fades to an inconsequential memory in the presence of the receptive, brilliant, warm-hearted people who contributed much to the making of history. Men with nearly total recall and no desire to embellish on their accomplishments, frequently have been found to be imbued with a nearly parallel desire to contribute to the record. They can turn research into joy.

Following publication of my semi-coffee-table book, *The Lockheed P-38 Lightning – "It Goes Like Hell" ... Kelsey,* more frequently referred to as "The Definitive Story," many people came to believe I was heavily biased toward that great WWII fighter. And hardly any readers of the book ever realized I am also WIDEWING PUBLICATIONS chief executive officer. Bias is out. The key to any of these books is a pervasive desire to present the clearest, most thoroughly researched material humanly possible, hardly a simple task. But nobody can ever be certain all bias has been screened out. After all, what was in the mind of your information source? Picture this situation: You are not a pilot but your information sources on the subject of aircraft compressibility are irrefutable. In the domain of a pilot with thousands of hours of world-girdling flight and test flying, he tells you he was the first man to exceed Mach 1 during a dive, in a P-38. You cannot possibly let that go unchallenged, but how do you go about telling him it never happened? It *couldn't* happen. Well, Jimmy Mattern was a gentleman of substance, taking my corrective information gracefully, not throwing me out of his home. Bias can be anywhere, not just with an author. Truth is the real goal.

Actually I wrote an extensive, multi-part magazine series about the P-47 Thunderbolt before moving on to the Lockheed P-38 story, done in a similar vein in the very early 1970s. My Lockheed P-38 series and the P-38 Symposium followed in the middle of the decade, and I authored a soft-bound book on the Thunderbolt at the end of the decade. Strangely, these airplanes were the most evolutionary (P-47) and the most revolutionary (P-38) piston-engined fighters of World War II. Perhaps it is true, opposites attract.

The late General Ben Kelsey had little connection with the P-47 project, so his influence in the beginning was minimal. My respect for the work and perceptiveness of Major Alexander P. de Seversky plus some association with C. Hart Miller put me on the track to the truth about the Major and the Thunderbolt. I soon learned the relationship between Ben Kelsey and M/Gen. Marshall Roth was a strong one, and the Kelsey-Cass Hough partnership was even stronger. Within a short time I was communicating with such luminaries as Hough, Gen. Roth, Col. George Colchagoff, Col. Robert Baseler, Dr. Walter Beckham, L/Gen. William Kepner, Col. John W. Keeler, Col. Jack Oberhansly, Col. Chester Sluder, Col. William Tanner, Alexander Kartveli, Lowery Brabham, Carl Bellinger and Gen. Mark Bradley. Material and informational inputs from all these men varied greatly, but it was substantial.

Evidently I was the first history researcher to contact Hart Miller and Col. Colchagoff despite their lengthy careers and heavy contributions to aviation. That was also true of my contacts with Gen. "Mish" Roth (Ben Kelsey's brother-in-law), Brabham and Bellinger. I visited Carl Bellinger and we seemed to strike up a good friendship. Col. Bob Baseler, former commander of the 325FG in the MTO, became a not infrequent visitor to my home. Of more than passing interest, we were, at the time, both Aircraft Service Engineers. We lost him rather suddenly, it seems.

Contact with the following former military personnel brought in important story and photographic material over a lengthy period of time:

Ray Bowers, Col. Harold E. Comstock, Col. Harry J. Dayhuff, Wayne Dodds, Robert Forrest, Sen. Barry Goldwater, Walter Grabowski, Col. James Harp, Jr., Col. Robert S. Johnson, Col. Fred Le Febre, Donn Madden, Stuart Moak, Charles D. Mohrle, Al Meryman, Robert Powell and Jack Terzian.

The importance of knowing, and gaining intelligence from, Col. Hubert "Hub" Zemke must not be underestimated. At age 80, he is as sharp as a razor. His soft side is heavily masked by a sabre-like wit, tongue and memory. He does not suffer fools gladly, so it helps to have your memory well stocked with facts and fast retrieval capability. Never underestimate the man. His Thunderbolt mounts must have known they had met their match. To have gone through life without knowing Hub Zemke would have been a tremendous loss to anyone.

If there are two more capable aviation history researchers than Frank Strnad and Edward Boss, Jr, I have never met them. After many years of hearing about Frank, we finally met in 1992. As a former Republic Aviation technical instructor, he had gathered more Seversky Aircraft and Republic records and photographs than you could ever envision. His intelligent and friendly wife, Helen, was a pretty secretary in Flight Operations at Republic during the war and afterward, and what Frank does not recall, she does. Frank took me all over the old Republic facilities and even to Major Alex de Seversky's former home on Long Island. Both Strnads seemed to ignore their own medical problems to make my visit ultra-productive. Frank's Seversky photo collection and dozens of P-47 photographs were made available for my use by that truly wonderful team. Great hosts. Add to them the quiet support of "Jim" Boss. He provided unflagging support in research work, always with a gentle smile and self-deprecatory attitude. The "Boss" seemed to quickly detect my seriousness about the project, producing a flood of very important material previously unknown to me. I had never viewed photographs of the P-44-IV mockup; I had never even heard of them, nor had others. A photo of a P-47 landing over the Ranger hangar was like the clang of an alarm bell to me. I knew it was another "special" I had never heard of, not even from Carl Bellinger. It proved to be an early-series P-47D with the complete XP-47J powerplant installed. If there is such an award, Jim Boss is a research hero.

Exciting Jeffrey Ethell and I are forging a new team which should be very upscale in publishing. I constantly caution him to be careful when flying around in P-47Ds, B-17s, Supermarine Spitfire XVIIIs and what-have-you. Daughter Julie, wife Bettie and Jeff took on the task of editing and designing the Thunderbolt book, and they will most likely never look back. Such is the manner in which new careers bloom. In age, we are years apart. However, we think almost as twins. His generous contribution of stunning Kodachrome color pix and information will never be diminished in my memory.

The name Kenn C. Rust on a book for a few decades

guaranteed it to be a well written, superbly researched publication. In a most cooperative spirit, to minimize my research time, Kenn granted me the rights to utilize any and all material from his long-defunct Ninth Air Force book. He has turned completely away from military history, concentrating on civil aircraft of the world. That is an extremely broad-based subject. Thank you many times over, Kenn.

Bob Boyd...not Robert...is a real Georgia Cracker. Constantly "hurting" from an athletic injury, he is a big bundle of good humor and talent. I spotted his artistic talent in 30 seconds. This big Boyd is loaded with it, all labeled "with Southern Hospitality." As you will see in this book, I have done more than praise him. I trust we shall remain friends as long as I, like Timex, keep on ticking. (Again, we are a generation apart.)

Valued contributors of information and data include Richard Sanders Allen, octogenarian engineer Casper Bowman, Walter Boyne, Mario De Marco, Leonard Godbold, William Green (yes, *That* Green) and Gordon Swanborough.

One of the most valuable assets I could have is the mythical Fellowship of Aviation Photographers. In the forefront of that group is none other than Col. Fred Bamberger, one of the longtime bestest. If Fred had not been an Intelligence Officer in the Twelfth Air Force in WWII, you would probably have gone through life without ever seeing air-to-air color photos of brilliantly marked 79FG Jugs at work. The picture selections, taken under adverse circumstances, are the gems collectors spend their lives chasing. Fred and I had lost touch for years, but I treasure the new connection. An old friend of his, Howard Levy, was helpful in coming up with some pictures of uncommon Seversky airplanes. Larry Davis was extremely helpful in providing some excellent color transparencies.

A. Kevin Grantham is really "Jack Armstrong" in disguise. He may be younger than my son, but we are really pals. Kevin and Jackie make me feel like I fathered a second son who married well. Oh, yes. he does excellent photo lab work on my behalf, and I have made certain he knows I appreciate his talent, honesty, intellect, and Jackie's cooking. Dan Hagedorn is such a nice guy, he makes me feel like an ogre. How can anyone be on such an even keel? It was Dan who connected me with Bob Boyd – and for that and the loan of some excellent color and b&w photographs, I will be eternally grateful.

Poor Jake Templeton. It had to be God's work that brought us together less than two years ago. That skinny little bundle of talent survived World War II and the Georgia ANG to gravitate to Hiawassee, GA, after I was magnetized here by Lake Chatuge. We shared common interests, and one day he produced nearly twenty original 4x5 negatives of Georgia National Guard F-47Ns in a moth-eaten photo wallet. Jake had taken them from a B-25 over these North Georgia mountains. "They're yours," he said. All I can say is, look at the endpapers and elsewhere in this book. Wow! Jake took those Jug shots in the 1950s and nobody I know has ever seen the originals in print. Gorgeous. A few weeks later and two months after his wife died, Jake's life came to an abrupt end in that "Storm of the Century" which caught us flatfooted in March 1993. Yes, it was a terrible storm. But the real miracle of this: How did Jake and Bud (Bodie) come together in the North Georgia mountains, after retirement, in this tiny corner (with one traffic light) of the United States? We are all beneficiaries.

At this juncture I have to prove a point which is certain to be debatable. This book has the strong potential for being the last great P-47 Thunderbolt book ever written. Republic Aviation is long gone and Nassau County (Cradle of Aviation sponsor) has no funds at all to catalog company records, publications or photographs, much less make them available to the public in any way. So they remain vulnerable in one or more scattered warehouses. Worse yet, United Technologies, guided by lawyers, has locked all historical material away, essentially unavailable to anyone. There are no staff members at all. The most important cog in the Pratt & Whitney history unit, Harvey Lippencott – a WWII field service rep – retired years ago and was only able to trace a fraction of engine photographs I was seeking from the New England Air Museum. But Harvey did manage to uncover a few rare photos for me, and he provided valuable information relating to operations of the Brazilian squadron in Italy. Contemporary researchers will be hard pressed to obtain anything at all from those and other sources.

In the 1970s, more of my words and pictures relating to the P-47 and Lockheed's P-38 appeared in the twin magazines, *WINGS* and *AIRPOWER*, than same-subject illustrated articles in all other aviation magazines combined. Publisher Joseph Mizrahi must be given credit for backing our common interests, something few publishers in history have ever done. In those "salad" days, Peter Bowers, Walter Boyne and I were strongly encouraged to do some of our best work. Bowers and I were the two veterans of WWII, and the "kid" of the trio was Colonel Walt Boyne, who needs nary a word of introduction. Two ranking aviation historians and one aviation history writer from Ohio were extremely helpful in lending valuable photos for use in this book. Robert Cavanagh and David Menard have been as close to the Air Force Museum Research Group (headed by Charles Worman) as blood cells, and that spans a number of years. Both are paragons of research in their allied fields of interest. Writer Larry Davis has been very helpful by supplying several excellent color slides.

Finally we arrive at the organizations which were of paramount support in providing research access, photographic copy resources in some cases, manpower support and considerable encouragement. Some of the latter was provided just because the sources were there. The Air Force Museum at Wright Field is monumental and inspirational. Up in Research, Charles Worman, Wes Henry and David Menard have been cooperative beyond expectations. Smithsonian's National Air & Space Museum proved to be so helpful to many they far exceeded current and future budgetary provisions. Monetary considerations will impact true historians the most. Almost daily the changes in access can be seen. *There ain't no more free lunch*. What occurs in the remainder of this decade will likely take NASM out of the true educational processes. Historians will suffer; the thriving home entertainment industry will benefit because they can afford the tariffs. Financial and staffing constraints hurt the New England Air Museum and the Glenn H. Curtiss Museums, but dedicated members and employees usually find a way to be of assistance. Robert Stepanek, in my opinion, is one of the finest museum builders of our time. And virtually every bit of his lifeblood contribution to NEAM is on a volunteer basis. The GHC Museum staff is frequently overwhelmed by new requirements, so its research value is severely limited.

A certificate of meritorious service goes to my USAF son, MSgt. Brad L. Bodie, together with a few other perquisites, for logistics support he provided during many months on both my P-38 and P-47 book projects. However, his greatest contribution to me has been computer science training and support, a field he knows well.

Of course somebody should be encouraged to be truthful about other sources of historical lore. Most disappointing from my point of view has been the San Diego Aerospace Museum (SDASM) since the departure of Brewster C. Reynolds. Any expectations about offers of assistance from the American Aviation Historical Society (AAHS) administration for a *real* history project disappear exponentially with the commercial success of the historian. SDASM has become a successful tourist stop. The AAHS just wanders along, hoping to find itself. If so, it may not like what it finds.

If I have forgotten somebody, you know I didn't do it intentionally. What can you expect from a septuagenarian?

INTRODUCTION

California and Washington played such a significant role in producing airplanes during five decades that began with explosive force in 1940 we tend to forget about the major role played by the Northeastern states. In fact the nucleus of America's aviation, aircraft engine and propeller industries was narrowly spread within a 200-mile radius of New York City when war erupted in Europe in September 1939. More than half of the aircraft procured by the U.S. Navy in the decade preceding the massive expansion program inaugurated by President Franklin D. Roosevelt in Fiscal Year 1940 were designed and produced in the Northeast – primarily in New York.

All of the naval fighters and torpedo bombers, except the SBD, together with a third of the Army fighters that fought a global war for America, originated in New York and New England. Although there was hardly a soul in the 48 states on New Year's Day 1940 who would have foreseen that Long Island would spawn such progeny, it was a fact by the end of 1943. Sadly, the great name of Curtiss, an aviation byword for three decades, eroded into nothingness even as two manufacturers born of the Great Depression surged forward with magnificence to fame and some fortune. The names Grumman and Republic had not even been a speck on the horizon when the stock market crashed in 1929, but that all changed as a result of their wartime performances.

For fully a decade after the New York Stock Exchange averages plunged to intolerable depths, Republic Aviation Corporation did not exist. Its beginnings were seeded by a Russian immigrant in the dark year of 1930, even as dozens of leading aircraft manufacturing firms crumbled into dust. Alexander P. de Seversky created the company bearing his name on the basis of a projected airplane design and a handful of patents. By sheer force of his entrepreneurship, the aviation firm known as Seversky Aircraft Corporation grew to significance in the dispiriting decade that followed. While finances were always shaky, they became grievous and threatening in the final year or so of the decade. As a result, the Major – as de Seversky was known far and wide – was overthrown by a dissident stockholders group and the newly created Republic Aviation Corporation supplanted the original Seversky organization.

There is considerable evidence to indicate that the War Department exerted tremendous pressure on Seversky management to oust the Major and get the corporation on track to cope with future growth in the industry. The finances of Seversky Aircraft were such that the organization teetered on the brink of bankruptcy, and financier Paul Moore was about at the end of his patience. His very large investment in the corporation was really in jeopardy; the War Department was not dissatisfied with Seversky products, but they needed a manufacturer who was not in such dire circumstances that a total production shutdown could occur at any time. Planners at the highest levels may have been far more aware of the threats than politics would allow them to reveal even to insiders. In retrospect, for example, it is almost a certainty that technical people at Wright Field looked upon the Seversky AP-4 contender as having better potential as a fighter than the Allison-powered Curtiss XP-40. But Curtiss-Wright was a large and powerful industrial firm; they had great production capacity and tooling while Seversky was ensconced in tiny, antiquated facilities; General H. H. Arnold was, for various reasons, a ferocious advocate of the Allison (and other inline) engine(s); and, almost finally, the influence of General Motors-Allison can hardly be overlooked.

Through an obviously evolutionary process, Republic would create one of the most important weapons in history and accelerate from virtual obscurity to become one of the greatest producers of aircraft in the entire world within a matter of months. That was accomplished without producing simple training aircraft or liaison types any time in the war years. In essence, Republic produced "Ford" or "Chevrolet" workhorse fighters based on a single well-defined concept. (In those days, the world seemed to be almost centered on Fords and Chevies as far as the automotive buying public was concerned.)

Rather curiously, the Seversky and Republic histories have been pre-loaded with inaccuracies, distortions of fact and pure falsehoods over a long period of time for a variety of reasons. Wartime publicity influences are not among the least of these. Witness the unwarranted widespread belief that Republic P-47 Thunderbolts dove to magical supersonic (!) speeds of 725 to 780 miles per hour – not just once but many times. It never happened. For lack of funds and the pressure of all other activities during Depression years, the Seversky model designation system, if it could be termed as such, was anything but definitive. In fact, it would be more logical to term it as hodgepodge. Alex de Seversky was not one to wastefully expend parts and assemblies. Something removed from one airplane would almost certainly appear on another, months or even years later. One single airplane, always bearing the legal four-digit registration number 2106, appeared in so many forms that the true number of configurations may never really be known.

Half a century is a long, long time. For anyone born after the Nazi invasion of Poland, it must be as difficult to envision what life was like in the 1930s and '40s as it would be to have the feel of lifestyles in Civil War days. Your chronicler feels fortunate in

Staff Sergeant Warren M. Bodie in 1944.

having been an indigenous part of history when the Republic P-47 Thunderbolt type first came on the scene in 1941. It may be an illusional concept, but I believe that a momentary return to at least a fractional aspect of the scene immediately before and after the attacks on Pearl Harbor and the Philippines can be helpful in making a brief transitional tour of those terribly trying but nevertheless exciting days. That "fractional aspect" relates to the burgeoning aircraft industry as it attempted to meet the Presidential call for 50,000 airplanes per year and to a road called "America's Highway," officially Route 66. Another affectionate name for the transcontinental (from Chicago) road was *The Mother Road*.

Initially completed – for the most part – in the mid 1920s, the popular Route 66 was undergoing major reconstruction in 1941-42. Perhaps some key people had the foresight to know how important the road would be to America inasmuch as we were certain to be drawn into the growing world conflict. Who could have guessed that major routes like US66 would soon allow new, massive trucks to transport major assemblies – such as complete B-24 center wing sections – from Fort Worth, Texas, and San Diego, California, to the new Ford Willow Run bomber plant in Wayne County, Michigan?

Catherine (Kay) Larson of West Palm Beach, Florida knew Warren Bodie when he was based at Morrison Field, Florida. She later became his wife.

Just about six short weeks after the XP-47B made its initial, unpublicized flight from Farmingdale, Long Island, NY, I graduated from high school. Having been enrolled for the better part of three years in the aeronautical curriculum at Cass Technical High School in Michigan, I had a great desire to get into the rapidly escalating aircraft industry even though I had been awarded one of those rare scholarships to the General Motors Institute. The Willow Run bomber plant did not yet exist and I had no real desire to work for Stinson on liaison aircraft. One large barrier stood in my way at that moment as far as going to work in any defense plant: I would not be 18 years old until July 11.

With the aid of some well-placed people, I had managed to obtain copies of eye charts ostensibly used by the Air Corps and the Navy/Marines in testing for cadet pilot training. It seems that an early childhood accident had seriously impaired the vision of my right eye, but I had a years-long dream of being a career military aviator. How unsophisticated, even naive, one could be in those days even as the eighteenth birthday loomed on the horizon. Both the Navy medical technician and the flight surgeon doing the testing for the newly created Army Air Forces detected my imperfect eye before testing was even completed. I was shown the door in both cases. Well, so much for that prospective career.

July 11 found me in line leading to the employment office at a local aircraft parts and tool & die company. Most of the 100 or more men (no women then) were certainly senior to me, so I did not really expect to be hired in those immediate post-recession days. However, when the employment manager noticed that I was a Cass Tech alumnus with that G.M scholarship, I was employed in less than five minutes. In a matter of days, I was fully involved in inspection of aircraft and machine tool parts. It soon became evident that this fairly large corporation had rather significant contracts with the U.S. Government and the British Purchasing Commission – probably because of the key part that company had in the growing Arsenal of Democracy. Incredibly, here was an 18-year old, suddenly making a phenomenal $60 a week on the midnight shift at a time when a foreman at Republic might have been very fortunate to be making near an equal amount. No more than three months after being hired, I traded my used 1937 Ford sedan in for a brand new 1941 Ford V-8 convertible coupe! Total price for that "spiffy" car was $1077, taxes and accessories included. It was near the end of the 1941 model year, so I was able to extract a trade-in value of $302 for the car I had purchased only 8 or 9 months earlier for $260. Much to my joy, the 1942 Ford and Mercury cars that came out only weeks later were not nearly as good looking as the 1941 models (but I was upset with myself for not spending an extra hundred dollars or so to buy the Mercury instead of the Ford). No parental assistance was asked or received (or needed).

For the record, in those immediate post-Depression/Recession days it was most uncommon for high school students, other than those with wealthy parents, to own anything better than a 1936 Ford or Chevrolet.

Theater newsreels depicting the destructive air raids on England soon ate at my soul. A month before December 7, I drove my new car across the Ambassador Bridge to Windsor, Ontario, in Canada and made a really serious attempt to join the RCAF for pilot training. I knew they did not give depth perception tests, so I believed I had a fighting chance to pass the eye examination (by cheating, of course). That 20/400 vision limit on my right eye did not fool their eye doctor one bit. No, I did not wish to enlist as an aircraft or engine mechanic in the RCAF. I had other ambitions. So, somewhat chastened, I beat a hasty retreat back to the United States via that international bridge across the Detroit River. It was quite evident that I was never going to be a fighter pilot. Period.

After working the midnight shift one Sunday morning – we had been the only department singled out for 7-day per week overtime in December – in near freezing temperatures because a plant expansion left only tarpaulins between us and the outdoors, I went home and fell asleep in a very comfortable chair. A couple of hours passed before I awoke. Turning on the new "No Stoop, No Squat, No Squint" Philco radio (no television in 1941), I found the airwaves jammed with reports of the Japanese attack on Pearl Harbor. Having been an avid reader of Buzz Benson stories in numerous issues of a pulp magazine (*Flying Aces*) and a believer in an inevitable attack on the U.S. by Japan since grade school days, I found that I was not taken by surprise. (The Benson stories always centered on a prophetic war situation with the Japanese Navy, with "Buzz" saving America – or something to that effect.) However, I was absolutely aware that my world, like that of millions of others, was destined to be turned upside down. That valuable scholarship

Having failed eye tests for RCAF, USAAF and USN flight training, Warren Bodie was heading west on icy Route 66, near Flagstaff, Arizona, in February 1942.

to the General Motors Institute was essentially gone with the wind.

Because 18-year-olds were not yet required to register for the draft, together with the inevitability of hundreds of thousands of young men rushing to enlist, I chose to wait for a while before making my move. But a small group of co-workers did decide to head for California with the idea of getting into airplane manufacturing. Within a few weeks that group had dwindled to one co-worker and me. Jack Dougherty and I quit our rather lucrative jobs in the chaos and excitement of the time, packed our luggage into that new Ford convertible and headed for the magic attraction of the Golden State.

U.S. Route 66 was mapped out for us by the AAA from Chicago to all points west. Of course it was mid winter and only natural that we would encounter ice-coated roads as we neared the Illinois state line. At that moment we knew little about Route 66 or its historical status, and we had no idea at all that we were really pioneering in a small way. The decade and a half that slowly passed since U.S. 66 was christened had – because of the Great Depression and subsequent Recession – progressed only moderately from single-lane dirt road status to mostly two-lane asphalt paving. Much of that transition had taken place only after Franklin D. Roosevelt created several massive programs aimed at dragging America out of the worst financial crisis in its history. As the country moved into a wartime economy, we plunged unknowingly into many miles of the transcontinental highway which had been reduced to corduroy dirt surface under the impetus of huge work programs sponsored by the federal government. Believe me, with the country still reeling under Japanese Navy and Army assaults and largely uncoordinated attempts to consolidate a disorganized, unprepared nation, this youthful pair was pioneering. Most of middle America was still functioning as if it was still 1940. Only on the West Coast and the East Coast was there any real wartime atmosphere. We felt like we were in a time vacuum.

For a youngster who had never ventured further west than Kalamazoo, Michigan, territory on the far side of the Mississippi River seemed terribly strange and remote. Midwinter road traffic was virtually nil as we passed southward from Springfield, – or was it Carthage? – Missouri. Out on the open highway we were soon aware of a new Ford sedan that looked rather official. Indeed, it was occupied by a pair of state troopers wearing campaign hats, and a short blast of the siren caused me to halt rather quickly. Heck, we hadn't been doing anything wrong; as it turned out, the troopers were merely curious about what two "kids" who appeared to be under the age of sixteen were doing in their state with a brand new convertible bearing Michigan license plates. It did not take long to prove we were not car thieves and we were soon moving west once more. But we could almost read their minds as they turned and drove away: "Damned rich kids!" They were almost right in a way. While neither Dougherty or I came from homes that could be considered as anything above bottom middle class, we had, in recent months, probably each been making more money in ten days than either patrolman was making in a month. After all, this was farm country and the impact of the "Arsenal of Democracy" economy was essentially missing except as a news item.

After crossing into Oklahoma, not far west of Joplin, Missouri, we found ourselves on a new concrete strip that, in retrospect, might have been the Will Rogers Turnpike. (Uncertainty now prevails because U.S. 66 still exists a bit west of the present turnpike.) We were suddenly brought bolt upright when the concrete ended abruptly, with absolutely no warning, out in the middle of nowhere. Quick reaction kept us from flying off the end of the paving. A minuscule sign diverted us to a pot-holed asphalt road, allowing us to proceed into territory that was strange appearing and remote to the Nth degree. Not a building or a sign appeared anywhere on that treeless plain.

When the "Okies" loaded every useful possession onto their already ancient cars – vehicles that were long since as tired as their passengers had become in the terrible drought years – for the trek west on Route 66, this colorless area seemed to have lost all vestiges of hope.

Amarillo, Texas was more exciting, but the lasting impression was hardly on the positive side of the ledger. It was cold, grey and miserable in a small city without much character. We roosted overnight in a downtown motel, heading off at 15 to 20 mph the next morning on sheer ice in what can only be described as zero-zero fog. It went on for miles, peppered with several wrecked or ditched and disabled cars. I admit to being afraid...for my new car. Many miles later we descended on that long grade into a more friendly looking Albuquerque, New Mexico. This surely was a new and different world for two "tenderfoot" types from Michigan.

Construction progress on the "mother road" had been dictated more by weather and slim finances than by the traffic load. And, although America was now at war, not a single military vehicle was ever seen by this wandering pair. Here were two young men traveling a route that became famous nearly a quarter century later when "Tod" and "Buz" rode a series of Chevrolet Corvettes to fame and some fortune in the TV series, "Route 66." That's twenty-five long years; a lifetime for many. In February, 1942, U.S. 66 was essentially "Main Street" in Albuquerque, being the main drag right through downtown. It might well have been called Motel Lane. We stayed overnight in one of those pseudo-adobe flat-topped buildings.

Never in our short lives had we seen straight highways that disappeared over the distant rises as a narrow pencil line. I have never been able to define the feeling I had on arriving in Gallup, New Mexico, in midwinter. Desolate, cold, drab, poor, depressing – all of those words applied, and then some. Could we have lost our way and stumbled into Siberia? Or it might have been

"Lower Slobovia" from one of Al Capp's cartoon strips. I recall thinking, "My gawd, why would anyone want to live here?" Dougherty and I had taken some pictures with my 35mm Argus A2F camera not long after leaving Albuquerque, the last "big" city we were going to see for many, many hours. And a note was made the Ford V-8 was averaging 20.9 miles per gallon of gas. The highest price we paid for fuel during the entire 2700-mile venture was a hefty 22 cents per gallon for Ethyl (premium grade). I can't recall using anything but Ethyl or what the lowest price was that I paid, but in California the price for Texaco Sky Chief (premium) was a memorable 14 cents. (Rationing had not yet been imposed. In fact, many months passed before ration stamps were issued to limit driving.)

We arrived in Flagstaff, Arizona, late in the evening on one icy cold weekday. Despite the late hour we were able to rent a cabin, a standard unit in any typical motel or, more frequently, motor court in that era. How cold was it? So cold that some 50 years later it still has the aura of fantasy. Even though we left the heat turned on (I presume) we arose in the early morning to find the water in the toilet bowl frozen! That is the truth! Fortunately for us the Ford started on the first try. Equally fortunate, I had somehow learned to use the relatively new Prestone anti-freeze product when I was only sixteen.

Heading west, once more, in brilliant sunshine that reflected blindingly off the pure white snow, we hardly made contact with pavement at all. We were driving on glare ice. No chains aboard and no place to rent or buy them. By good fortune, I had learned to drive in midwinter, in Detroit, on ice and snow. To say I was not nervous would be a lie.

Memories: We rounded an ice-coated curve on the mountain road in that high country to be confronted by a very big 18-wheel semi-trailer truck athwart the roadway. The driver was faced with a monstrous problem: the rear end of the trailer was against the face of the mountain where the highway had been carved out, while the front wheels of the tractor were dangling over the edge of the shoulder. Help eventually arrived and we were able to proceed after a lengthy delay. Very cautiously to say the least.

Another memory, hardly exciting but just as vivid, was the road paving seen frequently in Arizona, especially on the stretch (as I recall) near the way station named Seligman. It was not black asphalt or grey concrete. It was a reddish-maroon sort of surfacing, never before seen by Jack or myself. Wherever the aggregates came from, they must surely have been plentiful.

Before we arrived in Kingman, I recall there were stretches that gave a wild, sort of roller coaster effect as we cruised rapidly into the many dips designed into the road to allow passage of water from rare flash floods. Mountainous stretches were flush with sharp curves and switchbacks offering shoulders (sometimes) that could not have been more than a foot wide. Beyond that was great nothingness. Just a long drop off to oblivion. In those great stretches west of the Mississippi River we had encountered detours, narrow railroad underpasses with blind approaches, and more corduroy-surface dirt than we care to recall. Worst of all was the nighttime driving in desolate areas where we looked forward to hearing some good music on the radio to help keep us alert. What we wanted was Glenn Miller, Tommy Dorsey, Jimmy Dorsey or Woody Herman. What we soon learned, way out there, was that you could get about two stations and that was that. It was either Guy Lombardo (not a favorite) from Chicago or something much worse out of a Godforsaken place being touted as Del Rio, Texas. It was so depressing we preferred silence.

In and around Flagstaff we were at the highest altitude that I had ever been exposed to in my life, even in an airplane. The high point had been somewhere around 7000 feet. Kingman was so tiny at the beginning of WWII I don't even recall stopping there. If we did, it was probably for food and gasoline.

Needles, California, came up all too suddenly. Here was a town where temperatures of more than 100 degrees F. can be encountered at midnight. Heading across that vast, nearly featureless Mojave Desert was bad enough in wintertime; it was surely murderous in August. Gasoline and water stops like Amboy appeared as oases on the gritty sand. One lasting impression: the shoulders of the road across that wasteland, and out into the desert itself, were dotted with discarded tires and the rotting hulks of ancient cars that failed in their attempts at crossing.

Barstow, San Bernardino and Pomona faded behind us as we gazed admiringly at unfamiliar palm trees backlit by a beautiful red sky. Tomorrow was going to be wonderful. And, yes, the top had been powered down many, many miles before we had reached Foothill Blvd. In the dark, in Hollywood, we eventually found our way onto the new (under construction) Cahuenga Pass Freeway, that last part being a totally new word in our vocabulary. In later years, it became a part of the Hollywood or Ventura Freeway systems. It ended around Barham Blvd. then, and we soon found we were on Ventura Blvd. We checked into a motel for a week, I think, and the next morning we had a meal (lunch) at Bodie's Pago Pago Restaurant (no connection at all with the chronicler, but a rare coincidence).

Our 2700-mile transmigration came to a logical conclusion with a leisurely cruise down Santa Monica Blvd. to Pacific Highway and the ocean beaches. In retrospect, that was just something less than sixteen years after Route 66 was christened. Many improvements and changes would take place in the next 42 years before U.S.66 was relegated to the proverbial scrap heap by decertification in 1984.

As production of Republic P-47Bs became a reality on the East Coast, my career in aviation was into its initial phase on the West Coast, something over 3000 air miles away. When 1st Lt. Ben Kelsey flew the Lockheed XP-38 from California's March Field to New York's Mitchel Field in just a tick over 7 hours (flight time) exactly three years earlier, he had recorded the initial phase of the 400-mph fighter plane as equipment for the air arm of the United States. In actual fact, for the entire world.

Certainly those were exciting days, so is it any surprise that the wondrous Route 66 left a most indelible impression on this young, aviation-minded person – and most likely hundreds of thousands of others. Besides being known as the *Mother Road*, it was also named the *Main Street of America*. As musician Bobby Troup's song goes, "Get your kicks on Route 66."

Yes sir, that's the way it was – at least as Americans shook off lethargy and shock so that they could get going on the road to winning a major war.

A good storyteller is a person who has a good memory and hopes other people haven't. – I. Cobb.

This chronicle will make the most clearly defined attempt of this century to clarify and correct the Seversky and Republic stories published over several decades. Many of the histories were based on inadequately documented Seversky stories that appeared in abundance in the 1930s. Even worse, most public relations releases from (or for) Republic Aviation were distorted to purge Major Alexander P. de Seversky from any actual relationship with the P-47 Thunderbolt, primarily because of some bitter litigation that resulted after his expulsion from corporate management. An overabundance of in-house documents were loaded with such inaccuracies as to be comical. For example, a new company journal featured a cover picture of a formation of Air Corps P-35 fighters and had the gall to caption them as Republic P-35s. There has never been one shred of evidence to support such a statement. But it was indicative of many such imperfect statements to be issued for years. As this narrative is being written, a British magazine has published a feature story about Seversky aircraft. Written by a well-known American photo-

Warren Bodie when based at the Rattlesnake Bomber Base in Pyote, Texas, in 1943. He was in the Air Service Group 3rd Echelon.

journalist, it is peppered with many of the oft-repeated errors that sorely need to be buried forever. This P-47 history book is intended to reveal the truth, so – hopefully – it will be read by those who persist in avoiding research like the plague.

Truth crushed to earth shall rise again. – Bryant

To this stated end, key personages involved have been interviewed over a period of at least 25 years. Their words and personal documentation have served as a reservoir of provable facts. Resources (deep pockets) of the Cradle of Aviation Museum have been tapped. Surviving and long-gone engineers, pilots, executives, technicians, instructors, military personnel and respected historians have contributed a wealth of material and personal remembrances with enthusiasm.

Anyone who did not live in the decades of the 1930s and 1940s really lacks any factual understanding of what that period in history was like. Distorted views of the way people talked and acted frequently show up in motion picture films. A good example of that appeared in the movie "Memphis Belle." The language and actions were caricatures of life in those days. Many of us who survived the Depression and World War II have something akin to total recall, most certainly closer to the real thing than a script concocted by a 30 or 40-year-old "whiz kid." Even for those of advanced age it is mandatory that documentary records be studied with diligence, and if things fail to match up care must be taken to uncover corroborative evidence.

No one who had personal relationships with those decades wishes to see future students of history immersed in a sea of distortion. It is to them that we are deeply indebted.

Truth is the highest thing that man may keep. – Chaucer.

Catherine Larson Bodie with author's brand new 1947 Buick Super Convertible. (A rare commodity in the Autumn of 1947.) "Kay had great legs," Bodie notes,"also a valuable commodity."

PREFACE

In the first two decades after hostilities came to an end with the total sublimation of Germany and ultimately Japan, there was barely the slightest interest in publishing books about WWII aviation. Only in the 1970s came a real resurgence of interest in the airplanes of that momentous war. To this day, in fact, automotive books outsell aviation subjects by a very wide margin. In an era when idiotic and even badly acted plotless films can attract hundreds of thousands of people to part with their money, it is almost impossible to sell 10,000 brilliantly produced aviation books.

Part of the problem lies in the reality that only a fraction of the world population reads non-fiction books. In fact, even as the population multiplies newspaper readership declines. Well, just as you can't force a horse to drink water, there is little you can do to convince people it is imperative they read non-fiction books about aviation history, if for no other reason than to expand their basic understanding of world events.

Three important factors on the horizon do not bode well for the future of accurate, incisive authorship. If sales of beautifully illustrated, well researched books did not increase dramatically in the fiftieth anniversary years of World War II events, there will likely be little incentive to write and publish them in the next four decades. Secondly, good researchers have delved so deeply into the subjects at the half century point the trees of information have been stripped bare, especially true of the photographic coverage. Definite limits exist on the practicality of uncovering fresh photographs that have not been viewed numerous times, although long-forgotten material does surface from time to time. In just one more decade, logic tells us, most WWII participants will have departed the worldly scene or will be unable to accurately recount wartime events and experiences. As an old saying goes, don't chew your cabbage twice (or thrice) and expect to get an audience.

And (almost) finally, most WWII manufacturers of the predominantly important combat aircraft have now either disappeared from the business world or have been subordinated within conglomerates that see no sales value in aircraft produced 50 years ago. With the wartime designers and engineers long gone for the most part, even now, who speaks for the accuracy and importance of past history? Current competitive pressures and technology dominate the scene where profits are the issue, so history becomes an issue of minimal interest to management.

What about the impact of such thinking? Literally thousands of valuable documents and photographs have been destroyed or otherwise disposed of without conscience. Can anyone believe Republic Aviation Corporation – or what remained of that proud manufacturer – only had three or four color slides and transparencies taken during five years of war, the total of what has been passed along to local and national museums. Did Republic's public relations people send up six or eight fighters and a photographic airplane many times just to take two or three pictures? It would be more logical to believe that Elvis Presley still lives. When Fairchild acquired all of Republic Aviation's assets, did any executive care what a P-47 Thunderbolt was? Events would indicate the response would be a resounding NO!

To clearly illustrate just how deep such disinterest was and is, beginning immediately after VJ-Day, not a single manufacturer of the top WWII AAF fighters invested even a measly fee of $1200 to purchase one brand new example of their great money-making, world-beating fighters from the War Assets Administration. It seems heritage had no place in their vocabularies. Had anyone in charge learned any lessons at all from World War I, they would have known at the very least it would have been a good investment for the future, a memorial to workers and the military personnel, an invaluable asset. At that time, nobody could purchase even the cheapest new car for just $1200. Profit motive, it has been demonstrated, will pulverize heritage every time.

As a veteran of WWII, fathered by a veteran of World Wars I and II (plus the Mexican Border Campaign), I have concluded both profit and heritage warrant an equal place in my life. Therefore, this book and its predecessor, *The Lockheed P-38 Lightning – The Definitive Story*, have garnered my complete attention and dedication to excellence – at least to the best of my limited abilities. Hopefully, those dedicated efforts plus a massive investment of time and capital (truly, risk capital) will be regarded by a virtual handful of people as a worthwhile memorial to all who contributed so much of themselves in that major confrontation to overcome those who strove to dominate free world nations. In the winter of my life, I will feel fulfilled.

CHAPTER 1

DE SEVERSKY, FOREFATHER OF THE P-47 THUNDERBOLT

Which World War II American Fighter was the very best?

That is a question which raises the ire of many, although much of it is usually hidden behind a smiling face.

Can anybody in the world actually provide an irrefutable answer to that question which has refused to go away in more than a half century? Can existing written history be a substantive basis for such an answer? Most certainly not. It is more than slightly evident that in the initial postwar rush to capitalize on any interest in World War II aviation history, research in depth was hardly necessary. Newspaper and magazine articles were in abundant supply, and every person who flew a fighter in combat, worked on such aircraft in the factories or maintained them in the field was more than happy to relate his personal stories upon receiving the slightest encouragement. But once the initial rush was a thing of the past, interest in such subjects faded as new families were created, veterans were striving to expand their knowledge and capabilities in college, and the desire for home ownership soared over most other tangible items on the "want" list.

A significant increase in leisure time came into the picture for Americans in the late 1960s, leading to a renewal of the desire to learn more about World War II and the weapons employed in that major conflict. A massive growth in model building interests was accompanied by many technological improvements in kits just as the so-called "Baby Boomer" generation reached a stage where the resources were available to invest substantially in exciting hobbies. History was suddenly back in fashion. The growth of the Experimental Aircraft Association (EAA) became a phenomenon of the age,

It may be hard to believe, but there can be no doubt that the genesis of the Thunderbolt had its beginnings in Czarist Russia. Pictured in the back row, second from the right, the boy Alexander Nicolaiovitch Procofieff-Seversky was in pre-1914 military school where he gained his grounding in aircraft engineering. In adulthood he most certainly was perfectionist, overachiever and entrepreneur par excellence, and he would not suffer fools gladly. His sharp tongue and even sharper pen gained him scores of friends, perhaps more enemies. It is an absolute certainty that without Alexander P. de Seversky there would never have been a P-47 Thunderbolt. **(C.A.M.)**

***Major de Seversky, evidently overly anxious to respond to the mid-1930s competition for the Army Air Corps' first basic trainer, flew the demonstrator SEV-3 amphibian with a 420-hp Wright R-975 (J-6-9E) engine to Dayton. Of course the AAC did not want an amphibian and the crew configuration was all wrong. But Air Materiel Division assigned Project Number XP-944 (one of the last) to the airplane. The P-number series dated back to WWI at McCook Field, predecessor of Wright Field.* (WF/AAF)**

almost in lockstep with a surge of interest in warbirds, those surviving WWII aircraft hulks that were allowed to deteriorate for decades at airports and other less visible sites. Suddenly, in the 1970s, their value began to escalate exponentially, and once valueless aircraft became the targets of tireless trackers.

Restoration projects, recovery efforts and entire new industries surged ever upward to supply a boundless market with "Warbirds." That appellation has even embraced such once-denigrated types as Cessna UC-78s and Brewster Buffaloes. With all this came a real desire to learn all about those relics of a global conflict without equal. The route to recording and publishing such history, in large part, was by way of that early postwar material that was based on the weakest of research tools, not the purest of documentation, photographic and written evidence, and probing interviews with the kingpins of aviation in the prewar and wartime days.

With the passage of time, fading of memories and the loss of the most reliable and informative documentation, has come a new and disturbing pastime of controversy. Any gathering involving people ranging from the best combat pilots in the war to the newest neophyte history buff now leads to heated arguments about "Which Was the Best Fighter of WWII?" or "Which Was the Best American Fighter of the War?" Add to the unending arguments about religion or politics or sports an ongoing debate that can never be resolved – that "greatest" fighter debate.

Common sense dictates that we disregard the subject of which fighter might have been the all-time best in the world simply because patriotism and national pride obliterate any logical conclusions. In fact, logic and defined facts seem to be the least recognizable cards played in such debates. Concentration then homes in on the five most recognizable U.S.-built aircraft that could be contenders for a quasi title of "Best American Fighter Plane." That is the easy part. The contenders have to be the Grumman Hellcat, Lockheed Lightning, North American Mustang, Republic Thunderbolt and Vought Corsair (in alphabetical order of course).

With one outstanding exception, all of these aircraft started with a so-called clean sheet of paper. Therefore, the *best* written histories that have been published relative to the five types provide reasonably comprehensive, accurate and reliable historical paths to the ultimate truth. Unexpectedly, the least innovative and minimally advanced fighter type in the group of five is the one that has left the most incomplete and inaccurate "paper trail" of evidence relating to its origins and conception. That airplane is the Republic P-47 Thunderbolt, a fighter that could do battle in any war theater against anything the enemy was able to send against it in combat. Few of the world's most outstanding fighter aircraft had that capability.

Unlike the curvaceous and charismatic Lockheed XP-38 that burst forth on a totally unsuspecting public with a record-setting transcontinental flight when world peace still prevailed, the Republic XP-47B prototype moved into the limelight amid a flood of derisive comments about its bulk. The British press was especially unflattering when comparing the petite, svelt lines of their outstanding performer, the Supermarine Spitfire, with the gargantuan "avoirdupois" built into the Thunderbolt. After all, a mere "handful" of highly maneuverable Spitfires and Hawker Hurricanes had unilaterally defeated a Nazi *Luftwaffe* onslaught, the likes of which had never been observed in history. Could there be a more substantive display of attributes? Not only that, it could justifiably be claimed that it was the world's first major battle in which infantry, naval vessels, tanks and artillery played no significant role in defining the victor and the loser. It was almost completely a battle of aircraft against aircraft, pilot against pilot.

Without any portfolio of recorded accomplishments to divulge, this new U.S. Army Air Corps P-47 interceptor-fighter had been born and nurtured in one of the then smallest, oldest and least profitable plants committed to construction of military aircraft for any major world power. (As the British viewed it, the new American fighter was a competitor of sorts with their beloved Spitfire, but it was as vast as their badly mauled Fairey Battle bomber or the largely ineffective twin-engine Bristol Blenheim bomber. After the Battle of France and the Low Countries, the RAF's Battle simply disappeared from any combat role, relegated into oblivion, and the Blenheim type was not far behind.)

At the time of its conception in 1940, Republic's XP-47B was the product of an industry centered to a great extent in New York state and New England. In spite of oratory to the contrary, master plans for decentralization of the aircraft, engine and propeller industries had remained largely ignored until the Nazi war machine swept like a powerful scythe around the highly-touted Maginot Line of fortifications. France and Britain had launched an unbridled spending spree in the spring of 1940, attempting to buy thousands of tons of non-existent airplanes, engines, military vehicles and guns. Although temporarily distracted by Adolf Hitler's apparent preoccupation with North Africa, Balkan countries and Norway, the Allies instinctively knew that the so-called "Phony War" (a period of buildup and concentration on matters of more immediate concern) presaged an eventual assault of some kind on France. Any student of continental European history could have seen what was coming, even if the form it was to take could not be foretold.

Just to familiarize the reader with the prevailing situation in America on that day in 1939 when the *Wehrmacht* and *Luftwaffe* swept across the Polish Border and introduced the world at large to *Blitzkrieg* (Lightning War), major military aircraft builders in New York-New Jersey-Connecticut were:

Bell Aircraft
Brewster Corporation
Curtiss Airplane Division of C-W
Grumman Aircraft
Seversky Aircraft (subs.,Republic Aviation)
Vought-Sikorsky Aircraft Div. of UAC

(Not very far to the south was Martin Aircraft in Baltimore, Maryland.)

Concentrated aircraft engine and propeller manufacturers in the same region included:

Curtiss Propeller Div. of C-W
Wright Aeronautical Div. of C-W
Hamilton-Standard Propeller Div. of UAC
Pratt & Whitney Aircraft Engine Div. of UAC

The remaining bulk of the aircraft industry was, of course, located along what was then the rather remote West Coast region. Remote?! California, Washington, et al? Absolutely. Travel to a European destination was hardly slower. Even with expensive airline transport, that West Coast was at least 13 hours away from the Eastern Seaboard. By car, bus or train (the fastest of that trio) it was still more than a 48-hour trek. A civilized trip by car from Los Angeles to New York City was about a five-day venture. Picture this: the most important highway from the midwest (Chicago or St. Louis) was US Route 66. Most sections of that federal highway consisted of asphalt two-lane paving; very little in the way of concrete highway had even been partially completed. Many sections of road in Oklahoma, Texas and other states were in the form of corduroy dirt as construction progressed. Of course there was the Lincoln Highway headed off in the general direction of San Francisco, but you really had to be a pioneer to head west on that from Chicago. America did not have anything approaching a German Autobahn traversing the nation at that time. The new, relatively short Pennsylvania Turnpike was probably the only comparable roadway in America.

Republic Aviation, successor to the Seversky Aircraft Corporation at Farmingdale, Long Island, NY, was considered so unimportant in 1939-40 during a transitional move from the verge of bankruptcy that neither France nor England would purchase any Seversky or Republic product on the eve of World War II or even as the Battle of France was about to begin. Sales attempts launched by Seversky throughout Europe early in 1939 as heavy war clouds gathered had only resulted in some fairly substantial sales to Sweden, a nation that remained neutral and essentially unthreatened throughout the war. France, incredibly, had fallen from a fairly secure position as an aviation power leader to the status of an obsolete, floundering giant under the Socialist government that rose to political leadership during the

***With Major de Seversky at the controls, the Seversky SEV-3L landplane is shown in flight in the Dayton, Ohio, area, probably in September 1934. While you cannot see the bronze color paint job, which identified the SEV-3 airplane, your attention is drawn to that oh so familiar wing planform that no German was happy to see some 10 years later. That wing was the brainchild of the Major and his valued assistant (and chief engineer) Michael Gregor. It had been designed before Alexander Kartveli was hired as Gregor's assistant. That is historical accuracy.* (AAC via S. Hudek)**

***Dashing back to Farmingdale in X-1260, the Major had "Pop" Provo's team remove the amphibious landing gear and, for the first time, install the interchangeable panted fixed landing gear. Then it was back to Dayton for additional negotiations. A speed of 200 mph was attainable with only 420 horsepower, meaning that this landplane (redesignated SEV-3L according to Executive Committee notes) could outspeed all Air Corps pursuits except for the latest Boeing P-26 type and Consolidated PB-2As.* (WF/AAF)**

1930s. Of all countries facing a desperate shortage of warplanes of the types that could even offer a potential for combatting the new German Air Force, France was obviously in the position of being the most hopelessly unprepared. The hidebound military hierarchy was composed almost entirely of over-age generals so comfortable in their positions of power that they had failed to even study the role that aviation was about to play. Their intelligence agency had either been disbanded, ignored or compromised.

Hiding behind a facade of fearful false pride and misguided by a political organization that could not even recognize its own overwhelming lack of perspective and technical knowledge, France would not even consider purchasing the rights to manufacture the superior Rolls-Royce Merlin engine from its British ally. The idea of displacing the weaker and less reliable Hispano-Suiza and Lorraine-Dietrich engines or at least giving designers a powerful alternative unit was repugnant to the Frenchmen. When a readily available Seversky fighter (a development of the first-line U.S. Army Air Corps P-35) that was at least the equal of the Curtiss Model 75 Hawk was offered to the *Armee de l'Air* early in 1939, they much preferred to figuratively turn their collective back and limp along as before. Even as their decades-old war machine concept faltered, then fell apart some sixteen months later and their purchasing commission literally swamped the Arsenal of Democracy with last-second orders, they steadfastly refused to even consider buying from Seversky *nee* Republic. And so it goes.

Curtiss H.75C Hawks, purchased in some quantity by the French in 1938 before America joined the fray, played an almost insignificant role in the defense of France and the Low Countries in the spring of 1940. America had placed its security confidence in the Army Air Corps P-36A version of that same type pursuit (fighter) aircraft when they ordered no fewer than 210 of them in 1937. The order constituted the largest peacetime procurement of fighter-type aircraft ever placed in the U.S.A. to that time. In that same era, Boeing's Model 299X prototype "Flying Fortress" had outperformed the Douglas DB-1 (XB-18) only to have Boeing executives see the performance ignored in the award of a huge contract to Douglas Aircraft, a tragic, politically motivated action. Seversky Aircraft's AP-2 entry in the 1937 Pursuit Airplane Competition had technically outpointed the Curtiss entry, but the power of the Curtiss-Wright Corporation and a growing animosity against Major Alexander P. de Seversky within the War Department negated any performance advantage. Now, there was a mouthful of name to get your attention. It sounded vaguely French, strongly Russian, but what did he have to do with the American aircraft manufacturing industry? Possibly, quite probably, to the same extent that the 1920s firebrand, General "Billy" Mitchell had been involved about fifteen years earlier.

If we have learned nothing else in life, we have all seen that many of our greatest inventors, soldiers, politicians, engineers, industrialists, entrepreneurs and what have you could hardly be categorized as patient, politically conscious, tactful diplomats.

Quite the contrary. They are usually outspoken, impatient, quick-thinking – and they were unlikely to "suffer fools gladly." More often than not they have been targets of slanderous statements, scathing false accusations, and sometimes punitive actions on the part of their opponents. Firebrand names that come to mind include Lindbergh, Chennault, Patton, Mitchell, Gen. Frank Andrews and, in the 1990s, Perot.

The Seversky Aircraft Corp. was based in New York City from 1931, but in 1934 the vacated 1920s ex-Fairchild factory at Farmingdale, Long Island, was leased from the Aviation Corporation. Part of the facility, at the former Fulton truck manufacturing plant (lower left), was vacated in 1937 by the Grumman Aircraft Corporation. The small subassembly building (Ranger, circa 1936) immediately above the main plant was temporary home to Kirkham Engineering after it had defaulted in 1935 on manufacturing Seversky SEV-3M-WW military amphibians for Colombia in (then) newly leased Bldg.5, the main factory unit. **(Mfr. via S. Hudek)**

Alexander "Sasha" P. de Seversky was not the least outspoken or less pointedly "on target" in such a group. As a result, he was the quarry of a well-choreographed disinformation/misinformation campaign during the World War II years, involving some of the most respected names in the War and Navy Departments. In wartime there were few important people who were willing or able to take the time to boldly refute statements published as "facts" in some highly regarded national publications, primarily magazines and conservative newspapers. Many of the published statements and quotes attributed to highly placed political appointees and even military and naval officers had to be downright laughable in the opinions of men at war.

The Major stood up to the blows and, as might be expected of such a man, virtually all of his words of warning and wisdom were borne out by eventual facts.

Now comes the time to reveal the facts behind the connection of Major de Seversky and that great World War II fighting aircraft, the Republic P-47 Thunderbolt.[1]

The United States of America has always been an amalgam of various types of refugees and expatriots from virtually every other land on this planet. The men who pioneered in the growth of the nation in its earliest form all had rather shallow roots in America. Then, along came, not Jones, but Alexander Procofieff-de Seversky, soon to be regarded as a patriot, prophet, remarkable pilot, inventor, entrepreneur and, in the eyes of some, firebrand and misguided dreamer. Perhaps he was a devilish saint. He most certainly was a visionary aircraft designer with a harmful blind spot at times, impatient, intolerant and as good at the controls of an airplane as any contemporary.

He was born Alexander Nicolaiovitch Procofieff-Seversky in Tifflis, Russia, on 7 June 1894 to Nicholas and Vera Procofieff-Seversky. At an early age he was enrolled in the Imperial Naval Academy not far from a site where Igor Sikorsky was designing aircraft. His intense interest in these new vehicles led him to spend most of his off hours around the airplanes, and evidently he was even allowed to fly in them as "ballast" at times. Following graduation, he took postgraduate work at the Russian Military School of Aeronautics. Upon the entry of Russia into the world conflict, young Seversky was soon attending flying school and credited as an aviator. Commissioned in the Naval Air Service, Baltic Sea Area, he was assigned to the Second Bombing Squadron. In combat, his aircraft was shot down on 2 July 1915; his observer-bombardier was killed and Seversky lost a leg in the crash. It was almost a certainty that his flying career had come to an early end. Undaunted, he convinced his superior officers that he was not handicapped and was soon back in the combat arena wearing an artificial leg. Confirmation exists that he was credited with as many as thirteen German aircraft shot down and was awarded numerous decorations. As a matter of fact, he became chief of all Baltic Sea area pursuit aviation, attaining the rank of Lt. Commander.

Appointed by the Czarist Regime as a member of an aeronautical commission to study aircraft production and design in the United States of America, he arrived by way of France. In the process, French officials had issued his passport in the name of Alexander P. de Seversky, perhaps with the help of a few choice words from the recipient. With the end of the war late in 1918 and the Communist revolution succeeding in his country of birth, the very youthful de Seversky applied for U.S. citizenship. It is obvious that he had mastered use of his artificial leg and the English language because he was soon a test pilot and consulting engineer on the payroll of the United States Army Air Service in 1921. Still not a U.S. citizen, he was appointed as a special assistant to Brevet General William Mitchell on a project to design and perfect a bomb sight that could ensure drastically improved accuracy. Within three short years, this immigrant had filed no fewer than 364 patent claims. The general obtained funding to buy all rights to the new gyrostabilized bombsight design for $50,000, no small amount in the mid 1920s. This entrepreneurial engineer promptly founded the Seversky Aero Corporation. In 1923 he had married the extremely attractive American society girl, Evelyn Olliphant; by 1927 he was a naturalized citizen; in his spare time he gained a commission in the United States Army Air Corps Reserve with the rank of major. This was obviously no man with whom one should trifle. Proof of that was soon to come.

Like so many other aviation ventures, his small aeronautical firm was essentially bankrupted by the great stock market crash

[1] Every effort has been made to uncover and validate material pertaining to Alexander P. de Seversky's life. That is especially true with regard to his connection with the Seversky Aircraft Corporation and its successor, Republic Aviation Corporation. Extensive research has been performed by the author, drawing heavily on the documentation accumulated over a period of decades by a small number of aeronautical pundits, and especially material retrieved from every possible source by Nassau County's (Long Island, N.Y.) Cradle of Aviation Museum. Located at old Mitchel Field on Long Island, it is the repository for a large percentage of Major de Seversky's personal records. Information contained in several personal collections and in other government museums and libraries was made available to the author. Over a lengthy period of time, key personnel connected with Seversky Aircraft and Republic Aviation, either in civilian or military capacities, were interviewed.

In the summer of 1936, Seversky, having won the pursuit plane production contract, apparently purchased the 250,000-sq. ft. plant and the 127-acre airfield. They also got a 30,000-sq. ft. seaplane facility at Amityville on the Great South Bay. By the time this picture was taken in 1938, Kirkham Eng. & Mfg'g. Corp. had long since departed from Bldg. 5 amid much bitterness, and Grumman ultimately moved to their own factory a few miles away at Bethpage. Seversky was in the aircraft manufacturing business. Fairchild's Ranger Engine Div. had reoccupied their small engine building (No. 55), forcing Kirkham out in 1934. One production P-35 was sole occupant of the sod airfield, and others were inside Bldg. 5, the main factory. (Steve Hudek Coll.)

of October 1929. It did not take long to learn about the character of this man. He did not jump out of any window as the value of stocks plummeted to earth. In fact, he formed a new business, Seversky Aircraft Corporation, with $1000 capital to supersede Seversky Aero with assets consisting of six U.S. patents and three foreign patents. In consideration of the value of his patents, de Seversky received 2,000 of the 10,000 shares of capital stock issued with no par value. Three others, including Edward S. Moore, invested $35,000 in Seversky stock. The corporation came into being on 17 February 1931. Eleven days later, the Major was elected president, while two of the three incorporators – E.W. Poindexter and Miss R.V. Thomma – became Treasurer and Secretary, respectively.

Here we have some very "gutsy" people, putting their extremely valuable dollars (you could purchase a fine Pierce-Arrow limousine from a used car dealer for around a hundred dollars in 1932 and 1933) into a business venture of dubious value. In an exclusive section of Long Island or out in Beverly Hills, California, the $10,000 invested by Edward Moore would have bought a person any one of dozens of manors – with no mortgage to be paid off. (Within four decades, the real estate investment would have returned millions of dollars on the original payoff, and the investor could have lived there for forty years without making any payments other than low taxes.)

But it was in New York City that the seeds of the Republic P-47 Thunderbolt were planted in that bleak, depressing marketplace on a very cold winter day. Fact: Without Major Alexander de Seversky and his patents and mind, there never would have been a Republic Thunderbolt. Quite frankly, his direct and indirect contributions to America's last clearcut victory in war – and it most certainly was our biggest and toughest – were numerous and significant, if not monumental.

For those who may remain as "Doubting Thomases," even after the Major's actions of February 1931, consider the following: In that same month, but a year later and some 3000 miles to the west, Lockheed Aircraft Co. had, perhaps, two employees. They would have been Carl Squier and Harvey Christen, essentially holding hands to keep Lockheed as an entity. On the 6th of June in Los Angeles, Robert E. Gross and a few other venturesome aerophiles scraped up $40,000 to recapture the assets of Lockheed in order to form Lockheed Aircraft Corporation. Back on the East Coast, down at Dundalk, Maryland, a marginal aviation manufacturer by the name of General Aviation Mfg. Corporation – out of mergers unto mergers involving General Motors, Fokker, a North American Aviation that even owned parts of Douglas Aircraft and B/J (Berliner-Joyce) Aircraft – tried to keep afloat by building one last airplane, the GA-43 airliner. It flopped. Out of that mess, "Dutch" Kindleberger forged a viable, but poor North American Aviation relocated to Inglewood/Los Angeles, California, at Mines Field.

What do we draw from all that? Three "poor-as-churchmice" aircraft manufacturers with a combined production output of about five airplanes by 1933 eventually went on to produce the U.S. Army Air Forces' top three fighters of World War II![2]

Seversky Aircraft established its offices at 570 Lexington Ave., New York City, in April 1931. The Major soon hired his first chief engineer, a stubby little Russian by the name of Michael Gregor. This wily little fellow was an engineer of no small talent, having designed the Brunner-Winkle Bird biplane. They immediately began design work on an extremely advanced all-metal stressed-skin monoplane with interchangeable land/amphib-

[2] Yes, we concede that Major de Seversky had been "given the boot" in 1939 so to speak, but few of his associates/employees departed from Farmingdale unless it was to stay by his side. All that was needed was a firmer hand on business controls who would not bark back at the War Department.

ian landing gear. Out in California a designer by the name of John Northrop was busily engaged in design and construction of an all-metal monoplane of roughly contemporary configuration (except for the landing gear). Both of these airplanes were "pushing" the state-of-the-art in that they had metal cantilever wings featuring multicellular construction.

Michael Gregor's wing was actually far more advanced than the Northrop component in that it had a higher-speed airfoil (much thinner) and all fuel was carried in the centersection without having separate fuel tanks. The de Seversky-Gregor wing featured sealant to prevent fuel seepage/leakage. Exhibiting a classically beautiful semi-elliptical planform, it was probably the most attractive wing seen anywhere when it cut through the air in 1933. Here was a wing, essentially having the same planform, chord-thickness ratio and virtually all essentials that were used in 1944-45 on the first piston-engine airplane to exceed 500 mph (unofficially). That airplane was a modest derivative of the base Thunderbolt design that was "hotrodded" and intended for production as the XP-47J/P-47J.

The Major hired another expatriate Russian, trained in France as (initially) an electrical engineer and subsequently as an aero/ structures engineer. His name was Alexander Kartveli. His job: assist Gregor in all design assignments.

The first design to issue from Seversky Aircraft was, almost unbelievably, an expensive executive aircraft with possible military application in mind. Publicity and sales documents generated by Seversky Aircraft referred to the new, advanced airplane as a "Sport Amphibion" (sic) in 1933. But there was nothing to indicate that, even in the internal documentation created by de Seversky. It was essentially a 2-seat (even 3-seat) amphibian featuring retractable wheels and a hydraulically controlled float-positioning device. This feature allowed the amphibious floats to be angled for the best landing position on water and on land, but permitting repositioning for the best low-drag flight condition. The airplane, identified as the SEV-3, featured a low-profile, low-drag canopy and windshield. It carried the Bureau of Air Commerce license number X2106, a number destined to be around for several years. The Bureau of Air Commerce license stated that the aircraft manufacturing serial number was 301 with a manufacturing and licensing date of April 1933.

Without any factory facilities, the Major had approached the Edo Aircraft Corporation at College Point, Long Island, NY, sometime in 1931 to make arrangements for construction of the SEV-3. Edo was the prime supplier of metal floats in the United States and their facility was located on the water. It was an ideal setup for de Seversky to be launching an amphibious airplane since he also lacked so much as a hangar at any airport. Exotic tooling consisted of wooden 2x4s so arranged as to create a fuselage assembly fixture. No records are available to define whether the employees all worked for Edo or if some (or all) worked for Seversky Aircraft. Work progressed very slowly over a two-year period because money was not at all plentiful, even with the Wall Street backing that was in hand. At the same time, the Major was concentrating on a series of inventions in Manhattan, so much of the "bird dogging" fell on the shoulders of Gregor and Kartveli.

Originally equipped with a Wright R-975-ET engine rated at 350 horsepower (c/n 12602), the SEV-3 took off from water on its first flight in June 1933. Although the aircraft was skinned in Alclad metal, the flamboyant showman had the entire airplane painted with bronze lacquer, evidently supplied by the Murphy Varnish Company. Not much time passed before the amphibian was equipped with a Wright J-6-9E engine producing some 420 horsepower, and in this form the Major succeeded in establishing a new official world speed record for amphibians, averaging 179.76 mph over a 3-Kilometer course. The date was 9 October 1933.

Eventually that speed record was eclipsed by a U.S. Coast Guard Grumman JF-2 biplane amphibian manufactured by the Grumman Aircraft Engineering Corporation. In that tight-knit industry on Long Island, the company had been constructing their FF-1, SF-1, JF-1 and F2F-1 biplanes in a leased factory that would soon be occupied by Seversky Aircraft. When the Fairchild Aviation Corp. constructed a new plant at Farmingdale in 1928, they soon expanded operations into the former Fulton Truck plant alongside their main assembly building.

Just prior to the stock market crash, Aviation Corporation (Avco) purchased the entire plant. When Avco ceased operations in 1932 the old Fulton building was leased to a newcomer, Grumman.(Charles Kirkham, an extraordinary engine and aircraft builder who had been a Curtiss employee in earlier years, had moved into Avco's former engine manufacturing building a few hundred yards east on Conklin Street. American Aircraft & Engine, an Avco subsidiary, had operated in that building for a time without success. Kirkham Eng'g. & Mfg.

Wright Field, adjacent to the pike outside Dayton, Ohio, looked like this in 1937 when the first Seversky P-35 was flown in for acceptance testing. Much to Lt. Ben Kelsey's chagrin, it did not get passing grades and was rejected. Among aircraft to be seen on the pavement were an obsolescent Keystone bomber, a brand new Douglas B-18 "heavy" bomber and the very large Douglas YOA-5 observation amphibian. (AMC)

Seversky's engineering design department probably did not exceed one dozen designers and draftsmen in 1935 when the BT-8 trainer was in production. One blueprint on wall appears to be an inboard profile of the SEV-2XP pursuit plane that was the initial company entry in the 1935 Pursuit Competition. (Mfr. via F. Strnad)

Corp., was soon to play a key role in Seversky's future.)

The key financial backer for Seversky Aircraft was Edward Moore, but when he died his brother Paul took the reins and generally supported Major de Seversky in whatever plan of action he proposed. Paul Moore's investment in Seversky Aircraft as a single investor would hardly be insignificant six decades later, after some considerable effects of inflationary periods.

Something inspired "Sasha" de Seversky to start a high-powered sales effort in the summer months of 1934, to be quickly followed by a flying visit to Wright Field, Ohio. It may have been issuance of a Circular Proposal for a newly reinstated trainer category, the basic trainer. The SEV-3 amphibian, still powered by a Wright R-975 engine but fitted with a 3-blade adjustable propeller, was flown to Dayton by the Major. Demonstrations

Responding to an Air Corps circular invitation for a new basic trainer of advanced concept, de Seversky flew the SEV-3 amphibian to Wright Field, hoping to beat all other contenders to the punch. Air Materiel Division did assign a project number XP-944, but officials were emphatic about not wanting an amphibian. Perhaps a bit crestfallen, he flew back to Farmingdale for rework (installation of a normal land gear). In Board of Directors notes, the airplane was redesignated SEV-3L, a point missed by nearly everyone through the years. (Army Air Corps)

followed but he was very disappointed in some comments. Flying back to Farmingdale, he had the work crew install a new 2-blade propeller and possibly a different version of the Wright engine.

Not long afterward, the Major flew the amphibian back to Dayton with the idea of showing Materiel Division people that his airplane could far exceed written specifications for a trainer. Someone soon pointed out that the Air Corps had no need at all for any small amphibian, no matter how advanced it was in concept. "What kind of performance can we expect from the SEV-3 if it is equipped as a landplane?" was the question from a ranking official, emphasizing that a landplane trainer was demanded by the circular. Although the company brochures claimed that the landing gear arrangement could be changed in an hour, it took at least a week or so before de Seversky could fly back to Wright Field. One very interesting sidelight reported at that time was the fact that Col. Charles Lindbergh visited the Major at Farmingdale. Lindbergh, who was about to become a father for the third time (reported in the papers), was briefed personally by de Seversky on flying the just-converted SEV-3L (redesignated as such by the Executive Board) landplane. Apparently they flew over to Glenn Curtiss Airport together.

Shortly thereafter, the Major made the trip back to Wright Field in the demonstrator. There he appeared in a blue blazer with the sartorial touch of a handkerchief in his breast pocket. At one point he flew around Dayton with the low-profile canopy open, not wearing helmet or goggles. Everyone must have been forced to acknowledge the fact that this Seversky could speed past most of the first-line pursuit planes then in Air Corps service. A good Curtiss P-6E biplane might have given the low-powered but sleek training type a fair tussle for speed honors, but Boeing P-12s would have fared badly against the SEV-3L. And if the rather tightly knit military and civilian staff of the Materiel Division did not generate some excitement about the very strongly constructed SEV-3L that could give a new Boeing P-26A pursuit

With flaps extended, the SEV-3XAR was beautifully photographed at Wright Field in the autumn of 1934, ready to challenge any and all competitors. The question was: Were any Air Corps people ready for a quantum leap forward in the realm of training airplanes? This was a real high-performance airplane. On that day, the Seversky could have beaten every French fighter (and probably most American and British types) in a speed contest. Like all Severskys, it had only 3 degrees of (lower surface) dihedral, zero on the upper surface, so it could be "tricky" to land. (WF/AAC via Ed Boss, Jr.)

a scare during a speed dash (with only two-thirds of the pursuit's power), they had no real interest in their careers. They also lacked any forward vision.

Major de Seversky's brainchild had already proven (before FAI officials) that it could fly at more than 180 mph while mounted on big amphibious floatgear and with no more than 420 hp on tap. Suddenly, the Boeing P-12E as a first-line fighter was certain to be in trouble from this segment of the aviation world. Whatever transpired at Wright Field is poorly documented, but subsequent commentary sent the Major scooting back to Farmingdale – possibly with a smile on his face.

Seversky's Executive Committee approved allocation of $2975.00 on September 8 for modification of the powerplant and installation of full dual controls, complete military radio equipment and a higher-profile canopy that would be far more compatible with military pilot attitudes in an era when open cockpits were the norm. When the Sev-3L had been reworked and flown to Wright Field it was designated officially as SEV-3XAR (XAR most likely referring to Experimental Army).[3] By the time leaves were falling from the trees, the SEV-3 had been transformed into a contest winning beauty. "Sasha" (de Seversky) had evidently worked his magic again and borrowed another engine in the R-975 series from Wright Aeronautical for installation in the SEV-3L-*cum*-SEV-3XAR.

This bronze hustler, still bearing the ***SEVER/SKY*** logo on each side of the aft fuselage and the X-2106 registry, was so far advanced over the North American NA-16 competitor that it would be almost impossible to believe they were competing for the same contract. (The NA-16 prototype, licensed X-2080, came into view somewhat after a contract was awarded to Seversky. It was not even photographed at the factory until springtime 1935. At that time it had open cockpits and the entire fuselage aft of the accessory compartment was fabric covered. The design was years behind the Seversky airplane.) As an aside, it should be mentioned that the Executive Committee had agreed that if Seversky was awarded the contract for basic trainers, Kirkham Engineering & Manufacturing Co., already in the process of building aircraft for Seversky in response to its initial military contract (foreign), would manufacture those basic trainers in Kirkham's leased facilities at Farmingdale.

Just a few short months after Major de Seversky demonstrated the colorful SEV-3XAR in Dayton, the Army awarded Contract AC-7348 in the amount of $878,000 for thirty BT-8 basic trainers, but somewhat strangely they were to be powered by Pratt & Whitney R-985-11 engines rated at 400/420 horsepower. (All Seversky news items referred to a contract for 35 airplanes, but this was blown up to that number by the expedient means of including spares.) In the normal Lexington Avenue operational style, the SEV-3XAR was eventually referred to as c/n 35 in company records! But hold onto your hats. This story has some exciting ramifications.

We are uncertain about the timing and details of the following events, but there is no doubt that they occurred. When the Major was out peddling airplanes, Mike Gregor took it upon himself to move out to Farmingdale – most likely with the approval of Paul Moore, but just possibly on his own decision – into some corner of the former Fairchild/American A&E main assembly hall. When the hot-tempered Major returned to learn of this situation, he promptly gave Gregor his "walking papers." Since Gregor also had a temper, but was a stockholder, he cashed in his stock and departed for (eventually) Canada. (Reasons for the move to Farmingdale will be explained in their proper context because all of these events were overlapping in many ways.) The tragedy in this departure is that in the summer of 1936, following by a year or so the contract award for BT-8s, Seversky Aircraft arranged to purchase the plant on Conklin Street from American Airplane & Engine Corp. for $880,000 on the installment plan. Included in the price were the 250,000 sq. ft. factory buildings, a 127-acre flying field and even a 30,000 sq. ft. seaplane base at Amityville, Long Island. Gregor was a real asset to de Seversky and his design talents might have had resoundingly good effects on the pursuit airplane that was to make its appearance. Those flaring tempers must have cost Gregor a small fortune in the long run.

On the surface it would seem that Seversky received its first military aircraft contract from the Army Air Corps in 1935, but the Republic of Colombia beat them to the punch by several months. Neither internal corporate papers nor Bureau of Air Commerce records that were reviewed provide a date on which Colombia contracted to buy three (never six) amphibian military airplanes from Seversky, but the contract was signed before the SEV-3XAR had its day in the sun. (Seversky's contract with Kirkham was dated 16 May 1934.) Early elation about the contract was soon to fade, friends became enemies, and contract default and bankruptcy for at least one company had its fate determined.

Jumping to the conclusion that he might be declared winner of the competition by having an "off-the-shelf" airplane, Major de Seversky flew the SEV-3L back to Dayton for a second try. He was given specifications that would require certain additional revisions, especially in the cockpit arrangements. The most obvious changes required were in the canopy and powerplant installations. There is no proof that any Materiel officer ever flew (or even flew in) the prototype. In subsequent weeks, the airplane was converted into the SEV-3XAR. The Major stands by the SEV-3L at Wright Field. **(Army Air Corps)**

[3] Of all the aircraft manufacturers, Seversky must have taken "honors" for having the most disconcerting and illogical model designation and constructor's numbering system in the business. Vought, North American, Boeing and Grumman, by comparison, were epitomes of logic. Lockheed, by 1930, had developed a numbering system and design drawing numbering system that worked well for decades.

CHAPTER 2

AND INTO THE MILITARY ARENA

From Alexander de Seversky's and Paul Moore's viewpoints, their first true military aircraft contract had come from, of all places, the Republic of Colombia in South America. Apparently the Colombians – in 1934 – had acquired (or were acquiring) Curtiss Hawk IA fighters and Bellanca 77-140 twin-engine bombers to fight one of the many wars stemming from border disputes. According to *fable*, Colombia purchased six Seversky SEV-3M-WW type amphibians for use by their navy. They were expected to be operating from a new seaplane base located on the northwest coast at Barranquilla as reconnaissance airplanes with the capability of flying from any reasonably quiet body of water or an airstrip. The tale about procurement of six airplanes, kept alive in the U.S. for years by some, is pure legend.

History reveals that no more than three amphibians were ordered from Seversky Aircraft upon the recommendation of noted aviator Roger Q. Williams. The Colombian War Ministry had hired Williams as their technical advisor to a purchasing commission. When a contract was awarded to the New York firm, he was assigned to monitor aircraft construction and contract compliance in every detail. It is evident that the Colombian military purchasing commission placed an order for no more than three militarized versions of the so-called sport amphibian on the basis of its performances in 1933-34 and its obvious special capabilities. The contract called for each aircraft to cost $38,086. It is important to know that a surety bond was issued to cover an advance payment of $45,988.95 plus a 10 percent performance bond. Of course the financier Paul Moore endorsed that.

Without any production capability and no factory, Seversky was more than happy to turn to Kirkham Eng'g. & Mfg. housed, for the moment at least, in that original 1928 Fairchild factory on Conklin Street. Charles Kirkham was glad to get a subcontract for three aircraft (less amphibious floats) at $25,000 each plus spare parts. With the Depression hovering near its depths in 1934, the subcontractor had to post a $33,000 performance bond. At this point it seems appropriate to recall that a September 1934 Executive Committee agreement stated that Kirkham was to build all basic trainers if Seversky was awarded the contract.

Enlightened as to specific trainer requirements, the Major flew his SEV-3L back to Farmingdale for serious rework. Evidently a borrowed 420 hp Wright R-975-E2 engine was installed, with everything from the wing leading edge forward duplicating a SEV-3M-WW. A new "tacked on" proper canopy was installed; it did not blend well with the curved upper fuselage, but it was adequate. Here we now have the actual SEV-3XAR prototype for the BT-8 trainer, still being c/n 301 with the original registry. It was more than ready to take on North American's NA-16 (ultimately the BT-9) and all other competition. At the time, it was truly a wolf in sheep's clothing, being far more technically advanced than the new Boeing P-26 pursuits. (C.A.M. via Ed Boss, Jr.)

Charlie Kirkham was probably one of the most innovative and expert designers and builders of aircraft hardware in America in the early days of aviation. It then becomes logical to inquire about his lack of recognition and rewards. From what we can gather, his personality was abrasive beyond comprehension, and he must have been terribly impatient with most people. As for his attributes, he designed and built – his talent as a precision machinist was top echelon – the famed Curtiss K-6 and K-12 water-cooled engines, and most knowledgeable aviation insiders give him top billing for creation of the fabulously popular OX-5 engine built en masse by the Glenn H. Curtiss organization.

It is appropriate to mention that when the Curtiss OX engine of WWI days was unable to perform up to specifications, Charles Kirkham was made Chief Engineer on the program to save it. By the time he was finished and had exceeded specification requirements with reliability, it was known as the OX-5. Curtiss OX-5, that is, but Kirkham's reputation as an engineer and technician in the company's Garden City (N.Y.) plant was well established.

An airplane design, fully attributable to Kirkham, that should have led directly to strong aviation leadership in the U.S. in the early 1920s remains largely unknown and virtually unheralded. That was the Curtiss-Kirkham Model 18B biplane and its derivative triplane known as the 18T. The remarkably clean design featured the same kind of laminated wood fuselage that allowed later Curtiss R-6, CR, R2C and R3C racing planes to become world beaters in the first half of the 1920s decade. Originally designed as a 2-seat fighter biplane to outclass the Bristol Fighter of WWI days, the Curtiss-Kirkham 18B was the victim of the Armistice with its attendant ending of development financing. Navy triplane versions set world altitude records and were strong contenders in seaplane racing categories. After America ended participation in the great Schneider Trophy Contests following the great 1923-1925 triumphs, something that was entirely attributable to "penny-wise, pound-foolish" budget cutters, aircraft racing in this country was decimated. But Charles Kirkham got back into the act by using some remarkably creative techniques to convert a 1926 Curtiss R3C Schneider racer design into a semi-private contender for the trophy.

He redesigned the fuselage and wings to handle more power and then installed a borrowed Navy-Packard experimental 24-cylinder, liquid-cooled X-configuration engine. This radical narrow-X arrangement had a cubic-inch displacement of no less than 2774 c.i.– that was in 1927(!) – and was designed to deliver about 1250 horsepower. When you realize that the displacement was not significantly smaller than the powerplant ultimately used in the Thunderbolt but that frontal area was less than half that of the famed P&W R-2800, the Packard 1A-2775 might have been the most spectacular aircraft engine of its time. It was no longer than any contemporary 6-cylinder engine. The Detroit-based motorcar company did not have a well financed project, and their aircraft water-cooled engines were not particularly reliable. Metallurgy and bearings were most likely the culprits. (With the onset of the Depression, Navy funding essentially disappeared.)

However, the landplane version of this Packard-Kirkham racer that was scheduled to be flown in the Schneider contest by Lt. Alford Williams did something in 1927 that should have attracted much worldwide recognition. Using a landing gear that most certainly did not feature air-oil shock struts (not yet developed), skinny rubber tires of doubtful integrity and operating from a sod airfield, the tiny biplane was committed to fly and then land without benefit of any sort of flaps! Yet it managed to record an unofficially witnessed record speed of 322.6 mph. in the vicinity of Mitchel Field on a very windy day. As the old saying went, *Holy Toledo!*

Of course the brave pilot was Al Williams. Up to that time, no airplane had ever exceeded the magic 300-mph speed mark, even with that downwind condition. Kirkham's shop had initially put the racer on some new Naval Aircraft Factory floats to prepare the aircraft and pilot for the 1927 Schneider event to be held in Italy. While we suspect that with only two (borrowed) engines in existence and a dearth of development running time logged, engine reliability may have been a problem. There is far more evidence of float problems on the water and in flight. Inclement weather, restricting flying time, was a serious factor that summer. Worse, after Americans had cooperated by granting a postponement on behalf of the Italian entries in 1926, that host nation refused to return the favor in 1927. However, they received their just rewards when their three racers failed to finish the event and the British won handily. Time ran out for the Kirkham-Williams team effort.

Having survived the Great Depression's worst years, Kirkham surfaced in Farmingdale, New York, his home state in 1932. As previously stated, he had negotiated a contract for three Seversky SEV-3M-WW (for Wright Whirlwind), put it in his safe and went about the task of manufacturing Seversky Aircraft's first military airplanes. As every actor – especially James Cagney, Edward G. Robinson (alias *Little Caesar*) and Clark Gable – was prone to say in 1934, "Gee, that's swell."

Unfortunately, it wasn't swell. Kirkham was soon in trouble on the project, having overrun the budget and schedules. He

Flush with record-breaking successes, racing against special-purpose types in the Thompson Trophy Race, Cleveland and many demonstration flying activities, the Major flew this SEV-3M as a landplane. Performance was so impressive that it blinded him and Kartveli during the winter months to the realities of an upcoming Pursuit Competition in April 1935. Wanting to win with minimal expenditure of scarce funds, they virtually duplicated the SEV-3M-WW/SEV-3XAR as a land-plane two-seat fighter. Wright Aeronautical sold them on using a brand new SGR-1670 Twin-row Whirlwind engine. A new, smaller tailplane was fitted along with a controllable-pitch three-blade propeller. But, alas, neither top designer had done his homework. **(Mfr. via S. Hudek)**

blamed the Major, saying that drawings were delivered late and reflected changes that required rework and other added expenditures. (What's new?) What with all the bonds still being in place but endangered by impending contract defaults all around, Paul Moore provided the funds out of pocket to preclude disaster. But, as might be expected, the contract was abrogated as 1934 was ending and Charles Kirkham was out. And he did not depart on friendly terms with de Seversky. On 14 December 1934, the Major had sent C.B. Kirkham a "Christmas Greeting." He chastised the latter for allowing Sherman Fairchild to "examine and inspect" the Seversky SEV-3M-WWs, a thing that "constitutes a breach of our understanding," etc. He threatened to hold Kirkham liable for any "irreparable damage that we may sustain." It did not bear a rubber stamp signature and it did not say Merry Christmas.

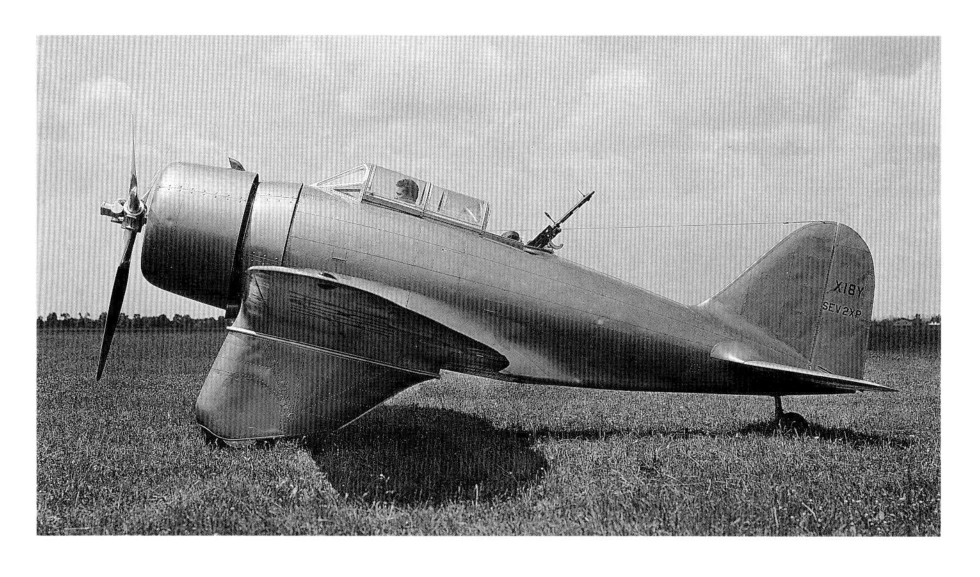

The Major and Chief Engineer Kartveli seemed unaware that the new Hawker Hurricane 1030-hp, retractable landing gear interceptor was about to fly in England. Bad espionage to say the least. Didn't Lt. Ben Kelsey's Circular Proposal specify a single-seater? Nobody else sent a 2-seat entry; in fact the Consolidated entry was a ridiculous single-seat version of their current in-service PB-2A. There were two other major anomalies in the new Seversky SEV-2XP design: (1)It was the only new fighter in the group that did not have a retractable landing gear, and (2) the engine chosen to power the pursuit plane was a badly configured, totally undeveloped Wright unit that failed dismally. **(Mfr. #145)**

Suddenly, Alexander de Seversky was in the aircraft manufacturing business, and that was not all bad. In fact, all things being considered and, especially Charles Kirkham's business batting average, it was probably a good thing. Seversky Aircraft took over the lease on the original Fairchild factory building facing on Conklin Street and hired as many of the ousted Kirkham employees as he could. It was now 1935 and Paul Moore was anxious to complete the Colombian contract and not be faced with problems that could cause the Air Corps to cancel the newly won contract for BT-8 trainers. Fabrication and final assembly space suddenly became a priority, so the three partially completed SEV-3M-WWs were moved out of the factory building and into one end of the red brick hangar on the airfield east side. A long year would drift

Already having proved to be the world's fastest amphibian in 1933 with a Wright Whirlwind J-6-9E engine of 350 hp, in SEV-3 form, the SEV-3XAR was reworked during the winter of 1934-35 to a SEV-3M configuration. New amphibious floats, a big Wright Cyclone R-1820-F engine of 750 hp and a new controllable-pitch propeller were installed. Result: the airplane established a new World Speed Record of 230.03 mph at Wayne County Airport, Michigan, on September 15, 1935. **(Seversky)**

Granted the SEV-2XP was the best-looking Seversky produced to date, but it was half a decade behind the Detroit-Lockheed XP-24 design submitted to the Air Corps for test in 1931. Reflecting America's decimated aircraft industry, the Seversky and Northrop entries never made it to Wright Field for the competition, and those that made it were not able to outperform current service pursuits, especially in reliability. Northrop's 3A disappeared into the Pacific during its first test flight and the SEV-2XP suffered an accident or other indignity. Fortunately for Seversky, the Curtiss 75 Hawk was powered by the same Wright Twin Whirlwind SGR-1670 type engine rated at 750 horsepower. (File)

by before all three airplanes were entirely acceptable to Roger Q. Williams, representing the Colombian government.[1]

Americans were beginning to see the light at the end of the tunnel in 1935. The song "Brother Can You Spare a Dime" was giving way to "Happy Days Are Here Again" on radio and wherever there was musical entertainment. Speaking of musicals, they were very big in films in that same year, with Fred Astaire and Ginger Rogers leading the field. Inflation certainly was not rampant, and if you wanted to buy off-brand gasoline (and many did), there was plenty of it offered at 10 gallons for one dollar. Any number of new car models were available in the $500 range, and Henry Ford came out with just the right combination, the 1936 Ford V-8. A few changes were made to the 1935 model, creating a great looking car with excellent performance. Anyone who could afford the 4-door phaeton convertible was to be envied by all, and the price was under $1000. The major flaw in hardtop models was that they did not have the one-piece "turret top" featured on General Motors cars. No, that was not the major flaw. The Fords did not have hydraulic brakes.

In the meantime, things were happening in rapid-fire order out in the hinterlands referred to as Farmingdale or, more accurately, East Farmingdale. Seversky Aircraft was offering several other airplanes to potential customers. One of these was the proposed SEV-5 Executive 5-place amphibian that was equipped with side entry doors to accommodate persons seated in the aft cabin. It was to have a wingspan nine feet greater than the SEV-3, and the fuselage length was to be increased by more than seven feet. Sea level top speed was stated as 200 miles per hour on power from a 645-horsepower radial engine.

Looking at the water-based air transport market, the company proposed a 10-place single-engine amphibian Transport identified as the SEV-7. Overall length increased by six feet and wing span increased to 49 feet. Company brochure data said that it would carry the ten passengers at 190 mph top speed on one 675-horsepower engine. A convertible landing gear could be installed in place of the float gear, and that seemed to be the point at which the semi-retractable landing gear came into the House of Seversky. The system was not one of the best Seversky "improvements," being a relatively high drag arrangement. It was also unnecessarily complex and complicated. In retrospect, the Major would have been light years ahead if he had but hired somebody like Alfred Verville to design an optimum flush-retracting landing gear. Neither the SEV-5 or the SEV-7 were able to attract any orders.

Fabrication of the three SEV-3M-WWs moved at a snail's pace for all of 1935, but even the Colombians did not seem to be upset by that. Following expulsion of those airplanes from the newly leased (subject to purchase contract) main assembly building (henceforth to be known as Bldg. 5), the Seversky team went to work refurbishing the plant for production of the BT-8 basic trainers. A proper engineering office was established so that work could proceed on several projects. It was destined to be a year fraught with problems, but there were to be some victories as well.

Seversky's pretty SEV-3XAR flew back to Farmingdale after being tested at Wright Field by several pilots, including Capt. Claire Chennault. It is a known fact that Chennault resigned his commission in the Army Air Corps primarily because he was openly opposed to procurement, tactical and strategic policies embraced by (then) Col. H. H. Arnold and Capt. Ira Eaker, among others. Here we have another "curmudgeon" who was friendly with Major de Seversky. An international aircraft sales representative, A. L. Patterson who spent years in China, claims that Chennault was a key factor in the evolution of the Seversky P-35. Since Chennault was no advocate of multi-seat fighters, he may have been instrumental in prodding the Major into making some of the key decisions leading to the final format of the

---

[1] For purposes of historical accuracy, it must be noted that the SEV-3XAR prototype for the BT-8 basic trainer probably did not differ more than 10 percent from the configuration of the Colombian SEV-3M-WWs if and when the latter had the amphibious floats removed and standard Seversky wheel/pant assemblies installed. Externally the only differences were the absence of a protruding carburetor air intake on the top cowling and the SEV-3M/-3L cockpit fairing extension which remained on the SEV-3XAR.

SEV-1XP prototype pursuit airplane (which will soon enter this picture).

Upon arrival of the SEV-3XAR at the factory on Long Island, the Wright R-975E engine and cowling were removed and the wheel/pant landing gear was replaced by an updated version of the amphibian float gear. Then, revealing the versatility of the original SEV-3 design, a new Wright Cyclone SR-1820F-3 series engine rated at 730 horsepower[2] at sea level was installed in a new, pugnacious looking cowling. They not only nearly doubled the cubic inches of engine displacement, they doubled the available horsepower. Although the engine was moved back somewhat to adjust for greater weight, overall length did not vary much (if any) from the original 1933 SEV-3 amphibian dimension. Contrary to some published data this was not a "new" airplane. The XAR suffix became history in April 1935 and that airframe was redesignated SEV-3M to reflect its new mission. (It surely must have been obvious by that time that not too many sportsmen were about to spend $50,000 for an "executive" aircraft in the modest recovery period following the Depression.)

The Seversky SEV-3XAR as seen at Wright Field and flown by Capt. Claire Chennault. **(USAAC)**

The Major applied for relicensing the airplane in April 1935 as plans were developed to head off on a demonstration tour of South America. An extremely large D/F loop antenna was installed during testing, but the long-distance flight was postponed in the face of other more pressing business at home.

Close on the heels of the basic trainer contract award, the Air Corps Materiel Division announced, through issuance of a Circular Proposal, that a new pursuit (eventually changed to fighter in 1942) airplane competition would take place 27 May 1935. It may have been the influence of Capt. Chennault that set de Seversky on the road to submitting a pursuit plane entry, but somehow the signals must have gotten crossed or were totally misconstrued. While everybody, including the tiny Wedell-Williams Flying Service, seemed to understand that a single-seat pursuit airplane was the objective – as a replacement for the Boeing P-26 and the Consolidated PB-2A (since the bi-place concept was dying or dead by then) – that point obviously eluded de Seversky and Kartveli. Everyone wonders how in the world that could have happened.

Well, entirely overlooked/ignored in the scenario was the fact that a fourth SEV-3M-WW fuselage had been partially or completely built by Kirkham Engineering. With money extremely tight in the face of the bailout financing furnished by Paul Moore when Kirkham nearly caused a default on the Colombian contract, a decision was made to complete the aircraft as a 2-seater with the SEV-3M-WW canopy and equipment. It is quite possible that a revamped BT-8 landing gear design was employed, being slightly shorter than the one used on the SEV-3XAR and not having the same rake. Wright Aeronautical was offering a brand new twin-row radial engine, the GR-1670 Twin Whirlwind to its Curtiss affiliate and it is a fair bet that the Major "cut a deal" with Wright in getting the SR-1820F-3 engine for the amphibian. Seversky's new pursuit was designed to accept the GR-1670 in a cowling that was first cousin to the one seen just weeks earlier on the SEV-3XAR.

If this all seems difficult to follow, just remember that the Great Depression monster was still roaring loudly in 1935 and but for the intravenous feeding by Paul Moore, Seversky Aircraft could have gone belly up. It was incumbent on the Major to save wherever possible, and he was not inclined to even waste a squeal. In that, he had lots of company. Money problems could also explain the failure to design and employ a retractable landing gear. Some wing structural redesign and development of a brand new landing gear system probably loomed as expensive and non-essential items.

When the new 2-seat pursuit was rolled out of the Farmingdale factory (behind schedule, unfortunately), it was evidently painted with that same bronze varnish that was being

---

[2] Depending on which publication the data appeared in and the year of publication, ratings could vary from 710 to 900 horsepower, with the highest rating being the takeoff figure at 2350 rpm. Critical altitude was about 7400 feet where the engine was rated at 715 horsepower. Wright Aeronautical's full-page ads stated that 730 hp was being delivered.

In a tight knit industry, Curtiss-Wright's Ralph Damon – later to lead Republic to success and fame – had hired Donovan Berlin from Northrop to design a new monoplane fighter. Expecting to have a 1000 hp engine, he designed an airplane larger than the XFT-1. It was only natural that Curtiss would place faith in Wright's new twin-row R-1670 Twin Whirlwind, but the engine can only be described as a "turkey." The Donovan-designed Curtiss pursuit was designated Hawk H-75. Standard blue fuselage, insignia yellow wings and tail prevailed on every entry in the 1935-36 competition. It was a last hurrah for that color scheme, except for trainers. (Mfr.)

used on the amphibian. Therefore, this new SEV-2XP would look very familiar to those in Dayton who had become well acquainted with the SEV-3XAR. But responding to some outside influence, they apparently re-painted the entire tail chrome yellow before flight trials began at Long Island. There was a definite change in color. The few markings on the airplane consisted of the X18Y registry painted in large black letters on the lower surface of the left wing, upper surface of the right wing. The rudder displayed X18Y followed (below) by SEV-2XP. Rather strangely, the manufacturing serial number assigned (c/n) was No.2 in spite of the existence of the three SEV-3M-WW airframes already in various advanced stages of construction when the 2XP was started. This "pursuit" airplane was flown at Farmingdale to verify that everything functioned properly.

***Starting late, the Northrop team worked endless hours to convert all drawings of their Navy XFT-1 (with SEV-2XP type fixed gear) into a higher-powered Air Corps pursuit with retractable landing gear and controllable-pitch 3-blade propeller. That unusual cowling enclosed a P&W S2A4-G (R-1535) Twin Wasp Jr. rated at 700 horsepower in 1935. Some 90 percent of the airplane was built with their XFT-1 tooling, but the new landing gear was the epitome of fine design. Lt. Kelsey had already postponed the start date of the competition to accommodate Seversky and Northrop, so when the 3A dove into the Pacific during a high-speed calibration run in the final days of July 1935 he was loathe to order a further postponement. John Northrop asked Curtiss to agree to a delay and was rebuffed. He promptly sold the design, tooling and parts to Chance Vought.* (Bruce Burns via Bodie)**

Young Lt. Benjamin Kelsey, the one-man Fighter Projects Office, received his indoctrination with this 1935 Pursuit Airplane Competition, and within days he must have felt snakebit. What few people there were in the Fighter Aircraft Branch of the Engineering Section, Materiel Division, became more disenchanted with the aircraft and engine industry with every passing day. Although the Wedell-Williams XP-34 entry never did appear at Dayton, design drawings procured under Contract AC-8392 revealed some good design features. The fully retractable landing gear had a wide track and complete flush landing gear doors. What it seemed to lack was adequate empennage surface area and there is every indication that the engine selected was anything but the latest design. The cockpit evidently was very small and construction details were not to military design standards. From California, a message was received that the Northrop 3A entry would not be ready before July.

The Northrop plant was reasonably busy in those days building attack aircraft versions of the commercial Northrop Gamma series for Gen. Chiang Kai-shek's army in China and a reasonable number of orders were in hand for commercial mail/passenger planes. On 1 March 1935 the Air Corps signed a contract for no fewer than 110 Northrop A-17 attack planes. The big 2-seaters were to be powered by P&W Twin Wasp Junior engines. Northrop's Air Corps pursuit entry was to be a heavily updated version of their U.S. Navy XFT-1 experimental fighter that was not encountering great success. Committed to using the small 740-horsepower P&W Twin Wasp Jr., an early version of which was to power the Hughes (as in Howard) H-I racer as it broke the official world speed record for landplanes in 1935, Jack Northrop made his airplane as small as possible, putting his faith in the premise that the 1535-cubic-inch engine would be producing something on the order of 900 horsepower by the time the first production model was ready to go. In fact, most of the XFT-1 tooling was employed to construct the wing, fuselage and empennage. The Model 3A did have a beautifully efficient and simple retractable landing gear that would have made Major de Seversky's fixed gear look archaic.

Boeing had tried in 1934 to sell their XP-29 pursuit – in various formats – to the Air Corps with little to show for it. With every dollar committed to the XB-15 and Model 299X (XB-I7), Boeing did not try at all in 1935. Lockheed was out of the picture, especially after the bankruptcy of Detroit-Lockheed and the almost simultaneous loss of their XP-24 prototype in 1931.

From Buffalo, New York, Curtiss Aeroplane Division submitted what might well have been the most advanced competitor of all, designed by newly hired Donovan Berlin. He had left Northrop in the depth of the Depression to become chief engineer at Curtiss.

Many features of the new Hawk 75 configuration reflected his work at the El Segundo factory, but the airplane was larger and heavier than Northrop's 3A because Berlin believed he would have more horsepower available. His airplane was equipped with the same Wright XR-1670 type engine that was installed in the SEV-2XP, although the cowling and installation on the Seversky was far superior to the Curtiss installation. Fitted with a 3-blade Hamilton Standard propeller, the XR-1670 engine was supposed to develop 775 horsepower, barely more than the then current P&W R-1535. (Published figures rated it at 705 horsepower.) The big difference was that the Wright engine was totally unreliable.

The immense damage done to American industry by the Great Depression is incalculable. Just one facet of it can be seen in the aircraft appropriations for Fiscal Years 1930-31 (which actually began in July 1929) and FY 1933-34. In the earlier year, the Army and Navy were allocated approximately $71 million; the comparable budget for 1933-34 was only $56 million. It may have only been a 20 percent drop, but it certainly was not likely to provoke manufacturers into spending much for developments that would possibly appear on the scene in the next few years. Everyman's view of the future was pretty bleak, and the H.G. Wells novel, "The Shape of Things To Come" was unlikely to prompt many smiles.

Just to add to Lt. Kelsey's feeling of impending doom, Seversky's new SEV-2XP suffered the indignity of an accident. It

was returned to the Experimental Shop, but details have been cloaked in some sort of secrecy and failed memories. Two important competitors were already late and Curtiss was having so many powerplant problems that people in Buffalo remained relatively quiet. The Wedell-Williams airplane just seemed to evaporate. Kelsey took what seemed to be the only logical step and managed to postpone the competition until August. We presume, in the face of no disagreement, that the statement

***ABOVE AND BELOW:** Lt. Ben Kelsey, with approval from his superiors, delayed start of the 1935 competition for about 30 days. Northrop and Wedell-Williams were "no shows" and Boeing simply did not respond, even though their earlier YP-29 models were far more advanced than Seversky's SEV-2XP. Farmingdale crews worked feverishly to revamp the crestfallen 2XP, installing a semi-retractable landing gear of elementary design and a windshield/sliding canopy borrowed straight from the Curtiss 75 entry. Certainly it was a 1-seater, but with a barn-size luggage compartment. It was "forward to the past."* **(Seversky/Author's Coll.)**

***ABOVE AND BELOW:** Wright's Twin Whirlwind engine was a disaster, forcing Seversky and Curtiss to seek other power. The Major almost instantly installed a Wright R-1820F engine (as in the SEV-3M), doing the conversion at Dayton, while Curtiss' Donovan Berlin unexpectedly went to the Pratt & Whitney R-1535. Both airplanes needed 1000 horsepower, and both were struggling to see 750 hp. Curtiss then turned to a Wright R-1820, believing their own corporate claims. Alex de Seversky hired highly recommended pilot-engineer C. Hart Miller to manage installation of a new P&W Twin Wasp. (WF/AAC)*

"Oh well, back to the old drawing board" was born at that moment.

Somewhere, sometime with all the confusion as summer came on, Curtiss removed the disappointing Wright Twin Whirlwind engine from the Hawk 75 and went to work installing a Pratt & Whitney R-1535 in its place. Results were disappointing because the fairly large Hawk 75A was unlikely to even match the Northrop 3A. In fact the Curtiss airplane was more like a moving obstacle course.

By the time Seversky's SEV-2XP was back in the shop at Farmingdale, the single-seat pursuit requirement had gotten through to the Major and to Alex Kartveli. Everybody went into action as if their lives depended on it. The engine went to overhaul, the wings were removed for installation of a retractable landing gear and the fuselage cockpit area was reconfigured for one-man operation. Because engineering design work had already been completed in connection with the SEV-5 and SEV-7 proposals for use of a retractable landing gear as an interchangeable alternative to the amphibious floats, the design team chose the flawed simple aft retraction system. While it added drag to an otherwise clean airframe it did give the Severskys unforgettable character. Kartveli could not even rotate the wheels to lie flat because Curtiss-Wright held a patent on that system which was designed by Walter Tydon. With the gunner's position deleted, the newly configured pursuit had what somebody described as "one helluva baggage compartment." It surely would have delighted the hearts of certain highly placed Army officers. (A long-standing formal requirement for U.S. military airplanes was in all specifications. It established minimum standards for a luggage compartment.)

With work completed rapidly, the revamped airplane was reidentified as the SEV-1XP (but still c/n 2). Having been apprised of another overlooked general requirement, there is evidence that the manufacturing team painted the fuselage a metallic blue and the wings and empennage received a chrome yellow Air Corps paint scheme. Wright's R-1670 Twin Whirlwind was still occupying that space between the firewall and the propeller, but by the time the designated pilot flew the pursuit back to Wright Field the engine was a worrisome thing. It must be presumed that the Seversky team learned of Curtiss plans to install a single-row Wright R-1820 Cyclone in the Hawk 75 because a spare Wright Cyclone (ungeared) that was probably earmarked for the SEV-3M soon found its way to Dayton. They quickly discarded the twin-row Whirlwind at Wright and adapted the SEV-1XP to accept the extra 150 cubic inches of engine displacement with its greater weight. For this and later installations the cowling was moved closer to the propeller arc and cowling length was increased by a couple of inches. Not only did the 54-inch diameter engine look out of scale on the SEV-1XP when compared to the 44-inch diameter of the Twin Whirlwind, it made the pursuit almost unmanageable because of the small empennage. A teletype message was dispatched with all haste to Farmingdale.

We can guess that the Major was not at all pleased with the message because he was preparing to enter the SEV-3M amphibian in – of all things – the 1935 Thompson Trophy Race at the National Air Races in Cleveland, Ohio. The highly regarded closed-course event for unlimited (powerplant) racing planes had a revised format, up from 100 miles (12 laps) in 1934 to 150 miles and a total of 15 laps. The typical entry had a wingspan of less than 20 feet, but the SEV-3M had a span of 36 feet and was mounted on two floats, each as large as the fuselage on all but the more powerful racers.

Neither snow, nor sleet or rain... There was the Major, wearing his typical Stetson "helmet" running up the newly modified Demonstrator, hoping to see near 800 horsepower output, but lucky to be getting 715 at any time. That large hoop in front of the windshield was a radio D/F (direction-finding) loop antenna for navigation. Rarely (if ever) mentioned, the heavier engine required new, longer floats and significantly revised water rudders. (Mfr.)

Before the summer was even one third over, Lt. Kelsey's plans took yet another blow. In the early part of July, Jack Northrop and a team of engineers and builders that included Ed Heinemann (destined for design fame after WWII) completed work on the new 3A pursuit plane at El Segundo, California. It was wheeled out of a hangar in the complex across Imperial Highway at Mines Field, the major Los Angeles airport, and test pilot Arthur H. Skaer, Jr., prepared to take off on his initial test flight on July 10th. The commonplace fog had burned off by then and Skaer made a westbound pass with the tailwheel barely four feet from the surface of the runway. As the Model 3A passed over the beach highway, it banked southward and was never seen again. Although it was determined that the Northrop crashed into the ocean, no parts of it were ever located.

The 1935 Cleveland National Air Races began on 30 August with arrival of Bendix Trophy race competitors. Unfortunately for the racing plane pilots and the air race fans, the weather at Cleveland on Thompson Trophy race day, the final day of the Labor Day weekend, was dismal. The pilots flew in a drizzling rain and low ceiling conditions. Harold Neumann won the event in Benny Howard's DGA-6 "Mr. Mulligan," a conventional-looking cabin monoplane that was really a wolf in sheep's clothing. It was powered by a cleverly disguised P&W R-1344 Wasp radial rated at a ferocious 830 horsepower. Major de Seversky had no pylon racing experience, so the popular Lee Miles flew the SEV-3M and managed a creditable fifth place in the event. The incredible thing is that nobody had ever entered any kind of amphibian in the Thompson event, let alone succeeding in beating out a true blue racer like the Brown "Miss Los Angeles," which it did. Quite frankly, whipping that big 5500-pound floatplane around pylons in tight formation with the likes of famous Steve Wittman and highly regarded Roger Don Rae took real skill. In the rain. It was akin to running a stock car in the Indy 500 race. It was astonishing.

Just two weeks later, on the weekend of September 14-15, Major de Seversky made four consecutive passes over a 3-Kilometer race course at Wayne County Airport, Michigan, during the scheduled Michigan Air Circus weekend. Flying the same SEV-3M amphibian which was powered by a Wright R-l820F-3 engine rated at somewhere between the published 715 and 750 horsepower figures, he broke the existing amphibian speed record by about 39 miles per hour.

His average speed for the four passes virtually "on the deck" was an officially timed and homologated 230.03 miles per hour. His speed absolutely equaled the best recorded top speed of any Boeing P-26 pursuit plane, the backbone weapon of America's first line of defense.

John Northrop sent a request to Materiel Division asking for further delay in the competition beyond August 1935, affording him time to construct a replacement for the Model 3A. Engine performances all around were giving the manufacturers palpitations (or worse), so it did not seem unrealistic to ask. In a totally unexpected response, Curtiss Aeroplane Division filed a firm objection to such action. Northrop was furious. He pulled out of the competition. Absolutely by coincidence, Chance Vought's top management had decided that they needed to move into the realm of all-metal aircraft. They then lacked such expertise since even their latest XSB2U-1 scout bomber was built with the same fabric and fasten-on metal plate technique as North American Aviation was employing on the BT-9 basic trainer. Within a few days Northrop and United Aircraft/Chance Vought worked out details for the East Coast manufacturer to purchase all design rights, basic tooling and drawings from Northrop.

What with Curtiss and Seversky both in the throes of major powerplant revisions to their entries, Lt. Kelsey – most likely with the approval of B/Gen. Augustine W. Robins or his Materiel Division staff – announced the decision to postpone the 1935 competition until April 1936. Lieutenant Kelsey's political education was in the process of showing that it would be tougher than obtaining his degree at M.I.T. had been. People did not always play by the rules.

That big single-row Wright R-1820 Cyclone looked better in the third iteration of the Curtiss Model 75, designated 75B, than it did in the second SEV-1XP, but it still failed to produce an honest 750 hp, let alone the promised 1000. A huge myth existed about the superiority of American radial engines by 1936, and both major engine manufacturers were so close to failing that design engineering was incomplete at best. Those problems were exacerbated by Air Corps desires to produce great inline engines for projected aircraft despite limited budgets and by an overabundance of engine projects within each company. The Curtiss 75B was on hand at Wright Field (on the new apron) in the spring of 1936 in this form; it failed to outpoint the Seversky SEV-1XP. (AAC/WF)

CHAPTER 3

QUICK SUCCESS AND A RUDE AWAKENING

Charles Hart Miller – far better known as C. Hart Miller in all aviation circles – was destined to become a powerful figure in aeronautics although for a few years during the seemingly eternal Depression era there wasn't a glimmer of that. Hart would have been happy to see even the faintest glimmer of a job. So, one might ask, how did he come to be such an important cog in the P-47 saga? Miller was so important to the Thunderbolt story that he might be considered akin to the football place kicker who wins the championship games, one after another, in the last few seconds before time runs out. Just for the record, he had a framed award hanging on his Laguna Beach condominium wall confirming it was he who gave the P-47 its name. That was but one of numerous contributions.

Following attendance at Millersburg Military Institute in Kentucky, Hart Miller gained a degree in engineering before joining the Army Air Corps cadet program during the Hoover presidency. After assignment at March Field, California, for a time, he was sent to Kelly Field, Texas, for flight training. By June, 1930, he had graduated, only to crash headlong into a situa-tion nobody had ever foreseen. It was destined to change his life pattern.

"I was caught in the Hoover economy program," Miller recalled, "and not one member of our June graduating class at Kelly ever received a Regular Army commission." Grinning broadly, he said "At the end of sixteen months the Air Corps ran out of money and gave us a brief kiss of death by saying, 'Well we're sorry boys, but goodbye.' and we were out in the cold." That's the way it was with millions of people in the Great Depression days.

It was in that same period that Major Alexander P. de Seversky was struggling to get the prototype SEV-3 built. He had to be one of America's all-time optimists to have incorporated a business based on a "sportsman's" expensive amphibian. The rollcall of aircraft manufacturers seeking protection under bankruptcy laws was like a reading of Who's Who in Aviation. Alexander, Detroit Aircraft, Fokker America, Hamilton, New Standard, Buhl, Verville, Emsco, Inland, Huff-Daland, Kellett and Pitcairn were but a few of the familiar names appearing on the lists. Many other famous names merely disappeared via the merger route. As for those that remained, the margin between survival and foreclosure was very slim for most.

Miller searched for a job in a totally jobless market, failed to find anything in months and decided that it was logical to return to school – at New York University – to get his graduate degree in aeronautical engineering. Seversky's SEV-3 airframe was moving toward completion just a few miles away from the college inside Edo Corporation's small plant over at College Point, Long Island. The paths of Miller and de Seversky were not yet fated to cross. But they were getting close, very close. After Miller obtained his advanced degree, a close friend, Donald Putt (who was later to rise to the rank of a general officer), called in 1935, to tell him about a possible job with the Seversky organization. Reserve Lieutenant Hart Miller availed himself of the opportunity to go on active duty for two weeks at Schoen Field, Fort Benjamin Harrison, Indiana. A couple of cross-country flights to Wright Field in the summer produced some very interesting results. The Seversky SEV-1XP with its big Wright Cyclone engine sat forlornly in a hangar, unable to even come close to attaining the 300 mph speed "guaranteed" by Major de Seversky. He explained to Miller that the ungeared R-1820 engine was way down on power. The pursuit plane was directionally unstable and had other poor flight characteristics according to the Major.

Pratt & Whitney's R-1830 Twin Wasp joined the successful, but smaller displacement, R-1535 Twin Wasp Jr. to anchor the division's position as one of the top engine builders in the world. That twin-row radial type engine, in various versions, powered the P-35s, P-36s, P-43 and B-24 types with great success and reliability. This early version, Model S-BG of the type used in Seversky P-35s, was rated at 1000 hp, produced slightly less in service form. It was generally known as an R-1830 "B" type; the more powerful "C" types had downdraft carburetors. (NEAM)

OPPOSITE: While Curtiss chased its tail, trying to find a proper engine, a team (still at Wright Field) led by C. Hart Miller worked feverishly to install a brand new P&W R-1830 Twin Wasp. Intended to produce 1000 horsepower, it was struggling to provide 850 in August 1935. Curtiss got a taste of their own attitudes when P&W could/would only provide an R-1535 engine rated at about 700 hp at that time; C-W railed against a second, multi-month delay in July, forcing Northrop to abandon the competition. A furious John Northrop sold the Model 3A design and tooling to United Aircraft (Vought). Pratt & Whitney was also a part of UAC! It was a bitter story, and Curtiss was saddled with a sorely underpowered Hawk 75A. Hart Miller's version of the SEV-1XP is pictured as it appeared on 25 August 1936 at Farmingdale, obviously showing it was Thunderbolt's grandpappy. Almost exactly a year earlier it had looked essentially the same at Wright Field. (Mfr.)

The reserve lieutenant related this story about their brief meeting at Wright. "I saw de Seversky and gave him details of my background. He hired me as a sort of pilot-engineer and told me to report to the plant (Farmingdale), but I was to call him just before I left Lexington (Kentucky). I forgot about the time on the night before I planned to leave and didn't call until 11 o'clock in the evening." Miller could not stifle a chuckle as he recalled that "I woke him up." It was not the best way to make a good impression. "But as we went out the door the next morning, he called me and said to come by way of Dayton. I got to Farmingdale two months later."

Many business arrangements tended to be a bit informal in those slow-paced peacetime days, but it eventually became clear that Hart Miller was the Engineering Assistant to the President – at the ripe old age of 26. The reasons for diverting the young man to Wright Field involved the cantankerous performance of the Cyclone engine, the overall poor handling of the SEV-1XP at that stage of its development, a need for the Major to be seen at the National Air Races in Cleveland. Since arrangements had already been firmed up for the record attempt at Wayne County Airport near Detroit, Michigan, it played a significant part.

But even more important, de Seversky had somehow managed to obtain a brand new Pratt & Whitney R-1830 Twin Wasp 14-cylinder radial development engine for use in the pursuit plane. (It must be presumed that since Lt. Ben Kelsey was pressing Curtiss to install one of the Twin Wasps in their Hawk 75, civilian registry X17Y, he also suggested that Major de Seversky do the same with his SEV-1XP. Curtiss rejected the idea, but by that time the Major was willing to try anything. The influence of Materiel Division and possibly B/Gen. A. Warner Robins is probably what turned the trick.) The assignment given to Miller: Install the new R-1830, provide mounts for the required machine guns (up until then had been overlooked), and correct the handling deficiencies. It was time for the young engineer-pilot to do what had to be done.

On the day that Miller arrived in Dayton he found the following situation. "They were building up the new P&W R-1830 powerplant to replace the Wright engine," he recalled, "but it turned out to be a much bigger job than anyone expected." Once again the engine moved forward a bit and a completely new cowling had to be fabricated to accommodate an updraft

***Half a decade before the Pursuit Competition was convened, Britain's Air Ministry issued a specification F.7/30 for a new interceptor-fighter. Supermarine responded with a Type 224 Spitfire (unofficially) monoplane powered by an experimental Rolls-Royce Goshawk III inline engine rated at 660 hp. Completed in 1933, it had an inverted Corsair-like wing and Severskyish landing gear fairings. It was no faster than the SEV-3M amphibian, but it led to a new design with the same name...Spitfire.* (Vickers-Supermarine)**

carburetor. They had to increase the rudder and elevator areas, which involved changes in the fin and horizontal stabilizers. All of this work was accomplished in an Air Corps hangar at Wright Field, although some of the component fabrication effort was performed with metal-forming equipment back at Farmingdale.

Time had run out, or at least patience had in the Materiel Division offices at Wright Field. The Curtiss airplane, now designated Hawk H-75B. was plagued with somewhat the same powerplant problems that the Seversky people had encountered with their third configuration of the SEV-2XP/1XP. Northrop's Model 3A had crashed and so had Jack Northrop's patience and temper, normally very mild. Chance Vought sought the opportunity to enter their version of the Northrop airplane by the following spring, and it was obvious that Wedell-Williams was unlikely to be well received. Kelsey was seeing impending doom. Curtiss' Hawk was performing more like a tired dove, encountering great difficulty attaining a speed in excess of 280-285 mph when a minimum of 295 was required. As of August 15, when the near identical Cyclone was still powering the Seversky, the maximum speed attained had been 289. The Major had promised no less than 300 mph speed.

Nobody seemed to really know what power was being produced by the Cyclones, but reliability appears to have been the greatest problem. As might be expected, the development-series Twin Wasp was unable to produce the performance needed from an engine being touted as being a 1000 horsepower design. American engine manufacturers evidently had run into a solid wall in their efforts to obtain more than one-half horsepower per cubic inch of displacement. The Germans were having comparable problems, but the British were proving to be the exception. Their Rolls-Royce Merlin was not, at that moment, producing 1000 certifiable horsepower consistently but it was turning out well over the half-horsepower per c.i. figure.

After the competition was rescheduled

***Isolationism and the Great Depression certainly shaped design philosophies relating to American Army pursuits and Navy fighters by retarding development of liquid-cooled engines. Curtiss gave up on any follow-on to the Conqueror engine, while Allison development of the V-1710 lagged. The Rolls-Royce Merlin C was an immediate 1000-horsepower success, leading to the prototype Hawker Hurricane (K5083) regularly attaining 320 mph in late 1935. (Sadly, more than 3 years would pass before Allison and Curtiss could get a production prototype to even match that speed.) Flush landing gear was effective and optimum in simplicity. Blame for the Seversky 1XP/P-35 landing gear perversion has to be placed equally at the feet of the Major and Kartveli.* (Hawker Aircraft)**

***OPPOSITE: Somewhat amazingly, Seversky's advanced-design SEV-3XAR beat out the very conventional North American NA-16 design for the first real Air Corps contract to procure basic trainers. Thirty of the sleek BT-8 model were on order, and by the time Seversky was declared winner of the Pursuit Competition, the trainer was in production. The first one (AC34-247) is shown in flight sometime after New Year's Day, 1936, with the Major (as usual) at the controls.* (S. Hudek Coll.)**

Carrying full Colombian Air Force markings and a temporary export license number, one of the three SEV-3M-WWs built (not 5 or 6) poses in winter snow at Farmingdale. Major de Seversky is manning the big single-lens reflex aerial camera in this publicity picture from the mid-1930s era. Notice that the water rudders were changed significantly from those used on the original SEV-3.This was first of three airplanes manufactured, but all three were shipped aboard the Nichi Maru to Colombia on July 15, 1936. Although unconfirmed, pilot at controls may be Roger Q. Williams, hired by Colombia to oversee construction and delivery. **(File)**

for April 1936, the Chance Vought/United Aircraft Corp. received a terrific shot in the arm. On September 13 out at Santa Ana, California, Howard Hughes flew his P&W Twin Wasp Junior-powered H-1 racing plane, designed by Richard Palmer, over a 3-Kilometer course to establish a new international speed record for landplanes. His FAI-homologated average speed was 352.388 mph.[1] The R-1535 powerplant was essentially the same as the one that was to be used in Vought's version of the Northrop 3A pursuit and A-17 attack planes.

Following cancellation of the 1935 competition announced by Lt. Kelsey sometime in August, the SEV-1XP was flown back to Farmingdale, most likely by the Major or Hart Miller, with the new P&W Twin Wasp doing the honors. Evidently the Experimental crew at Wright Field had performed a near miracle, revamping the airplane in a matter of weeks. But that is exactly why Miller had been hired. The pursuit plane looked pretty good and de Seversky seemed to be pleased with it. But after he fractured the existing amphibian speed record at Wayne County Airport in September he evidently had a change of heart (or mind). The following scenario facts were misunderstood or ignored for more than a half century, but the appearance of new photographic evidence led to questioning of former members of Seversky's Experimental organization. Department of Commerce records and answers to the questions provided proof that the SEV-1XP military prototype went through two more major modification events before returning to Wright Field in April 1936.

The successful record and racing efforts of 1935 must have convinced Major de Seversky that the geared successor to the Wright R-1820F-3 or F-5 would boost the performance of the SEV-1XP up to 300 mph as promised. In a small corner of the hangar, cor-

[1] Dogged stubborness and inflexible minds cost America a truly great opportunity in 1936, possibly as early as 1935. Howard Hughes contacted the top officers in the War Department and at Materiel Division, hoping to sell his H-1 racer to the Air Corps as a pursuit plane. It really was not outlandish to consider it, but Hughes wanted to force his way in and the Wright Field people naturally resisted on the basis that "your beautiful racer has to meet specifications." (It featured a wooden wing which would not meet contemporary requirements. That also may have been the key to failure of the Wedell-Williams to appear. If only Jack Northrop and Howard Hughes/Dick Palmer had combined forces, by 1936 the Air Corps might well have been procuring a 330+ mph pursuit with superior growth possibilities. In a matter of months, Hughes raced nonstop clear across the Continental U.S. in the same airplane at an average speed of 327.5 mph!)

had more or less enjoyed great success in the fields of bomber and fighter aircraft. The U.S. Navy's carrier-based fighters were unchallenged in their performance by naval aircraft of any nation. The Air Corps Martin B-10 bomber was without a peer anywhere, and the Boeing P-26A "Peashooter" pursuit was technically, if not in performance, advanced over the latest Royal Air Force and French *Armee de L'air* types in service. The multiseat Consolidated PB-2A pursuit was far ahead of any competition except that Lt. Kelsey and others considered the 2-seat fighter to be a ridiculous concept at best. The Bell XFM-1, first flown in 1938 (less than a year before Lockheed's 130-mph faster XP-38 took off from March Field for its initial flight) proved Kelsey's premise. Both airplanes had the same Allison V-1710 supercharged engines of essentially identical power. The Kelsey-inspired XP-38 carried as much firepower (even more concentrated) and had greater range from the start. As for the PB-2A, it must have been somewhat embarrassing to Air Corps people to know that in 1935 the big float-equipped Seversky sport plane had shown a fast lap of 235.96 mph, outdoing the best recorded speed for the pursuit plane by about 4 miles per hour.

Before the first production BT-8 was wheeled out of Seversky's Bldg. 5 production facility (about the size of Lockheed's out in Burbank) early in 1936, it was photographed with others on the assembly line. The two men at far left appear to be the Major and his "financial angel" Paul Moore. (Mfr. via Hudek)

doned off by a wall of draperies to maintain some element of secrecy, a team of engineers and technicians performed extensive modification work on the airplane in the middle of the 1935-36 winter. Production of BT-8 basic trainers was progressing satisfactorily in the main factory building (Bldg. 5) and design activity was under way on a new trainer, the SEV-X-BT.

An 830-horsepower version of the Wright Cyclone, the GR-1820G-5, was installed and this time the empennage was revised extensively. They had created a monster which took honors as the ugliest Seversky ever built. Most members of the modification crew posed with the airplane when it was rolled out onto the sod airfield at Farmingdale. The legendary Hal "Pop" Provo, who led much of the modification work, posed with the group, but down on his knees. Was he asking forgiveness? It is more than likely. The naked (it had been stripped of all paint in the rework process) SEV-1XP – *sometimes* referred to in official documents as the SEV-4 – was also pictured with one SEV-3M-WW, the first production BT-8 and the SEV-3M on wheels. Since the single-seater was due at Wright Field for the 1936 "running" of the Pursuit Competition, test flying had to proceed with haste. Results were so bad that nobody even suggested obtaining photos in flight.

Practically overnight a P&W R-l830B engine rated at 850 hp, unless that was one of the numerous documentation errors that dotted the records, was reinstalled and the tailplane reverted to the one installed by Miller's team at Wright Field. The pursuit airplane was painted in the blue and yellow military scheme (soon to become history). Then it was off to Dayton for "finals."

The Depression's damage to the aircraft and engine industry must have been far greater than contemporary publications reported. Up to the time of the 1935 competition failures, America

Meanwhile, British and German designers were about to pitch any superiority of American fighters into the trash bin, and make spectacular inroads on America's dominant technical position. While the American press seemed to virtually ignore the aviation advances in those European nations, the Spanish Civil War was soon to reveal this country's real status. Back in 1933, British designer Reginald Mitchell – who had created the Schneider Trophy winning 400+ mph Supermarine racers – and Vickers-Supermarine, for which he worked, had produced a low-wing interceptor to Spec. F.7/30 with some of the features exhibited by the Seversky SEV-2XP. The Type 224, defeated in evaluation trials by the technically inferior but more maneuverable and faster Gloster Gladiator biplane fighter, had a "panted" landing gear that was a close relative of those used on the SEV-2XP and the BT-8.

However, Mitchell's monoplane was big and was powered by an experimental inline Rolls-Royce engine. To his everlasting credit, the British designer recognized the flaws in his design. Inspired by the sleek lines, wing shape and retractable landing gear of a German Heinkel He 70 employed as a test bed for versions of the Rolls-Royce Kestrel and prototype Merlin engines, he laid out the lines of a new fighter. Vickers-Supermarine agreed to backing a private venture interceptor which the designer proposed, the beautifully sleek Type 300 Spitfire. The Royal Air Force assigned a valid RAF airplane number, K5054.

Arrangements were made for Vickers chief test pilot, Capt. J. "Mutt" Summers, to make the initial test flights. The interceptor lifted off from Eastleigh Airport on 5 March 1936. Performance was so spectacular that the British Air Ministry awarded a production contract for no fewer than 310 examples just one month later. If the American Air Attache in London failed to send an

immediate teletype to State Department officials in Washington, with a copy to M/Gen. Oscar Westover, Chief of the Army Air Corps about published reports on this new British interceptor, he was certainly unqualified for his job.

Exactly four months earlier, on 6 November 1935, the similarly powered (Rolls-Royce Merlin) Hawker Hurricane prototype had flown for the first time with Group Captain P.W.S. Bulman at the controls. The RAF soon revealed that it had attained a speed of 320 mph even though the initial engine was not providing the normal power expected. In the same month that Vickers-Supermarine was awarded the contract for 310 examples of the Spitfire, Hawker received a huge Expansion Scheme order for 600 Hurricane Mk.Is. Hurricane and Spitfire prototypes had fixed-pitch wooden propellers which compromised their speed capabilities in comparison with the 3-blade Hamilton-Standard constant-speed propellers then available in America. It did not take long for information to come out that the Spitfire had flown at an astonishing 349 miles per hour at 17,000 feet altitude. All subsequent indications were that nobody with any authority over military aviation in the U.S. raised a finger to stop the near farcical contest that was about to be renewed at Wright Field in early spring, 1936.

At least one person in the Air Corps learned a major lesson from the 1935-36 competition. That was Lt. Benjamin S. Kelsey. He proved that in 1936 by drawing up the radical specifications, essentially on his own volition, leading directly to the 1937 design competitions for single-engine and twin-engine fighters. The two winners of the 1937 *design* competition were Bell's XP-39 and Lockheed's XP-38, both being extreme departures in virtually every way from anything the Air Corps had ever specified. Kelsey was risking his military career in generating formal requirements which deviated so radically from what had been looked upon as standards.

Fate decreed that nobody would issue a logical order to cancel the 1936 competition at Wright Field, and perhaps that seemingly innocuous inaction was the key to the eventual appearance of the Thunderbolt almost exactly five years later. Seversky Aircraft would almost certainly have failed but for the events that took place as the baseball season got under way in

The Experimental Shop crew posed with the terribly disappointing Cyclone "G"-powered SEV-1XP at Farmingdale in late winter/early spring 1936. Talented "Pop" Provo, a flight line fixture at the plant, is shown kneeling in front of the propeller. Tall young man in overcoat is W. Howard Ehmann with Walter Hoenes by his left arm. Brother Erwin Hoenes and Judson Hopla were the Major's most valued mechanics. (Hal Provo)

1936. Preparing to walk off with the production contract for pursuit airplanes and Wright engines, Curtiss' Donovan Berlin had concentrated on polishing the Wright Cyclone SRG-1820-G5 engine installation in the Hawk H-75B, probably not knowing that Seversky had quickly abandoned that powerplant after intensive work to install it in the SEV-1XP. Having speedily returned to the P&W R-1830 Twin Wasp in spite of that engine's failure to produce the promised 950-1000 horsepower, de Seversky and Hart Miller were committed.

After observing the debacle at Wright Field in 1935, Major Reuben Fleet, as president of Consolidated Aircraft and close friend of Air Corps chief M/Gen. Benny Foulois (who had just retired in August 1935) ordered rework of a "bailed" PB-2A. The newly relocated company was so cheap in creating their entry that the designers did nothing more than eliminate the gunner's seat and equipment from the aging 2-seat pursuit so that they could add a metal turtledeck, slightly improving the aerodynamic shape. They hardly reduced weight; they did not increase horsepower over their 1933 specifications; they even kept all of Detroit-Lockheed's 1930 external dimensions.

Frankly speaking, the airplane that arrived at Wright Field in time for the 1936 rendition of the Pursuit Airplane Competition was an anachronism. It was 20 mph slower than the RAF's Gloster Gladiator biplane that had been placed in production in England in 1935. Reuben Fleet should have been too ashamed to appear in public. Before autumn made an appearance, his entrant had been written off. (Through a strange twist of fate, the Consolidated PB-2A spinoff that was entered in the 1935 competition *a year late* had stemmed directly from the Lockheed XP-900/XP-24 of 1931. This writer has stated in previous works that even Detroit-Lockheed would have been far ahead of the game if they had merely scaled their XP-900 down to the size of the 1924 Verville-Sperry R-3 racer they had plagiarized in the first place. Mercifully, the Consolidated entry just faded away.)

Major Reuben Fleet had status and insider information available to him for the asking in those days, but he certainly failed to use those attributes to his advantage. When it became a known fact that Britain's Rolls-Royce Merlin I was developing a fraction more than 1000 horsepower in the latter months of 1935, perhaps Fleet or some Materiel Division officer had access to status data about America's own General Motors-Allison V-12 engine. Even if they could only obtain 1000 hp for a few minutes but knew that the prognosis for later success was excellent, Fleet (with a little vision) might have obtained one on bailment for installation in a lightened and improved PB-2. Any performance in excess of 300 mph would have probably won him a contract. Such an award to that manufacturer would surely not have been in the best interests of Americans.

Consolidated's top management did not spend one legitimate penny on improvement of the breed. They simply "greased the pig" to see if they could slither into the winner's seat. In retrospect, that is exactly what that same company had done in taking over Detroit-Lockheed's XP-24 advanced design (for 1930) to develop into the PB-2A after Lockheed Aircraft (California) had been "milked dry" by expansion-crazy New York stock speculators. In the roaring twenties it wasn't illegal to pass on insider information, and having friends in high places is still associated with perquisites ("perks") which, we all know, are virtually impossible to control.

Abandoning the cowling design which U.A.C.'s Pratt & Whitney had pushed so hard for Northrop to use on the 3A, the corporation's Chance Vought Division built a complete monocoque fuselage for their new V-141 Army Air Corps pursuit plane in just a month and a half. A conventional NACA cowling was fitted around the R-1535 Twin Wasp Jr. and they simplified the landing gear fairings, but otherwise the airplane was just about what Jack Northrop had envisioned late in 1934. Considering that experience with all-metal stressed-skin structures was minimal at Vought's East Hartford plant, they had performed a remarkable task. The V-141 was left in natural metal finish but it did carry the basic Air Corps markings and no civil registration number.

It did not take much flight time for the test pilot to encounter directional instability and spin recovery traits that were unacceptable, in fact very dangerous. With that short Northrop XFT-1 fuselage moment arm and the tiny tailplane, any damping action that the original large landing gear pants probably offered was suddenly missing because of the flush gear installation.[2] Vought's immediate solution was to increase rudder area by a couple of square feet. It proved to be only a bandaid on a more serious problem, and performance of the airplane on less than 750 horsepower was not likely to win them a contract. (It has been previously exposed by this writer that many facets of the Mitsubishi A6M1 "Zero-Sen" fighter of WWII stemmed directly from sale of the V-141-*cum*-V-143 and all design rights to the Imperial Japanese Navy by 1938.)

Nobody had to be a genius to know that Hawker Aircraft and Vickers-Supermarine in Great Britain were actually in poorer financial condition than America's Consolidated Aircraft, but that hardly deterred them from doing their absolute best to design new interceptors with private venture capital paying the bills. Fairness dictates that all of the home-grown entrants in the Pursuit Competition be directly compared to the Hurricane and Spitfire contemporaries designed in England to properly evaluate America's competitive determination. It was quite obviously of little importance to industry.

There is an interesting sidelight to that speculative vision of the PB-2/Allison aircraft/power-plant combination. General Mark Bradley, an officer who was in time to be a key player in the Republic P-47 program, had been involved in testing Allison engines when he was a lieutenant. He related a story about the many hours he spent fly-

Totally unknown to all who have published Seversky history in the past, and virtually undetectable in company records – possibly because it was so ugly – a fourth iteration of the SEV-1XP was rolled out of the shop (Hangar 1?) before winter officially ended in 1936. The Twin Wasp had been removed and a geared Wright G Cyclone (R-1820-G) was installed. That big 54-inch dia. engine was supposed to deliver 815 hp at 13,500 feet, but it was hard pressed to show 700. Mammoth cowl and short fuselage forced the installation of a vertical tail 50 percent larger than that on the original SEV-1XP. It was photographed with the BT-8 and the SEV-3XAR in front of the main factory. **(Seversky)**

[2] If stability problems were so quickly encountered with the V-141, it is not at all unlikely that Northrop test pilot Art Skaer fell victim to the same deficiency while flying at very low altitude over the ocean in July 1935. In fact, the even larger cowling may have exacerbated the instability. The knowledge that Skaer was probably distracted momentarily to switch fuel tanks enforces the theory that he may have lost control and had no time to attempt recovery.

ing an Allison-powered PB-2A at Wright Field in the late '30s when the engine was undergoing acceptance testing. As he vividly remembered it, "I used to fly one (a PB-2A) that was powered by a turbosupercharged Allison. When I was sent down to the Flight Test Section in the summer of 1938, that was my first job assignment. Old Umstead (Major Stanley Umstead, chief of Flight Test) said, 'Alright, Bradley, see that thing out there' – what it was was one of those damn Consolidated PB-2s fitted with the first Allison, not a production type but probably a Y (service test) model. 'Well, put time on it.' They had put a lot of lead shot in the tail, and the prop was too big for the landing gear," said the retired former chief of the Air Force Logistics Command in our 1972 interview.

That brings up some interesting "Ifs" about the V-1710 in 1936. If it was ready for turbosupercharging (available in the PB-2A type) and if it could fly for even a couple of hours in that form, the PB-2A Special might have been faster than the new Hurricane, especially above 16,000 feet. However, Gen. Bradley had also remarked that turbo regulator failures were at the top of the list of flaws in the program, especially in 1938.

Nobody would have been at all surprised if some upper-level officer – most appropriately Gen. Arnold or Gen. A. Warner Robins – had canceled the pursuit competition early in July 1935 in view of the performance of competitors. By early 1936, knowledge of the Hurricane's performance was certainly global in scope; it would have been more than logical to cancel the reopening of the competition in May. But fate was to decree otherwise. Logic was surely not at the top of many lists in Depression years. If anyone had been foolish enough to even suggest purchasing manufacturing rights to the Merlin in that decade, derision and various unsavory epithets would have been heaped upon him.

When Douglas Aircraft won a huge contract for B-18 bombers by defeating the Boeing 299-X (XB-17), there was little, if any, doubt the main deciding factor was "how many for how little." The pursuit competition was not a matter of which airplane had the performance superiority; it was more a matter of which had the least dismal performance. The fact that Curtiss Aeroplane had underbid Seversky by $29,412 to $34,900 in quantities of 25 airplanes and $14,150 to $15,800 in lots of 200 evidently could not overshadow the fact that the SEV-1XP was a trifle faster and was far more reliable. Another factor may have been that Berlin and his superiors refused to embrace the P&W R-1830 and steadfastly remained faithful to their own corporate Wright R-1820 in the face of some pressure from Materiel Division.

So it was that Seversky Aircraft was chosen as winner of the competition for 77 pursuit airplanes to be designated P-35 after de Seversky promised to incorporate substantial changes in the landing gear, flap system, powerplant installation, and to totally revise the cockpit/canopy situation. All that remained of the SEV-1XP design when the initial production item was delivered was the famous wing. The contract provided $1,636,250 to cover the 77 actual aircraft procurement plus spare parts equal to another eight airplanes.

Remembering that 1934-35 can likely be rated as the bottom of the Great Depression, it is interesting to note that Alex de Seversky was serving as president and a director of the company. His salary for the year beginning in May 1935 was $17,000 and in the following year even before they had received the contract for P-35 – airplanes – his salary rose to $25,000. Of course his stock must have been going up a bit with the fortunes of the company, so the Russian immigrant was rather well off in a period when a family of four could live very well on $30.00 to $50.00 a week.

Later in life, retired B/Gen. Ben Kelsey confided, "Probably the worst thing we could have done in 1936 was to give Seversky (or anybody for that matter) the production contract for pursuit aircraft. If we had given them a service test order for the more commonplace thirteen airplanes, old de Seversky might have developed a much better airplane...." (Based on information about civil servant and military pay rates in the period, there is very serious doubt when Ben Kelsey sent in his official recommendations about award of that contract he was making more than $150.00 per month including flight pay and allowances. Yet, this young 1st Lieutenant was the Pursuit Projects Officer for the Army Air Corps. Reuben Fleet probably made as much in a day. Alex de Seversky surely had to work a day and a half or even two days. His company was poorer.)

As a matter of fact, that is approximately what occurred. Even with two fat contracts awarded by the War Department in the safe – for 30 of the BT-8 trainers and 77 pursuit airplanes, designated P-35 at the time of award – the Major initiated a pet project for a racer-type pursuit airplane to be designated Seversky AP-2. Ben Kelsey, after four decades of activity, apparently did not have anything to trigger his memory about that creation. After all, it never had a real chance to prove itself and, in 1937, when it arrived on the scene America was falling headlong into what has been referred to as the "Roosevelt Recession." In many ways the country was suffering more than it had in the darkest days of the Great Depression. However, there were other matters of more immediate interest occurring in that period

Vought's V-141 served not only as their entry in the Pursuit Competition (1936 phase) but to bring their own manufacturing and engineering staff technology up to modern levels. They lagged far behind industry levels, even in their newest Navy XSB2U-1 dive bomber which did not feature metal stressed-skin fabrication techniques. Poor spin recovery forced Vought to quickly increase V-141 rudder area by about 33 percent, but it still ranked third behind the Seversky and Curtiss entries. A half-hearted try by Consolidated only won them the "cheap skate" title. (AAC/WF)

In one unprecedented forward surge, the Royal Air Force adopted fast monoplane multi-gun fighters to replace biplanes that had dominated two decades. First flown on March 5, 1936, before the winner of the USAAC Pursuit Competition was announced, the prototype Supermarine Type 300 Spitfire was able to exceed 360 mph despite the limitations of a 2-blade fixed-pitch wooden propeller. The straightforward, uncomplicated landing gear remained virtually unchanged during years of production. In direct contrast to the Seversky SEV-1XP and P-35 landing gear designs, the Spitfire and Hurricane systems were lightweight and created very little drag. (Author's Coll.)

Three well-designed, beautifully fabricated (only partially by Kirkham Engineering) SEV-3M-WWs for the Colombian Air Force were lined up on the Seversky airport immediately prior to delivery. Although the construction of these first-sale military airplanes caused a tremendous amount of friction between the Major (seen here) and Charles Kirkham and nearly bankrupted both firms, the Colombian government was happy to get the amphibians. (Mfr.)

around Farmingdale. (The AP-2 and the disturbing and incomprehensible Seversky model designation system will be examined at the proper chronological time.)

Photographer Luis Azzaraga was known to "doctor" photos quite well, and his camera work often led to the mistaken belief that pictures in this 1936 series were "cut and paste" pictures. Using a large format (Speed Graphic) camera, he captured the Major cruising in his competition-winning SEV-1XP over Manhattan Island's man-made Alps peaks. His lens must have had a remarkably flat depth of field to get the 3-D effect evident. The airplane fuselage was a metallic blue color; wing and tail surfaces were military yellow-orange. This is the Hart Miller version of the SEV-1XP, done under pressure. A factory attempt to improve it failed miserably, forcing overtime effort to get back to this status. (Luis Azzaraga/Mfr.)

American Army and Navy intelligence systems must have gone completely haywire in 1936, or the people who were responsible for long-range defense and war planning were incapable of understanding what was happening at flank speed in Europe and Great Britain and possibly uninformed. If they had done nothing more than read *Aviation* or *Aero Digest* magazines (especially the editorials by cutlass-witted writers like Cy Caldwell and Frank Tichenor) they would have understood that America was rapidly falling behind the real progression in Great Britain and Germany. How could we be allowed to permit years of design breakthroughs to wind up in the old "circular file"? The British produced the Hawker Hurricane and the Supermarine Spitfire despite their own battle with world depression. Even without the benefits offered by some of the best American propellers, those British airplanes could outrun our leading prototypes by 40 to 70 miles per hour in 1936. The Nazis were not yet doing as well because, like U.S. aircraft manufacturers, the Germans were still slowed by serious powerplant development problems. However, objective viewing by even the most vehement anti-Nazi military planner would have revealed that German talent for building effective weaponry was something soon to be reckoned with. Unfortunately, the intelligence types or the planners evidently had taken sedatives. The "2-minute warning" went unheeded.

The War Department, through the good offices of Materiel Division, awarded contract AC-8892 to Seversky Aircraft for seventy-seven P-35 pursuit airplanes plus spares. For all concerned at Farmingdale, the timing was even quite good. Plant expansion initiated to handle production of thirty BT-8 basic trainers was fortuitous, and a work force of skilled production people was certain to be a fine cadre for an even greater number of workers who would be hired. The two "Sashas," de Seversky and Kartveli, combined their talents to ensure that the production P-35 would be a significantly different airplane from the SEV-1XP that won a TKO over vaunted Curtiss-Wright. It is logical to presume that Curtiss' loss to a bunch of White Russian expatriates who spoke with distinct accents goaded them into pushing very hard on Don Berlin to get the Hawk 75 act together. Unfortunately for all Americans, Curtiss-Wright did not really learn what it was to be a hungry fighter. The P-36 was, perhaps, their finest hour. If so, they did not have a contender for any figurative crown. A corporate arrogance persisted throughout World War II, destroying the brain and the heart of a once-great Glenn H. Curtiss dynasty.

CHAPTER 4

AMERICA–THE GIANT GOES TO SLEEP

When the war department authorized the Air Corps Materiel Division to award that large production contract for 77 pursuit airplanes to Seversky Aircraft in the spring of 1936, economic conditions in the United States seemed to be on the upswing. The contract was issued on June l6. The company was in full production on their contract for thirty of the BT-8 basic trainers, an aircraft that would eventually revolutionize U.S. military training programs.

The first production BT-8 made its initial flight on 29 January 1936, and the Major was at the controls – as usual, since he also wore the chief test pilot hat. In appearance it wasn't much changed from the SEV-3XAR prototype, although it was not quite as good looking. Certain military requirements dictated a few of the physical changes, those most evident being in the engine/cowling area, the cockpit canopy and in landing gear details. The fin and rudder looked a bit taller, but that may have been an illusion because of the color and normal Air Corps tail striping. There was no streamlined fairing on the tailwheel which was also redesigned. A tall radio antenna was mounted forward of the windshield. One thing generally overlooked in connection with the BT-8, and a bit later with North American's BT-9, the Air Corps had basic trainers featuring enclosed cockpits and cantilever monoplane wings at a time when the standard single-seat pursuit airplane in service (the Boeing P-26) had an open cockpit and wire-braced wings. All but a couple of the BT-8s were assigned to the sparkling new training center at Randolph Field, Texas. The base was such a contrast to older fields which had preceded it that it was very frequently featured in Hollywood films.

Rather strangely, the BT-8's wheels were carried on oilite

One or two flights proved that the SEV-X-BT lacked adequate vertical tail area, requiring rework prior to its appearance at Wright Field. Alex de Seversky is shown at the controls of the new trainer which had a P&W R-1340 engine derated from 550 to 450 horsepower. It all turned out to be "overkill." A slightly improved North American NA-16 was declared winner of the BT-9 contract With fixed landing gear and simple tubular fuselage structure, it was certainly far cheaper per unit. (Attention is called to the SEV-X-BT registry, X189M, for subsequent reference.) **(Luis Azzaraga/Mfr.)**

When Seversky Aircraft received the order for $1.64 million on June 18, 1936, for 77 pursuits to be designated P-35, it forced an immediate expansion. At the same time, a SEV-X-BT prototype was being designed and built for a new (1937) basic trainer competition. However, it was closer to what the Air Corps hierarchy (should have) wanted for advanced training. Unfortunately, the Major or Kartveli was frozen in place on that cursed semi-retractable landing gear. Lockheed had used flush gear retraction systems since 1931, and North American was about to employ a comparable design on trainers. (Mfr./F. Bamberger)

sintered-bronze plain bearings rather than ball or roller bearings. They often contributed to brake grabbing as they wore. Many pilots blamed a "narrow-tread" landing gear for a serious tendency to groundloop. Actually the tread was probably greater than that on most contemporary monoplanes (especially the P-26). The real cause was in movement of the tail-wheel aft on a shortened strut as compared to the SEV-3XAR. This changed the angle of attack on the zero-dihedral wing. In a proper 3-point landing touchdown attitude, the tip stall was sudden and unexpected. A ground-loop was often the result. Seversky P-35s had a similar problem, and it showed up in the later Republic YP-43s and first production P-43s, leading to the tall tailwheel strut seen on most Lancers. If the pilot of a BT-8 made a perfect 3-point touchdown and the brake grabbed on one side, a groundloop was a virtual certainty. With an increase in weight, the airplane was definitely underpowered. The P&W R-985-11 cranked out some 400 horsepower and actually had a military rating of 450 horsepower, but de Seversky was probably correct in his arguments that the airplane really needed a 500-hp engine. By 1939, all BT-8s were withdrawn from active training and were placed in Class 26 as ground training machines or were scrapped.

Long ignored and almost overlooked in the hectic advances of 1935 and 1936 were the three SEV-3M-WWs intended for delivery to Colombia. It seems evident that Roger Q. Williams, who was overseeing the project for the South American country, did not press Seversky Aircraft to complete all three airplanes on any particular schedule. He did fly each one on a final check flight as customer inspector. Finally, the episode came to an end as the three amphibians (temporarily licensed X15391, X15689 and X15928) were flown a few miles west to North Beach Airport to prepare them for export. On 1 July 1936 they were on their way to Cartagena, Colombia, aboard the Japanese freighter *Nichi Maru*. Sailing from a Brooklyn berth aboard the vessel were Williams, his wife and an inspector who was to be responsible for final assembly and checkout. For a project that began with much optimism, it is probable that everyone involved was delighted when the freighter put to sea with the SEV-3 spinoffs safely aboard and paid for.

As Americans suffered through the Depression into 1935, all was not quiet in the world outside. Italy had flexed its Mussolini muscles in October 1935 by invading Ethiopia, a primitive Afri-

In brilliant summer sunlight, the new SEV-X-BT showed its good lines to advantage. Increased fin and rudder area was required to compensate for the larger-diameter engine that was a bit too far forward. Northrop had made a similar mistake on their 3-A pursuit that evolved from the the Navy XFT-l. Without those large fixed gear pants near the c.g. a larger empennage or longer aft fuselage was required. (Air Corps Materiel Div.)

can nation. With more than 300 aircraft operating in support of the ground troops against ragtag volunteers with little more than ancient rifles to defend themselves, it was akin to giant boxing contender (world title) Primo Carnera attacking an emaciated beggar whose hands were seriously injured.

The economy of the United States was showing some recovery strength in 1936. For example, while only 459 military aircraft were produced in 1935, deliveries increased to 1141 in 1936. (Recovery was jolted in 1937 with the start of the "Roosevelt Recession" that proved to be every bit as devastating as 1935 had been. Aircraft production for the Army and Navy slipped back to 949 units.) Civil war broke out in Spain in 1936, serving as a realistic proving ground for aircraft from Nazi Germany, Mussolini's Italy and Communist Russia. A handful of American types that slipped through international screens were generally second-hand commercial aircraft converted to hopefully perform in the role of bomber or attack aircraft. It was an ugly, frustrating affair.

As the Nazis and Fascists gained strength in Europe, the Japanese went to war with China in the Far East. It seems only fair to look at such wars as the most realistic war games and proving grounds conceived on the face of the earth. The U.S. government, encountering a strong isolationist movement, did everything possible to avoid foreign entanglements. The toothless League of Nations was decaying rapidly without seemingly recognizing the facts. If the American military and naval planners were learning anything from those wars overseas, it was not being reflected in their buying policies. Procurement of those large numbers of Curtiss P-36s and Douglas B-l8s in 1937 and even larger quantities of Curtiss P-40s (simply a 1935 Hawk 75 with an Allison V-1710 unsupercharged engine up front) after the 1939 competition revealed that military planners were either asleep or so painfully suppressed by Congress that they dared not even raise an eyebrow. (Probably the most disheartening example of this occurred after 16 January 1939 when M/Gen. Frank Andrews, Chief of the GHQ Air Force, gave a speech in which he rated the U.S. a 5th-or 6th-rate air power. By February 9 he had been demoted, transferred and replaced by B/Gen. Delos Emmons, promptly promoted to Major General. Andrews was actually one of our best generals.)

***After beating the powerful Curtiss-Wright organization at its own game, the Major retired to Farmingdale to enjoy his victory and make new plans. The SEV-1XP was reworked to remove armament equipment, reminding us that no fixed machine gun provisions were anywhere to be seen on any version prior to the arrival of Hart Miller in 1935. The reworked area is quite prominent. The 1XP began to perform as a demonstrator and racer.* (Seversky)**

Seversky Aircraft prepared to produce P-35 pursuit airplanes for the Air Corps in their "new" factory while they were in the process of completing the Colombian SEV-3M-WWs, manufacturing the thirty BT-8 trainer. The designers appeared to have a host of new projects under way. The SEV-1XP prototype was reworked extensively in October 1936, taking on many of the features soon to be seen on the production P-35s. A new P&W R-1830SB-G Twin Wasp engine rated at 950 horsepower was installed in what was to become the production-type cowling, and a completely new production-design tail was installed as well. Not yet ready to install a redesigned landing gear, the shop crew hammered out a rather crude set of fairings that provided a shape similar to those which ultimately graced the first production airplane. The 1XP was then ready to serve as a development "mule" to test production items, and it was also to assume the mantle of demonstration aircraft, frequently as a racing airplane. There is evidence that orders from Frank Fuller, Jr., and Jimmy Doolittle (for Shell Aviation Products) stemmed directly from the revised 1XP. It was relicensed in January 1937 in this new form, with the X18Y license to expire 9 October. Contrary to some published stories, it never received any other designation.

The two Alexanders, de Seversky and Kartveli (actually Kartvelichvili) – who was by that time a vice-president in charge of Engineering – got their heads together to conjure up the trainer that they believed the Air Corps should have. As time was eventually to prove, they were in the process of telling the military leaders they needed what eventually was to become the AT-6 advanced trainer, an airplane familiar to virtually everyone after 1940. Unfortunately, de Seversky and Kartveli were too far ahead of the planners[1]. North American Aviation turned

[1] Some will contend that military planners in Washington were absolutely "on top of the problems" in prewar days, and that our view is overly critical. If plans

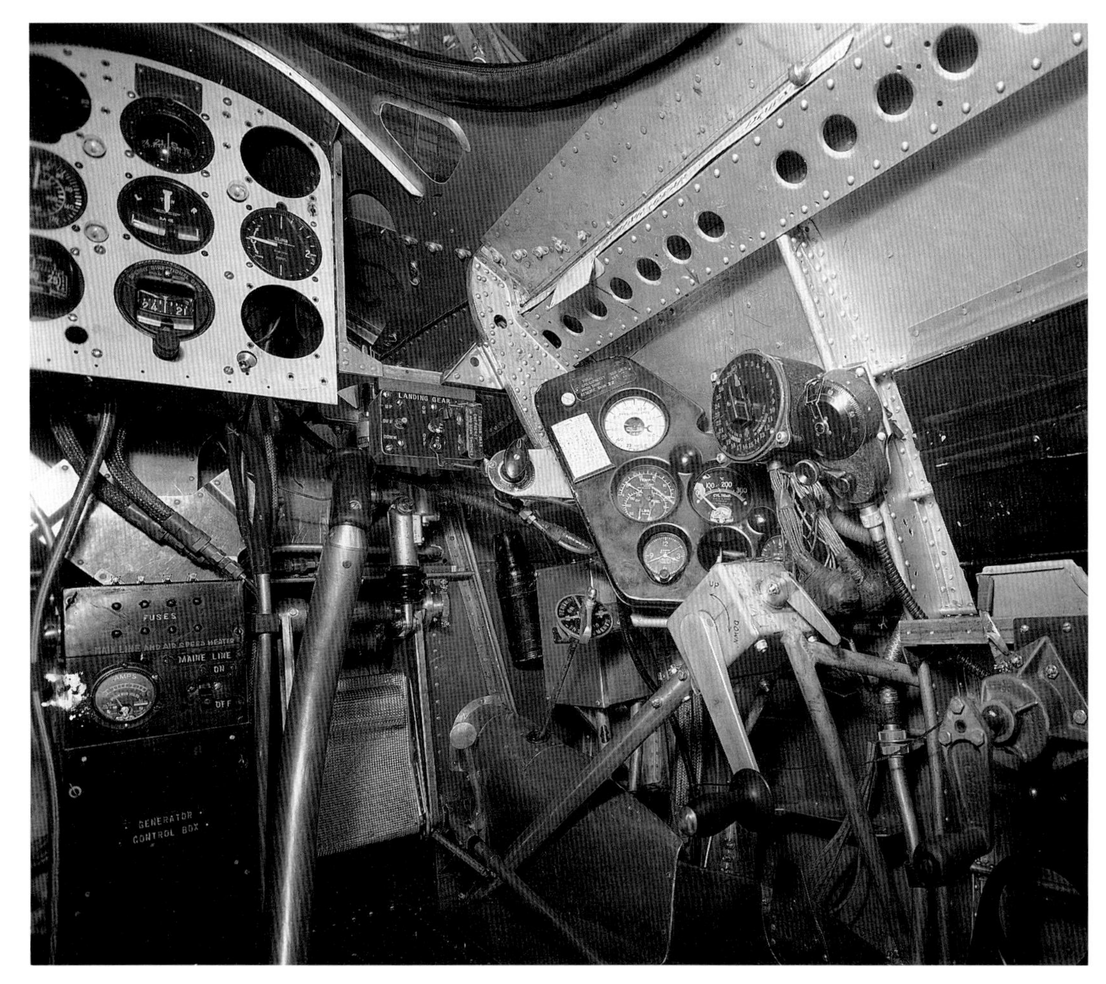

***A rare insight to the cockpit area of what was essentially the second Seversky airplane built, and probably the first airframe built by Seversky Aircraft personnel. The first Kirkham-built SEV-3M-WW was not far behind, and the SEV-3 had been built at an Edo Manufacturing facility, mostly with their employees. This photo was taken on 30 May 1936 immediately after the armament was removed.* (C.A.M/Ed Boss, Jr.)**

out the BC-1-*cum*-AT-6 at the right time, just as the military awakened to the need. The trainer that Seversky Aircraft created for rollout on 16 August 1936 was the SEV-X-BT (c/n 6), not built to fill any government-specified role. It was truly a private venture. As the manufacturing date reveals, it most assuredly was not the predecessor of the SEV-1XP as some have claimed. In fact, it seems to have acquired certain parts fabricated for and used on the 1XP at an earlier date since the Major was loath to discard any useful part as the company continued to lose money.

Their SEV-X-BT was an all-metal stressed-skin monoplane powered by a P&W S3H-1 Wasp radial rated at 550 horsepower. That suited the designers just perfectly. It was, in fact, a modernized, more powerful SEV-3XAR-styled trainer with a retractable landing gear and an innovative wing that could be fitted with long or short outer panels having different lifting areas, thereby altering the performance and landing speed. Still smarting from the Air Corps constraints that limited trainer horsepower to something in the 450 horsepower range, the design team included throttle blocking to restrict normal power to that figure. It was convenient to go through the blocks and pull 550 hp providing the airplane with outstanding characteristics. Nation-wide demonstrations by the Major elicited very favorable press reports and a great deal of interest, but the military failed to show one iota of interest. Rather significantly the outer panels were designed to be attached with a few degrees of dihedral, long missing from Seversky airplanes. This design improvement was applied to a new airplane almost exactly one year before Materiel Division's Experimental Engineering Section pilots reported on the deficient flight characteristics of the first P-35 which were attributable to lack of dihedral.

That factor, the company's financial condition and the generally chaotic situation everywhere on earth were soon to direct the attention of de Seversky to foreign markets and create a widening rift between him and the War Department.

Seversky Aircraft's commitment to improve the P-35 production models was valid and remained so over the entire contractual period. The first P-35 (there was never any real XP-35) was a very different airplane from the SEV-1XP. It should have been possible for the Air Corps Materiel Division to ask for and get special changes (including flush-retracting landing gear installations), even if they had to reduce the number procured in order to obtain funding for such changes. As a matter of fact, in the same springtime period in which the first production P-35 was flown to Wright Field, the Major had personally created his new Seversky AP-2[2] (under a revised in-house designation system) to compete against Curtiss' Y1P-36 trio which were procured as a "booby prize" after the giant corporation gave in to the Army urging they try the P&W Twin Wasp, but that comes a bit later in the story.

Meanwhile, bacK at Farmingdale,one of the least understood episodes involving the original SEV-3 airplane was quietly taking place, or at least putting events in motion. With the Spanish Republicans having virtually nothing in the way of airpower with which to face Franco's Nazi-Fascist supported

made years in advance were so useful, one wonders how the Pearl Harbor attack ever was launched and why was it such an overwhelming success. Why was the Air Force, in mid-1941, without even a prototype of any heavy-lift transports, relying on modified private-venture airliners (DC-2, DC-3, Electra 10, and soon to come, DC-4 types) as the closest thing to it? Why was the best photographic aircraft a Beech Model 18 with some camera ports in it? Why was it necessary for a mere lieutenant colonel (George Goddard) to suggest to Kelly Johnson at Lockheed a manner in which P-38s could become excellent high-speed reconnaissance airplanes? This fertile cell led to the outstanding F-4/F-5 series. Why were the best replacements the AAF could provide for ancient P-26s in the Western Pacific in mid 1941 second-rate Republic EP1s which had been ordered by Sweden? Our pilots were committed to possible combat with integral fuel tanks, absolutely no armor plate, and instruments using the metric system (then ignored in America) and Swedish-language callouts. Was there any planning at all???

2 Company records, Board Minutes, and even Bureau of Air Commerce records relating to the AP-2 seem to have either ignored the existence of the airplane or have cross-pollinated it with the first P-35 (which the AAC rejected) which became the AP-1 in 1937. If we are to believe certain records, the AP-2 actually was completed before the first production P-35 rolled out. It premiered many of the features soon to appear in the SEV-S2 and SEV-DS executive/racing airplanes, but was even more advanced. It used the flush landing gear that Materiel Division should have required on the P-35s, and it even featured the addition of wing dihedral which division pilots and chiefs demanded for the service P-35s.

While military and naval officialdom in the United States bent over backward to avoid publicity about new aircraft, Britain was happy to exploit its perceived (sometimes illusional) dominance in new military aircraft. The RAF displayed its brand new prototypes in every category at the 1936 Hendon Air Pageant. Numbers 1 and 2 were prototype Hurricane and Spitfire, respectively, while No.3 was the Vickers PVO-10. All had flush-retracting landing gear installations; No.4 was the ill-conceived Fairey Battle light bomber prototype – made worse by having the Seversky-type wheel retraction system (partially retractable). Nearly as big as the prototype Vickers Wellington (K4049) which had two engines, the Battle had only one engine essentially identical to that powering the prototype Hawker interceptor. (Author's Coll.)

As dozens of pilots were soon to learn, Seversky BT-8s (and the first-line P-35 pursuits) were exceedingly prone to ground loop. The touchdown angle of attack for a 3-point landing, caused by the short-coupled fuselage, resulted in a lightning-like wingtip drop at a crucial point. Unwary pilots, caught short, ground looped, frequently with this result. Charles Lindbergh was not happy with the P-35 for that reason. (Author's Coll.)

Once again the original, much revised SEV-3 popped up in a new guise in December 1936. As a landplane SEV-3XAR, this airframe had a Wright Whirlwind engine, but here it was (as a SEV-3M) equipped with a big direct drive Wright R-1820-G5 Cyclone engine (reworked to G2 standard) rated at 810 horsepower. And, lo, there was the vertical tail last seen on the SEV-1XP in August 1935 at Wright Field. A new bronze paint job had been applied overall. (Bodie Coll.)

aviation forces, a global effort was energized to obtain any aircraft that could even conceivably be converted into a military machine. In the U.S.A. such actions were ostensibly blocked by a Roosevelt-invoked embargo against such sales and shipments. The record-breaking SEV-3 with its big Cyclone engine (even with the Wright's failure to produce a promised 950-1000 horsepower) was undergoing substantial conversion to wolf's clothing. The amphibious floats were removed and the original vertical empennage was replaced by the tailplane taken from the ugly version of the SEV-1XP in that airplane's final conversion. The Major's personal mechanic, Judson Hopla, has stated the resulting landplane (which, in falsely arranged records was stated to be a "BT-8") was designated correctly as the SEV-3M and could attain a speed of 285 mph. That might have been possible if they could get the Cyclone to produce 950 hp for two minutes or more, but it was probably never even recorded as an airspeed.

Certainly the experimental team at Seversky was working diligently on modifications and restoration of the SEV-3M to like-new condition in Hanger 1 in the final quarter of 1936. It appears that they just added that old extension to the fin and rudder of the landplane SEV-3M to compensate for the forward float area. The well-used demonstrator was then coated in a beautiful new gold or bronze paint scheme and the amphibian landing gear was reinstalled. It was soon acknowledged that the Major's pride-and-joy personal airplane had been sold to Col. Roberto Fierro, former Chief of the Mexican Air Service. Gear changes were made in the Wright R-1820-G5 engine to convert it to a G2 Cyclone for long-distance flights. Although the Cyclone did have a takeoff power rating of 1000 hp, the normal rating for flight at 5800 feet was down to 850 horsepower.

The Christmas season of 1936 was not going to pass quietly for Major de Seversky. With the engine being altered as needed from G2 to G5 configuration and landing gear changes at will, the Major and Frank Sinclair proceeded to establish four intercity

Major de Seversky exhibited the updated SEV-3M, in December 1936, to some well-heeled friends at Farmingdale. A Pierce-Arrow limousine stands out in the background, giving us a clue. With no record to guide us, we presume the man at left is Col. Roberto Fierro from Mexico, while the man at de Seversky's left may be financier Paul Moore. (Seversky via S. Hudek)

records and one new 100-Kilometer closed-course speed record (amphibian). At the start, probably on 19 December, Seversky flew the amphibian some 1200 miles in 5 hrs. 46 mins. from New York to Miami, Florida, followed almost immediately by southbound and northbound flights in 1 hr. 4 mins. to and from Havana, Cuba. The closed-course record speed was 209.4 mph, established at Miami. At the time, negotiations with Col. Fierro were under way. The U.S. State Department insisted that Fierro guarantee that the airplane would not be shipped to Spain. His response was evidently satisfactory and the sale was authorized. The Major flew back to Farmingdale where the SEV-3M (registered as NR2106) was converted to landplane format. Frank Sinclair then dashed 1250 miles from New York to New Orleans, Louisiana, averaging 250 mph for the distance. (Few service-type fighters were likely to have done that well.) The Seversky test pilot then made a record-breaking two-stop, 1100-mile flight from New Orleans to Mexico City to deliver the airplane to Fierro for an agreed sales price of $45,000. As an aside, Judson Hopla reported that the floats and machine guns were in a truck that was hijacked on the way to Mexico.

Two years before Maj. de Seversky and Alex Kartveli committed an unpardonable sin by trying to foist the SEV-2XP 2-seat pursuit plane on the Experimental Engineering Section at Wright Field, the USSR had a prototype fighter with retractable landing gear. (Some European journalists claim it was the first fighter with retractable landing gear to enter squadron service anywhere. They forget the P-30/PB-2A.) Powered by copies of the Wright Cyclone, various versions of the Polikarpov I-16 were entering service as the SEV-1XP appeared at Dayton. On 775 hp it was as fast as the 950 horsepower P-35 two years later. Finish was marginal, and it was tricky to fly. By March 1939, some 475 I-16s had been sent to fight in the Spanish Civil War. **(William Green Coll.)**

At some time in this process, Col. Fierro told Major de Seversky that the Spanish Republicans had a tremendous gold reserve and they were more than anxious to purchase Seversky products to combat General Francisco Franco's Fascist revolt. Some reports indicate that de Seversky was willing to establish a factory in Mexico or Spain to produce a hundred SEV-1XP "type" fighters. Those 1937 or 1938 airplanes would have very closely matched the fighters that eventually became first-line military types for Sweden. All of this occurred just prior to the initiation of the 1937 Pursuit Competition, so it is nearly impossible to establish with certainty who provoked what actions and when in the Air Corps vs. Seversky prologue to confrontation. Wisely, Kartveli kept his distance from the impending battles.

With a tremendous international political undercurrent of intrigue behind the scenes, the major manufacturing deal involving Seversky was finally cancelled in March 1937. In the meantime, the SEV-3M had been provided with a Mexican civil registration of XA-ABG, flown to Vera Cruz where it was evidently shipped on the Spanish freighter S.S. *Ibai* on 24 December 1937. After a reasonably uneventful voyage, the freight arrived at Le Havre, France almost exactly one month later. Somehow the SEV-3M arrived in Spain between 17 March and 13 June 1938, but inaccurate reporting and the need for secrecy leave great gaps in the war zone utilization of the first Seversky airplane in the Spanish Civil War.

When Barcelona fell before a Nationalist onslaught early in 1939, every Republican aircraft based at Vilajuiga was ordered moved to an airfield at Banolas in anticipation of an immediate move to France. Starting the Cyclone engine after a period of inactivity delayed departure of the SEV-3M for perhaps 15 minutes, so it arrived on the scene late. Treachery ensured that the entire group of aircraft would arrive at the airfield immediately before an attack by the Condor Squadron. Every arriving airplane was destroyed on the muddy base. The Seversky's pilot arrived late, so his fate was not sealed. However, in landing on the only strip of real estate that was not cluttered or holed, the heavy monoplane dug its wheels in and flipped over. The pilot

And yet another old tail, this time from the ugliest version of the SEV-1XP, reappeared to adorn the final version of the long-lived SEV-3M. Mounted on the big amphibious floats, the airplane needed additional fin area when it was sold for $45,000 (U.S.) to Fierro on Dec. 23, 1936. Mexican records show the sale date as Feb. 27, 1937, being registered as XA-ABG in place of the final NR2106 applied at Seversky Aircraft. Shipped from Mexico to Spain on December 26, it was destroyed in a landing accident there on Feb. 6, 1939. Old Severskys never died; they just – travelled away. **(C.A.M./Ed Boss, Jr.)**

Nicely fabricated, the first production P-35 (AC36-354) was delivered early in 1937 from expanding facilities. On April 15, Seversky took over the former Grumman factory immediately west of Bldg. 5; total plant area expanded to 250,000 sq. feet. As delivered to the Materiel Division, the P-35 had an odd windshield, complex landing gear fairings and the 0-degree wing (upper surface) dihedral that proved annoying to test pilots. The airplane was rejected and the Major (evidently) brought it back to New York. Those actions and record keeping manipulations created a situation wherein the Air Corps received a replacement P-35 per contract, via spares and/or "black" parts. The company retained this "reject" under the designation AP-1. (AAF/WF)

and mechanic survived that crash landing on February 5th, but the long-time favorite of Alex de Seversky was never to appear again.

As for the SEV-X-BT, perhaps it was just as well that the Air Corps procurement types remained at arm's length. There is no record that anyone at Wright Field or Patterson Field ever flew the civil-registry airplane or if they even witnessed flights with de Seversky in the pilot's seat. What is certain is complete lack of any expression of the slightest interest in his concepts. When the SEV-X-BT was photographed at Wright Field, the vertical tail shape changed one more time. It is not known if the extension was added at Farmingdale or at Dayton. The Major flew it back home, most likely thinking very hard about what the next step would be with this and other airframes. To his credit, the airplane was not allowed to sit and collect dust or corrosion.

Either before the SEV-X-BT was flown to Dayton in 1936 or upon its return, photographs were taken to demonstrate the use of interchangeable wing outer panels to provide either 200 square feet of wing area or 250 square feet. By 8 July 1937 the experimental shop personnel had removed the Wasp engine and installed an 875-hp Wright Cyclone GR-1820-G3 engine. Single

With the pilot seat removed (or tilted back, Seversky style), a fine view of the P-35 cockpit was captured on film. For a fairly complex modern airplane, instrumentation was hardly overwhelming. In some ways it seems elegant. (F. Strnad Coll.)

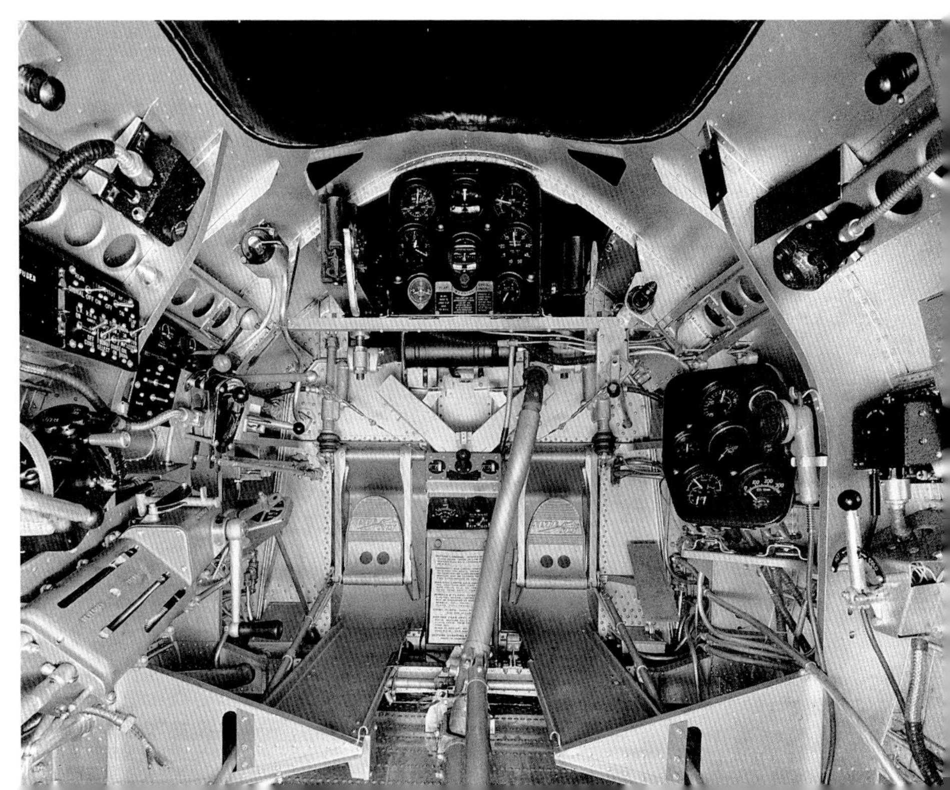

.30-caliber and .50-caliber machine guns were installed to fire through the propeller arc, while a .30-caliber flexible-mount weapon was installed in the rear cockpit. Short-span (38-foot) wings were used and, despite some erroneous licensing data that was entered into the record, the airplane was revealed as a Model 2PA bearing the correct license number R189M. As of 10 September 1937, this "Convoy Fighter," as it was named, re-registered NR189M for a demonstration tour of Brazil and Argentina in South America.

Although the first test flight in this 2PA configuration was made in July 1937 with Frank Sinclair at the controls, it was also flown by Major de Seversky at Farmingdale while planning went forward for the demonstration trip. Judson Hopla, Major de Seversky's personal one-man support team, was called back from Mexico to go along with Sinclair to support the 2PA "Convoy Fighter" on its travels[3]. The 2-seat fighter was shipped to Buenos Aires, Argentina, that fall where it was off loaded. Suddenly, they were faced with the problem of getting the aircraft off the dock and over to the airport. The unusually wide center section of the wing would not allow it to pass through the city's streets, and Sinclair chose not to remove the outer wing panels. The dock was found to be approximately 200 feet in length; however, Sinclair and Hopla (with the permission of appropriate authorities) got into the airplane, taxied to one end of the dock, took off in that short space and proceeded to fly to the nearby Army airport. Hopla stated that the Wright Cyclone installed in the 2PA was unreliable and had failed them three times during the Argentine demonstration. As a result, a Curtiss competitor defeated Seversky Aircraft for a contract award. The airplane was ferried to Brazil, but suffered yet another failure. Hopla recalled rebuilding the Cyclone under difficult conditions so the pilot could demonstrate the airplane to a panel of military evaluators. Eventually Hopla was called back to Farmingdale to prepare for support of a new EP1-68 and 2PA-202 demonstration and sales tour throughout Europe. (Judson Hopla stated the Seversky 2PA was flown back to the U.S.A., but indisputable evidence reveals the fighter remained unattended in South America. No report of its ultimate fate has been discovered.) Many others have written that the Seversky went to Russia and was destroyed there in an accident, but such reports are the result of poor data analysis. A few reports even placed it in Spain. The "Russian Connection" is discussed in some detail elsewhere in this chapter. (Incidentally, it is common writer practice to this day to credit the late 1930s Wright Cyclone-series engines with a 1000-horsepower. By 1939 a Cyclone was hard-pressed to produce up to 900 horsepower with any chance of operating reliably, but supposedly produced that one-grand figure for takeoff. Pratt & Whitney's R-1830 Twin Wasp proved to be far more reliable and a bit more powerful.)

On April 30, 1937, Alex de Seversky was viewed in the P-35/AP-1, AC36-354, on the ground and in flight. This real P-35 prototype featured yet another windshield that was close to the final configuration. It would serve as a clue to this particular aircraft's identity, as would a pair of small landing gear wheel fairings added a bit later. One thing was certain then as it is now; that AP-1 airplane had charisma. (Luis Azzaraga/Mfr.)

Production of 77 Air Corps P-35s got under way in 1937, and the first airplane of the series was flown by the Major on 4 May. The only thing that looked even remotely like the SEV-1XP prototype was the wing planform. Whereas that competition winner had begun to look like a homebuilt racer for the Thompson Trophy Race because of various crudities inflicted upon it during 1935, the initial 1937 airplane (first of a "proofing" batch of three) had a certain charisma. The airframe could have – and should have – been reduced in size and weight without excessive engineering effort, but some tooling design work would have been compromised. It would also have required approval from Lt. Ben Kelsey's office and most certainly, much additional flight testing would have been required. Therefore, the P-35 was saddled with all the excessive cubic footage and unwanted avoirdupois that was a hangover from the use of tooling created earlier for 2-seater SEV-3M-WW/SEV-2XP projects. Although the P-35 showed many signs of unexpectedly good production engineering, it was still a true 2-seat airplane dimensionally. That was proven less than two years later when the fuselage of the AP-4 accomodated a big turbosupercharger system. It was built in the same jigs.

However, a touch of evil permeated the Farmingdale premises. Somebody who remains unidentified to this day was

[3] Hopla had joined the Major as his personal support engineer/mechanic in 1933 and was still with him up to the time of de Seversky's death on 25 August 1974. When Hopla was interviewed years later, it was clear that he had valuable memories, but as much as two-thirds of his information has been shown to be flawed by failing memory attributed to his advanced age. Because his personal associations with many company activities are invaluable, data that seems to be accurate has been used in this narrative.

Major de Seversky posed for publicity/advertising photographs with that No.1 P-35, this being done in connection with Shell aviation fuel. Minutes later, the Major was shown drinking from a Coca Cola bottle (more ads). It is a good bet that the AAC people were upset later to see those military markings used with commercial ads, if they did appear in print. **(F. Strnad Coll.)**

responsible for redesign of the prototype landing gear, and a great many man-hours must have been expended on that effort. Virtually everything except the concept went out the window; yet, that is the one thing which should have gone first. The new installation was unnecessarily complex, created a great deal of unwanted drag and, with the landing gear extended, there were two great fairings-in-reverse that created unexpected airbrakes. It is an unmitigated fact that excessive pitch changes during takeoff were caused by those wind-grabbing pockets. Even worse, after acceptance testing of the production airplane was conducted at Wright Field, the landing gear had to be redesigned once more. (The story of that controversial landing gear is told in detail in a subsequent Gallery.)

The contract required Seversky to deliver those three early production pursuit planes to prove full interchangeability of parts. Therefore, the first months of 1937 were devoted to tooling up for mass production. Factory expansion efforts ensured that all 250,000 square feet of space was functional in early spring to support a production rate of one airplane per day.

While the P-35 engine cowling and new empennage showed superior design work, nobody did a thing to simplify the canopy/turtleback structure and reduce overall canopy drag. Even worse, somebody designed an entirely new windshield that can only be described as an abomination. It was ugly; it distorted the pilot's view; it was far more complex to manufacture than the original. Fortunately, that was something which the Air Materiel Division pilots refused to tolerate. (In direct contrast, the 77th airplane on the P-35 contract was produced as the XP-41 with a cockpit enclosure which was nothing short of superior for 1939.) In addition, average fliers found the absence of dihedral unpleasant to cope with in weather minimums and in formation flight.

Another action which was, at the time essentially unprecedented, saw the Fighter Projects Office (which we might very well read as Lt. Ben Kelsey) reject the first P-35, serial AC36-354 (constructor's No.44). The airplane was flown back to Farmingdale, returned to the manufacturer, and was quietly absorbed by Seversky Aircraft, which in itself, created another unprecedented reaction. As a result, historical fact has been badly distorted until now. All variations of the theme that have

It seems obvious that the Major's reach exceeded his grasp. One design could not be all things to all people. In an era where the Navy wouldn't even think of adapting an Army airplane to a carrier-based function, Seversky installed a Wright R-1820-22 radial engine on yet another "diecast" airframe, attached the big, bulky landing gear fairings and that blunt windshield removed from the No. 1 rejected P-35 and installed an arresting hook, thus creating a naval fighter. Someone added an illegitimate designation of NF1, but a civil license number X1254 was inscribed on the airplane. It flew to NAS Anacostia, D.C. on September 24, 1937, to compete for a contract. **(Author's Coll.)**

Always leave them guessing. That must have been a de Seversky motto. When the first P-35-cum-AP-1 was seen at Langley Field, where it had unexpectedly arrived for testing, it was a conundrum in the making. In a virtually unprecedented happening, the P-35 carried full Air Corps markings PLUS a new civil registry X1390.There was that "one-off" windshield peculiar to the No.1 P-35 and a small pair of wheel fairings never seen on any other Seversky airplane. Gone were the bulbaceous fairings; in their place were new production fairings that became a trademark. Not only that, the wing had suddenly gained 4 degrees of dihedral (measured on the upper surface) vs. 0 degrees. (NACA 13875)

appeared in print just ignored the facts about AC36-354. Although the written paper trail barely exists – especially in the stockholders Annual Report for any year or in minutes of Directors meetings – the facts prove that Seversky Aircraft always retained that No.1 P-35 as the AP-1 until it was destroyed in a hangar fire at a later date. At the same time, there was never a murmur from the Air Corps about not getting the total quantity of 77 airplanes ordered. The reason: they got their 77 airplanes consisting of 76 new P-35s and one XP-41 in accordance with original plans. Even though the SEV-1XP had been updated considerably, it was not an optimum demonstration airplane. The AP-1 fit the bill much better.

But let us be pragmatic, even skeptical, about something. First flown as the P-35 in May 1937, the AP-1 was destroyed in a Miami, Florida, hangar fire less than a year later. Isn't it quite likely that the AP-1 was rather fully insured? In any event, a new Demonstrator was manufactured very promptly, designated AP-7 and registered NX1384. Incredibly, it made its first flight about 30 days after the AP-1 was destroyed. (But remember, there was never even a whisper of suspicion.)

The first thing changed on the P-35 No.1 was its oddly crafted windshield which may well have been inspired by a misguided Bureau of Air Commerce directive issued in the mid 1930s requiring the use of "birdproof" windshields on commercial aircraft. Variations of the idea appeared on Vultee V-1s and Boeing 247s, but no company other than Seversky tried the idea on military aircraft.

Major de Seversky then made another serious mistake in trying to capitalize on the fast (for its time and place) pursuit airplane he had in hand by utilizing it for Coca Cola and Shell Oil advertising campaigns. This could not have pleased War Department and Air Corps officials at all. In fact, the campaign seems to have been aborted. A third element was the announcement, in June 1937, that Amtorg Trading Corporation had, on behalf of the Russian Government, placed an order with Seversky Aircraft in the amount of $780,000. The Russians were to obtain two SEV-3/2PA -based amphibian aircraft and manufacturing rights for the sum of $370,000; the balance of $410,000 was to cover an option for two additional planes together with complete tooling to manufacture as many as 10 amphibians per day. When the contract was an-

Major de Seversky was a conservationist well before it became fashionable; he seldom wasted a part, component or entire airframe. When it became evident that the Army was not about to allow sales of P-35 types to foreign buyers, he merely doubled the SEV-X-BT's horsepower and reworked the "trainer" into a 2-seat "Convoy Fighter" by July 1937. Powered by an improved Wright Cyclone GR-1820-G3 rated at 875 horsepower and swinging a 3-blade Hamilton-Standard controllable-pitch prop, it would have flown circles around the comparable SEV-2XP of 1935. Relicensed as R189M, this neat package went on a sales demonstration tour to Argentina and to Brazil. It never returned to the U.S.A. Unexpectedly, a duplicate of the AP-1 windshield replaced the trainer windshield. (C.A.M./Ed Boss, Jr.)

nounced, the conversion of the single SEV-X-BT to "Convoy Fighter" format as the 2PA had not even been completed. It seems obvious that Seversky had already designed an airplane to meet Russian needs, and that the 2PA was an expedient means for demonstrating capabilities of such an aircraft to many other potential foreign buyers. The company expected to be involved in supplying many manufactured parts for a lengthy period before the Russians could get such a high rate of production fully operational.

The Farmingdale factory seemed to have no trouble producing a variety of enigmas. One of the most perplexing was the potentially successful, high-performance AP-2. Built for the new 1937 Pursuit Competition at Wright Field, it lost out to the potent Curtiss-Wright lobby. The Hawk 75 "that should have been" in August 1935 won as the P-36A. Curtiss had installed derated P&W R-1830 "C"s in three YIP-36s to get performance about equal to production P-35s. The flush-gear AP-2 with basic "C" series powerplant was faster, but C-W could build for half the price per unit. When the AP-2 landing gear failed on September 1 as the Major was landing at Floyd Bennett during preparations for the Bendix Trophy Race, he became depressed. Damage was obviously reparable on R1250, but it never flew again. (H. Levy/C.A.M.)

In the meantime, Seversky engineering went about the tasks of increasing P-35 wing dihedral (measured on the wing upper surface) by 4 degrees, simplifying the landing gear fairings and correcting other deficiencies. Lo and behold, their first P-35 showed up at Langley Field, Virginia, in the hands (temporarily, at least) of the respected NACA for recommendations to improve performance. In yet another precedent-setting situation, the airplane had full military markings and a new civil registration number X1390 at the same time. Nobody seemed to be upset by the situation. A close look at the cockpit area reveals that the windshield appeared to resemble the ultimate production windshield, but it was substantially different. Nothing comparable was seen on any other Seversky airplane. Wing dihedral had been incorporated, and production-type landing gear fairings were installed with one major exception. Some very odd extensions had been applied to the fixed landing gear fairings near the trailing edge of the wing. No other P-35 was ever seen with those extensions. After completion of tests at Langley, the airplane went back to the factory. It was then redesignated AP-1 and all military markings were deleted. Without the benefit of publicity about its "underground" life and all facts suppressed in company files, it is hardly any wonder that so-called official records about the entire affair are terribly flawed.

During the following winter, the AP-1 cowling was modified to accommodate a massive propeller spinner. It must be presumed this was an outgrowth of the Langley Field episode, delayed to some extent because the NACA was still trying to develop their special drag reduction cowlings for radial engines. Still another subtle change was made to reduce the cowling air entry slot as much as possible without causing engine overheating. As mentioned earlier, the airplane was destroyed in a hangar fire at Miami. There is no record pertaining to the AP-1 cowling configuration at the time of destruction. Seversky experimented with large-spinner cowlings on several different airplanes without much ini-

A crew of U.S. Navy firefighters based at Floyd Bennett Field, N.Y., extenquished a small blaze that erupted in the engine compartment of the Seversky AP-2 when the routine landing turned into a crash event as Major de Seversky flew in from Hartford, Conn., on September 1st. He had planned to enter the new pursuit airplane in the 1937 National Air Races. This rare photo has never appeared in any book or magazine. (F. Strnad Coll.)

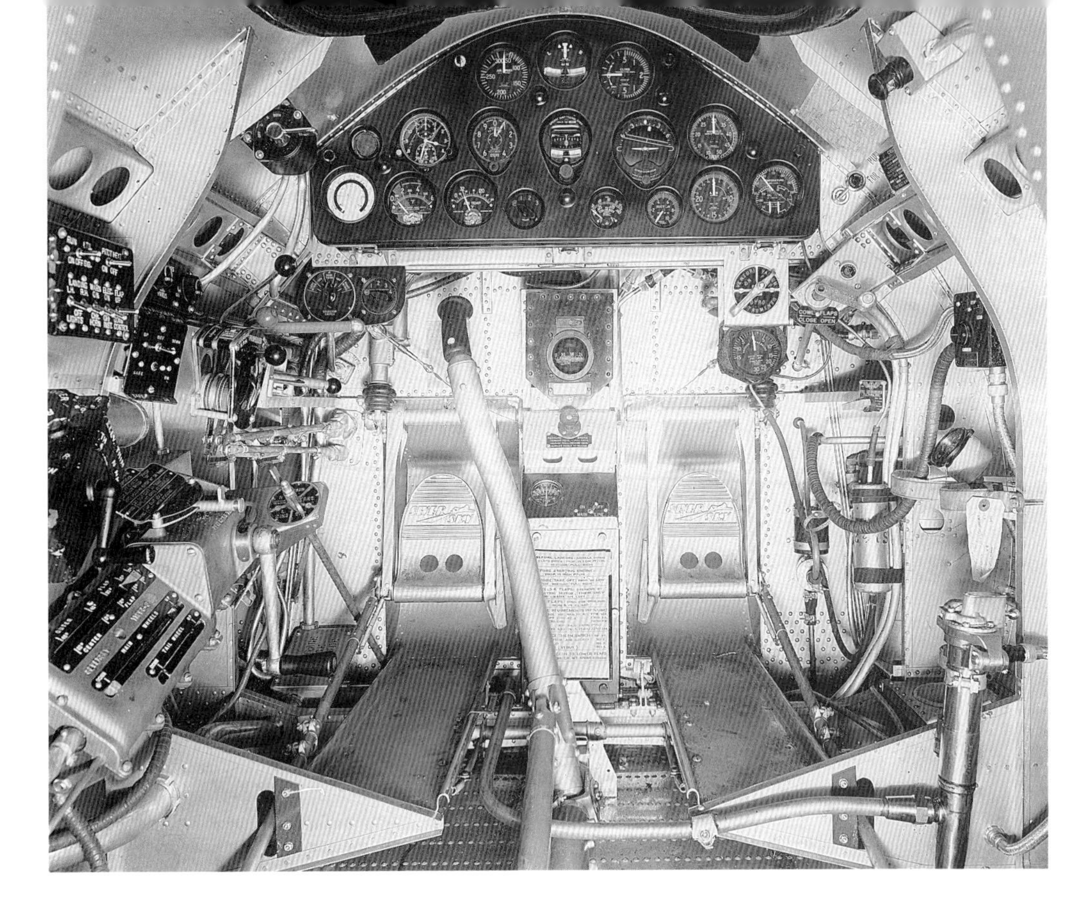

The neat racing cockpit of Frank Fuller, Jr.'s Seversky SEV-S2 surely did not concentrate on weight reduction. This looked more like the office of an executive transport of that era. Fuller evidently did not feel comfortable with pylon racing. Ray Moore flew it in the 1937 Thompson event, was handily outclassed even by Steve Wittman's backyard "Bonzo" and finished a distant sixth. Clark Gable "flew" here when the airplane was called the **Drake Bullet** *for the great film,* **Test Pilot.** **(Mfr. 892/F. Strnad)**

tial success. However, such early experiments must have had an impact on design of later units. The aircooled-engine cowlings fitted to the XP-72 fighters and the XF-12 reconnaissance airplane were among the best ever produced to enclose radial engines.

The Seversky AP-2 saga is not a highlight of the overall Seversky Aircraft story. The airplane could have led to significant performance improvements and an increase in overall orders for the company's products. Well prepared for the new 1937 Pursuit Competition, the neat AP-2 (c/n 39, temporary registry X1250) was everything that the P-35 should have been and could have been. It had a flush, wide-track landing gear, a lower-profile canopy – less weight and less drag – and a newer SC3-G type Twin Wasp with the more efficient downdraft carburetor that most often was fitted to the Seversky racing/development airplanes. Depending on the state of tune and stage of development, those engines produced 1050 to 1200 hp for takeoff. According to some published reports, the AP-2 was substantially faster than the new Curtiss Y1P-36 test airplanes. It traveled 340 miles from St. Louis to Dayton at a speed of approximately 340 mph, unofficially of

A close look at the abominable landing gear design that never seemed to go away is presented here. Seversky test and racing pilot Frank Sinclair posed with the blue and yellow prototype with refinements (?) for racing on September 1, 1937, at Burbank's Union Air Terminal. Trim changes must have been horrendous with the gear extending. Just look at those scoops! **(Frank Sinclair)**

While all of the individual development work proceeded at Farmingdale, the Major never lost sight of the need for record breaking and speed contests. To this end, the original SEV-1XP was reworked in January 1937, bringing it close to a standard P-35 configuration. Major differences were in the original 1935 windshield and zero wing dihedral, a new P-35 tailplane and cowling, and retention of the original landing gear with bulky fairings added. Unexpectedly, it had metallic blue paint on the fuselage and yellow wings and tail. Test pilot Frank Sinclair and aviatrix Jackie Cochran (shown here) flew the SEV-1XP racer (#63). Only Sinclair flew it in racing competition; Cochran for record breaking. It was written off after an accident in Miami on 13 Dec. 1937. **(Author's Coll.)**

course. Even allowing for tailwinds, it was certainly faster than the Curtiss competitors in the June 1937 "competition" because none of the YIP-36s could exceed 294 miles per hour. There is every probability that the AP-2 was a 320 mph pursuit plane, ensuring that it was as fast as a Hawker Hurricane Mk.I and, most likely, with three times the range of the RAF interceptor. In appearance, the AP-2 was essentially a militarized SEV-S2 racer with flush retracting landing gear.

Certainly the Curtiss Y1P-36 entrant in the 1937 competition was dead slow by comparison, leading to complete redesign of the cowling seen on all eventual production versions. In contrast, facts about the new Seversky AP-2 were stated in a classified test report – which was endorsed with a handwritten comment to the effect it was not to be used in whole or part for publicity – sent by Capt. Stanley Umstead to (then) Lt. Col. Oliver Echols on 8 June 1937. Test results showed that at 900 hp, 2500 rpm at 10,000 feet altitude, the new AP-2 had been clocked at 307.0 mph with manifold pressure showing 34.6 in. Hg. on the P&W R-1830-C engine indicator.

Compared to the contemporary Hawker Hurricane Mk.1 with 1030 hp on tap, knowing the newest Twin Wasp was soon to be producing up to 1200 hp, it is more than probable a developed AP-2 could have matched or exceeded the speed of the RAF fighter. Some facts about Umstead's report A-19-698 are of critical interest. It was classified in 1937; on 11 May 1938, the acting chief of Flight Test, Materiel Div., a major, made that handwritten suppression entry. The Curtiss Y1P-36 won the largest peacetime pursuit contract in history for the Curtiss-Wright Corporation, although the airplane was obviously some 20 mph slower than the Seversky AP-2. The smoking gun aimed at Alex de Seversky could hardly be more evident.

Although the Major put on a "hot" flying exhibition in his AP-2 on Memorial Day weekend at St. Louis in 1937, the Army forbade him to present the airplane in the airshow static display. (He flew to St. Louis with his beautiful wife Evelyn safely ensconced in the cavernous "baggage" compartment.) Even more likely, they may have been so perturbed by his very spectacular flight demonstration that they refused to even consider purchasing more of his airplanes. As it was, Curtiss walked off with big order for pursuit aircraft, some 210 airplanes of a basic P-36A type. (The only way that the War Department could prevent the Major from demonstrating his private venture airplane was to threaten exclusion from future procurement of anything. It is a matter of record that the Air Corps never ordered any more Seversky airplanes after the P-35 model. An old military axiom: We can't make you do anything, but we can make you

All three competitors for a Navy fighter contract in 1937 were concentrated almost within sight of each other in New York. They were Grumman, Brewster and Seversky. Unexpectedly, Brewster won out over Grumman with its XF2A-1, while Seversky was a distant third. The NF1 is shown here with a very large test thermometer mounted on the wingtip. Seversky's engineers seemed to lack any real knowledge about BuAer requirements, so they lacked any chance for serious consideration. **(Howard Levy)**

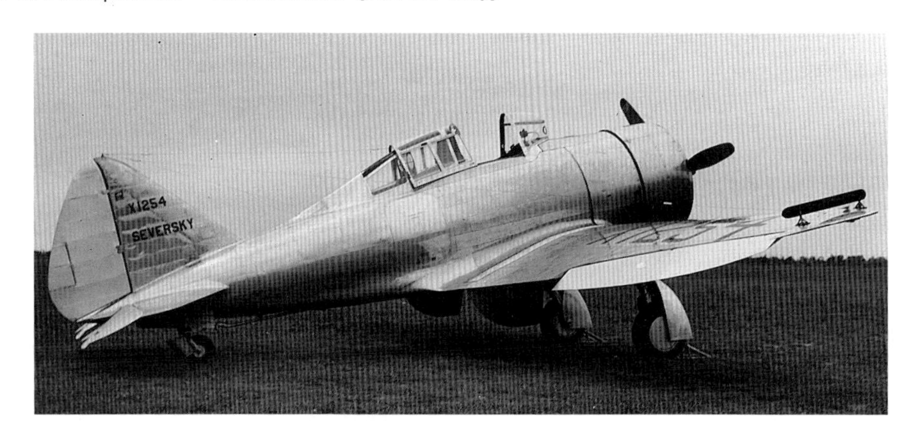

damn sorry you didn't do it. The clearest illustration of that was presented in James Jones' novel *From Here to Eternity*. Jones drew more on fact than on fiction. It has to be obvious to all that decision makers in civil government and in the military did not always wear white hats.)

Shortly thereafter, Major de Seversky was preparing to fly that same AP-2 in the 1937 National Air Races at Cleveland in competition against company test pilot, Frank Sinclair, who was entered in the much revised blue and yellow SEV-1XP showing race No. 63. (Sinclair raced in the Bendix Trophy event, coming in fourth; he then placed very well [fourth] in the Thompson Trophy Race – a 200-mile closed-course event – against pure racing airplanes that included the new Turner-Laird Special which proved to be only 1.4 mph faster.) However, the Major chose that moment to make one of his few piloting mistakes. On 1 September he landed hard and out of shape at Floyd Bennett Field in New York. The new landing gear failed, possibly as a result of bad design or metallurgy, and the AP-2 was damaged too badly to be repaired in time for the races. (There were no spare landing gear parts.) Despondent to the extreme because of the 1937 Pursuit Competition results, knowing that he had a very fast airplane with good potential, and having bent his prize speedster at the most inopportune moment, he evidently "punished" himself by committing the AP-2 to purgatory. It was never seen again in the United States.

And then there was the brief flirtation with the U.S. Navy's Bureau of Aeronautics. Seversky's experimental shop was very busy in 1937 on efforts not related at all to the P-35 production program. Among the single-seat types was the one and only SEV-S2 racing airplane for paint tycoon Frank Fuller, Jr. and a slightly different rendering of the same theme for Jimmy Doolittle and the Shell Oil Company that bore the designation SEV-DS. The Navy Bureau made a determination in the latter part of 1936 that a proposed Grumman XF4F-1 biplane was not likely to exceed operational performance of the new F3F-3 offered as an upgrade of the first-line F3F-2s then in service, they cancelled the effort and ultimately awarded development contracts to Grumman and Brewster for new carrier-based monoplane fighters. If Grumman's "Iron Works" was unable to produce a high-performance XF4F-2 type to meet or exceed requirements, Seversky's staff of engineers with no USN experience at all was reaching for the moon. Brewster's XF2A-1 design was an obese airplane powered by the Wright R-1820-22 Cyclone engine, the same type powerplant that just refused to produce advertised power in the latter half of the 1930s decade.

With some slick salesmanship, Seversky Aircraft got their foot in the door by promising quick delivery of a suitable fighter. Grumman was unable to fly their prototype before September and the Brewster XF2A-1 was not to fly until December. As it turned out, the XF4F-2 was delivered to the Navy at Anacostia, D.C., just before Christmas 1937 and the Brewster was not turned over to BuAer until a year later. The Farmingdale company simply built a duplicate of the No.1 Army P-35, even to the installation of that abominable windshield that the Air Corps so quickly rejected. Major differences from the P-35 were installation of a Wright R-1830-22 radial engine and a standard Navy tailhook. But the airplane was ready to go by 3 June 1937. It did pass all initial tests for the Navy, but was unable to show a top speed in excess of 270 mph.

By the time the Seversky – given a company designation of NF1[4] (for Navy fighter 1) which sounded somewhat like a formal USN designation – could be fully navalized, the gross weight was likely to further reduce performance. As of 31 March 1938, the NF1 was in the hands of Navy test pilots at NAS Norfolk or Anacostia. Ultimately, the NF1 did not meet BuAer requirements, but on the other hand it took three years to bring the XF4F up to the standards for carrier operation, and the Brewster F2A series left a taste in the mouth that nobody ever forgot. In its few WWII combat encounters, it was trashed by Japanese fighters.

In one interesting sidelight of the NF1 program, C. Hart Miller conducted all company flight testing on the airplane at Seversky Field without benefit of having a paved runway. He had to go through the series of zig-zag taxiing trials, false takeoffs in which the throttle is cut just after leaving the ground, and then a standard Navy routine of flight tests followed. That is all capped by the necessity to dive the airplane thousands

Jimmy Doolittle required installation of numerous additional test instruments and controls because the Seversky SEV-DS was to be used extensively for testing of higher-octane aviation fuels. Those fuels gave the Allies a valuable powerplant asset in WWII. That was a pretty "busy" cockpit for 1937. (Mfr. 961/F. Strnad)

[4] In Navy designation terms, a formalized designation of NF-1 would apply to a Navy training airplane built by Grumman (N for trainer, F being assigned as the code letter for all Grumman airplanes.)

It was a case of the Major and the Major (not the Minor). Major de Seversky talks to a bemused Major James Doolittle (AC) seated in his newly purchased SEV-DS. Doolittle was on reserve status then and was manager of Shell Oil Company's aviation products division. He planned to use the SEV-DS for fuel developments and for business travel. Although it closely resembled a special racer designated SEV-S2, it differed in having a Wright Cyclone R-1820-G5 engine in place of the usual P&W R-1830. (Seversky)

of feet to terminal velocity, ending with a mandatory 7.5G pullout. Miller was reminded that test pilot Jimmy Collins dove into a nearby cemetery during such a test in a Grumman fighter on 22 March 1935. Collins had performed nine of a series of ten dive and pullout tests that day, but on his last dive test the airplane was seen to pull out abruptly, whereupon the wings and engine broke away from the fuselage and the remains spun into that cemetery.

Alexander deSeversky was, without a doubt, an entrepreneur, salesman and speed merchant. Back in 1935 when the Pursuit Plane Competition had a rather bumpy ride on the way to a contract award, the Major made a rather remarkable decision to enter a National Air Race event. His experience at pylon racing was nil. Therefore, as described in an earlier chapter, he was wise in choosing experienced race pilot Lee Miles to guidethe big SEV-3 amphibian to a most respectable fifth place in the closed-course Thompson event. In 1936, after the SEV-1XP won the Pursuit Competition at Wright Field, the Major may have actually considered entering the fighting plane in the Bendix cross-country race from New York City to Los Angeles (the Nationals had been moved to California that year). But the Air Corps supposedly turned him down, if he actually presented such a proposal, with the rejoinder that it was a Classified airplane. To the media in those days, that always meant "Secret," even though the SEV-1XP was not even classified "Restricted" on photographs taken at Wright Field. Logic, the financial condition of Seversky Aircraft at the time, and a daunting workload were far more likely to have kept him from racing. Actually, in view of Air Corps policy about the XP-38 just two years later, it is too bad the SEV-1XP did not participate in 1936. It most certainly was capable of "running away" from any of the fastest competitors. The Air Corps could have gained something from a win by the Seversky, and even if that prototype had then crashed, it would have been no real loss. The military competition was history.

However, 1937 was an entirely different matter. The un-

The single Seversky SEV-DS (Doolittle Special), registered as R1291 and bearing the Shell Oil emblem on the cowling, is shown on landing approach to Burbank. Wright Aero had upgraded the R-1820-G5 Cyclone from 815 horsepower to 1000 in 1938. By August 1940, a total of 584 hours had been clocked. Just prior to the opening of hostilities with Japan, the airplane was exported to the Ecuadorian Air Force, being redesignated C-1. (Author's Coll.)

With only the letter "R" appearing on wing and tail, the machine gun armed Seversky 2PA-L for Russia was test flown over New York in November 1937. The pilot on its first flight on November 2nd was evidently C. Hart Miller, then assistant to the company president (de Seversky). There is no record of any CAA registration being issued; therefore it was not supposed to be flown, especially with armament. Both the 2PA-A and 2PA-L were exported under Export Lic. #4420. (S. Hudek Coll.)

happy stories about the Seversky AP-2 have already been related, and de Seversky's popularity in the highest echelons of the War Department was disintegrating. Memories of his close association with and support for Gen. Billy Mitchell rose to the surface among military men in high places who did not share the Major's high regard for the general. Almost lost in the mists of time are some relevant facts. Americans were very much in an anti-war mood; the military had a hard time even surviving in that environment, and military/naval personnel were not really held in the highest regard; and, after Mitchell's trial there was little tolerance in the services for outspoken dissidents. (In just a bit over a decade, America's greatest hero, Charles Lindbergh, ran afoul of the governmental hierarchy to such an extent that he resigned his commission in the Air Corps.)

But it was more likely that de Seversky's sharp tongue and outspoken views appearing in the press rather frequently created greater animosity. Dealings with the Amtorg Corporation representing Russian interests, and others of doubtful character, did much to annoy Washingtonians (as in D.C.). With the AP-2 written off, at least the Major had some racing/record-breaking potential in the updated SEV-1XP, and Frank Sinclair was scheduled to participate in the Bendix race from Burbank, Calif., to Cleveland. For some months prior to the National Air Races, de Seversky had been honing the talents of aviatrix Jackie Cochran through an agreement giving her considerable flight time in the 1XP, renamed "Executive." At least as a publicity designation. It appears that the 1937 plan had called for the company president to enter the AP-2 in the Thompson Trophy Race while test pilot Frank Sinclair would fly the SEV-1XP in the Bendix. Following the Nationals, Cochran was to fly one (possibly both) of the speedsters in attempts to establish new speed records. As the major 1937 racing season came to an end, Frank Sinclair had managed a respectable fourth place in the X-country Bendix event and managed to give the hottest racing planes in the country a run for their money, placing a close fourth in the Thompson Trophy Race in that rather tired SEV-1XP.

Additionally, some salesmanship and other negotiations had been initiated with Fuller Paint's top executive, Frank Fuller, Jr., on behalf of Seversky Aircraft, the object being to provide him with a new P-35 clone (primarily a botany term in 1937) upgraded with a few in-house secrets. Fuller Paint would derive a lot of good publicity from a racing and record-breaking program just as oil companies had in the first years of the decade. It appears that this new racer, designated SEV-S2, was made available at a most attractive price. Fuller willingly joined the team and was about ready to ramble. (See Chapter 5 for a 1939 photograph.)

At the same time, Jimmy Doolittle had joined Shell Oil in their aviation products division and was in need of a high-performance replacement for his Lockheed Orion "Shellightning." Tests with high-octane aviation gasoline were in progress, so they purchased a near duplicate of the SEV-S2 except that it was powered by a Wright Cyclone (R-1820) engine and instrumentation was quite different. Both "Executives" featured the cut-down cockpit canopy first seen on the AP-2. That Doolittle Special was logically designated as the SEV-DS[5]. Although the SEV-S2 belonging to Fuller relied on a Pratt & Whitney Twin Wasp (R-1830) like its military brothers, it featured a "C" series model with a downdraft carburetor. At most times in its racing years, the Fuller racer could count on having an engine developing anywhere from 50 to 250 horsepower more than "B" engines in P-35s were producing. Any commercial or military airplanes in the country that were faster than those Severskys had to be experimental types (including Roscoe Turner's Meteor racer).

Seversky Aircraft garnered a tremendous amount of first-rate publicity in that 1937-1939 era from the performance of those three SEV airplanes. On 3 September 1937, the SEV-S2 with Frank Fuller at the controls won the Bendix Trophy Race, continuing on to New York to set a new West-East speed record. On September 21, Jacqueline Cochran broke the 3-Kilometer world speed record for women at Detroit while piloting the SEV-1XP at an average speed of 292.271 mph, making her the fastest woman in the world. In the final quarter of the year, Fuller, de Seversky and Cochran established new inter-city speed records

[5] For several decades there has been warranted confusion about the Seversky designations for aircraft. During a period of transition, even some disarray, certain models had the SEV prefix while others didn't. Company model designations appearing in this book are based on the best company internal documentation that has been uncovered by top volunteer researchers working at Nassau County's Cradle of Aviation Museum.

***Essentially identical in most respects to the landplane version of the 2PA-A, the initial airplane on the Amtorg contract was the 2PA-L Convoy Fighter. The change to a Wright Cyclone GR-1820-G7 geared engine boosted top speed to 287 mph, and with overload fuel carried internally it had a range of 2020 miles. A comparable change in dihedral (to agree with Air Corps requirements) was being made on all Seversky types, ie: 7 degrees 20 minutes measured on the lower surface. Constant-speed propellers were being fitted as normal equipment. Nobody ever explained how company (and other?) pilots could fly around with no CAA registry.*(Author's Coll.)**

***BELOW: With a top speed of 240 mph at an altitude of 15,000 feet or more, the Seversky 2PA-A could not have had much competition from other types of amphibians. Amazingly (for 1937), the fighter had a "maximum range of 2300 miles," assuming takeoff from land at its overload gross weight. The Russian government had a fairly expensive option to manufacture many of those "Convoy Fighters" on a production basis with Seversky Aircraft support. It was never exercised.* (C.A.M./Ed Boss, Jr.)**

at the controls of Seversky Executives. That trend was to continue through 1939.

When the company made that press announcement about the Russians purchasing a pair of amphibians based on the SEV-3 and 2PA prototype designs, facts were (as usual) a bit distorted. The buy was really based on a design upgrade of the 2PA which was a landplane showing four fuselage-mounted and two wing-mounted machine guns plus the rear gunner's weapon. It was the real Convoy Fighter design. Combining that aircraft configuration with amphibian floats as used on the SEV-3M-WWs for Colombia produced another evaluation prototype. The pure landplane version (with fewer guns at that stage) was designated 2PA-L, the amphibian became the 2PA-A and differed only in the landing gear and cowling airscoop arrangements.

Rather strangely and remarkably, only the amphibian was given a Bureau of Air Commerce license although both aircraft flew over American soil – with guns installed! The 2PA-A, registered NX1307 (hand lettered rather crudely on the wings and rudder after the airplane was painted overall prior to shipment

Accented by a Seversky P-35 coming in for a landing, this single Seversky 2PA-A amphibian Convoy Fighter was built for export to Russia via the Amtorg organization in 1937. Touted as being developed from the P-35 pursuit, it was really a cross between the prototype 2PA and the original SEV-3M. Powered by a direct-drive Wright Cyclone 9 (G series) rated at 1000 hp at S.L. (850 at 15,200'). A civil registry NX1307 was initially applied somewhat crudely to expedite test flights. **(Seversky)**

overseas) was shown on its alternate retractable landing gear that could be interchanged quickly. It was ultimately flown with the amphibious landing gear installed. The 2PA-A had to be a very fast amphibian with its Wright R-1820 Cyclone engine probably matching the unit installed in the SEV-DS in the same timeframe. That Cyclone was then rated at 815 horsepower. Evidently the Major was not around, so the first flight of the 2PA-A was made by Hart Miller, by that time a v.p. in the organization.

Miller also piloted the landplane Convoy Fighter, the 2PA-L, when it went up for the first time – sans any registry except for the letter R. If the media had learned of that situation, the roar from Washington might have been deafening. Why? Well, here was an unlicensed armed fighter (no bullets) built for Communist Russia flying over Long Island, New York. At the time, the Spanish Civil War tempo was rising in crescendo as Nazi, Fascist and Communist weapons of war were being widely used with maximum aggressiveness. Several U.S.Government agencies would have been hard pressed to explain the flights. This export airplane was probably capable of outrunning America's newest first line pursuit aircraft just being assigned to Air Corps units. Performance figures with the Cyclone engine credited the 2PA-L as having a top speed of 290 mph at 16,500 feet. On internal fuel alone, it had a cruising range of 2600 miles. If Walter Winchell or some other firebrand commentator had roped that story it would not have taken many distortions of facts to inflame the population of the entire country. It was hard enough to determine what was fact within the company. It seems pretty obvious that at the very least the Bureau of Commerce had no handle on the situation. An export license (No. 4420) had been issued for both aircraft on October 23rd, but that couldn't have been the shipping date. Hart Miller did not fly the 2PA-L for the first time until November 2nd. Seversky may have avoided the poison pill when they turned the airplanes over to Amtorg representatives, but the multi-million dollar project never did materialize. Amtorg did not exercise the option, most likely because somebody recognized that toughened embargo laws could cut off the supply of necessary components needed for mass production.

One slightly disturbing notation appeared in Seversky's stockholder report for 1937, issued on 31 March 1938. It stated clearly what the company had designed, built and delivered to Amtorg and the USSR, namely one Convoy Fighter (obviously the 2PA-L), and that "a second plane of its amphibian design is undergoing its final test and will probably be delivered during the first week of April (1938)." That had to be the 2PA-A airplane. The export license was supposedly related to that airplane. How did the unlicensed 2PA-L leave the country? (Sometime later, in the story *"America, Too Young to Die!"* written by de Seversky, he added fuel to flames. The story said the Russians had ordered two long-range fighters and he referred to them as AP-3! Nothing else ever confirmed construction of any airplanes known as AP-3s.)

But worse was yet to come. The squalid story of selling Convoy Fighters to Japan in their days of aggression against China and at a time when the USS *Panay*, a river gunboat operating in waters near Shanghai, came under attack by Japanese airplanes was squelched. Seversky management went so far as to issue totally misleading stories about any such sale to their own employees. In the face of irrefutable evidence that has never been suppressed, many former employees believe the company's misleading information to this date. Why would an otherwise honorable panel of executives authorize a disinformation release to their own people? In 1938, Seversky Aircraft was in dire financial straits, somewhat camouflaged by the successes in record breaking and in winning speed contests. Paul Moore and his wallet seemed to be imbued with the patience of Job, but even he was feeling the pinch and had to insist on some financial recourse.

The 1937 (bad recession year) loss from operations was $1,210,516, a significant fortune in 1937. Anybody with just $1500 could buy a brand new Wiley Post biplane with a Model A Ford engine. Can you imagine purchasing a factory-fresh Beechcraft

***One feature of the single 2PA-A that was rarely observed was full interchangeability between the amphibious landing gear and a conventional retractable landing gear. Evidently the float gear could be removed fairly easily and a typical Seversky landing gear fitted. A close comparison between this configuration and that of the one and only 2PA-L reveals differences in the landing gears and the air intake arrangements.* (Frank Strnad Coll.)**

B17 staggerwing biplane with a 225-horsepower radial engine for less than $9000? A beautiful 3-place Waco YMF biplane picked up at the Troy, Ohio, factory would not cost you more than $8000. An export model of the Curtiss Hawk 75 would not set you back more than $27,000, while that same check would put you into a Bel Air (California) mansion, possibly as a neighbor to Gloria Swanson. Or you could live close by star actor/flyer Wallace Berry, just off Sunset Boulevard on North Alpine Drive for

***Just about 6 months after the SEV-2XP failed its first assignment, Hawker's F.36/34 prototype Hurricane flew for the first time on 6 Nov. 1935. Even with a wooden fixed-pitch propeller set for optimum climb, this 1,000 horsepower interceptor attained 318 mph at critical altitude. During the summer, Curtiss and Seversky were unable to see 275 mph on the clock with their new prototypes. By 1936 the RAF had ordered 600 Hurricanes. The first production Hurricane Mk.1 (L1547) is shown at time of delivery in October, 1937. It had eight guns!* (British Aerospace)**

ABOVE AND BELOW: *Utilizing some new NACA data, Seversky installed a revised cowling and large spinner on the enigmatic AP-1 (formerly P-35, AC36-354). It remained the only actual P-35 airplane to have the odd wheel fairing extensions and different windshield. The company used it as a flying testbed and demonstrator before it was destroyed in a Miami fire.* (Frank Strnad Coll.)

less. Ford Motor Co. came out with a new V-8 60 sedan in 1937, and you could drive it away for less than $500. A Pontiac 6 was not likely to take more than $700 out of pocket. Seversky Aircraft could not afford to upset valuable employees by letting them know that a major contract for at least twenty Convoy Fighters was coming from the Japanese. The type of airplane to be procured was designated 2PA-B3, and the price per airplane was to be about $62,500 plus spares and engineering data fees. (See Chapter 5 for details.)

Throughout the latter months of 1937 and on into 1938, production of P-35s for the Air Corps moved along nicely. The pursuit planes were being completed at the rate of about one every other day as deliveries came to an end in August. Seversky was never to see another production contract from the U.S. Army Air Corps, and every action by one or the other brought about a further reaction. Aside from those machi-

nations at the highest levels, P-35 integration into the GHQ Air Force was accomplished with no more than normal teething problems. If nothing else, the airplanes were quite popular with the public, especially in the midwestern states after virtually all of the Seversky pursuits were stationed at Selfridge Field near the town of Mt. Clemens, Michigan. The base was on the shores of Lake St. Clair. Personnel assigned to Selfridge had seen many significant changes there, primarily in the disappearance of a plethora of World War I wooden buildings and hangars. Dozens of Williamsburg-style brick buildings had been constructed under WPA/PWA authority following election of Franklin Roosevelt. Other welcome additions included new hangars and a very large concrete parking area for the fighters. Pursuit aircraft duties were shared with some Consolidated PB-2A two-seaters, a breed that was extremely controversial to say the least and the last of their type to serve. Selfridge was part of the First Air Force headquartered in New York at Mitchel Field. Curtiss P-36s were soon to join the P-35s in Michigan, especially since attrition would deplete the Seversky ranks by about half as the decade of the 1940s arrived.

Showing that a battler can be beautiful, the Supermarine Spitfire lived up to its great appearance, attained greatness. Designer Mitchell made few mistakes as exemplified by this stirring photograph of an early Mk.1 in the series. It even proved to be adaptable to carrier operation. No wonder nearly ever pilot wished to fly any Spitfire, and was rarely disappointed. (Author's Coll.)

Back at Farmingdale, that same No.1 airplane delivered on the P-35 contract – complete with its peculiar windshield and those odd, pointed wheel fairings protruding – appeared with Bruno Associates publicist Dick Blythe in the cockpit. Air Corps markings had been deleted and a revised registry number was assigned by the CAA, and the designation became AP-1. Several production P-35s came back for incorporation of 4 degrees of dihedral in the wing (measured on the upper surface) and, remarkably, one of those was marked AC36-354 – again. Now, how do you suppose that happened? It was not mentioned in any board meeting minutes or any other known records except in one of Mr. Meyerer's production reports. Hmmmm! The AP-1 was destroyed in a hangar fire at Miami, Florida, on 2 April 1938. (C.A.M./Ed Boss)

When the Seversky 2PA (NR189M prototype) did not return to the U.S.A. following Frank Sinclair's and William Klenke's 1938 demonstrations in Argentina and in Brazil, the company manufactured a new test/demonstration airplane, the AP-7. It was in most respects a duplicate of the new EP1-68 demonstrator, but it was the only one to eventually have a one-piece clearview canopy on the turtledeck. At one stage it had a massive prop spinner and narrow-inlet cowling, almost certainly the same installation seen earlier on the AP-1 (which the AP-7 also replaced). Testing was in connection with the projected 4-place high-speed executive transport being designed. (Author's Coll.)

Operational Seversky P-35s assigned to the 27th Pursuit Sqdn., 1PG, are shown lined up for review with the squadrons commander's airplane at the front. The P-35 was pleasant to fly, beautifully built and had excellent range. But Gen. Mark Bradley, closely associated with P-47s later on, remembers the Seversky as being a profound groundlooper. For the unwary, it could groundloop while taxiing at considerably less than 30 mph. (Author's Coll.)

Seversky P-35 pursuit planes of the 27th Pursuit Squadron are shown in formation flight over Michigan circa 1938. A major event at Selfridge Field in many Depression-era years was the John Mitchell Trophy Race event. The weather always seemed to be perfect. No particular pursuit type seemed to survive more than four years. One or two Boeing P-26s were always on hand long after the P-35s arrived. (S. Hudek Coll.)

Posing nicely in the skies over Selfridge Field, Michigan, the 17th Pursuit Squadron's commander showed off the classic lines of a new Seversky P-35. Virtually every P-35 manufactured was assigned to the First Pursuit Group, long based at the Mt. Clemens area field. This photo exhibits the major difference between a production P-35 windshield and that seen on the P-35 No.1 (AC36-354) when it was returned to the company. It soon became the AP-1 demonstrator which was evidently covertly replaced by a new airplane with that same AC number. The Army did get their 77 airplanes per contract. (via Jack Terzian)

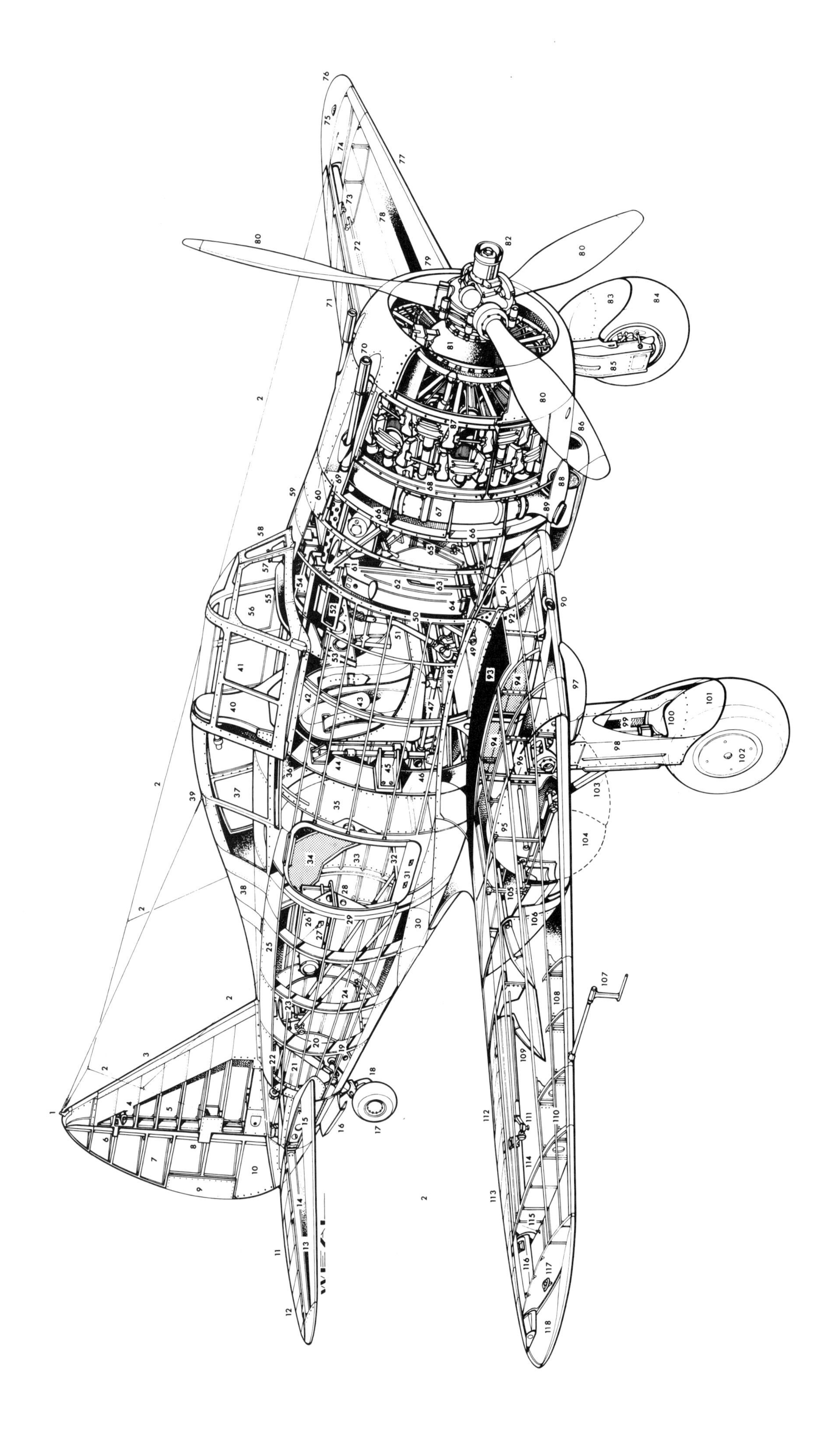

Seversky P-35 Cutaway Key

1. Antenna attachment
2. Antenna wires
3. Vertical stabilizer leading edge
4. Aft navigation lights
5. Vertical stabilizer structure
6. Rudder upper hinge
7. Rudder structure
8. Rudder center hinge fairing
9. Rudder tab
10. Rudder assembly
11. Elevator tab
12. Right hand elevator
13. Right hand horizontal stabilizer
14. Elevator hinge
15. Horizontal stabilizer attachment
16. Retractable tailwheel door
17. Retractable tailwheel w/pneumatic tire
18. Tailwheel strut assembly
19. Aft fuselage lifting point
20. Fuselage frame (typ.)
21. Tailwheel shock absorber
22. Control rods
23. Elevator control rod linkage assembly
24. Fuselage longerons
25. Semi-monocoque riveted Alclad fus. structure
26. Radio receiver-xmtr. equipment
27. Radio antenna lead-in
28. Radio equipment support tray
29. Baggage compartment aft beltframe
30. Wing/fuselage fairing fillet
31. Access door latches
32. Baggage compartment floor structure
33. Baggage access/pilot escape door
34. Baggage compartment w/alternate provisions for second crew member
35. Forward fuselage aft beltframe
36. Sliding canopy track
37. Canopy fairing w/fixed Plexiglas windows
38. Dorsal spine fairing
39. Radio antenna lead-in
40. Headrest/turnover structure
41. Sliding cockpit canopy
42. Emergency tilt-back pilot's bucket seat assembly. (Rotates 90 degrees backward for emergency egress.)
43. Oxygen cylinder (behind seat)
44. Seat track assembly
45. Seat support mounting bracket
46. Bellcrank assembly for aileron and elevator control rods
47. Aileron/elevator control rod
48. Wing root intermediate fillet
49. Rudder pedal assembly
50. Firewall and bulkhead assembly
51. Control stick
52. Expended cartridge case ejection chute
53. Instrument panel assembly
54. Two Browning machine guns: one .50-cal. and one .30-cal. (Specified)
55. Instrument panel shroud
56. Windshield assembly
57. Gunsight mount
58. Optically flat windshield glass
59. Removable access panels
60. Engine oil tank
61. Engine mount struts
62. RH ammunition magazine (.30-cal. and .50-cal.interchangeable)
63. Cooling vent slot
64. Ammunition magazine retaining clip
65. Engine controls
66. Cowl flaps (shown in phantom)
67. Stainless steel exhaust collector
68. Cowling former rings
69. Machine gun blast tube
70. Machine gun barrel troughs
71. LH aileron
72. Aileron control rod
73. Aileron bellcrank and hinge
74. Antenna wire attachment
75. LH navigation light
76. LH wingtip assembly
77. Wing leading edge
78. Wing spar
79. NACA cowling nose ring
80. Hamilton-Standard controllable-pitch propeller
81. Engine reduction gear case
82. Propeller hydraulic pitch change mechanism
83. MLG wheel/strut fairing
84. MLG wheel and tire assembly
85. MLG strut fairing
86. Engine oil cooler air intake
87. Pratt & Whitney R-1830-9 Twin Wasp 14-cylinder aircooled radial engine
88. Wing leading-edge fillet
89. Collector ring exhaust stack
90. Cockpit air inlet
91. Lower engine mount attachment fitting
92. Wing center-section front spar
93. Wing root non-skid walkway
94. Center-section integral fuel tank
95. MLG wheel well
96. MLG strut hinge fitting
97. MLG strut forward fixed fairing
98. MLG wheel/strut fairing
99. MLG strut assembly
100. MLG strut axle forging assembly
101. Wheel and tire assembly
102. Wheel coverplate
103. Retraction strut and gear drive assembly
104. MLG wheel/strut fairing (retracted position)
105. Split flap actuating rod
106. MLG trailing edge wheel fairing
107. Pitot tube head
108. Multi-spar cantilever wing structure
109. Trailing-edge split flap extended
110. Wing leading-edge ribs
111. Aileron control rod and bellcrank assembly
112. Aileron trim tab
113. RH aileron
114. Aileron balance weight
115. Antenna wire attachment
116. Aileron outer hinge
117. RH navigation light
118. RH wingtip assembly

CHAPTER 5

EXPORT OR PERISH

Perhaps nobody can fault the Major for many of his actions subsequent to his personal writeoff of the Seversky AP-2, and maybe it had something to do with a mid-life crisis or some such thing. He *seemed* oblivious to the feelings of ranking officers and civilians in the War and Navy Departments, and he must have detected some undercurrent of bad feelings toward him in various Air Corps offices. But even if he did not sense it or hear about it, the wealthy investor Paul Moore should have learned something about such matters from other investors or stock exchange members. In the final third of the 1930s decade, de Seversky seemed intent on distancing himself from the War Department, and the NF1 debacle at least made him aware that no "kluge" airplane was ever likely to win a contract from the Bureau of Aeronautics...unless, like LeRoy Grumman, communication lines to Navy Department insiders were well established. He would have done well to hire somebody for analysis of Grumman contacts and organization, thereby providing himself with the necessary tools required for opening the proper doors.

Instead, de Seversky and his pretty wife Evelyn (who had become a proficient aviator in her own right with the Major as her instructor) seemed content to concentrate on establishing new speed records between cities or in breaking old ones. He may have been convinced that the Government wasn't going to let him win many contracts, especially if giant Curtiss-Wright was in the bidding. One wonders if anyone, other than Evelyn, was ever close enough to him to learn what he knew or suspected. Even if he had his mechanic/associate, Judson Hopla, as a confidant there is little or no chance that we can ever learn the truth. The passage of time has led to a blending of fact and fiction, making it all but impossible to rely on the validity of Hopla's information. Therefore, only the man, himself, really had any ideas about the reasons for his actions. Knowing that he was capable of producing some excellent products, he turned toward the export markets of a world in turmoil. For every action, there seemed to be a more-than-equal reaction on the part of some people in government offices.

Although documents relating to initial procurement of twenty Seversky 2PA-B3s have not surfaced, many of the later details of the con-

***At first glance, the 2PA-B3 version of the "Convoy Fighter" was identical to the Russian 2PA-L. All of those airframes obviously came right out of the P-35 jigs, with a revision to the cockpit and canopy. However, 2PA-B3 cowling details, the P-35 main landing gear, vertical tail, and a fixed tailwheel were significant differences. In 1937 it was surely unpopular to be selling military aircraft to the Japanese, but this one marked F-1 (later X1321) was the first of 20 shipped to that nation.* (Seversky)**

tract are a veritable fountain of specifics. Amazingly, at a time when almost every record flight by Seversky airplanes was listed in detail in the company's Annual Report, along with minute details about Mrs. Odlum's (Jacqueline Cochran) financial contribution of $30,000 to the company, there was not one word about the 2PA-B3 contract with the Aircraft Trading Corporation. It should be mentioned that the president of Aircraft Trading was a Mr. S. Megata. When some details about the transaction began to appear in Seversky records, it quickly became closely held information because they were often dealing with a Mr. Yamamoto. But within the company, personnel were informed that twenty fighters had been sold to Siam. There was nary a word about Japan.

***Back into the shop for rework once again, the AP-7 (NX1384) was hotrodded with a 1200 hp P&W S1C-G Twin Wasp engine, and licensed as a 2-place airplane. It was completed on 22 August 1938 and flown for the first time on that same day. As with all Severskys, there was plenty of room for one or two in the baggage compartment. And, presto, there were those bulbous landing gear fairings duplicating what had been created for that initial P-35/AP-1. Perhaps we should refer to them as "curvaceous" since they helped Jacqueline Cochran outrace Frank Fuller's SEV-S2 by 23 minutes in the 2043-mile Bendix Trophy Race in 1938. She went on to Bendix, N.J., to establish a new west-east transcontinental woman's speed record. Severskys placed 1-2 in the racing event. AAC generals weren't amused.* (William Yeager)**

Rather surprisingly, the first "complete information" drawing of the Seversky 2PA-B3 reveals it as, in most essentials, a production P-35/AP-1 with a 2-seat canopy. Features included the zero-dihedral wing, P&W R-1830 engine and a production P-35 landing gear except for a non-retractable tailwheel. When the first airplane on the contract (c/n 126) was rolled out in the latter quarter of 1937, a Wright GR-1820-G3B engine had been substituted. Most people believe the Russian 2PA-L and the 2PA-B3 were near duplicates, but tails, engines, landing gears and several other features were quite different. To avoid ongoing misunderstandings about the 2PA-B3 airplanes, the following details are provided:

c/n 126 Long-range wing, ordered by A. deS. Reg. X1321
c/n 64-1 Short-range wing. 1 of 9. Reg. NX1391
c/n 64-2 Short-range wing. 2 of 9. Reg. NX1388
c/n 64-3 thru 64-9. Short-range wing. Seven airplanes
c/n 66-1 thru 66-10. Long-range wing. Ten airplanes

Seversky's Export Shop was thrown into panic on April 23rd when crating of 2PA-B3s on Project 64 was ordered to stop. The Japanese suddenly realized that those airplanes with short-range wings only carried 200 gallons of gasoline. Long-range versions had an additional 75 gallons in each outer wing, and the customer insisted on 350 gallons internal capacity at that late date. On the 27th, the shop got orders to ship c/n 64-2 and -3 by rail and another order required assembly for flight to the West Coast. They were to try and catch the *Hokkai Maru* at Los Angeles harbor.

Three Seversky 2PA-B3s were licensed for U.S. test and ferry flight to the West Coast, but the rest were crated and shipped from New York to "Siam," if you want to believe the cover story issued by the company. The first 2PA-B3 flew only once because of very bad weather (but ATC accepted it), was dismantled and shipped on the *Hokkai Maru* on 12 April 1938. It appears the total price for aircraft and spares was $1,623,267, giving Seversky a net profit of $402,580. In the Imperial Japanese Navy, Seversky 2PA-B3s were given the designation A8V1 as a Type S airplane.

***Featuring a rather massive one-piece windshield and a cowling design absolutely predicting those used on the Republic XF-12 years later, the 1938 Seversky Executive was upstaging the Spartan Executive designed in the same general timeframe. It was to have a 950 hp engine, flush landing gear and would carry four people and full luggage. Looking remarkably like the considerably later XP-47J in this Heaslip rendering, the Executive absolutely forecasted the Thunderbolt outline and size.* (Frank Strnad Coll.)**

In the Depression and Recession days of the 1930s, everybody found enough money to take in the movies each week and buy a copy of Life or Look magazines for escapist reasons. It only took one thin dime to get the magazine, or maybe attend a Saturday matinee at a neighborhood theater. In Detroit, a towering four-dip Thompson ice cream cone could be had for a nickel in 1935 when the first Seversky pursuit was fighting for a contract. Whenever a film announcement stated that Gable and Loy were "together again," you could be certain it would attract big crowds. Here they are seen with that SEV-S2 in makeup as the infamous "Drake Bullet." (Frank Strnad Coll.)

As such, they were intended to operate as fighter escorts for twin-engine Mitsubishi bombers over China. After the first few aircraft were completed, the Japanese requested use of long-range wings. Seversky even supplied kits to rework delivered airplanes. With only twenty aircraft of the type procured, it could hardly be looked upon as a serious operational program. It is a safe bet that successful operational use would have led to licensed or unlicensed production in Japan.

Neither the State Department nor the Commerce Department raised even the slightest objection to this sale. In fact, de Seversky contended they heartily endorsed it. When details of the sale hit the War Department, the Major's goose was cooked.

According to Board of Directors (BOD) Meeting minutes and other documents, conditions at Seversky Aircraft were tottering on the brink of disaster. Example: Paul Moore sent a letter to the BOD on 22 December 1937 from which the following opening paragraph is quoted. "This company is still operating as if it were a going concern. Financially it would have been bankrupt had it not been for my personal credit which has been extended to the limit." Yes, and a very Merry Christmas to you fellas too. (No, he did not add that comment.)

Unfilled orders as of 15 March 1938 totaled $2,913,959, but where was the money to fulfill such orders? One "bright" spot highlighted in reports was an order in the amount of $80,000 from the Air Corps for conversion of the 77th airplane on the P-35 contract during construction. That turned out to be no little job. A major change involved providing a fighter of significantly revised design, featuring a vastly revised fuselage with an entirely new canopy and a smooth, flush-riveted skin. That was also true for the wing. This was supposed to be available for a new Pursuit (High Altitude) Competition in September 1938(!) as initiated by Circular Proposal CP38-390. We have to assume that, as usual, it did not go according to schedule and was actually the event conducted in March 1939. The airplane mentioned was to become the XP-41. Somehow, against opposition from Alexander Kartveli and others, the Major talked the BOD into allocating $50,000 to build a new (undefined) demonstrator for that same competition. Strangely, $30,000 of those funds came from Mrs. Floyd Odlum, better known as Jacqueline Cochran, in settlement of a claim against her relating to the SEV-1XP. (Now there is an interesting item. Did Jackie wreck the airplane? If so, there was never even a rumor about it.) This new and unidentified airplane was to be made available to her for entry in the forthcoming Bendix Trophy Race and two other competitions. Are we to believe that this was the AP-7 and that it was going to be in the pursuit plane competition? Well, not likely because minutes were recorded toward the end of April. The AP-7 was completed in May. It happens to be the airplane she got to fly. The airplane mostly likely funded underhandedly was

Virtually every Junior Birdman of America (thank you Mr. Hearst) must have envied Clark Gable for his racing skill with the Seversky clone known as the Drake Bullet. By golly, that sure looks like Frank Fuller's SEV-S2 to us! Even the wonderful Myrna Loy probably knew that. Just about everyone must have seen the movie Test Pilot, starring Gable, Loy and Spencer Tracy, at least once. (Frank Strnad Coll.)

Although 1938 was a banner year for new concepts and prototypes emanating from the halls of Farmingdale, recession times had descended upon Seversky Aircraft in a big way. The company was on the verge of bankruptcy, and the profitable sale of convoy fighters to Japan did nothing to enrich the Major's status with the War Department, or others. Production contracts had been all but completed by year end, and expectations about large new contracts were not uplifting. A contract to totally revamp the last P-35 production airplane into the XP-41 was not a moneymaker, and the AP-9 (Project 68-2) was built to try out features to be embodied in that new pursuit. A mechanically supercharged P&W R-1830S3C-G engine was installed, as was a flush landing gear and a modified airfoil shape (with a sharper leading edge). Delivery of the XP-41 to compete against the XP-40 was weeks late, so the AP-9 was flown to Wright Field by an Air Corps pilot on 17 January 1939 as a "stopgap" entry. **(S. Hudek Coll.)**

the de Seversky dream fighter, the AP-4, direct ancestor of the future P-47 Thunderbolt.

The loss of the AP-1 in the Miami hangar fire on 2 April 1938 was serious at a time when the company had virtually no production contracts of any kind. A high-performance demonstrator was needed at once, and it was rather fortunate for de Seversky to have authorized construction of eight or more airframes constructed around a fuselage/wing/empennage concept with great commonality of components. Although the AP-4 was assigned constructor's No.144 and would seem to have been next in line, it was a radical departure from others under construction. Scheduled completion was still a long way off. An airframe (c/n 145) which more closely followed the basic AP-1 concept was designated as the AP-7. In its early format, it closely resembled the standard P-35 military airplane, but it featured a special molded Plexiglas turtleback behind the canopy and was some 20 inches longer than the mass-produced pursuits. That was the result of a plan by which the single-seat types and the 2-seaters would have the same fuselage length, oddly a compromise which worked to the advantage of the fighter types by reducing the great tendency of the earlier models to groundloop. In its early appearances, the AP-7 (licensed NX1384) had standard P-35 landing gear fairings and a very large spinner matched with the NACA special cowling. The initial R-1830-BG engine of 950 hp was replaced by an R-1830 S1C-G engine of 1200 hp (for takeoff) as of August 26th. Although it looked just like any other single-seat airplane, it was licensed as a 2-place cabin airplane because there was full accomodation in the lower aft fuselage for a "passenger" who evidently had a set of flight controls. The idea was not ridiculous for instrument flying on long trips, but to land the plane from a nearly blind position in an emergency was certain to create white-knuckle conditions, even for an exceptional pilot.

With the change in powerplant came a regressive step involving installation of a new set of those bulbous, protuberant clamshell landing gear fairings. (Somebody had a love affair with those shapely covers.) Although dimensional data relating to the AP-7, as modified, is nowhere in evidence a 3/4-front view of the airplane would appear to confirm some rumors that the setup for Jackie Cochran's 1938 racing and record-breaking efforts included a set of long-span wings.

Miss Cochran's 1937 performances in Seversky's fast-tiring SEV-1XP as a racer – fitted out, like the AP-7, with full accommodation for a second person – won an International Harmon Trophy for her in January 1938. Her reputation was on a fast rise, thanks largely to the wisdom and talents of the Major.

Speaking of de Seversky, he flew the revised AP-7 from New York to Los Angeles in 10 hr. 3 min. on 29 August 1938 in the role of ferry pilot, breaking the existing east-west transcontinental speed record in the process. A few days later it was Jackie Cochran taking the same airplane back across the U.S. as a National Air Race participant. Her achievement was to win the coveted Bendix Trophy Race on the 2042-mile trip from Burbank to Cleveland by crossing in 8 hr. 10 min. 31 sec., then dashing on to NYC to establish another new west-east transcontinental speed record. The race No. 13 did not seem to be a bad luck omen. Frank Fuller's SEV-S2 was slower by about 23 minutes, but he did come in second. Speed merchant Paul Mantz, piloting the former Lockheed Orion "Shellightning" that Jimmy Doolittle had flown for several years, was more than an hour slower than

ABOVE: Evidently the Major either had a lot of "clout" left, or he was very lucky. While contract work proceeded on the Army's XP-41, de Seversky forged ahead with his own pet private venture, a thing that other company management not only failed to support, but a design that absolutely saved them from failure. Incorporating all the best updates of the XP-41, Alex de Seversky had more foresight than anyone except possibly Kelly Johnson and Lt. Ben Kelsey. He initiated Project 65 along the lines of the XP-41, and innovatively stuck a G.E. turbosupercharger back in the former baggage compartment, upside down. His new AP-4 (NX2597) did everything right that Curtiss and Bell did wrong. But, like Ben Kelsey, he was far ahead of his time. He may have lost the competition, but he got the Army Air Corps Guilt Prize: i.e., a Service Test quantity of a new YP-43 design. (Author's Coll.)

LEFT AND BELOW: Long before the 1939 Pursuit Competition was convened at Wright Field, Alexander P. de Seversky was off to Europe on a sales tour that had a share of successes and failures. While he was absent in behalf of the company, he had relinquished the presidency to Mr. Walter Kellett, recently brought on board as the vice-president. The Major was not in the U.S. when, at an altitude of 5000 feet, test pilot Frank Sinclair declared an emergency because the AP-4 was afire. Flames entered the cockpit on that 22nd day of March 1939, forcing Sinclair to bail out. The airplane crashed near Fairborn, Ohio, and these two pictures show all that remained that was recognizable, including the turbosupercharger. The NX2597 registration was still readable. No matter that the sleek AP-4 was gone; the airplane had proven its point. You are allowed one guess about which company received the major fighter contract. You're right! Curtiss. **(Frank Strnad Coll.)**

Fuller. However, Mantz obviously had a problem because in 1939 he flew the same airplane over that route an hour faster. But tailwinds can do strange things for the record-setting crowd.

Superficially, at least, the AP-4 could be (and has been) a source of confusion with the XP-41, like fraternal twins. But a close study of the "twins" would reveal major differences not even readily evident to engineers. And it is a good thing the Pursuit Competition was not really held in September 1938. The AP-4 was not rolled out the door until everyone was saying "Only 2 more shopping days until Christmas." Far worse, the XP-41 wasn't on time even for the opening of "festivities" at Wright Field. It was finally flown to Dayton on 4 March. In desperation, the company had dispatched the AP-9 as a stand-in, but even that airplane was not completed until the eve of New Year's eve, more or less.

Even some of the most "official" military records list the XP-41 as the "prototype for F-43 series" (sic) and "procured for high altitude performance tests" (sic) although its maximum speed turned out to be recorded at 15,000 feet, just like the Curtiss XP-40. Leaping ahead, the real facts about Seversky's XP-41 were that at 323 mph (best) it was no faster than the Major's AP-2 of 1937; the actual cost of the airplane modification work was $61,963 (remarkable) according to a Procurement Division report; the 1939 armament, like that of the XP-40, was one .30-cal. and one .50-cal. machine guns. Evidently the XP-41 was delivered with a large propeller spinner and the low-drag NACA cowling for its P&W R-1830-19 engine equipped with a 2-stage (and possibly 2-speed) mechanical supercharger. (At some date, a -31 engine was substituted for the -19 to be used with the NACA low-drag cowling and the very big spinner.)

Revisionist history tells us that the Major's pet AP-4 hardly left a shadow, but it was the true prototype for the Republic YP-43 service test quantity of fighters. It was the first single-engine fighter in the world to have a turbosupercharger for its aircooled radial engine located inside the aft fuselage. Fitted with the mammoth spinner and the narrow-inlet cowling, it was a handsome airplane. If you search hard for the details, it is easy to see only this AP-4 had the ability during the competition to perform well at more than 23,500 feet altitude. But who do you suppose got the big production contract for "high altitude" fighters? Of course, it was Curtiss Airplane Division at Buffalo, awarded an initial contract for 524 pursuit planes. The XP-41 was shunted off to the NACA at Langley Field for further testing, never to be acknowledged again.

OPPOSITE: Here, dear friends and countrymen, is the genesis of the P-47 Thunderbolt! A baby Thunderbolt if ever there was one. And something that nobody has been able to refute: this was purely an Alex de Seversky design concept, being the first fighter ever built with a midship supercharger installation. In fact, the "loyal" opposition distanced themselves from it with some zeal. The supercharger, Seversky-Gregor wing, flush landing gear and the near-final P-47B/D windshield and canopy are all there. This is the AP-4 as it appeared just days after the 1939 competition flyoff was completed. Had Lt. Ben Kelsey enjoyed "clout" in 1936, the EP1-106s for Sweden (P-35As for the AAF) might have looked more like this. But fate was not kind in the next day or two. **(AAF/Materiel Command)**

Although Kartveli was most certainly involved in development of the XP-41, records and affidavits made public in later litigation initiated by de Seversky against Republic Aviation firmly established proof that the AP-4 was almost entirely the Major's own project. Even the BOD minutes substantiate that. In order to save funds, the XP-41 and AP-4 made use of tooling and any design features that could be shared. As will be shown later, the AAC contract for thirteen service test airplanes and Wallace Kellett's news releases referred to the airplanes (at least initially)

At a glance, the AP-4 and the XP-41 (AC36-430) could pass for identical twins, but close examination shows they were closer to being fraternal twins. The carburetor air intake on top of the XP-41 cowl and mechanical supercharger intakes, a Curtiss Electric prop, completely different flush landing gear, and no sign of a turbosupercharger aft become identifying points. Most published material has wrongly stated that the XP-41 had a turbo unit. One significant feature of both pursuits was use of flush riveting. **(Author's Coll.)**

as AP-4s. Most assuredly, the Kartveli design team was responsible for the aerodynamic configuration of the two key Seversky (not Republic, yet) airplanes that were the main participants representing Farmingdale manufacturer at the 1939 competition. The AP-4, then, was the *incontrovertible* prototype for the YP-43 service test airplanes, and through the following chain it was the direct linkage to the P-44:

PROGRESSIVE EVOLUTION – P-44 INTERCEPTOR PROPOSALS

AP-4B	P&W R-2180	1400 hp	P-44-I
AP-4C	Wright R-2600	1500-1700 hp	P-44-II
AP-4D	Wright R-2600	Alternate proposal	P-44-II #2
AP-4J	P&W R-2800	1850-2000 hp	P-44-IV*

*This is believed to be mockup designation. Nothing firm on an AP-4L, but it may have been a later, final design study. All were to be interceptors based on YP-43. (See Bob Boyd's profile illustration in color sections.)

In a test flight after the competition was history, the AP-4 caught fire in flight and test pilot Frank Sinclair got a bit singed before he went over the side. The cause was either turbine wheel failure or (more likely) fracturing of the Inconel exhaust piping. The plunging airplane crashed near the town of Fairfield adjacent to Patterson Field. Somebody was smart enough to recognize the potential in the airplane's configuration, leading to the award of a service test contract for thirteen AP-4s (YP-43s) on 12 May 1939. Contract W535-ac-12643 called for these airplanes to be delivered in 1940, along with one skeleton airplane! A feature eventually to be seen (to a degree) in the P-47 called for the second airplane to have the G.E. Type B-2 turbosupercharger to be "completely buried" within the fuselage structure (as opposed to being partially exposed).

The autumn of 1938 could have also witnessed the collapse of Seversky Aircraft but for Paul Moore and his wallet or deep pockets.

Suffice to say, the Major had authorized construction of several airframes for sales/demonstration purposes. Changes in direction and plans would appear as if by magic. Of course we have faulted de Seversky for not fully adopting flush landing gears on all new aircraft, but he was not a magician. Even as president of the company, his hands could be tied by the BOD or by Moore. And in the Board Minutes of 26 September 1938 we find that is exactly what happened. His operating budget was suddenly slashed from a proposed $514,000 to $183,000. The weekly payroll (shop employees, etc.) was cut from 472 to 367. The entire business may not have been worth the $2,020,000 which was owed to Mr. Moore. It was at about this time de Seversky effectively lost any control he might have had.

Wallace Kellett, the autogiro man, became Managing Director with full power and authority to conduct company operations. More or less in desperation they agreed to proceed with the Major's plan to take two of the new demonstrators to France on a sales tour, and also to then demonstrate the aircraft in at least five other countries. Sweden and Poland were not even mentioned. Revealing his business weakness, de Seversky evidently did not see how he was being set up. With the War Department lurking in the shadows and nudging Moore and others along with a big carrot, the die was cast. It is very possible that nobody on the BOD expected any orders to be forthcoming from France or Great Britain in spite of war clouds and chaotic conditions prevailing in France's nationalized aircraft companies. After all, the purchasing commissions for those two Allies had not even approached Seversky as they proceeded to increase purchases from a host of other manufacturers.[1]

Although the Curtiss XP-40 failed to outpoint its competitors – only the Seversky AP-4 showing real capability above 20,000 feet – Curtiss-Wright Corp. was awarded a contract for 524 pursuit planes on April 26, 1939. Their P-40 was, in every respect, a 1935 airframe mated to a 1050 hp Allison V-1710-19 engine. By that time, production P-36Cs were powered by P&W R-1830 radials rated at 1200 hp. With 150 less horsepower and a weight increase of 750 pounds, Curtiss and the AAC would have us believe it was 31 mph faster than the standard P-36C. Actual production P-40s (no suffix letter) gained another 340 pounds...but somehow it was then supposed to be faster by another 15 mph!! Just 4 months after that contract was awarded, no fewer than nine squadrons of the 362-mph Spitfires were ready for combat and the Luftwaffe had the definitive Messerschmitt Bf 109E in Gruppe strength. (Walt Tydon/File)

[1] In 1938, war clouds were thickening over Europe. The French aircraft industry had been nationalized by the Socialists, and it was a massive failure. The British were procuring whatever they could from the U.S., and the French were in the process of buying Curtiss Model 75 Hawks and other aircraft. For unknown reasons, the joint French-British purchasing commissions (avoiding Seversky) even bought

(Mention has been made of the confusing model designation and serialization of Seversky airplanes, and nowhere was it more misunderstood than in connection with the demonstrators and fighters/bombers involved with the European sales campaign of 1938-39. All designations and constructor's numbers used have been confirmed from factory manufacturing and contract records.)

Through some serious compromise, two demonstration airplanes authorized and constructed in 1938 utilized as much of the P-35 contract tooling as possible. One single-seat fighter, designated as EP1-68, was to have exactly the same overall dimensions as the 2-seat convoy fighter that evolved from previous products. The latter type was designated as a 2PA-202, soon to be commonly known as the 2PA-BX when the sales team realized light bombers were more likely to sell than escort fighters. The French concept of multi-seat fighters had the connotation of twin-engine as well, and the RAF was headed strongly in the direction (secretively) of turret fighters. All single-seat fighters were firmly locked into the interceptor role. With that in mind, Seversky Aircraft was offering both demonstrators with short-range and long-range wings, just as they had 2PA-B3 airplanes sold to Japan. Powerplants for the two airplanes only differed in the blower ratios and other details dictated by the missions, but the P&W R-1830 engines featured downdraft carburetors as previously seen on AP-2, SEV-S2 and the then concurrent AP-7. The power range was from 1050 to 1200 horsepower. (The 2PA-BX went to Europe with an R-1830-S3CG engine rated at 1100 horsepower; the EP1-68 was equipped with an R-1830-S1CG producing 1200 horsepower.)

At first glance, the EP1-68 was just one more P-35 but among other things it was 14 inches longer (matching the 2PA-202). One side benefit was a reduced tendency for the EP1 type to groundloop. Col. Charles Lindbergh had been victimized by that P-35 trait a few times, so he made little attempt to conceal his displeasure with having to fly that military aircraft from time to time. Provision had been made in the EP1 for installation of at least one machine gun in each wing panel. There was no problem at all in providing space in the 2PA for a rear gunner since every Seversky built to date had adequate room for a second (even a third) person in the aft fuselage. For the record, the 2PA-202 had been sequenced as c/n 146 and was registered as NX2586; the EP1-68 was c/n 147 and registered as NX2587. (Evidence shows that airplanes constructed in the experimental shop had three-digit numbers in the 100 series while production airplanes – as in the case of the Japan-contract airplanes – had a two-digit project number followed by a dash number in sequence.)

Initially rolled out on the turf of Seversky Airport on 15 August, 1938, the 2PA-BX appeared to be an elongated P-35. Almost exactly one year had passed since the last production P-35 was delivered to the Air Corps at Selfridge Field, Michigan. When partially dismantled and nicely crated by the Dade Brothers company, the closely related airplanes headed for European shores. In a letter to Alex de Seversky, Kartveli had suggested the company could guarantee a top speed of 325 mph for the EP1

Two of the rarest Seversky types were the EP1-68 (proper designation form, pursuit demonstrator, and its companion 2PA-BX, also correctly 2PA-202). This EP1-68 would likely be identified by most as a P-35, but the cowl leading edge for its P&W S1C1-G engine of 1050 hp had a rounded lip and the carburetor air intake was on top. Registered as NX2587, it generally flew in Europe with a set of long span wing panels to increase range. Before shipment, a set of AP-7 type bulbous wheel coverings were installed. Fuselage length on this fighter (not pursuit) was identical to that on 2PA-BX 2-seater, at 14 inches more than any P-35. EP1s were less likely to groundloop because of the added length. **(Mfr. via S. Hudek)**

if they switched to the higher-altitude-rated engine installed in the 2-seater. That change could have been accomplished before the airplanes were crated. According to details provided by Kartveli, the "BX" demonstrator was only about 2 to 9 mph slower than the EP1 prototype despite a deficit of nearly 100 horsepower.

Since the high quality of construction evident in Seversky airplanes was a real asset, a guaranteed speed of 325 mph in a readily available fighter should have invited a flood of orders, but in the long run only the Swedes had the perceptiveness to take advantage of the opportunity. Another criterion which should have loosened the purse strings was manufacturing and manpower availability at Farmingdale. Poland's situation was perplexing; they had desperate requirements for combat aircraft, but they did not have great wealth and they recognized time was not on their side. Documentary proof exists, however, proving they would have ordered 250 aircraft but for an attempt by Wallace Kellett to garner credit for the transaction through devious communications. He avoided reference to a request by the Polish government for demonstrations by Major de Seversky. Of course Poland would have been overrun by the time a contract could be signed. Therefore it is only important to understanding the power plays behind the scenes.

Accompanied by Major de Seversky, Evelyn de Seversky, test pilot George Burrell and service engineer-mechanic Judson Hopla, the two demonstrators were aboard ship and headed for France on or about 26 November 1938. Unexpectedly, when both airplanes were reassembled there the old, bulbous full-coverage landing gear fairings were installed on *both* airplanes. It seems they had been specially constructed for the demonstrators. (Thankfully, they were never used again.)

The Major and Evelyn flew together in the 2PA-BX while Burrell and Hopla traveled in the EP1-68. Without access to any de Seversky diary or an itinerary, details of the venture are unknown. The primary French target must have been a total failure for there is a dearth of words about it, anywhere. Great Britain was the last stop on the tour, although that may not have been intentional. Sweden, not even mentioned in the planning stages, turned out to be the saving grace of the entire cam-

militarized Lockheed 14 airliners. Engineers in Great Britain could surely see that Seversky AP-7s with flush landing gears, improved cockpit canopies and weight reduction could equal Hurricanes in speed, surpass them in range. But no orders came. If Major de Seversky was anticipating sales of aircraft to either country, he was fortunate in being allowed to visit those sites. Why? If there was no smoking gun as such, a knife was being wielded with a vengeance.

***So you thought all of those Severskys were single-seaters. Rarely seen was the interior of a P-35, SEV-DS or, in this case, the one and only EP1-68 European Demonstrator. All could be classed as 2-seaters (in black leather), and in a pinch a third seat could be installed. The pilot's seat, when released, flopped backward to allow the pilot an emergency egress through the large baggage door (more weight). The seat is shown here in the "escape" position.* (Mfr.No.1451/F. Strnad)**

paign. Responsible leaders recognized they might soon be cut off from obtaining combat aircraft from Allied or Axis nations, and their aircraft industry would be unable to supply the nation's own total needs. (Norwegians passed up the Seversky types and eventually ordered Vultee Model 48 fighters, but they never got them once the Neutrality Act went into effect on 4 November 1939.) In a series of contract actions, the Swedish *Flygvapnet* (essentially, their Air Force) ultimately ordered 120 very slightly revamped 4-gun fighters, designated EP1-106, and 52 redesigned dive-bombers under the designation 2PA-204. The latter aircraft, essentially in the same category as the unsuccessful British Hawker Henley light bomber, were lengthier, heavier and not at all intended to serve as fighters.

A totally unfamiliar airplane shape appeared over Southampton Airport, England, on 30 March 1939 creating quite a stir at that landing field, the home of Vickers-Armstrongs' factory where Supermarine Spitfires were being built. That "shape" was the Seversky EP1-68 fighter demonstrator flown by none other than Major de Seversky. He had flown down from Martlesham, near Ipswich, after leaving the companion 2PA-BX two-seat demonstrator with the Air Ministry for evaluation flying. In return, the director had arranged for the Major to visit the factory and fly a Spitfire. During a 4-hour visit, he was allowed to fly a new RAF fighter and Flying Officer J.K. Quill flew the EP1-68 in that same time period. The British test pilot was up for less than a half hour, but did make two landings and two takeoffs. The president of Seversky Aircraft was instructed on the flight operation of the Spitfire by Vickers-Armstrongs' chief test pilot (fighters), Flight-Lieut. George Pickering. He taxied around a bit to become acquainted with the narrow-track landing gear and then proceeded to fly for 30 minutes. The Major gave the Spitfire a good workout during that short flight. To get a proper perspective on the performance of the Spitfire, he used his entire repertoire of maneuvers. That episode undoubtedly placed the Major in a unique position of being the singular public voice who could speak from experience about the actual capabilities of at least one of the best European fighter aircraft. He was certainly aware the Spitfire and Hurricane were not only in production (with at least 1000 of the interceptors on order) but manufacturing facilities were extensive and the aircraft were in first-line service units.

It is important to remember that this unparalleled event took place almost exactly five months before Hitler's Nazi forces attacked Poland and brought on World War II. During his tour of all Europe and the Scandanavian countries, the sharp-eyed and technically astute aviation executive probably gained a better insight into European war machines than all War Department, Navy Department and Air Corps officialdom combined. (Contemporaries such as Charles Lindbergh and Al Williams were also outspoken aviation experts of the period, but they were generally only exposed to the German *Luftwaffe* aircraft with first-line capabilities. We must presume that de Seversky was not at all conversant with Messerschmitts, Dorniers or Heinkels.)

Interviewed by an aviation writer for a local newspaper, the Major had this to say – in the then-current vernacular so evident in American films – about the Spitfire:

"She is certainly a swell little bus," he said. "She handled very well, and I thoroughly enjoyed the time I spent flying her. She is very stable, and I am impressed by her speed and the way in which she answers to the controls.

"The visibility, too, is much better than I expected when I was sitting in the cockpit on the ground."

In further conversation, de Seversky confirmed that he had

ABOVE AND BELOW: *Two universally ignored views of the demonstration Seversky 2PA-BX, or Model 202, show NX2586 as it appeared at Farmingdale just before it was shipped to Europe. Despite all documents issued, a detailed memo from Alex Kartveli in 1939 states that this convoy fighter was powered by a P&W Twin Wasp S3C1-G giving 1200 hp for takeoff, but only 950 hp at 15,900 feet. Incredibly, the CAA memos for licensing say it had a Wright Cyclone. Factually, the EP1-68 and 2PA-BX were grouped in the last half dozen Severskys ever built under that name. Probably the XP-41 qualifies as the last one.* **(C.A.M./Ed Boss, Jr.)**

come to Europe to study progress made in aircraft manufacturing.

There was one small oddity about the EP1-68 demonstration fighter – it was marked with what could easily be taken for a racing number, but it had a different significance. The number 68 painted on both sides of the aft fuselage obviously refers to the manufacturing sequence number used by Experimental as part of the model designation. As flown in Europe, the airplane was virtually a clone of the Seversky AP-7 flown by Jackie Cochran when she won the 1938 Bendix Trophy Race in a bit over 8 hrs. 10 mins. for the 2043-mile distance. Contrary to information contained in a few books, that airplane (race no. 13) did not have a flush landing gear until 1939. It had the bulbous-style wheel enclosures of the type used on the first production P-35 (later, AP-1). As flown in Europe, both the EP1-68 and 2PA-BX demonstrators displayed those fairings. In retrospect, de Seversky may well have won bigger contracts for his airplanes from European nations if Manufacturing had done nothing more than install flush landing gears on both airplanes. Certainly money was short and so was time, but that sales effort was a "life or death" program for the company. And by 1938-39, that semi-retractable landing gear was a 1935 anachronism. Because the various fairings tended to disguise the situation, it is not generally recognized that only about 25 percent of the wheel was actually within the wing contour. (In all fairness to the Major and everybody concerned, he had ordered Manufacturing to turn out four EP1 type fighters and four 2PA type machines. At least one each was to have a flush type landing gear installation. Man-hour expenditures exceeded all budget allocations, and the Factory Manager protested successfully, causing

Quite unexpectedly, the EP1-68 and 2PA-BX showed up in Paris with landing gear fairings like those on Jackie Cochran's winning AP-7 (actually owned by Seversky Aircraft). The Major must have loved them. Both demonstrators were to be flown in France, England, Sweden and other European countries, but the 2PA was seriously damaged in a poor weather landing at Croydon Aerodrome near London on 1 April 1939. Total flight time was only 56 hrs. 55 mins., so the tour was shortened considerably. If the airplanes had been fitted with flush landing gears and AP-4 type canopies, orders might have come from France and England, unless politics turned out to be the damning factor in Great Britain, too. (Author's Coll.)

deletion of many features known to be useful additions. This reveals, once again, how crucial it was to introduce such features in the midst of the AAC contracts already in place to produce P-35s. But for the want of........)

Test pilot George Burrell and the Major went on to make appropriate sales pitches and flight demonstrations in Sweden and Poland before lack of familiarity with airport conditions led to a 2PA-BX accident. After the "BX" was demonstrated in France, Sweden and at the RAF's Aeroplane Armament Experimental Establishment at Martlesham Heath, the Major was faced with unexpectedly bad weather during a landing approach at Croydon Aerodrome. Landing long, the BX ran into an iron fence and was seriously damaged. That ended the tour for the 2-seater, but the EP1-68 was to continue. Nobody was injured in the 1 April 1939 crash landing, and Sweden was ready to order 52 upgraded dive-bomber versions (*Flygvapnet* designation, B6). Somewhere along the line, the name "Guardsman" was adopted.

Darkening war clouds, events at Seversky Aircraft (Kellett's duplicity for instance) and other factors ended the 1939 tour before spring ended. However, the trip was a success in that the Major's efforts had led to what became the largest aircraft orders ever received from any customer, far exceeding any Air Corps procurement of that manufacturer's products.

Before de Seversky had departed for Europe, he evidently issued an order for conversion of the AP-7 demonstration airplane to feature a revised, sharp leading edge wing and a flush-fitting landing gear duplicating the one scheduled to be installed on the AP-9. It is highly unlikely he would have ordered the conversion from Europe and even more unlikely Kellett would have approved it. Jackie Cochran had submitted her entry for the 1939 Bendix Trophy Race and there was an exceptionally good chance that she could beat Frank Fuller's beautifully painted (metallic blue-green with yellow and black trim) SEV-S2 by a good margin. She would be at the controls of the modernized AP-7 with its "fast" wing and the completely retractable landing gear. The metal skin had been polished to a mirror finish. Apparently she took off from Burbank's Union Air Terminal in foul weather and promptly aborted the flight.[2]

Cochran's plan to make an attempt on the 1000 kilometer international speed record was not denied. That Seversky AP-7 was up to the job, recording an official speed of 305.926 mph over the course on 15 September 1939. (In April of the following year, she used the same airplane to establish international records of 331.716 mph for 200 kilometers and, more importantly, breaking the 2000 kilometer record held by Germany's Ernst Seibert when she averaged a bit over 324 mph. The point about flush retracting landing gears was proven in dramatic fashion. It is easy to visualize that the Major, flying that AP-7 (possibly with the large spinner and narrow inlet cowling, plus some Pratt & Whitney magic) might easily have matched the Spitfire's top speed over the

Just a month after factory rollout in July 1937, executive Frank Fuller, Jr., flew this Seversky SEV-S2 from Burbank to Cleveland in 7 hrs. 54 mins. to win the Bendix Trophy Race. He continued on to Bendix, N.J., setting a west-to-east record at 255 mph. As seen here in September 1939, he again won the Bendix event at a speed of 282 mph over the 2043-mile course. Once again he broke the coast-to-coast record in under 9 hours. The Fuller paint on this racer was metallic bluish green with yellow accents. Registered NX70Y, this airplane helped establish the Seversky name. Like the SEV-DS, it was exported to Ecuador in 1941, being redesignated C-2. (Author's Coll.)

2 Reed Kinert's book *Racing Planes* has always been regarded as an authoritative book on such things for years, but the only thing said about Cochran in the 1938 segment showing a picture of the '39 entry is that it was the version she raced in 1938 "with the flush retracting landing gear." (Sic) He even has her in the SEV-S2 in 1938 tabular data when she won the race with the AP-7 which had those big fairings on the landing gear. There is not a word in the 1939 pages about her.

standard 3 kilometer record course. One thing is certain. Jackie Cochran absolutely ranked at the top of all female pilots in America, if not the world in those pre-war days, and there were few male pilots around who could have bettered her performances in a spirited airplane like the AP-7.

Aside from the one speed record and Fuller's victory, Major de Seversky could not have found much to his liking in 1939. The 2PA-BX, with a total flight time of only 46 hrs. 55 min., was described in the engine log by Judson Hopla as having a damaged propeller but no serious injury to the engine. He provided no details about the aircraft other than to say it was a "complete washout." His final entry was a comment about dismantling the BX for shipment from an English port direct to the factory. Its ultimate fate remains unknown. Wife Evelyn became ill, dictating a delay in return to America by ocean liner. But that only delayed what was to become the most bitter pill in his life. (It has been assumed that the BX airplane was scrapped upon return to the States, but new evidence reveals issuance of an export license proceeded. Unfortunately, no destination was listed on the document.)

Aviatrix Jacqueline Cochran was seen at Union Air Terminal (shortly to be Lockheed Air Terminal) on August 25, 1939 with the considerably revamped AP-7 (NX1384), a Seversky-owned demonstrator. Engines were changed frequently for various events, but all were versions of P&W Twin Wasps. In August, Cochran was ready to break the 1000 km. speed record and to attempt a second win in the Bendix Trophy Race. She failed at takeoff attempts for a Bendix start at fogbound Burbank, but in April 1940 wrested the 2000 km. record from a Luftwaffe pilot. The AP-7 went to Ecuador in 1941 to be the C-3. (C.A.M./Ed Boss)

This brilliant entrepreneur, exceptional pilot, industrious inventor, charming salesman was – in the eyes of his best friends and his most bitter enemies – a lousy businessman. Money normally came to him with such ease that he never learned its value. If he wanted something, as he did the lovely Evelyn, he turned on the charm and it came to him. But his great weakness was never learning to duck. He was a dupe. He never really understood the Wall Street people, and Paul Moore was evidently content to balance the checkbook...if that ever had been possible. During at least one interrogation session in connection with the Major's lawsuit, Moore was visibly shaken to find he had been seriously misinformed by Kellett on matters relating to foreign sales. Polish officials wanted Seversky to demonstrate his airplanes near Warsaw with an eye to purchasing 250 examples, communicating that information to their Air Attache in Washington. Kellett concealed many facts about the request from Moore. However, the die had been cast with regard to Seversky's future. Marquis of Queensbury rules did not apply.

His return to America, his home away from home, his Utopia, was not going to be his greatest homecoming.

When Pratt & Whitney developed their 1830-cubic inch Twin Wasp in direct competition with Wright's R-1820 Cyclone, they must have made technical discoveries that allowed them to move ahead of the competitor and set a pattern that carried over into the masterful R-2800 Double Wasp. All SB-G models of the Twin Wasp had updraft carburetors, produced a maximum of 1000 hp for takeoff while rated at 950 hp normal. Seversky was quick to adopt SC-G through S3C-G versions that developed up to 1200 hp as a normal rating. Thousands of these radials powered C-47s and, in turbocharged versiond, the B-24 series. Earlier, all Seversky racers and varied demonstrators used the "C" models of the Twin Wasp. When Republic developed the world's first successful turbosupercharged radial-engine fighter, the P-43 from the Seversky AP-4, it set the stage for the evolutionary P-47 Thunderbolt. (NEAM)

CHAPTER 6

SEVERSKY IS SEVERED; SWEDEN IS SEQUESTRATED

Paraphrasing the Corporation Board Minutes of December 1938, W. Wallace Kellett had taken full charge of corporate operations and was soon promoted to vice-president even as Evelyn and "Sasha" probably prepared for Christmas in Paris, far removed from the disappointments that were to befall them.

George Burrell and the Major demonstrated the EP1-68 and the 2PA-BX rather successfully wherever they had the opportunity, but potential buyers remained strangely silent and unappreciative. Back in the States, the personable and impressive Hart Miller had no greater success. At the same time, the French ordered Vought V-156 scout-bombers (equivalents of the U.S. Navy's SB2U-3s which were never even a factor in WWII), were given White House approval to evaluate the very hush-hush Douglas 7B twin-engine bomber that had not even been evaluated by the Air Corps against the North American NA-40B and Stearman X-100, and Curtiss Airplane Division had inked heavy orders for export versions of the obsolescent P-36C under the designation Hawk H75-A1 (French designation H75-C1).

All hell had broken loose in the media outlets when the 7B crashed into the North American Aviation parking lot adjacent to Mines Field (Los Angeles Airport) with a representative of the *Armee de l'Air* on board. In Isolationist America it became a *cause célèbre* scandal. However, it generated big orders behind the scenes for an improved configuration known as the DB-7, evidently even before the Air Corps could order the A-20 version. Meanwhile, although the British Purchasing Commission

The EP1-106 that had crash landed with test pilot Herbert Hulsman at the controls was repaired and prepared for delivery to Sweden by February 1940. The Flygvapnet number 2121 and insignia had been applied in compliance with new approvals. The second EP1 (2122) in the background differed in that it had wing gun pods installed. Generally unrecognized in published histories is the fact that all EP1s and P-35As had fuselages lengthened by 14 inches (shown to effect on S/N 2122). **(Mfr. 1706)**

had not even intended to visit tiny Lockheed in Burbank, a message was dispatched to their offices in New York. It created enough interest for them to view a hastily designed and created mockup of the Lockheed 14 airliner during a visit to California. C.L. "Kelly" Johnson's design prowess and some salesmanship gave the British an opportunity to procure a much-needed replacement for the smaller, slower, far-inferior Avro Anson reconnaissance airplane for Coastal Command. Phenomenal orders for the Lockheed Hudson soon followed.

Criticism comes easier than craftsmanship. – *Zeuxis. The British press hammered at the "monstrous P-47" when it appeared on the scene, seeing it as the most gigantic fighter ever. What they ignored was the fact that Hawker's Tornado and Typhoon fighters were in the same vein, and the prototype Tornado (Type F.18/37) had actually flown in October 1939. Governmental secrecy and the war precluded any announcement of the type, but derogatory comments about the Thunderbolt at a later date were baseless. Main flaws in the design: Rolls-Royce's totally unreliable Vulture engine and a very thick wing section on a low-altitude fighter.* (Ministry of Supply)

Then the entire "behind the scenes" situation became crystal clear. A master move was initiated to "gain the confidence of the United States Government." Kellett sent a letter to the Secretary of State and the Secretary of War, informing them of corrective actions being taken to (obviously) save the Corporation. The BOD also considered closing the plant, a thinly veiled warning to the Government agencies that Seversky Aircraft was on the ropes. There were, at that moment, no orders on the books except for spares. The company had lost $553,819 in 1938. They attributed part of the loss to costs associated with the XP-41 contract. (That flies in the face of a much later government document which listed total costs at nearly 20 percent less than the $80,000 allocated, but who really knows?) The company said the two military airplanes (undeniably the XP-41 and the AP-4) built for the competition of 1939 cost something over $365,000. A reminder: The final project figures in an official USAAF document stated actual final Procurement Division Funds expended totaled (for the XP-41) $61,963.00.

As stated previously, the Major never learned to duck. He was in the "pit," and here came the pendulum! On 18 April, with de Seversky mourning the loss of the 2PA-BX, Mr. Kellett was elected President of the Corporation. Alex Kartveli was to remain as Vice-President. During the summer months, important changes were taking place in rapid-fire order. In lock step with those actions, the Consolidated Trading Co. signed a contract for 54 airplanes and spares for overseas shipment at a price of $3,210,000. Which 54 airplanes? Nothing in later contract documentation comes close to matching that number. Who did Consolidated Trading represent? (Could it be that Consolidated translated to *Kungl-Flyforvaltningen* in Stockholm, Sweden? Doubtful. Also the date is all wrong.) And the painful pills did not stop even then. The Army and Kellett jointly announced a $974,325 order for – *Fasten Seat Belts* – fifteen (sic) Seversky AP-4s plus a separate order (at a better price) for no fewer than 100 (sic) production AP-4s! The Army made its public announcement in May 1939.[1]

On 15 September 1939, the Seversky Aircraft Corp. was renamed Republic Aircraft Corporation, but that must have evoked some additional action on the part of the Major's lawyers. The organizational changes resulted in the second name change, this time to Republic Aviation Corporation. Evidently everything became official on 13 October 1939 when the smoke cleared. Most of the actual corporate structural changes occurred in October. Seversky's stylized logos on airplanes disappeared almost overnight after the 13th, and the change to Republic Aviation Corporation – with a logo not radically unlike the old one – soon was to appear on buildings, letterheads and elsewhere.

Meanwhile, the British and French Allies unenthusiastically declared war on the Axis nations. But the French could not even launch a minor-league bombing offensive against the Nazis at minimal range, thanks to the havoc wrought on the French industrial machine and the *Armee de l'Air* by Pierre Blum's Socialist government. The British evidently hoped to exhaust the Germans in picking up leaflets dropped by a few venturesome RAF bombers. As propagandists, the British were not in the same league with the Nazis.

Major Seversky (then, by his choice) made a darn good try at pulling the rug out from under the Corporation. He wanted lots of money; he wanted his patents back; he insisted that Paul Moore pay him $7 a share for all his outstanding stock, and he also had already forced them to change the name from Seversky so that he could have that right to the name for his own business ventures. The BOD, of which he was to remain a member for some lengthy time, refused the Seversky "offer." Thus, the battle was on. And it went on in the courts until settled in September 1942.

The smoking gun of the War Department, State Department or some official body or bodies was, by the end of 1939, clearly in sight. Even with last quarter shipments of nearly $740,000 worth of parts and aircraft in 1939, Republic's backlog had catapulted upward to $10,000,000. In that critical time, Sweden had only placed orders for 45 of the EP1s and 24 of the 2PAs. What about Consolidated Trading and the 54 airplanes? Wallace Kellett probably transposed the numbers or was counting on Swedish orders that were in the discussion stages. A later statement about the P-43 attributed to him was far worse. As for the

[1] In truth, the orders were for the thirteen AP-4/YP-43 service test airplanes and an order inked exactly 5 months later for eighty (not 100) P-44s. Yes, that is correct. The Air Corps issued a new Circular Proposal #39-770 which generated Republic Model Spec. #87. That evidently spelled out the P-44-IV design, equal to the company's AP-4J or AP-4L. (In retrospect, they were never to be seen as such; the contract was canceled in favor of a much upgraded design (Model Spec. #88) which appeared as the XP-47B.) As anyone can see, following the paper trails requires some extensive detective work.

***A revamped version of the first prototype Tornado did not take to the air until 14 long months later, delayed more by engine problems than any other factor. Vulture engines were rated in the 1750 horsepower realm, but their failure rate (especially in the Avro Manchester heavy bomber format) was disastrous.* (Hawker Aircraft)**

Major, and as many others have discovered, don't mess with the military unless they sanction it. And smile.

Everyone tends to refer to the Curtiss P-40 contract as "huge" and in many ways it was. But it must be remembered that about three years earlier, the British Ministry of Supply had authorized procurement and issued a contract for more than 600 Hawker Hurricanes in June 1936. In July, no fewer than 310 of the more complex (to manufacture) Vickers-Supermarine Spitfire interceptors were ordered. By contrast, the order for 524 Curtiss P-40s was hardly overwhelming. Three long years had crept by since the Hurricane and Spitfire orders were signed, most likely to accommodate development progress of the Allison V-1710 engine. In retrospect, it might have been of great financial value to this nation in the 1937 Recession days to initiate the program which ultimately saw the award of production contracts to Packard Motors in Detroit for mass production of Rolls-Royce Merlin engines. The British should have appreciated the opportunity to make money and at the same time reduce unit production costs significantly. In the long run, the Air Corps might well have gained as much as two years technologically and should have obtained far better fighting planes.

Alas, such thoughts were, in that decade, repugnant to the American psyche. U.S. planemakers could have benefitted from obtaining a 1030 horsepower inline engine as the basis for a really new single-engine fighter, er...pursuit. But the powers of General Motors (Allison) and Curtiss-Wright (Curtiss) were such that the idea could hardly have been considered, no matter how much merit it had. General Electric might have gotten into the fray as well. It is logical to recall how slowly that company responded to demands by the War Department for increased production of turbo-superchargers for the P-38 and B-17 programs.

***Republic Aviation's P-44 and XP-47B designs owed something to the Bureau of Aeronautics of the Navy Department for pushing development of the P&W R-2800 Double Wasp in the face of Army Air Corps pressure on United Aircraft to develop inline engines. That engine, first utilized in Vought's XF4U-1 (Design V-166B) as the XR-2800-4 was soon to become the heart of all P-47 Thunderbolts. First ordered on June 11, 1938, more than two years before a Change Order was issued to develop the XP-47B, the XF4U-1 Navy fighter first flew on May 29, 1940. It was soon claimed to have become the "World's fastest airplane, the first fighter to exceed 400 mph" and "holder of the World's Speed Record at 405 mph," and such claims are made in publications to this date (1993). Every such claim is pure hallucinatory garbage. The official speed record of more than 440 mph was achieved with an Italian Macchi-Castoldi seaplane in 1934, and the Germans had established an officially observed FAI record of 469+ mph with a Messerschmitt record type in 1939. Additionally, the XP-38 had attained speeds in excess of 400 mph in February 1939.* (U.A.C.)**

A cadre of ranking officers in the Air Corps took great umbrage with the plans and directions of H. H. Arnold and Ira Eaker, to name but a few. To illustrate how the wind blew in those days, consider the reaction to M/Gen. Frank M. Andrews and the speech he gave at the annual convention of the National Aeronautic Association on 16 January 1939. He was then Chief of General Headquarters Air Force, one of the main bastions of American defense. He told the members of NAA that the U.S. was a fifth, even sixth-rate airpower and that his own command had only slightly more than 400 fighting aircraft. By February 9th Andrews was out and reassigned (to purgatory, it would seem), replaced by a newly promoted M/Gen. Delos C. Emmons. (Those who forgot to duck were doomed to unpleasant fates.)

There were officers who, in the 1936-39 period, had come to extol the virtues of operations in the stratosphere, the one area in which Americans were not three years behind all others. Fighter Projects Officer Capt. Benjamin Kelsey was in the front ranks of those advocates, although the contract award to Curtiss would hardly seem to support that premise. Was any captain going to go up against those who were the real Army Air Corps decision makers? He knew the officers in Materiel Division procurement and engineering who had a positive interest in high-altitude fighter operations stemming from procurement of the experimental Curtiss XP-37, Lockheed XP-38 and Bell XP-39 trio of turbosupercharged fighters, all powered by Allison V-1710 engines. The XP-37 was just a supercharged flying testbed, and the thirteen service test YP-37s were far from being suitable combat aircraft, even in pre-war days. (Curtiss designers of that era did not think to deepen a fuselage to accommodate a supercharger. They just lengthened the fuselage, putting the pilot about 20 feet behind the nose.) By the time the Bell XP-39 was undergoing serious evaluation, Capt. George E. Price had been added to the Project Office and assigned responsibility for the Bell P-39 program. It did not take long to see he was not a believer in turbine supercharging.

Was Kelsey correct in his decision making? Only the P-38 Lightning type, of that trio or their offspring, was capable of going into combat against the best *Luftwaffe* and Nipponese fighters with any chance of success. And to gain the upper hand over such adversaries was something totally out of reach for Curtiss and Bell types. Although Kelsey could hardly be classified as a real advocate of Seversky pursuit airplanes, he had been up front in recognizing the importance of the AP-4's performance and configuration. He surely played a large part in getting a service test contract award for thirteen AP-4s, quickly to be properly designated as YP-43s. Just as the P-35 had evolved from the SEV-1XP prototype, the YP-43s showed significant design improvement over the lone AP-4. This time the Project Office dictated terms which assured the Air Corps people drag reduction was to play a large part in additional contract awards.

Only a short time earlier, few people could see the advantages of operating at altitudes above 22,000 feet. When Ben Kelsey studied the advantages of over-weather flight plus numerous other advantages in combat situations, he became a believer. That was reflected in the specifications he laid down for the competitions that produced the XP-38 and XP-39 in 1936. There had been other turbocharged competitors at Wright Field that spring of 1939. One was a Curtiss H-75R, and the other was the Curtiss XP-37. A test pilot who had spent several hours at the controls of the latter airplane was Lt. Mark Bradley (later to become commanding general of the USAF Logistics Command). He related his experiences with the XP-37, remarking it was flown frequently without benefit of a useful turbo regulator, creating numerous problems which did little to improve pilot confidence. Not the least of these faults was turbine overspeeding. Inability to control such conditions could lead to turbine failure or, at the very least, poor response to changes in power demands at critical times.

Something never mentioned before is the sudden desire on Major Seversky's part to utilize the General Electric supercharger in one of his airplanes. Had he suddenly become an expert on such matters? Why was he willing to risk so much in the face of opposition from his own chief engineer, Kartveli? Even if he was on the cutting edge of destiny leading to combat in the sub-stratosphere and stratosphere, he must have been enlightened by some specialist in propulsion who was a strong advocate of turbocharging. Unfortunately, that question was not put to Hart Miller in any interview or correspondence in the early 1970s.[2]

Herbert Hulsman, a young, newly hired test pilot, took the first production Republic EP1-106 (NX15687 for test) on its initial flight on 14 Oct. 1939, and flew for about 30 minutes. On final leg of his landing approach at about 100 feet altitude, the EP1 was seen to drop behind a clump of scrub oaks. It rolled about 300 feet before flipping onto its back. The windshield and sliding canopy were crushed, and Hulsman was killed (broken neck). There was little additional damage, and the airplane was redelivered in February 1940 with full Swedish markings as #2121. It became a J9 in the Flygvapnet. (Carl Bellinger)

Placement of the supercharger so far aft and far removed from the engine was not only innovative, it was remarkable. The Allison-powered Curtiss XP-37/YP-37 fighters and the turbocharged version of the P-36, labeled H-75R, all had the supercharger located aft of and below the engine. Placement of the unit aft of the cockpit, far removed from the engine, had never been tried on any American aircraft. In fact, it evidently was never on any fighter aircraft anywhere. Why did the Major choose that location? Well, that oversize luggage compartment was rarely filled and few passengers ever rode back there. Unlike the Curtiss XP-37/YP-37 airplanes where the pilot's posi-

[2] One 1990s decade "authoritative" book states the "bulky" turbosupercharger located low in the aft fuselage "placed it "near the center of gravity." One glance at any picture of the AP-4 or a P-43 shows that location would be aft of the trailing edge of the wing, a remarkably aft c.g. location!

tion was far aft, with forward vision extremely limited, the pilot in the AP-4 had a commanding position well forward. In the Curtiss H-75R, proximity of the turbine compressor to the pilot's feet most certainly had shortcomings.

The AP-4 airplane which had impressed Kelsey and others in its performance at Wright Field had been entered in the competition as a "private venture" by Seversky. And because it was there, the Seversky AP-9 (with a pair of pseudo machine guns) went as a temporary substitute for the late XP-41. Civil registry on the AP-4 was NX2597. If seen in formation flight with the XP-41, with or without the NACA large spinner cowlings, the airplanes would have appeared to be duplicates to all but the most observant viewers. But there were significant differences; most easily detected were decidedly different flush landing gear arrangements, the Hamilton-Standard Hydromatic propeller on the AP-4 vs. the Curtiss unit on the XP-41, the turbine drive unit on the G.E. turbosupercharger protruding from the lower aft fuselage on the AP-4 and subtle fuselage differences forward of the windshield. During the evaluation process at Dayton, it became evident the XP-41 was not a bit faster than the Seversky AP-2 speed recorded during the 1937 competition, but the AP-4 was about as fast as a Spitfire Mk.II. More to the point, its best speed performance was at a critical altitude of 23,500 feet where it outstripped the Spitfire and the XP-37. Apparently they took the Seversky AP-4 to 35,000 feet, far exceeding the winning XP-40's capability, but oxygen systems of the period were far from satisfactory or reliable.

In an interview with retired General Mark Bradley we learned he had flown one of the first service test Curtiss YP-37s into Wright Field for extensive testing at just about the time Frank Sinclair was forced to bail out of the burning AP-4 demonstrator. The total expenditure of funds to develop Dr. Sanford Moss' turbosuperchargers during the two decades following WWI was likely shockingly low. Oxygen system development funding was far lower. (The motion picture *Test Pilot*, made in the late 1930s, had a memorable scene in a development B-17 showing Clark Gable and Spencer Tracy with oxygen tubes stuck in their mouths. Pretty elementary equipment, to say the least. But the film was contemporary and very well executed for the period...the tubes were "state of the art" then.)

Capt. Kelsey and his close associates were instrumental in obtaining funds and the authority to purchase thirteen (not fifteen as stated by Kellett) AP-4s, shortly thereafter to be given the military designation YP-43. Contract AC-12643 was issued in connection with the earlier Circular Proposal 38-390, so it turned out to be a little reverse English on the earlier Seversky P-35/ Curtiss Y1P-36 situation but on a larger scale all around. The contract was dated 12 May 1939, calling for delivery of the first airplanes exactly a year later. However, the first engine was delivered by the Air Corps a couple of months late.

While Major Seversky was still in Europe, the factory lease agreement he had entered into earlier in the decade required completion of the purchase by 1 January 1940 and notice of intention by 1 July 1939. Fairchild-Ranger also had a similar obligation for the portion of the plant they had been leasing. The total purchase price of the Seversky portion was $730,000, but that did include the airport and the hangar. It would soon be realized the deal was spectacularly good for the company. With the outbreak of war in Europe just weeks away from the July due date, immense changes were to affect the company fortunes.

It seems incredible now, but Seversky Aircraft had only 176 employees on board as of May 1st. Alex Kartveli and Hart Miller may well have contemplated the need to make out resumes for submittal to Grumman or Brewster or...perish the thought...Curtiss. Just one year later the books would show about $14,000,000 in unfilled orders (backlog) existed. Should anyone wonder if Seversky Aircraft would have survived with the Major still at the helm, just recall the Swedish orders he alone had garnered – although he had been stripped of power before they were actually *de facto* considerations – would likely have proven successful in stabilizing the company. Additionally, it was *not* the XP-41 which led to the Thunderbolt but the Major's own AP-4 concept, the direct predecessor of the YP-43 and ultimately the P-47. Certainly he could have benefitted from managerial assistance; there is no doubt about it. Perhaps it was Paul Moore who failed; he was the money man and maybe he should have exercised more power in his important role to make the company work.

Alex Seversky, who was more and more inclined to discard the lengthy association with the "de" in his name, had planned to sail home from Europe on the French liner NORMANDY in May but the plan had been delayed at that crucial time by Evelyn's serious illness. However, they did sail together in June. Normally we think of sailing off together into the sunset and a more romantic lifestyle. In this case, Alexander Seversky sailed into the sunset of his unforgettable career as an airplane builder.

However, this story has a much wider scope about many more people than the Major. And, of course, the main topic has to do with the men and machines and World War II.

Misinformation or disinformation relating to the Republic (not Seversky) EP1-106 fighter for Sweden (and directly applicable to the AAC's P-35A) has been so commonplace as to obscure most of the actual facts. It became imperative to present here the core engineering data and salient de-

Although this photo's information statement identifies it as a P-35 production line, the photo sequence number and the second fuselage (at left) cast doubt on that. That second fuselage has a solid razorback structure, ensuring it was no P-35 or EP1. It may have been a YP-43 or a P-43. Photo number virtually ensures that an EP1-106/P-35A is in the foreground, a YP-43 in second position and the date most likely mid 1940. Total company floorspace was only 250,000 sq. ft. then. (Via Frank Strnad)

Prepared for duty in Sweden, the second EP1-106 was pictured in a winter setting on Long Island early in 1940 while the "Phony War" crawled on in Europe. America was, as yet, largely unaffected by events associated with war, although the nation was rapidly building toward being the "Arsenal of Democracy." According to commerce records, factory hourly pay at Republic had risen to about 30 cents per hour. A little over a year earlier, it had been about 20 cents, but layoffs were rampant then. (S. Hudek Coll.)

scriptive material relating to the type.

Basically the EP1-106 was an all-metal cantilever-wing fighter monoplane powered by a single Pratt & Whitney R-1830-SC3G Twin Wasp 14-cylinder radial engine. Republic General Spec. 100 applies to the type. The engine had a takeoff and military rating of 1050 hp at 2700 rpm, the latter rating being set at 7000 feet altitude. It mounted a Hamilton-Standard constant speed propeller having a diameter of 10 ft. 6 in.. In external appearance it was generally identical to the Air Corps Seversky P-35, but there were several distinctly EP1 features. The top cowl displayed an airscoop to feed the downdraft carburetor and the bottom side of the cowl featured a considerably larger oil cooler scoop than the one seen on P-35s. Although not as apparent to a casual viewer, the overall length of the EP1-106 was 20 inches greater at 26 ft. 10 in.. The *Flygvapnet* and AAF EP1/P-35A airplanes carried a pair of wing-mount guns in addition to the cowl-mounted units. Those *Flygvapnet* versions featured two 7.9-mm KSP M/22 machine guns and two 13.2-mm AKAN M/39 machine guns. These were essentially the equivalent of two .30-cal. cowling guns and two .50-cal. wing guns.

EP1 aircraft had exactly the same 36 foot wingspan as the production P-35 and even the SEV-1XP prototype, providing a wing area of 220± sq.ft. and a wing loading of 26.5±0.3 lbs./sq.ft. at a normal gross weight of 6118± pounds. Normal fuel capacity was 130 gallons, but an overload capacity of an additional 65 gallons could be carried internally. That gave an overload gross weight of 6965 pounds.*

(*All figures guaranteed, +/- 5% per Alexander Kartveli.)

In the performance categories, a speed of 320 mph was guaranteed (with the same allowances), but actual test speed was 306 mph. Had they been able to install the 1200 hp engine, it is quite probable a top speed of 320-332 mph would be registered. One actual performance figure exceeded the guarantee. The recorded rate of climb was 2930 fpm as opposed to a guaranteed figure of 2800 fpm. Normal maximum range was 750 miles, but that could be increased to 1122 miles with the overload fuel situation.

Meanwhile, the financial fortunes of the company soon to be reidentified as Republic Aviation began to change in rapid-fire order. On June 29th, the contract was signed with Swedish representatives, calling for delivery of the fifteen fighters identified ultimately as Republic EP1-106s. The order called for the first delivery to be made 120 days after receipt of equipment

Camouflage paint schemes were applied to all EP1-106s (J9s in service) and the two 2PA-204As (B6s in service) delivered to Sweden before assignments to operational units. The fighters went to the Svea Flygflottilj F8 to act as a defense force protecting Sweden's coastal cities. Those P&W R-1830 Twin Wasps delivered anything from 1050 to 1200 horsepower, depending on which official source you want to quote. (R. Dorr Coll.)

Test Pilot Lowery Brabham is shown piloting a Republic 2PA-204A Guardsman, built for the Swedish government, over the tiny Republic factory at Farmingdale. A circus tent erected for Dade Brothers (at bottom center) was used to prepare military aircraft for shipment to Sweden. Ranger's engine plant had tripled in size, but Republic was still not out of the woods. **(L.L. Brabham)**

from the Swedes. That quantity was increased by an additional 45 airplanes on October 11th, and the first 24 dive-bomber 2PA-204As were ordered on the same day. (In Sweden, the EP1s were to be known as J9s; the 2PAs would be known by the B6 designator.) As the "Phony War" slogged along in Western Europe, tiny Republic Aviation celebrated the New Year with another contract for 60 additional EP1s on 5 January 1940. That brought the total number of the fighter-type orders to 120 while 2PA bomber types increased to 52. Rarely mentioned, but extremely significant at the time, was the licensed manufacturing agreement, a spare parts contract and provision of tooling to manufacture both types of aircraft in Sweden. (Another factor long overlooked was an additional order for 36 of the Guardsman bombers signed on 12

Although the 2PA-BX prototype derivative of "Convoy Fighters" was virtually always written up as a fighter, the "BX" suffix shows that it was intended primarily as a bomber. And, to solidify that concept, the Flygvapnet provided the designation B6 to the 2PA aircraft they expected to receive. The production airplanes were longer, had greater wingspans and early publicity pictures showed the Guardsman loaded with bombs. Sweden only received two of the 2PA-204As; the USAAF eventually had the other 50 in service under the advanced trainer designation of AT-12. **(Via R.S. Allen)**

July 1940.)

In the meantime, at Republic Airport (Farmingdale) on 14 October 1939, the first of the initial fifteen EP1-106s was being readied for delivery toward the end of the month. Originally it had appeared the first airplane would be delivered well ahead of the scheduled delivery date – actually December 12 – but a "Murphy's Law" accident resulted in delay. The first EP1-106 completed was licensed as NX15687 for flight testing, and it was being flown by newly hired production test pilot Herbert Hulsman. On final approach after an uneventful flight the airplane dropped below the tree line according to a witness. It landed short, was tripped up after a 300-yard journey through scrub oak and flipped on its back. Hulsman officially died of a broken neck, although the aircraft was virtually undamaged except in the canopy area. The fighter was righted and repairs were effected within weeks. Then bearing the *Flygvapnet* number 2121, it was test flown, crated and shipped to Sweden.

If this airplane looks somewhat familiar to you, that should not be a surprise. The Reggia Aeronautica (IAF) was responsible for development of a slightly earlier design known as the Reggiane Re.2000, a conspicuous theft of the Seversky P-35 or – more likely – the EP1-68 prototype export fighter. One development that was placed in production was a slightly more powerful version designated Re. 2002, shown here in Luftwaffe markings. The Italian design team improved on Seversky's landing gear, powerplant and cockpit enclosure. (Author's Coll.)

Admittedly there is a great difficulty in understanding all of the ramifications which the Neutrality Act invoked when war broke out in Europe in September 1939. However, to blame that for non-delivery of at least 110 airplanes to the customer in 1940 – as virtually every published story has stated for more than 50 years – is a distortion of the facts. President Roosevelt quickly saw the Neutrality Act was penalizing the victims and aiding the aggressors, exactly the opposite of what was intended. By November 4th, just two months later, the law was repealed. As might be noted, many of the Republic airplanes were ordered by Sweden *after* the Nazis invaded Poland, primarily because Sweden was NOT a belligerent. The American law applied only to belligerents. Not long after contracts were signed in connection with the licensing agreements, the Swedes realized the supply of American-source components could be cut off for a variety of reasons. As a result, those licensing contracts were canceled. Perhaps that is why they had ordered additional 2PA bombers in July 1940 to bring the total to 88 units.

If all of this is fact, why did deliveries to Sweden stop so abruptly and why did the Swedish representatives cancel the final contract for 36 Guardsman bombers with hardly a murmur from anyone? Well, on 10 October 1940, the War Department, acting on a weapons embargo order, was authorized to seize all of the undelivered aircraft because of a so-called, and certainly controversial, "emergency" in Europe. Quite naturally the Swedish government protested, but the protests were dismissed by the State Department. Perhaps the action was taken because of some reports sent by Colonel Carl Spaatz and Captain Ben Kelsey subsequent to events in England which they were privileged to witness firsthand. Or it could have stemmed from secret communications between England's Prime Minister Churchill and our President. Agreements had already been made to grant access to America's latest combat types, and the Air Corps generals protested the authorized expansion programs were falling apart. At the time there were terrible deficiencies in the Philippine air defenses – despite the growing Japanese threat in the Western Pacific – so the seized EP1s were quickly redesignated Republic (not Seversky) P-35As and shipped to Clark Field as replacements for Boeing P-26s. Concurrently, the fifty 2PAs sequestered were assigned to a training role and designated as AT-12s.

That "brilliant" move by the War Department did little to aid in the defense of the Philippines, but there was a virtually unknown benefit to one of the Axis nations – Italy. In one of the most blatant thefts of an aircraft design ever recorded, "designers" Alessio and Longhi of the Reggiane S.A. copied the entire major structure of a Seversky P-35, installing an 18-cylinder Piaggio engine and a flush-retracting copy of the Curtiss P-36 landing gear. A low-profile Seversky-type canopy, as installed on the SEV-S2 and SEV-DS airplanes, was used by the Italians. This design, known as the Reggiane Re.2000, almost won a competition for fighters to equip the *Reggia Aeronautica* (Italian Air Force) on the basis of its performance, but it had some technical deficiencies (primarily maintenance of the "wet wing" fuel supply system). That loss precipitated the action which was detrimental to America. The Reggiane organization soon sold a small number of Re.2000s to Hungary, along with manufacturing rights which were exercised. Since the U.S. had deprived the Swedish *Flygvapnet* of sixty sorely needed fighters, they simply purchased an equal number of the Italian aircraft to supplement the Republic EP1 "first cousins". Kept alive by the foreign orders, Reggiane produced versions of the Re.2000 for the Italian Navy, and advanced Re.2002 versions probably served in the defense of Italy against the Allies. While the Republic EP1s, delivered earlier, served the Swedish military as first-line fighters (and later as trainers) until 1947, the Italian counterfeits were disposed of by 1945, attesting to the "quality" of the product manufactured by the pirates at Reggiane S.A..

This story points out the illegal Italian incorporation of a Curtiss-style flush retracting landing gear in the Re.2000 and

Re.2002 versions without any great difficulty, resulting in a top speed for the early version of 329 mph (Italian optimism?) on 986 hp at 16,400 feet altitude. According to official Republic tests, their own EP1-106 could not better 306 mph on 1050 horsepower. (If the Reggiane figures have any validity, this writer's later opinions/conclusions about the flush landing gear advantage are correct.) Any engineer working in that 1940 time frame should have recognized the advantages of a fully retractable landing gear. After all, virtually every contemporary modern combat airplane incorporated that feature. On-site representatives of the Swedish *Flygvapnet* should have gained access to performance figures or calculations from Alex Kartveli. The Swedish engineers could have evaluated current data and insisted on an appropriate design change. With their orders eventually reaching the immediate vicinity of 200 airplanes, they surely enjoyed a lot of leverage. Even if they were forced to reduce orders by five or ten airplanes to obtain speeds close to Hurricane performance at additional expense, it would have been worth it. They evidently did not even ask the questions.

As if to extol the virtues of the de Seversky-Gregor wing design and some other Seversky features, the Reggiane organization turned out one of the best *Reggia Aeronautica* fighters of the war with that same wing and empennage, mating those features to an outboard-hinged flush landing gear retraction system and a German DB605A liquid-cooled engine of 1475 hp! Armament was increased to two wing-mounted 20-mm cannon, two cowl-mounted 12.7-mm machine guns and the typical engine-mounted 20-mm cannon. Maximum speed turned out to be a spectacular 390 mph at 22,900 feet, an impressive figure if it is valid. This was the Reggiane Re.2005 Sagittario, with production deliveries beginning in 1943. Of the 48 airplanes delivered prior to the Italian capitulation, most went to the 22° Gruppo Caccia defending Rome and Naples. Some 300 Sagittarios were in various phases of construction when the factory was flattened by bombing.

Just to show where America was on the airpower scale in 1940, it is germane to point out some of the things occurring elsewhere in the world as well as in the U.S.A. at midyear. Over in England the prototype Hawker F.18/37 Tornado took to the air for the first time on 6 October. That large 10,250 lb. fighter was aimed at the 400 mph class with its 1800 hp Rolls-Royce Vulture engine, but the engine ultimately proved to be a disappointing failure. A very similar Hawker airframe was also being developed with a 2000 hp Napier Sabre engine in a parallel program. In fact the prototype Typhoon had made its maiden flight on 24 February 1940. The Sabre engine was also plagued by problems in development, resulting in a long gestation period for the Typhoon. As of mid 1940, the Army Air Corps did not have one single-engine fighter which could begin to approach 400 mph with a military load.

Fortunately, it was the U.S. Navy which would be instrumental in helping to rapidly close the gap between American fighters and foreign types. The Navy's Bureau of Aeronautics had sponsored development of a 2000 hp-class engine, the experimental Pratt & Whitney XR-2800-2 Double Wasp. By October 1940, a prototype engine was flying in the new Vought XF4U-1 fighter, a strong contender in the 400 mph class. It had flown for the first time on October 6th and broke through the 400 mph barrier a short time later. Less than 60 days prior to that flight, the Materiel Command issued Change Order #2, S/N 3248 against contract AC-13817 for design and development of the Republic XP-47B featuring a turbosupercharged version of the same

Certainly qualifying as one of the most unusual formations of the WWII era, a Republic AT-12 is seen leading a Curtiss P-36A and one of the rarely seen North American P-64s, probably near Luke Field, Arizona. The first two types in the group had P&W R-1830 Twin Wasp engines, and the P-36 was an obsolete pursuit plane. Few (one-half dozen) single-seat P-64s were taken over by the AAC after being ordered by Siam. With a Wright R-1820 up front, it was the slowest of the trio. **(A.C. Head via F. Strnad)**

basic R-2800 engine.

Major Alexander Seversky was, as would be expected, very unhappy about the corporate takeover situation, especially since many of the detrimental charges against him were the statements of non-aviation people brought in to deprecate the Major's performance and contributions. He most certainly resented the widespread published stories which always credited Kartveli and the XP-41 with being the direct predecessor of the YP-43s. (In actual fact, the XP-41 proved to be an embarrassment to the AAC, having a very short active life. Its major value was in revealing what the P-35 should have been at the beginning of 1937.) And since the Major had hired the Russian engineer and brought him along as a team player he had to be somewhat disturbed to see his former assistant become the corporation vice president. After he was forced out as president of the company that had borne his name, many of the harsh comments which Seversky made about Kartveli in written communications were valid. Included were the points about Kartveli being graduated as an electrical engineer and, not caring much for that activity, he later majored in aeronautical engineering. However, he specialized in structural design, not aerodynamics. In writing on legal matters, Seversky stated that "Kartveli had shown no innovation of any kind in aircraft design." Alex Kartveli, whose forte was mathematics, did not get embroiled in any battle with Seversky. He preferred to avoid the arguments and move on as a first-rate chief engineer. He had some very good engineers working for him, and such talent has served other chief engineers well over the years.

Showing its underside with full pre-war markings, the Republic AT-12 (canopy open) drifts lazily through western skies in 1941. Well built, attractive and pretty fast, it became a favored "hack" for officers of rank. (A.C. Head via F. Strnad)

A Republic "history" issued two decades later (in 1956) may have been on target as to financial operations, but was so technically flawed as to have been a joke. The writers totally ignored Major de Seversky's outstanding contributions in personally obtaining European contracts for more than 172 valuable airplanes, all developed with his personal inputs. Did he believe in his airplanes? Absolutely. He did more than half of the test and demonstration flying himself. Simply expressed, the Major may not have been all things to all people, but he certainly was a creative one-man band. He was surely a true American, although his flamboyant style offended some. Examples of at least a couple of other great Americans who ruffled the feathers of important people were Generals Frank Andrews, George Patton and Benjamin Kelsey; and then there was Abraham Lincoln.

As Henry Ford reminded his Willow Run employees, "Observe the turtle. He only makes progress when his neck is out."

They can because they think they can. – Vergil

Paul Moore settled $880,000 in debentures (money owed to him) for just $110,000 and sided with new management, having been convinced by Wallace Kellett that Major de Seversky was not competent as chief executive. Results generated one huge question mark. Why did it take eight years for Moore to come to such a conclusion? Every poor situation could have been managed without the ultimate route they took to change the corporate structure. A good chief executive officer with top management abilities was needed to run the business side of things. Kellett turned out to be the wrong choice. Well, by 1939 the die was cast. The depersonalized name of Republic Aviation was adopted in deference to pressures from Washington and from the Major himself.

War in Europe brought sudden and momentous changes to American industry, together with two new terms that had not been in the American lexicon, i.e., "defense industry" and "Arsenal of Democracy." Both terms were about to become as common as Abbott and Costello's "Who's on first?" It would be *Blitzkrieg, Luftwaffe* and *Wehrmacht* leading the parade to the fall of France and to the desperate events at Dunkirk which made those terms familiar to every U.S. citizen.

At Farmingdale, in the early months of 1940, there was no real sign of such things. Tiny Republic Aviation was assembling export EP1s and 2PAs on parallel assembly lines with a minuscule order for service test YP-43s, the outgrowth of impressive performance by the private venture Seversky AP-4 during the winter competition at Wright Field in 1939. That work was being done in a plant originally constructed by Fairchild in the 1920s. Factory design took a giant leap forward in just a decade and a half from the early 1920s. Very large clear-span factory buildings that could accommodate airplanes with wings spanning 150 feet were being constructed by firms such as Austin Company. Developments in air conditioning gave rise to windowless "blackout" plants, something impractical just a decade earlier. There was no way for Republic to be considered a part of the "Arsenal" buildup. After President Roosevelt made the momentous decision to expand the Navy and Army aviation arms by 50,000 airplanes a year, plant expansion for the major manufacturers surged forward at a tremendous rate. But no such expansion was in progress at Farmingdale until the last quarter of 1940.

Adding contrarotating propellers and an extension shaft to this 1939 Republic Rocket (AP-12) was looking for "a bridge too far" trouble. The design's best feature was the ventral fin, something rarely incorporated in fighters that could surely have benefited. But the P-39 absolutely needed it, and the P-38 had it. According to the crystal ball, this AP-12 design eventually evolved into Republic's stillborn XP-69, a fighter which was really designed around Wright's Tornado "cylindrical" engine. Known in plant as the AP-18, it was intended to fly with the 42-cylinder version of the Tornado. (Mfr./ E. Boss)

Production of the thirteen YP-43s would be but a brief flurry of activity. A Republic proposal for a 400-mph fighter in the AP-4 series resulted in a modest production contract for a "hotrodded" version of the service test YP-43, the new aircraft carrying the designation P-44. Had this P&W R-2800 powered interceptor (AP-4J or -4L) been developed to flight status, it surely would have been in a close battle with Vought's XF4U-1 to become the first single-engine 400-mph fighter in the world. (Refer to Chapter 7.) Spurred by the President's decrees about expanding the Air Corps as never before, the War Department ordered 80 of the P-44-IVs (company internal *nom de plume*) months before the first YP-43 was to take to the air. Giving some strong credence to rumors that the War Department was a key factor in the demise of Major de Seversky as president of the then existing company, the contract was awarded on 12 October 1939, coinciding with the takeover. It was also obviously aimed at keeping the company (as Republic Aviation) in business with a trained cadre of skilled workers. After a decade and a half of existing on the most meager rations – often with an entire strength of approximately 1500 airplanes – the Air Corps was suddenly exhorted to plan on having 10,000 to 20,000 airplanes a year. (Those totals often expanded within a few months as military disasters continued apace.)

Sweden's *Flygvapnet* ultimately operated sixty EP1-106 fighters as first-line types until 1947 when new operational types – particularly some North American P-51Ds – came into the inventory. On balance, they were regarded as delightful airplanes. Even to pilots who had been flying Gloster Gladiator biplanes, the Republic monoplanes were considered to be excitingly maneuverable. Despite the fact that only two of at least 52 Republic 2PA Guardsman dive-bombers ever reached Sweden, there is a gem of a story in their service lives. America took title to 50 of the 2-seaters, using them to train pilots in a sensationally expanding pilot training program in 1940-41. More than half a century later, a single example of the type still remains in the U.S. in private hands. One of the pair delivered to Scandinavia might well exist in nearly perfect shape except for a poor decision by someone in authority who refused to spend a trifling amount of money for modest repairs in 1953.

Flygvapnet officials had assigned serial numbers 7201 through 7252 to the bombers they had designated as B6s, all intended for duty with Vastoga Air Wing F6, based at Karlsborg near the Baltic Sea. When the seizure action occurred on 10 October 1940, only two of the 2PAs had been crated and shipped, numbers 7203 and 7204. (For a time, one Guardsman aircraft remaining stateside – No. 7219 – was flown by Swedish Purchasing Commission officials in the Washington, D. C., area, but it soon appeared in AAC markings as an AT-12.) Upon arrival in Sweden, most likely in company with at least several EP1 fighters, the two B6s were assembled at the Central Flying Works, Malmstatt.

After the third airplane of that type was taken from Sweden's Air Purchasing Commission, it became AT-12, serial no. AC41-17510. (Republic AT-12 Air Corps serial numbers were AC41-17494 through -17543, inclusive.) Although there had not been any airplanes classified as advanced trainers (ATs) since the 1920s, the incredible expansion of the Air Corps initiated in 1939 forced a quick change. A few North American BC-1As, a basic combat type, were redesignated as AT-6s in 1940. Almost overnight there seemed to be hundreds – then thousands – of AT-6s in the inventory. The powerful Republic AT-12s were really something special, enjoying about 75 percent more power. It appears most of the Republic AT-12s were based with Training Command in the Southwest U.S., probably at places like Luke Field.

Senior officers at many locations looked upon the AT-12s as ideal "hack" transports. If flown with normal fuel load of 130 gallons, maximum cruise range would not have exceeded 725 miles, although that could be boosted to more than 1000 miles by increasing the fuel to the overload capacity of 195 gallons. When flying out of the Arizona or California bases in Army Hot Day conditions, it would have made sense for the pilot to travel alone if that extra 65 gallons was pumped in. With two aboard, and some luggage, it would have made more sense to use a normal fuel load with refueling stops every 2.5 to 3 hours, cruising at about 275 mph on 650 hp at 2250 rpm in the neighborhood of 15,000 feet. Any pilot familiar with either the Seversky P-35 or the Republic P-43 Lancer would have felt quite comfortable with the AT-12 and its Twin Wasp powerplant. After all, the airplane had an additional 25 square feet of wing area to take care of the extra 650 pounds or so above normal gross weight. Under those conditions, the wing loading was a touch lighter than that of the Republic P-35A.

The fate of the two B6s in the *Flygvapnet* presents an interesting contrast. On 20 August 1940, #7203 was assigned to Air Wing F6. After the exciting new aircraft had accumulated only about 32 hours of flight time, novice pilot G. B. H. Lindstrom was permitted to fly the high-performance Republic B6. Through an error in somebody's judgment, Cadet A. G. Nystrom was also permitted to occupy the gunner's seat for Lindstrom's initial solo flight on 20 September 1940. On approach for landing, the young pilot decided to make a touch-and-go at the last second. In an uncoordinated action, he pulled back on the stick but failed to increase power on an airplane with a relatively high wing loading – at least for an inexperienced flyer. A wing dropped sharply, causing the Republic bomber to spin about a half turn into the ground. Lindstrom died at the scene, while Nystrom suffered severe injuries. Of course the airplane was totally wrecked.

By stark contrast, the second airplane, #7204, which had been assigned on the same date (August 20) to Flotilla Eight (F8) at Barkaby, was intended for use by members of the Headquarters Air Staff to maintain flying proficiency. Painted with dapple-style camouflage like its counterpart and all of the Republic J9 fighters, that airplane bore the large identity number 46 on the tail. Because it was considered to be a delight to fly, the airplane soldiered on as the only one of its type in all Scandinavia – and, in fact, in Europe – for several years. As late as 1951, General B. C. Nordenakjold, the Swedish AF commander flew #7204 on a tour of bases for the 25th anniversary celebration of the *Flygvapnet* (Swedish Air Force). A couple of years later, nearly 13 years after that single B6 airplane went into service at Barkaby, an unnamed pilot ground looped the machine rather viciously on 13 March 1953. Although the damage was not "fatal" to the airplane at the time, a decision was made to reject a request for expenditure of about the equivalent of $5000 to repair the B6 and return it to service. Sadly, the only example of the type in Sweden was committed to the boneyard after being written off on May 29th.

Airplanes built by the Seversky/Republic factory were extremely well-structured and fabricated, and their performances could easily have been improved by some changes that were hardly innovative. Perhaps the changes were proposed, being rejected only because of the financial situation prevailing at the time and place. In retrospect, any experienced aircraft engineer would wonder about the lack of a drag reduction program, just as others might wonder about failure to work with vigor on weight reduction. With 20-20 hindsight, we can only speculate on the reasons for failure to maximize the potential, but the most likely reason was the already overextended financial condition of the company as 1938 came to a close. The Major was far more technically proficient than he was in managing budgets. Sadly, he was among the last to recognize that shortcoming.

While the AP-10/XP-47 was struggling for recognition, another project group located in the former Grumman factory (Bldg. 1) was designing something resembling a bit of upsmanship over the Bell P-39. Tagged as the AP-12 Rocket, this escapee from fiction artist Frank Tinsley's drawing board had four nose and two wing machine guns plus a nose-mounted cannon. We can only assume that it was not designed to utilize the Allison V-1710 as some have contended. Nobody ever saw an Allison cannistered as shown, and no V-1710-engined airplene ever needed dual propellers. Especially in 1940-41. (Mfr./E. Boss)

Republic XP-47B Preparations for First Flight.

Opposite: Republic XP-47B – Death Dive of the Prototype.

Republic P-47D-3-RA "Daring Dottie III" 341FS, Fifth Air Force.

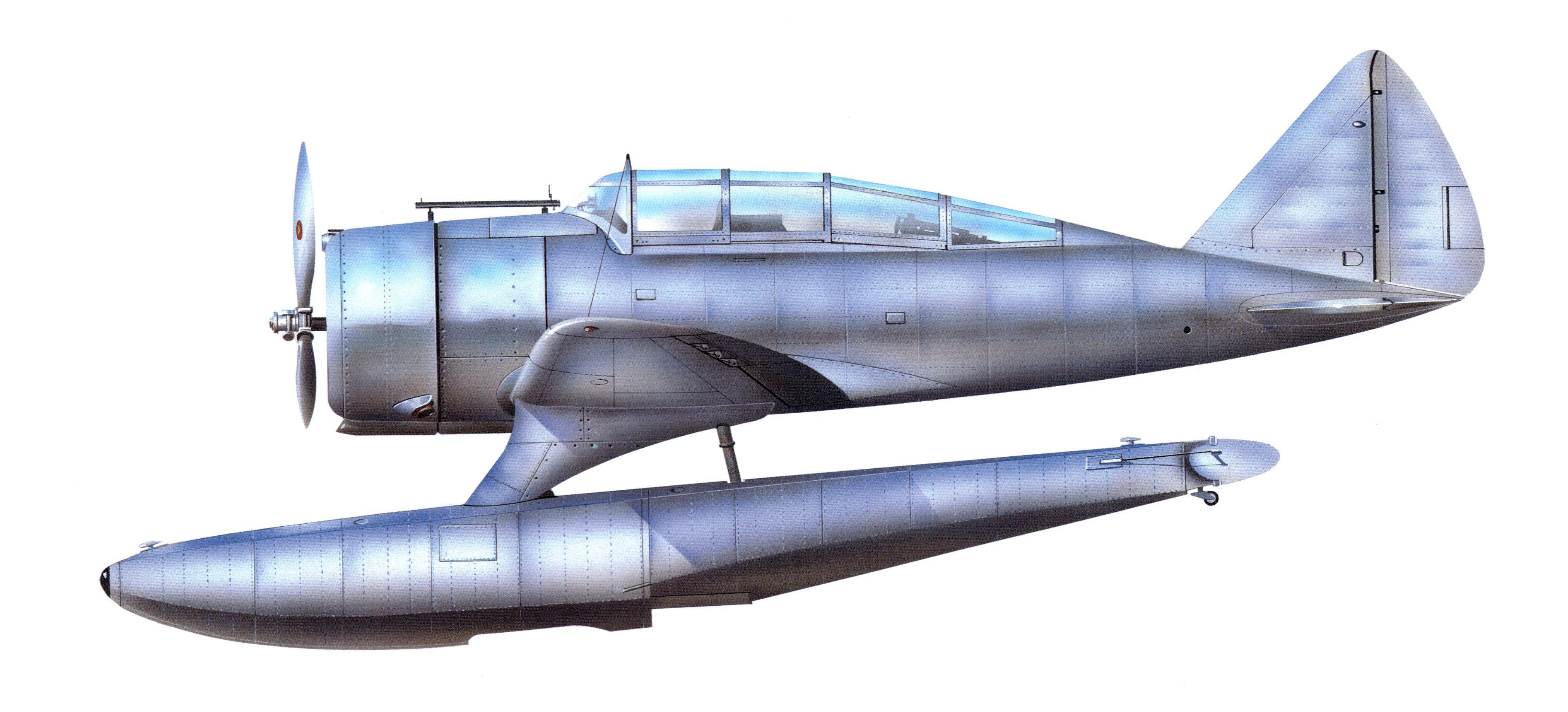

Seversky 2PA-A Amphibian Reconnaissance-Fighter for Russia.

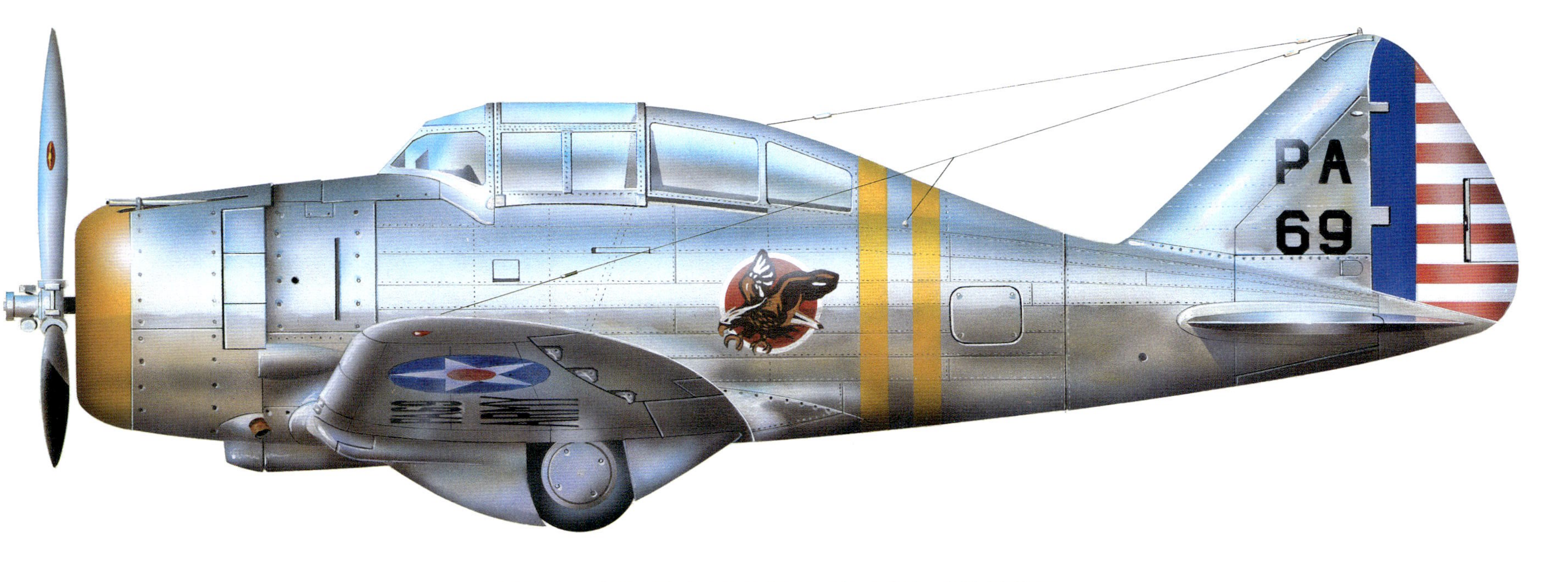

Seversky P-35: 27PS, 1PG, Squadron Commander, Selfridge Field, Michigan.

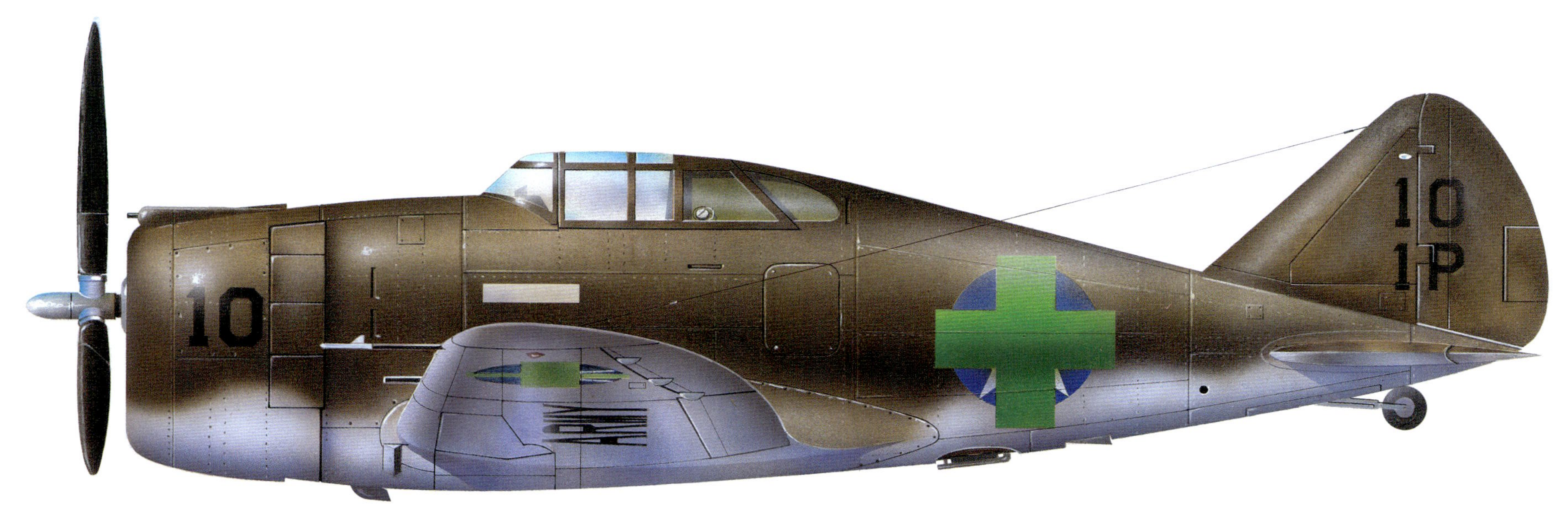

Republic P-43 Lancer: 1st Pursuit Group ,1941 War Games Markings.

Republic P-44-IV Mockup Design. Production Contract Awarded. Cancelled. Shown as 1PG, 94PS Commander's Aircraft.

Republic RP-47B : Lt. Col. Hub Zemke's Aircraft, Springtime 1942. New York/New England Defense. Zemke was 56th Fighter Group Commander.

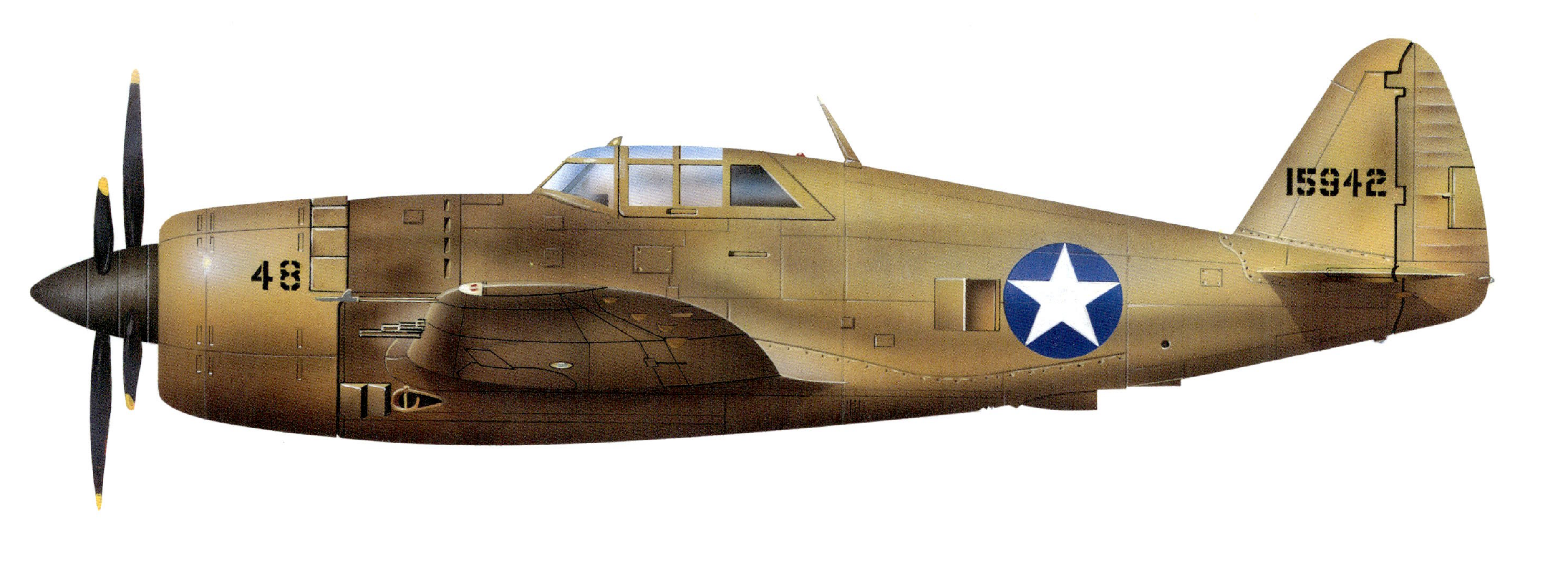

Republic P-47B "Double Twister" for Contrarotating Propeller Test. C. Hart Miller-Pilot.

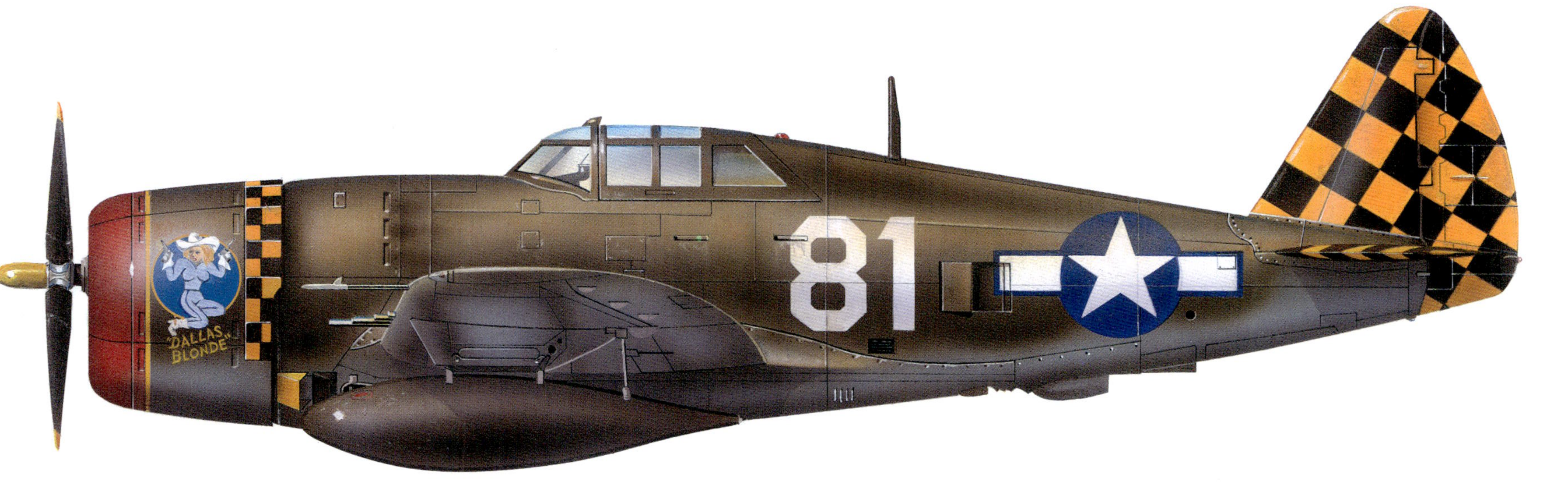

Republic P-47D-10-RE "Dallas Blonde," 319FS ,325FG, 15AF, Italy 1944. Lt. Don P. Kerns-Pilot.

Republic P-47D-16-RE "Touch of Texas": 510FS, 405FG, 9AF, with 54 Ground Support Missions. Capt. Charles Mohrle-Pilot.

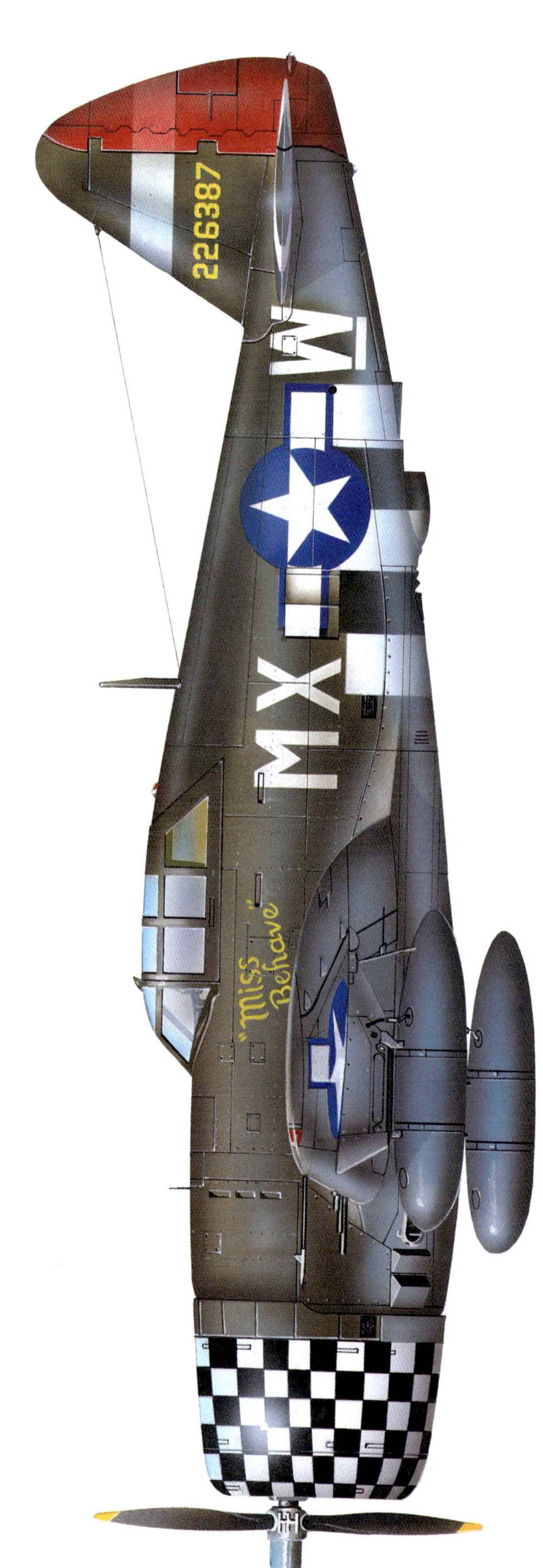

Republic P-47D-22-RE "Miss Behave": 82FS, 78FG, 8AF.

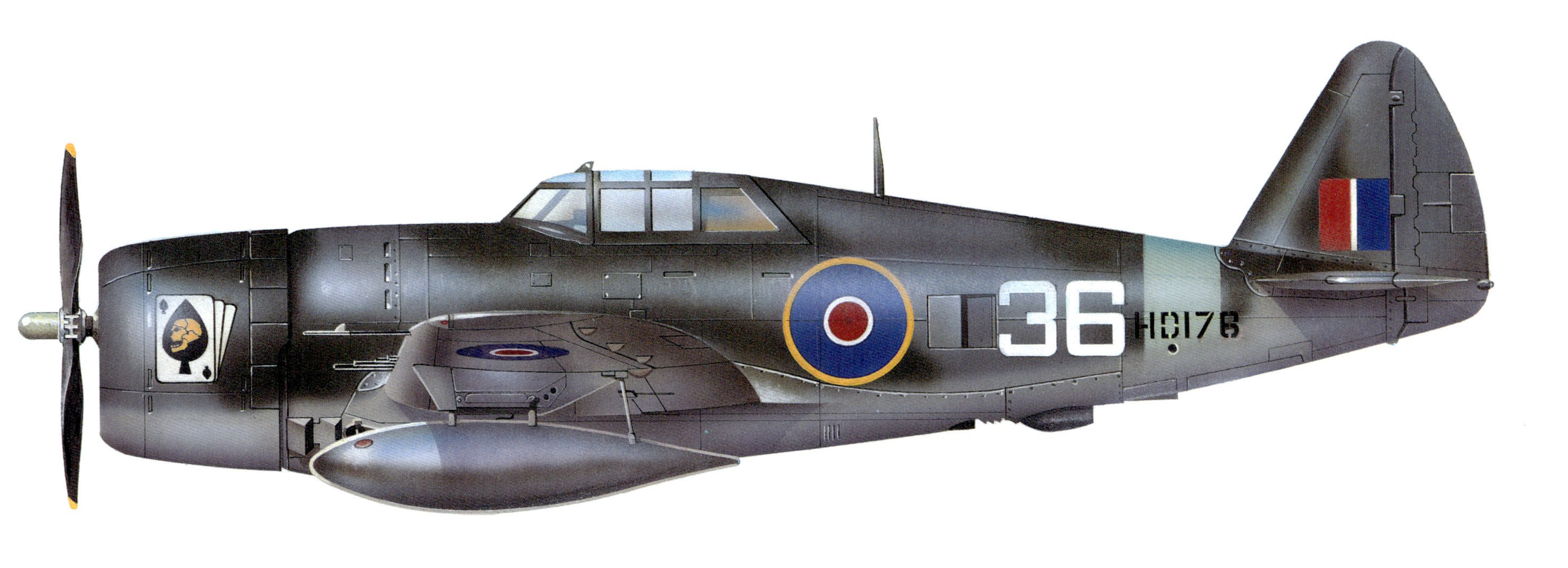

Republic Thunderbolt Mk. I, Royal Air Force, 73 0TU, MAAF, Egypt.

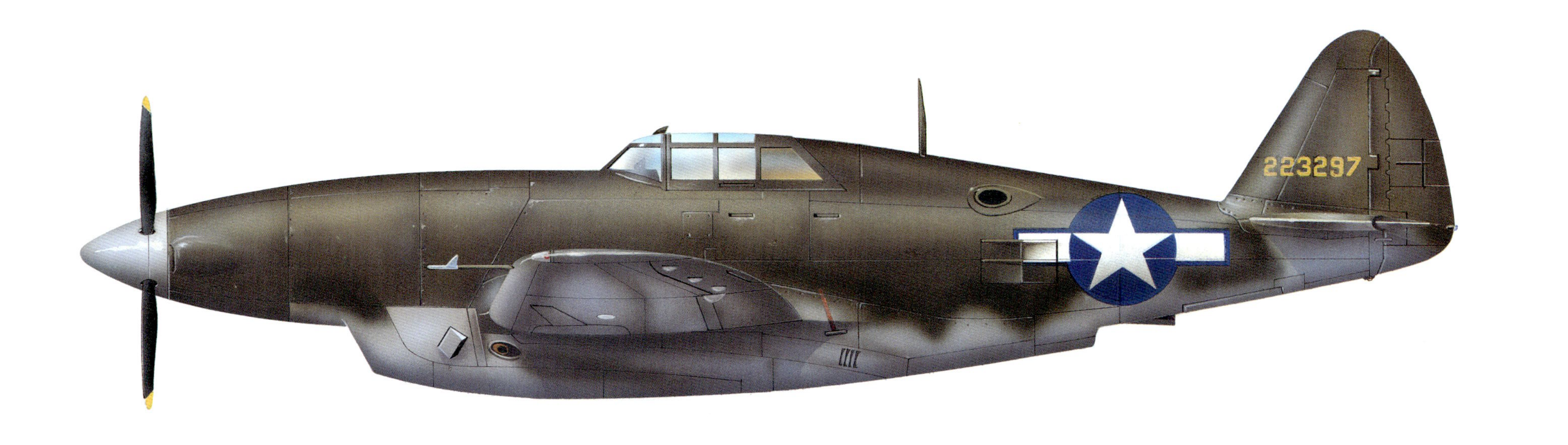

Republic XP-47H : Chrysler XV-2220 Development Engine.

Republic P-47D-27-RE "The Trojan Warhorse," 79FG, 12AF, Italy.

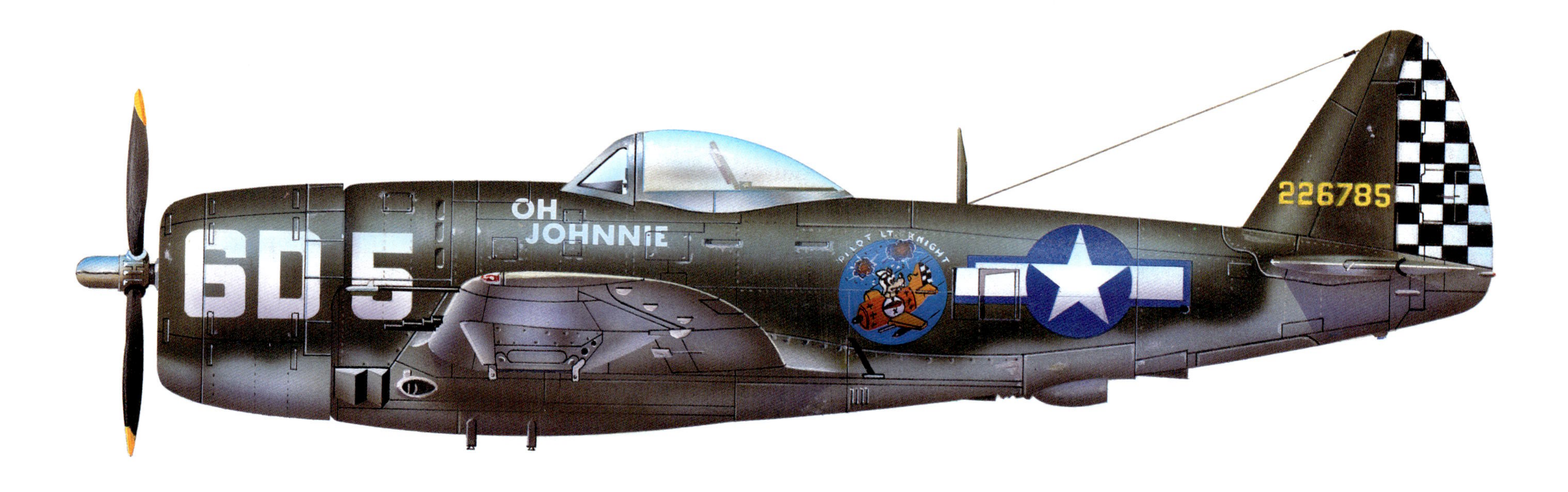

Republic P-47D-27-RE "Oh Johnnie," 346FS, 350FG, 12AF, Italy. Lt. Raymond L. Knight-Pilot, Medal of Honor, KIA.

Republic P-47D-27-RE "Angie," 512FS, 406FG, 9AF, ETO.

Republic P-47D-28-RA "Battle Baby," #80

Republic P-47D-30-RA "Stinky," 356FS, 358FG, 9AF, ETO. Lt. Don Volkmer-Pilot.

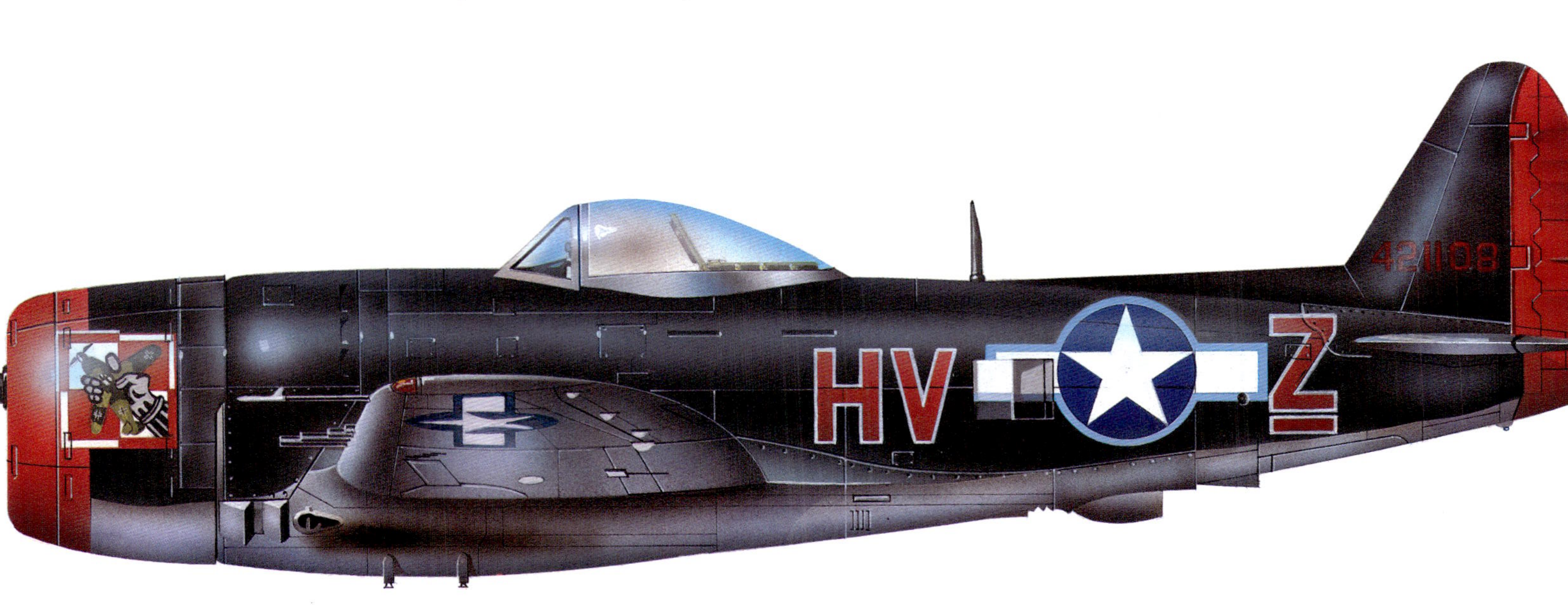

Republic P-47M-1-RE HV★Z , 61FS, 56FG, 8AF. Lt. Witold Lanowski-Pilot.

Republic P-47M-1-RE "Teddy," 62FS, 56FG, 8AF. Maj. Michael Jackson-Pilot.

Republic P-47N-2-RE "The Virginia Belle," 333FS, 318FG, 7AF. Central Pacific Area.

Republic F-47N-25-RE, Puerto Rico Air National Guard. Unapproved Markings Incomplete.

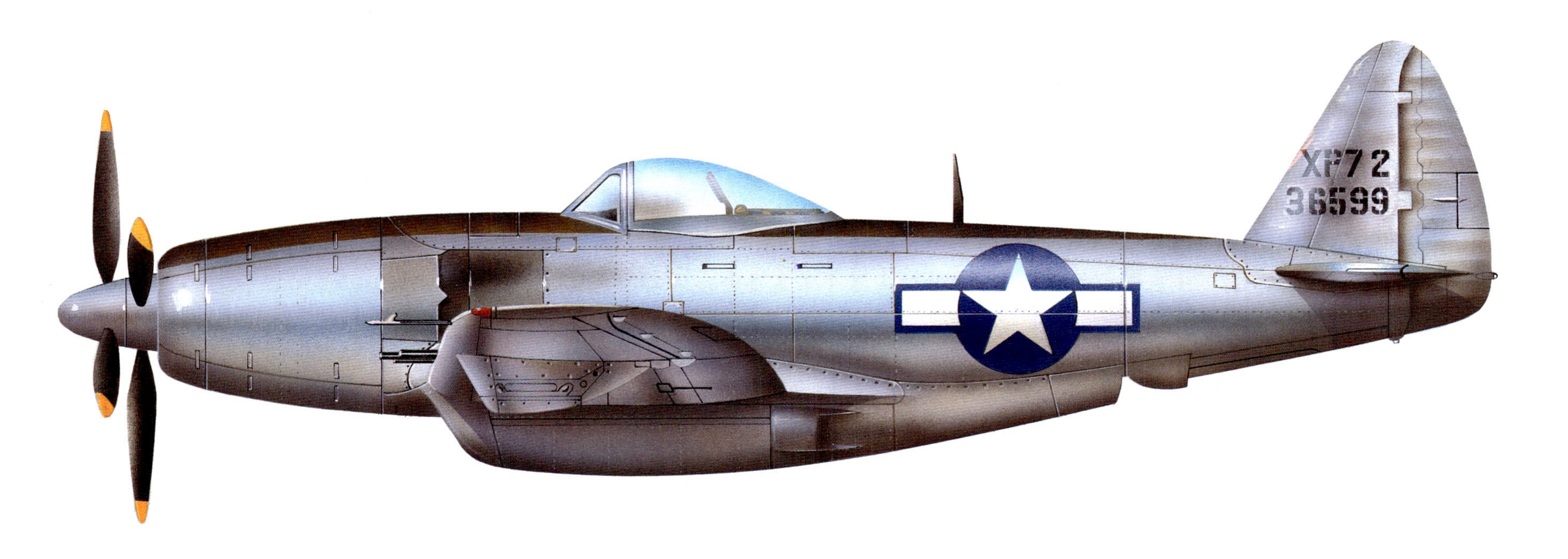

Republic XP-72 No. 2 Prototype. Pratt & Whitney R-4360 Engine. Contrarotating Propellers.

If it quacks like a duck....Not so! It was not a Seversky (anymore) and it was not a P-35. It may have looked the same, but there were multiple differences. Republic Aviation was the builder and the AAF designated all 60 of them as P-35As. Overall the P-35A was 20 inches longer than a P-35, and the P&W R-1830-45 Twin Wasp was company-rated at 1200 hp for takeoff. Forgotten in time is the fact that Vultee V-48 Vanguards had the same Twin Wasp engine at 1200 hp, the same span, were two years newer and the Flygvapnet had ordered 144, only to have the AAF take them over as P-66s. Supposedly they were 30 mph faster than P-35As and P-36Cs having identical power. Not likely, and armament was light. (Mfr. 2209)

CHAPTER 7

BATTLE HYMN OF REPUBLIC

If the decade of the 1930s started with emotional cries of pain and the thud of bodies crashing to the pavement on Wall Street, at least for people connected with every business and industry, who can say it ended in any better manner? Muffled cries must have been plentiful in the ruins of Poland, and it is said ships moan as they slip beneath the waves. The outbreak of war in Europe was attended by a quick, lightning thrust in Eastern Europe, some cries of rage from Great Britain and France, and then a relatively uneasy quiet for months. Aside from torpedoes exploding and taking an assortment of ships to the bottom with some of those unlucky enough to be aboard, the decade ended with murmurs about the "Phony War." British factories hummed with accelerated production of every possible war material, but the RAF could only launch small-scale bomber raids against the sworn enemy, and it was paper (leaflets) fluttering to ground that made the most noise.

Even as bitter accusations and distortions of truth flew freely in and out of the courtrooms after de Seversky was ousted from Republic Aviation, the YP-43 went into limited production in company with P-35As and AT-12s. The first YP-43 flew in March 1940, showing very close kinship to the Major's earlier pet project, the AP-4, even as the Kartveli-backed XP-41 faded. Test pilot Lowery L. Brabham is shown in the tufted Lancer. (Rudy Arnold Photo)

Out on Long Island, much a part of New York State, the imaginary sounds of reveille echoed through the old buildings at Republic Aviation Corporation on Conklin Street. It was time for everyone to get up and get going. In mid summer 1939, there had been a fairly radical change in the business part of the fading Seversky Aircraft Corp. when Wallace Kellett organized a new Military Contracts Department. Its major objective would be the management of a rapid increase in activities stemming from business with the U.S. Army Air Corps. There was going to be a new Director of Military Contracts, and he was hardly more than a kid...albeit a large-sized one. Charles Hart Miller was barely into his thirties, but had done a good job as Assistant to the President and he obviously knew many of the people in military aviation who were on their way up. At the time, Miller was not only the director of the department, he was the entire staff. Within weeks of his appointment, he was dealing with Major Bengt Jacobsson representing the Swedish Air Force Commission. Surely, they had not anticipated the foreign facet of the operation when the department was formed. Yet it was a welcome addition to Miller's assignments. In essence, he ranked No.3 in corporate management, immediately following Vice President Joseph McClane.

When production work on the YP-43s started, Republic Aviation had fewer than 1500 employees on staff. By the time this final YP-43 was ready for delivery on September 13, 1940, total employment had risen to 2322 crammed into the pre-Depression Fairchild plant (and some decrepit outbuildings). This Lancer (AC39-716) showed that Factory Manager G. Meyrer had overcome some trying times and his people were doing a commendable job. Project Engineer D.A. Tooley's staff was faced with devising a new tailwheel installation that would effectively overcome an annoying inherent tendency for Lancers to groundloop. (Mfr. via Author)

Hart Miller had to be a pretty busy person in those months when he had to create a department and still handle some flying activities as a test pilot, help sell airplanes, negotiate contracts and serve as a customer relations man. Just to illustrate how things were changing, net sales of completed aircraft, parts, manufacturing rights, etc. in 1939 totaled $923,559; in 1940 deliveries amounted to no less than $10,313,740, more than a tenfold increase. Against a net loss of $524,781 in 1939, the company showed a profit of $923,332 in 1940.

Project 79 on the company books was for construction of the YP-43s in old Bldg. 5. By contract, all were supposed to be delivered in 1940, but the Air Corps immediately fell behind in required deliveries of the engines. One rather uplifting and important event occurred at Farmingdale on 9 September 1940 when National Defense Coordinator of the Office of Production Management William S. Knudsen and Army Air Corps chief M/Gen. H. H. Arnold visited the factory. That was most certainly a "first." They wanted to inspect the plant facilities, take a close look at a new YP-43 interceptor. (The best interception job ever done by P-43s was keeping the sheriff from posting a bankruptcy notice or keeping the wolf from the door.) That interceptor project and the Swedish EP1-106 fighter contracts had been responsible for bringing the company back from total employment of fewer than 200 people in 1938 to a late summer 1940 roster of approximately 2300 people.

In the idiom of the period, Kellett made some announcements that were pretty far removed from fact, but they need to be presented if for no other reason than to show how far afield from fact were the pablum dishes being served up to the public. Some 99 percent of the public would have believed any or all of the pronouncements as they appeared in the New York News the next day. By the time they got into the aviation press, they may have become even more distorted. Airspeed and performance (usually one and the same) figures on the new fighting plane, described by "fighting pilots" as "sure poison to bombers," were said to be a closely-guarded military secret, but it "was taken for granted that the single seater would do better than 350 miles per hour at the 20,000-foot substratosphere level." (Guaranteed figures provided to the military will be presented following a few more "accurate" statements.) After saying the company would be expanding its then current plant "fivefold" and hiring 10,000 *men* – oh, how that gender comment was going to change – he proceeded to state that a new, high-altitude interceptor was going to be produced at the rate of four per day beginning January 1, 1941. That would be less than 4 months away, and in that crowded old Bldg.5. And on a holiday?

He was probably referring to the new P-44 contract for 80 airplanes.[1] According to the terms, only one P-44 was scheduled

---

[1] One 1990s book about Republic Aviation stated that on 19 July 1940 the AAC ordered 225 of the P-44-2s and they increased the orders to a total of 827. The P-44-2 was to have "extra built-in fuel tanks, making it the longest-range fighter of the period." What incredible statements! According to a Republic Military Contracts Dept. report book updated to 24 June 1940, a total of 80 of the P-44 fighters had been ordered, and C. Hart Miller's handwritten entry said "All is canceled." As will be seen in this chapter, nobody was in a better position to know anything about contracts at Republic. Regarding range, as data will show, the P-44 could never even approach the XP-38 which had already flown about 18 months earlier and YP-38s and P-38Ds then currently being manufactured. *Love truth, but pardon error.* – Voltaire.

to be delivered in the next month (October) and no more than six were to be delivered in *that* January. A production rate of four P-44s per day was not going to be achieved in 250,000 sq.ft. of factory space, and they had not even broken ground for the "1,000,000" sq.ft. of manufacturing space a new plant would provide on the other side of Conklin Street. "At least 1 million" was the number Kellett used when addressing the audience and reporters.

It is appropriate at this point to provide some technical specification data pertaining to the YP-43/P-43 series as provided to Hart Miller by Kartveli. Comparable data relating to the P-44-IV is provided, although those were the proposal figures supplied by Engineering (essentially from Chief Engr. A. Kartveli, P-44 Project Engr. R.C. Bergh and YP-43/P-43A Proj. Engr. D.A. Tooley.):

YP-43/P-43A Pursuit	P-44 Interceptor
Circular Proposal 38-390	Circular Proposal 39-770 (Rev.)
Mfr. Model Spec. 60-A	Mfr. Model Spec. 87
Number ordered: 13	Number ordered: 80
P&W R-1830-35/Type B-2 Turbo	P&W R-2800-7/Type D-2 Turbo
T.O. Power: 1200 hp	T.O. Power: 1850 hp/2600 rpm/S.L.
Mil. Rating: 1200 hp/2700 rpm/@ 20,000'	Mil. Rating: 1850 hp/2600 rpm/@ 25,000'
Curtiss Elect. Constant Speed 3 blade	Curtiss Elect. Constant Speed 4 blade
High speed, Mil. rating, @ 20,000': 351 mph	High speed, Mil. rating @ 20,000': 402 mph
Design Gross Wt. 7300 lbs.	Design Gross Wt. 8700 lbs.
Weight empty 5478 lbs.	Weight empty 6939 lbs.
Wing loading 31 lbs/sq.ft.	Wing loading 37 lbs/sq.ft.
Normal fuel: 145 gals.	Normal fuel: 162.5 gals.
Overload fuel: 73 gals.	Overload fuel: 98 gals.
Endurance*: 2 hrs.@75% pwr.	Endurance*: 1 hr.@ full throttle
Endurance**: 1 hr.@75% pwr.	Endurance**: 1 hr.@75% power
Armament: 2x.50 cal. cowl, normal; 2x.30 cal. wing/alt.	Armament: 2x.50 cal. cowl + 4x.30 cal. wing/normal

* = Normal fuel load. ** = additional for overload fuel. Neither design had self-sealing fuel tanks or armor plate. It is very obvious that the P-44 was to be purely an interceptor.

Just four days after General Arnold and Mr. Knudsen flew into Republic Airport in a Douglas B-18 bomber converted into an executive transport, Mr. Kellett announced the War Department's award of a contract to Republic in the amount of $56,290,316 to procure "pursuit aircraft and spares." Even in a later report to stockholders there was no clarification of which airplanes were ordered for that amount of money. However, the following procurement actions took place in due time and were authorized by the appropriate contracts or change orders there to:

773 Republic P-47B airplanes on Contract W535-ac-15850, dtd. 13 September 1940. There were two subsequent Change Orders.

54 Republic P-43 airplanes on the same contract, but C.O. 2 took care of the appropriate Change Order requirement.

There was an additional order on 10 December in the amount of $5,210,000 to cover the cost of building *500,000* additional square feet of manufacturing space. (Wasn't Kellett just quoted as saying that there would be 1,000,000 additional square footage? Trying to keep the facts straight when all that misleading information was being circulated could not have been a simple task.) All things considered, Republic would have – by early June 1941 – 700,000 sq.ft. of manufacturing floorspace, more or less, and the remaining 50,000 sq.ft. would be devoted to office space, storage, etc..

In anticipation of beginning a lot of procurement action with Republic, the AAC had dispatched Capt. Russell Keillor to Farmingdale to assume a new post as Air Corps Factory Representative to oversee the production of fighters in response to the immense orders being awarded to Republic.

Was the path to that $56 million order as clean, orderly and simple as it sounds? Did Alex Kartveli simply design the P-47 Thunderbolt in anticipation of what the Army Air Corps was going to need – as many books, magazines and publicity releases would have us believe? No sir, not on your tintype! So, what is the real, authentic path everybody had to follow leading to the XP-47B design and thousands of Thunderbolts? It was one tortuous, potholed road ending with nothing short of a firm statement detailing appropriate action expected. Furthermore it was supported by a well-defined set of ground rules, better known as specifications. Whether General Echols or General Kenney was in charge of Materiel Division at the time, documentation of that type was mandated.

According to a logical legend, Col. Henry H. Arnold and some of his mid-1930s cohorts had become thoroughly convinced liquid-cooled (Prestone, or ethylene-glycol) engines with small frontal areas could be buried in the wings of bombers. They would be more efficient than the reliable radial engines developed in parallel by Pratt & Whitney Division of United Aircraft Corp. and by Wright Aeronautical Div. of Curtiss-Wright Corporation. All development effort was largely accomplished with sponsorship from the U.S. Navy.

However at a most inappropriate time, a French Caudron C.460 racing plane powered by a Renault aircooled inline engine defeated all other contestants in the 1936 National Air Race Thompson Trophy contest. The long, narrow cowling caught Arnold's eye and he evidently mistook the Renault for a liquid-cooled engine. For unknown reasons, Col. Arnold was somehow convinced lightweight fighters, especially those with liquid-cooled engines, were the optimum answer to America's needs for efficient interceptors to defend against enemy bombers. Perhaps that is the way the long-range planners saw it, but at the time there was hardly a glimmer of such a long-range bomber threat far into the future.

Some misfiring could be detected in the Republic Aviation

Although this photo shows a relatively basic Republic YP-43 cockpit arrangement, it depicts some equipment that was foreign to your average fighter pilot. Immediately below the cockpit frame padding (top center) was a latch installation for automatic release of the instrument flying hood. Installed on the third YP-43 Lancer built, it may have only been for experimental purposes. (AFM)

corporate and military enclaves shortly after award of a contract for thirteen YP-43 interceptors. The company was to manufacturing thirteen YP-43 service test developments of the Seversky AP-4 entrant flown in the January 1939 Pursuit Airplane Competition. First, there was the not unexpected unpleasantness accompanying the corporate reorganization. An influx of orders from Sweden for fighter and light bomber aircraft which had flowered from decisions made solely by Major Seversky in 1938 tended to brighten the scene. However, the specter of bankruptcy hovered over the dingy red brick factory buildings. Fortunately the Swedes were cooperative and helped with some much needed cash not required by terms of the contract.

By then it became evident B/Gen. "Hap" Arnold was more or less enamored of the Allison V-1710 engine and some others being developed by Continental Motors and Lycoming. It did not seem to matter that none of those manufacturers had any great track record with production of high-powered aircraft engines. Notice the total absence of the producers that had carried American aviation to the forefront in the 1920s, namely Wright Aero, Pratt & Whitney, Curtiss and Packard. All of those companies had extensive experience and facilities used for production and test of powerful engines. Beyond some association with the Wright Cyclone and P&W Twin Wasp, the Army Air Corps was being led astray from their "bread and butter" engines, most likely by Lt. Col. Edwin R. Page, head of engine development at Wright Field. The only people who seemed interested biting on the hook for really lightweight pursuit aircraft were Bell Aircraft (surviving through production of biplane dive bombers developed by Great Lakes Aircraft, then defunct), Tucker[2], who was planning to use some Miller-designed engines, and Republic Aviation (standing with the aid of crutches). Miller, known for building race car engines, had never even designed a production aircraft engine.

This had to be after the time Howard Hughes tried unsuccessfully to get the powers at Wright Field to buy his H-1 racer. Oh, certainly it had a Twin Wasp Jr. radial engine of about 800 horsepower in it, but designer Richard Palmer certainly could have adapted it to the Allison. He had already proven to be innovative. They scoffed at the wooden Hughes speedster – already having proven it had great range capabilities too – but ultimately chose to let Bell try to develop the XP-77 lightweight fighter which, by the way, was a wooden craft. And guess what, the Tucker XP-57 (Mfr. Model AL-5) was to be flying on *wooden wings* on a steel tube fuselage covered with plywood. The Hughes airplane, even with a radial engine that could have been boosted to 1000 projected horsepower, would probably have been the best of the lot.

Of course that did not occur to anyone but the German engineer Kurt Tank who simply "borrowed" the concept and reworked it until it emerged as the Focke-Wulf Fw 190.

Arnold's wishes evolved in the form of a 1939 Circular Proposal No. CP39-770. The Curtiss Airplane Division at Buffalo, N.Y., responded with their Model CP39-13, essentially an Allison-powered lightweight rendition of the P-40. They were awarded a development contract for two airplanes, the XP-46 and the XP-46A, the latter being essentially a flying aerodynamic model of the XP-46 without combat-type equipment. Seversky Aircraft (at that time) responded with a remarkably similar design (designated as Model AP-10) but only weighing 4600 pounds, about two-thirds as much as the Curtiss fighter. The company's response to the Circular Proposal was sent to Dayton in August 1939.

By that time it was clear to everyone Major Seversky was no longer the man in charge at Farmingdale, so the AP-10 had become Kartveli's key project for the moment. One very unique aspect of the project was the use of a full-size combination wind tunnel model and mockup. In fact the proposed fighter also turned out to be a pacemaker for the revitalized corporation in that no other Seversky/Republic large-scale wind tunnel model had ever been tested in an NACA facility. In comparing the Curtiss, North American NA-73 and Seversky/Republic designs, it

World conditions, start of the draft in America and other factors forced rapid changes at Selfridge Field and other bases. Fine hangars and other buildings constructed by Pres. F. D. Roosevelt's administration proved to be mightily fortuitous as war threatened. Seven YP-43 and P-43 Lancers were being prepared for pilot proficiency and indoctrination flight. This scene was in stark contrast to remembrances of less than a decade earlier when the author's uncle, Capt. Harry Tunis (AC Res.), was thrown from his nearly new, uncowled, fabric-covered Boeing P-12 near Selfridge in a midair collision. Injuries ended his career. (C.A.M./Ed Boss, Jr.)

seems rather clear new NACA data about the location of main cooling scoops for liquid-cooled engines influenced all three manufacturers equally. The NAA design was the most advanced of the three aircraft, but by the same token it must be realized there would be no response from the Inglewood firm. The design had come along a few months later to meet requirements of the British Purchasing Commission. In its initial conception, the AP-10 was a very conventional airframe aft of the wing trailing edge. It really only had three outstanding features: 1) very small size, with a wing span of 31 ft. 1 in. and a length of 29 ft. 1 inch, 2) only two .50-cal. machine guns which could not even fit within the fuselage mold lines (being mounted at the wing-fuselage juncture within contoured fairings) and 3) unmistakable Seversky-Gregor wing shape. Other than the initial wing planform, it had no recognizable Seversky features. Both the aft-belly-mounted

[2] Yes, that Preston Tucker. He tried to sell an armored car with a 37-mm AAC cannon in a plexiglass turret to the Army. His proposed lightweight fighter had a wooden wing, unproven engine (and not even tested, if built at all). Suffice to say, Preston Tucker was no James Doolittle. Had the French *Armee de l'Air* been able to field 1000 of their best lightweight fighters against the Luftwaffe in 1940, all would have been destroyed in a week.

Almost a year before the XP-47B ever saw the light of day, Chrysler Motors produced their own Thunderbolt in 1940. A very advanced design car with a retractable metal top, it was designed by famed le Baron studios and anticipated many 1950s automotive concepts. (Author's Coll.)

coolant scoop and the empennage were uninspired. The engineers issued Model Specification 88 on 1 August, describing the salient features of the tiniest possible combat aircraft that could be wrapped around an Allison V-1710-39 engine. The little scrapper was designed for a gross weight of no more than 4600 pounds. With at least 1090 to 1150 hp on tap it was touted as being able to attain a speed of 415 mph at 15,000 feet altitude, which appears to be a bit optimistic since that was a couple of thousand feet or so above the engine's critical altitude.

To place the AP-10 in a proper perspective, here are some very contemporary pursuit/fighter airplane comparisons (Gross weights and horsepower ratings can vary by as much as 6 percent, depending on the information sources; data by manufacturer, military model, wingspan, gross weight, horsepower):

Bell XP-39	35'10"	6204 lbs.	1150 hp
Bell XP-77	27'6"	3940 lbs.	520 hp
Curtiss XP-40	37'4"	6870 lbs.	1090 hp
Republic YP-43	36'0"	7800 lbs.	1200 hp
Curtiss XP-46	34'4"	7322 lbs.	1150 hp
No. Amer. XP-51	37'0"	8633 lbs.	1150 hp
V-S Spitfire IA	36'10"	6317 lbs.	1030 hp

Formal evaluation of design responses to the Circular Proposals was in the hands of a Board of Officers, with points being assigned or subtracted to give weighted averages. The board's report stated the AP-10 was unsuitable for mass production (meaning it should not be adopted as a service type fighter). It was recommended that the design merited further investigation in the form of an experimental vehicle. Materiel Division's Experimental Engineering Section, commanded by L/Col. F.O. Carroll, generally agreed but wanted several minor design changes, but these required raising the design gross weight to 4900 pounds, including the addition of 5 square feet of wing area, up to 115 sq.ft., and bomb racks to carry six 20-lb. fragmentation bombs or two 100-lb. demolition bombs. (One concept of the period was the dropping of small bombs into bomber formations...but proximity fuses were still years away.) The Armament Section people, perhaps handcuffed by congressional demands for adhering to World Disarmament Treaty concepts and out of touch with reality, insisted on a two-gun armament while Spitfires and Hurricanes carried eight rifle-caliber machine guns and the Germans and French were installing cannons without question.

Nearly as soon as the name Republic Aviation had been adopted, Contract AC-13436 was written in the amount of $461,800 for procurement of two modified AP-10s under the military designations XP-47 and XP-47A. The paperwork went to Washington for approval, but was returned unsigned on November 16. Reasons for rejection included (a) insufficient firepower, (b) high wing loading at 42.6 lbs/sq.ft. and (c) inadequate high speed as compared to the competing 355 mph Curtiss XP-46 design. (Wasn't the design speed of the XP-47 some 60 mph greater? Didn't the XP-47 fit two of Gen. Arnold's two favored ideas – namely very light weight and an Allison engine? Ah, yes, the old Curtiss-Wright magic of the 1930s was still a phantom influence factor.)

Back at Farmingdale the wind tunnel model/mockup took on a most unusual appearance. The left side of the AP-10 exhibited an entirely new wing panel of much greater span and extremely conventional shape with a mildly tapered straight trailing edge. Provision was made for two .30-cal. wing guns and one .50-cal. wing gun on each side of the aircraft. The belly air scoop was closer to the design adopted by Curtiss for the XP-46 than to the shape of the scoop employed on the NA-73X being constructed for the RAF. At that point, the fuselage remained as before but a new contact was written. Aligned with AAC classified project M-48-40, contract AC-13817 was approved on 17 January 1940 with an estimated cost of $616,700. The Allison V-1710-39 engine requirement did not change.

Designed in response to Air Corps Circular Proposal CP39-770, Republic Aviation's first new effort was the AP-10, given the military pursuit designation XP-47. The lightweight pursuit design was submitted to the military Experimental Engineering Section at Wright Field in August 1939. With a 29 ft. 1 in. length and a weight of less than 2.5 tons, the Allison-powered pursuit was considerably smaller than its near twin, the Curtiss XP-46. This RH side view of the mockup evidently represented the earliest firm configuration with tiny wing. The mid-position liquid-cooling duct was NACA-fashionable then. (Frank Strnad Coll.)

While a full-scale model of the AP-10/XP-47 went to the Langley Field wind tunnel, a mockup reposed in the west end of Republic Bldg. 5. Close analysis reveals that the RH wing had the familiar Seversky planform and was close to half span of a Boeing P-26A, while the LH wing resembled the Bell P-39 planform (distorted here). This wing panel shows two .30-cal. guns and one .50-cal. gun. There were to be two fuselage guns as well. A developed canopy would have mimicked that on a P-43. As shown on this side of the asymetrical mockup on 12 August 1940, span would have been 41 feet. Col. Howard Z. Bogart, then Chief, EES Technical Staff at Wright Field, provided this information. (Mfr. No. 1959/F. Strnad)

While all of this light-weight, Allison-powered nonsense was going on as if no Americans had ever seen or heard of cannon-armed Messerschmitt Bf 109D/Es or speedy, well-armed Vickers-Supermarine Spitfires, a "sleeper" idea was establishing a foothold in the shadows. Slowly, perhaps for the first time, the influence of aerodynamicist Costas E. "Gus" Pappas was beginning to be felt at Republic Aviation. Having joined the company in 1938 after graduating from New York University, it appears Pappas may have been the key to the canopy and fuselage improvements which first appeared on the XP-41 and the AP-4 airplanes. Flush riveted, butt-jointed skins were being embraced rapidly by the most progressive manufacturers such as Lockheed and Bell. High speeds suddenly made skin friction an important drag factor, and the day of the brazier head rivet for such airplanes was nearing an end. As has been mentioned previously, Kartveli's team was doing everything possible to leapfrog ahead and produce something faster and better than the competition. Major Seversky's pet AP-4 design, upgraded to the firmly entrenched YP-43, was evaluated with every new radial engine Pratt & Whitney or Wright had on the test stands. Fortunately for all Americans, East Hartford's P&W had some talent that came up with the R-2800 Double Wasp at exactly the right time and with the sometimes elusive capability of accepting turbosupercharging as a bountiful partner. (Wright's contemporary R-2600 never did show any affinity at all for such boost.) Target horsepower for the Double Wasp was the 2000 hp range.

The engine selected as the heart of a proposal (AP-4J or -4L) was the R-2800-7 which had a single-stage integral supercharger and was set up for turbocharging. Initially it had a takeoff rating of 1850 hp at 2600 rpm and a normal rating of 1500 hp, 2400 rpm at no less than 25,000 feet. When the data was presented to Materiel Division officials they were sufficiently impressed to issue a new Circular Proposal 39-770, dated 11 March 1939, aimed directly at Republic's Model Spec. No. 87. Contract AC-13380 was issued on 12 October, the day before the changeover to Republic Aviation became a fact of life, officially. Of course nobody else could produce a design to meet the proposal by the deadline, so a contract was awarded to Republic for 80 new-design P-44s. (Within the company, this was generally looked upon as the P-44-IV concept.) It was all very confusing.

Originally the interceptor was to attain a speed of 406 mph at 20,000 feet but an armament change reduced that figure to 402. A mockup of the P-44 was created from a severely reworked YP-43 airframe featuring a cowling shape that carried over to the ultimate P-47B design, except for retention of the two .50-cal. machine guns in the familiar position. Cowling shape was not really dictated by engine size. Oil cooling requirements and increased airflow for the D-2 supercharger and intercooler established those criteria. Even without the benefits of flight testing, it did not take long for the engineering people to acknowledge that the large single oil cooler installed in the

So few engineers are involved with most preliminary design proposals and so few such proposals ever even get to the mockup stage that nobody retained solid information about them. When any proposal is abandoned early in a program, little if any information is retained. This illustration, unidentified in surviving corporate records, may be the AP-4A or B, otherwise known as the P-44-I. It most assuredly falls somewhere between the AP-4 and the P-44-IV (AP-4J), and it does not feature one of the larger engines. (Mfr. via F. Strnad)

The strong heart of the P-47 (and the projected P-44) from the very beginning was Pratt & Whitney's R-2800 Double Wasp engine. An R-2800-5 engine is shown because it was similar to the -17 version used on the XP-47B. A P&W chart rated it at 2000 hp (Military), but only 1625 hp for a Normal category (S.L. - 25,000') rating. The manufacturer said only one -17 was built. The P-44-IV mockup evidently had a borrowed X-series engine, was to have -11 unit as in North American XB-28s. (U.A.C.)

mockup was but a crude stopgap design. (Bob Boyd's profile rendering of what the P-44 would logically have looked like appears in the color profile section. Unlike the single distorted illustration released more than fifty years ago by Republic, this profile is based on study of the mockup and other information never previously seen in print.)

Despite every iota of pressure applied by "interested parties" with influence, the aircooled engine advocates seemingly had triumphed over the liquid-cooled engine proponents. Neither Republic nor Curtiss had developed designs with any significant growth potential for an unknown future and, as history would eventually prove, the people who ignored Capt. Ben Kelsey's foresight with regard to high-altitude combat requirements had strayed down a dead end path. At the very best, Curtiss' P-46 design in 1939 was about where England and Germany were three years earlier.

General Arnold's lightweight interceptor concept might have sufficed for the French *Armee de l'Air* or defense of Britain's cities, but nobody can win a war by remaining in a defensive posture. After Curtiss-Wright had run out the string on the Curtiss D-12 through Conqueror in the mid-1930s, they were not inclined to spend another penny to develop a successor. Ford Motor Company had made some halfhearted effort to develop an inline aircraft engine, but it was possibly more of a "testing of the waters" than any real desire to proceed with full-scale development. Old Packard had been burned pretty well with their 1200 hp narrow-X engine for the Navy in 1928-29. However, when they were asked to manufacture Rolls-Royce Merlins in World War II, they went quietly about turning them out in great numbers without any scandalous failures. Unlike great failures on the part of Curtiss with the Caravans, P-47Gs, SO3C observation types and the SB2C Helldivers or Wright's dismal failure to even produce a viable prototype of their Tornado X-1800 cylindrical engine, Packard was almost too successful. They were not memorable. That mystical Wright, according to corporate information, could be built in a 35-cylinder form (5 rows of 7 cylinders) with an 1800 cu.in. displacement for 1800 hp, or in a six-row XR-2160 form (2160 cu.in.) producing 2350

The oil cooler installation for the P-44 mockup looked like a "shade tree mechanic's bad dream." Fortunately for all concerned, it never had to be flown. Big horsecollar firewall behind the P&W R-2800 illustrates, for the first time ever, how a YP-43 fuselage had to be reworked extensively for the mockup. Wright engine for the AP-4C/-D was approximately 2 inches larger in diameter, providing evidence of the reasons for its rejection. There were no production airplanes ever powered by turbo-supercharged R-2600 engines since they probably did not respond well to exhaust supercharging. (Mfr. via C.A.M.)

horsepower. If anyone ever saw the engine, even in static condition, they never came forward to say so and Wright Aeronautical seemed content to just let the stories of its existence fade.

How many Army Air Forces veterans of WWII have any concept of how close America came to not having an operational Pratt & Whitney factory when Pearl Harbor was attacked? Without the United Aircraft team to rely on, the North American AT-6 might have been far less functional; Consolidated B-24s would have had to rely on some engine which may well have compromised their capabilities; Grumman F6F Hellcats might have been hamstrung with a less capable Wright R-2600 engine, never having the slim margin of success they enjoyed over the best Japanese fighters. That story is worth telling here because of the narrow margin by which our key engine manufacturers survived in the Depression days as the War Department planners gazed fondly at the Allison engine which never seemed quite able to catch up with the league leaders. Army Chief of Staff General Malin Craig was no friend of the Air Corps and simply ignored pleas to have provide development funds for sorely needed multi-stage, multi-speed mechanical superchargers. The Germans and British were the only ones to devote time, energy and funding to such projects.

Even though the advocates of aircooled radial engines were still around to proclaim a modicum of success, they could hardly be attributing it to the foresight of our War Department officials who were making the key decisions in the 1930s. Perhaps the Air Corps high command can be blamed for overstating their case in favor of the low-drag concept of inline engines. Or it might also have been their overreaction to the supreme success of the Rolls-Royce Kestrel and Merlin engine development story inasmuch as the latter type burst upon the scene so spectacularly in 1935. On the other hand, development of the Allison V-1710 had been actively pursued since the beginning of the decade when single-cylinder examples were produced and tested successfully, but that was a world apart from manufacturing production engines in large numbers.

Whether it was due to the right officials being in positions of responsibility or to some other fact-or, L/Col. Edwin Page's Powerplant Branch was relatively successful in launching an entire family of so-called "Hyper" (for high-performance) liquid-cooled engine projects in the midst of the Great Depression. Continental Motors was funded to develop their horizontally-opposed O-1430 engine in, of all times, 1932. The resulting engine somehow evolved into one of Allison's major competitors when the prime outgrowth was an inverted-vee configuration along the lines of the Daimler-Benz DB series which proved to be so successful in Germany. Perhaps the biggest difference was in the use of mechanically-driven superchargers which, according to Ben Kelsey, had been effectively stymied by General Craig. The 1430 cubic-inch Continental in its inverted vee layout was identified as the XI-1430. (Since it is a known fact American engineers hired to work in Germany were major factors in the design of the Volkswagen, it would not be hard to believe some Continental Motors engineers were responsible for the later near success of Chrysler Motors XIV-2220 sixteen-cylinder engines powering two Republic P-47 experimental airplanes.)

ABOVE AND BELOW: Following receipt of a stopgap contract for a service test quantity of YP-43s, the newly created Republic Aviation Corp. quickly moved to propose an airplane called P-44-I (AP-4B). That projected successor to the YP-43s (13 ordered) was to be powered by a P&W R-2180 Twin Hornet engine but estimated performance was disappointing. Succeeding P-44-II (AP-4C and -D) versions were to have Wright's R-2600 engine generating from 1500 to 1700 hp, but it was an unhappy fit. Turning to the new P&W XR-2800 Double Wasp, which was assuredly compatible with turbosupercharging, they created the P-44-IV (AP-4J). One service test YP-43 was to have a "buried" supercharger (others were exposed), but that idea was shifted to the P-44-IV. An "off-the-board" contract was awarded, based on this metal and wood mockup using a reworked YP-43 fuselage. The eighty P-44-IVs were to be powered by R-2800-7 engines rated at 1850 horsepower. Technical data shows clearly that it was to be an interceptor...nothing more. (See color profile.) (C.A.M./Ed Boss, Jr.)

Never before seen in print (applies to all P-44-IV photos), here is the cockpit of the Republic P-44-IV mockup, company designation AP-4J. But shades of WWI, there are the cocking handles for two .50-cal. machine guns! Notice the "exotic gunsight" complete with crosshairs. (Sorry about that.) (Frank Strnad Coll.)

Also involved in the program, Lycoming Div.of A.M.C. (located in Williamsport, Pennsylvania,) was proceeding with a degree of success with its own "flat" engine, the XO-1230 liquid-cooled unit intended primarily for buried installation within the wings of bombers. As engineers in the Materiel Division saw it, there would be bombers with buried engines which contributed virtually nothing to the drag, only thrust from the propellers. Fighters were envisioned as having the engines buried in the wings or within the fuselage driving propellers via extension shafts and small gearboxes. One Hyper engine, the Wright Tornado, had the faint odor of fraud. It was to be a liquid-cooled radial engine that could be built with increasing numbers of cylinders for higher horsepower. Eventually there were any number of airplanes scheduled to be powered by the Tornado, including Republic's AP-12 and an outgrowth of that type known as the XP-69, about which more anon. The Tornado was the flying saucer of its day: rarely to be seen, evidently never photographed. The biggest mystery surrounding it lies in the apparent inability of any ranking officer to question the funding of the project or to demand its cancellation unless Wright could prove it existed and could at least pass a 50-hour operating test. Even the wartime Truman Senate Investigation Committee failed to raise the flag. One wonders what part Colonel Page played in suppression of information about the Tornado. (The tax-paying public has been defrauded by too many projects like it, the Tucker [Miller-engined] XP-57 lightweight fighter and Ford Aeronutronic's "Sergeant York" anti-aircraft vehicle about 50 years later.)

Whatever the reasons for this preoccupation with liquid-cooled engines, the overall effects on the production heart of America's aircraft engine industry flirted with disaster. This country was a sleeping, indifferent giant – a fact even recognized by the Japanese before WWII. The sum total of all military appropriations for aircraft, engines and research activity for the new fiscal year beginning 1 July 1938 amounted to just $122 million, but even that pitiful amount was *150 percent greater* than total appropriations for the previous fiscal year! It was as if nobody had even noticed Hitler's rapid rise to power in central Europe from about 1933, or the expansionist threat of Japan in the orient, or that wars had been fought in Ethiopia and Spain and China during the decade.

No more than 36,000 people in every job level were employed by the combined aircraft and engine industries in 1938. Industrial assets come to an appalling total of only $125 million. That was a *combined total* for all aircraft and engine builders in the 48 states!

One major brewery or motion picture studio at that time could boast of greater net worth than two combined industries whose disappearance could precipitate world domination by one man. Pratt & Whitney Aircraft was in dire straits in the autumn of 1938. While its engineering and development staff was producing powerful and reliable aircooled engines which were the envy of every nation on earth, some of its inadequate resources were being bled off into a realm with which the company had no real familiarity, namely a bevy of liquid-cooled engine projects for the Army Air Corps and the Navy Bureau of Aeronautics. Important as that was, it paled into insignificance when compared to the dearth of production orders. With no more than 3000 people on the payroll – engineers, executives, mechanics, foundrymen and the ever important machinists and tool & die makers – their backlog of new orders (primarily for commercial aircraft engines) had fallen to $8,700,000!!

Unless some huge new infusion of capital could be found somewhere in the 1937 Recession, which was acting more like a full-fledged depression, the company's officers concluded they would be forced to close the factories and discontinue operations by 1 May 1939. P&W may have had more employees at that moment, but the Connecticut firm had farther to fall.

***Although the appearance of EP1-104s with USAAC markings indicates that this Bldg. 5 assembly line photo was taken on 8 January 1941 well after the Republic Aviation takeover, the timing also indicates that the other line is producing P-43As instead of scheduled P-44s. That is evidence that many P-43 pursuits were camouflaged after delivery. Painted and unpainted Lancers seen at Selfridge Field in 1941 tend to confirm that.* (Mfr./Ed Boss, Jr.)**

Those same executives were hardly invigorated when, in April, the War Department unexpectedly awarded a huge production order for P-40 fighters to Curtiss in the face of the unbelievable performance of the Buffalo-built airplane in what had been touted as a contest for high-altitude capability. Allison (read that as General Motors Corporation) was recipient of a $15,000,000 contract to supply engines for the Curtiss products. At that time, 1st Lt. Ben Kelsey and General "Hap" Arnold had just about sold Washington's congressional people on the idea of procuring a quantity of Lockheed's Allison-powered YP-38 after the unfortunate loss of the one and only XP-38 prototype. Bell Aircraft was expected to receive substantial orders for its P-39 Airacobra if the prototype met or exceeded expectations.

P&W's anxiety increased in multifarious ways. To observe the rejection of their powerful and reliable Twin Wasp engine was unbearable. Informal information leaked to UAC management was convincing enough for them to believe the R-1830 would come out the winner of the competition in some development of Seversky's XP-41 or the AP-4. They were stunned. What

***With the War Department forcing the issue to reorganize the troubled Seversky company, the military was obligated to maintain at least an effective cadre of experienced employees. When the P-44, XP-47 and XP-47A were all canceled, the production P-47B was not going to be rolling off assembly lines in a huge new factory building for about 1½ years after the last YP-43 was flown. Once the smoke cleared, 54 plus 80 Lancers had been ordered. Here, a P-43A (with early AC number) prepares to fly at Luke Field, Arizona.* (USAAF)**

could be behind these actions which could easily doom East Hartford? Here was a company that had been producing more than half of the military aircraft engines turned out by American factories in 1938 which, by June 1939, would be forced to close its doors, perhaps forever. And what company do you suppose was deeply committed to development of the first engine in the United States likely to produce a reliable 2000 horsepower? It certainly was not Allison or Continental or Lycoming...or Wright.[3]

Help was on the way, but it was not from those who would have been thought of as "best of friends." It came from the other side of the Atlantic Ocean and in a backhanded sort of manner. France was in big trouble in 1938 at about the same time United Aircraft Corporation – comprising P&W, Chance Vought Aircraft, Sikorsky Aircraft and Hamilton-Standard Propeller – had fallen on hard times. A series of weak governments, backed by archaic military thinking and planning, had seduced the French populace into the belief that nationalization of industry by the Socialists would bring prosperity, happiness and respect to all. Aircraft and engine design and production were partners in chaos. True, the aircraft industry was producing a few of the most efficient-looking and beautiful military aircraft in the world, but for the most part their products were among the leaders as the most ungainly, ugly and impractical warplanes of any period in history. They generally matched up well with the Maginot Line (fortress) concept. French-designed liquid-cooled engines were inevitably far behind those of any first-line power in reliability and power output. And although their best manufacturers had produced some exquisitely small-diameter, twin-row radial aircooled engines with high power output, the reliability factor bordered on the ludicrous. And productivity in the country was such that the flow of completed units was minuscule. The leading airpower in the world in the 1920s and early 1930s was in disarray and could hardly be expected to hold its own against the Italians, should it come to that.

Adolf Hitler had created an impressive monster just five years after his takeover of the German Government, threatening widespread takeovers of weak adjoining nations. Conditions on the European Continent were accelerating toward a state of crisis. French Premier Edouard Daladier was forced into being somewhat realistic upon his return from one of those peace missions to Munich. He had become a charter member of that famous old "If" Society.

"If I had possessed 4000 airplanes," he sighed, "there would have been no Munich."

That confrontation with the Nazis was the equivalent of a bully facing off with you as he stands on your toes. It is an untenable position to be in. Of course it is easy to deride the French and British for their lack of determination, but Americans were at more than arm's length away with a huge lake between ourselves and the bully. Jarred into reality by the events confronting him on a daily basis, Daladier ordered his aides to seek an immediate solution to the daunting problems plaguing the *Armee de l'Air* and the all-but-crippled French aircraft and engine industries. Better late, he conceded, than never.

By some magic coincidence, Pratt & Whitney's representatives in the Paris office had previously convinced French officials to authorize a formal type certification test of the R-1830 Twin Wasp engine. All agreed it could be rewarding. Against the best contemporary engines produced in France, the Twin Wasp was an outstandingly successful competitor. But those results had led to a seemingly insignificant contract for two million U.S. dollars. However, at the time, the contract constituted about 20 percent of the company's entire backlog. Furthermore, the Daladier aides were aware of a specific American presence. Then the Nazis precipitated a further crisis in February 1939, and alarm among the French grew rapidly. They may have finally realized what Hitler's real intentions were. He was going to continue pressing until he got exactly what he wanted unless he could be stopped. That was unlikely since Nazi power was growing more rapidly than Great Britain's and France's combined. The hour was late, but there was still a chance they could defend themselves even though it was far too late to deal from a position of strength. With time running out, the French element of the Joint Purchasing Commission in America signed a much larger contract in the spring of 1939, containing options for even larger procurements. All of those options were later exercised, and by the time Germany's *Wehrmacht* eventually crossed the border into Poland, more than $84,000,000 in orders had been booked by the East Hartford factory.

Faced with the Arms Embargo Act, Pratt & Whitney management committed themselves to an all-out expansion program, a risky business not backed by Washington. Back in 1938, their monthly production had averaged about 70,000 horsepower (a common measure of production in that industry). Early in 1939 a war plans committee formed by the company had calculated existing capacity at 450,000 horsepower per month. It was then necessary to calculate a commitment as to what was required to increase effective capacity to no less than 1.7 million hp per month. An unconstrained French Purchasing Commission agreed to finance construction of a modern quarter-million square foot addition to the East Hartford facilities. Fortunately for all concerned, the Embargo Act was soon relegated to history. P&W's shipments jumped to more than 400,000 horsepower per month, a six-fold increase by the end of 1939. The survival and viability of America's most important aircraft engine manufacturer had been assured, not by the War Department or Navy Department in Washington, but by the decisions of a Parisian at the head of the French Republic.

Those actions ensured a flow of R-1830 engines to Republic Aviation to meet schedules and produce P-43 fighters for the Army Air Corps, preventing the Farmingdale, N.Y., company from locking its doors, too. It also enabled P&W to continue some accelerated development of their experimental R-2800 Double Wasp engine, a powerplant certain to be ten thousand times as important to America's role in history as all the "Hyper" engines combined. Neither Wright Aeronautical nor Allison engines allowed the Army Air Corps to keep Republic Aviation from skidding into obscurity. How the War Department expected to keep that aircraft company viable while evidently ignoring the plight of United Aircraft Corporation can hardly be understood. No written record explains it, and any truthful explanation went with the demise of those responsible for the actions. All actions do not necessarily stem from the obvious. *Fortune* magazine published an article in wartime saying Alexander Seversky had far greater impact on aviation as a result of his published writings than he ever did as an airplane builder. And perhaps they were correct in reaching that conclusion because there is no way to measure what actions occurred because of his impact. But what about the simple creation of his company and development of airplanes leading directly to creation of the Thunderbolt and the part those fighters played in the defeat of the Nazis? That impact is beyond comprehension.

There can be little doubt about the significant role played in history by the Republic P-43. Its importance cannot possibly be measured by its military combat record, nor could any longevity record for the Seversky AP-4 be set as a fine example. But what if it had not been "invented" by the Major? Nobody can even point to a single successful predecessor. At this late date there are still rumors among former Lancer pilots about orders actually existing prior to Pearl Harbor or immediately thereafter which would have sent the men and fighters to the

[3] Should anyone have any doubt about such happenings, what became of such household names as Essex, Hupp, Graham-Paige, Pierce Arrow, Cord and Duesenberg in the auto industry? Aircraft companies which failed include Detroit-Lockheed, Verville, American Aircraft & Engine, plus a host of others.

Philippine Islands to participate in their defense. If a group of P-43s had been dispatched immediately after the "Battle of the Carolinas" was a fading memory, how would they have performed against the invaders? Would Air Corps logistics have handcuffed use of the Lancers as it did some other types? Is it possible the actual performance of the type and of young, inexperienced pilots could have stemmed the invasion tide and the fall of Corregidor? Probably not. Imponderables such as the lack of preparedness, inadequate early warning systems, poor maintenance, indifferent command, ammunition shortages were involved. Two decades of peacetime tactics and training were unlikely to aid in turning back a tide of an enemy already experienced with years of combat in China and Manchuria.

Performance of the Republic Lancers would hardly have been superior to the performance of an America which had slept through reveille in 1933. Or at least in 1938. Adequate tools had been there. The spirit was unwilling.

***Somewhere in Louisiana in 1942, a pilot learned that groundloops could not be outlawed, lengthened tailwheel strut or not. Well, that damage did not even come up to the "deductible" on the State Farm policy. Republic's contract stated that the P-43 had a guaranteed top speed of 351 mph at critical altitude, but that was most likely without camouflage paint.* (Norman Taylor Coll.)**

CHAPTER 8

THUNDER ARRIVED QUIETLY

When Adolf Hitler finally showed his hand in the waning days of summer 1939, the War Plans organizations of every major nation tossed their 1-, 5- and even 10-year planning documents into the trash bin. Figuratively, of course. Anyone can plan for war, plan for depressions, plan for earthquakes, but there have been few modern-day successes in knowing just how, when and where they will hit. Americans would really get a taste of that philosophy a bit more than two years after Hitler made his move.

Planners in England had counted on having more time to develop a new generation of aero engines in the 2000 hp class, but the Nazi machine activities threw schedules out the window. The rather smallish Napier company was using their favored concept of the "H" block format, essentially two horizontally-opposed flat engines stacked in parallel, connected to a common propeller shaft. Yes, it was complex but it was developing more than 2000 hp. They called it the Sabre. Over at Rolls-Royce they had been developing their Peregrine engine, essentially an upscale Kestrel. (The Peregrine was the engine selected for the RAF's Westland Whirlwind fighter, Great Britain's contemporary of Lockheed's P-38. Suffice to say, only 112 Whirlwinds were accepted by the Air Ministry. The rest were canceled.) The Peregrine engine was never a real success, and the R-R idea of mating two in an "X"-block layout to create a 1760 horsepower powerplant named Vulture was not a premium concept. An engine with that heritage could be expected to act like the bird of prey too. In essence it picked on the Hawker Tornado fighter and the Avro Manchester bomber projects until they expired. Looked upon with disdain by Hawker's design chief, Sydney Camm, and the engineers at Supermarine, the Bristol Centaurus aircooled, sleeve-valve radial engine was in a class with P&W's upcoming Double Wasp. It ultimately proved to be the best of the three British engines. So much for planning.

Backtracking a bit, we pick up the traces of the P-47 project in its lightweight fighter stages. Things had a bad habit of changing rapidly after Labor Day 1939 became but a memory. Some years later, retired L/Gen. Laurence "Bill" Craigie was good enough to relate his personal view of those changes.

"I had worked for (Major Howard Z.) Bogert in Experimental Engineering, primarily as Project Officer (like Ben Kelsey) on cargo transports, personnel transports and trainers. Things happened so fast as the war progressed, I was never even a colonel. Although I had been a lieutenant for years," he recalled with some affection, "I jumped directly from Lt. Colonel to Brigadier General." (Moves from lieutenant to Lt.Col. had been remarkably fast. Ben Kelsey went from captain to colonel almost in one jump.)

When the Contracts Section of Materiel Division (later, Command) refused to approve the paperwork for development of one XP-47 and one XP-47A in November 1939, Majors Howard Bogert and Franklin O. Carroll – the main players in Experimental Engineering – convened some conferences in an attempt to overcome the objections behind the rejections. Capt. Kelsey was no longer in that picture, if he ever was, because expansion programs could not possibly be handled by one officer anymore. He had his hands full with the P-38 project and other assignments. Capt. Marshall S. Roth had entered the P-47 picture at some point, but he had never been able to define his assignment at that time (his records had been destroyed).

Hart Miller, as assistant to Republic Aviation President Wallace Kellett, was at Wright Field to negotiate changes to the contract and to aid in resolving performance problems as the Thanksgiving Season neared. Those technical conferences resulted in specification changes, forcing a rise in gross weight of the XP-47 to 6150 pounds. An increase in wing area to 165 sq.ft. dodged the problem of unacceptably high wing loading. Actually the wing loading was reduced a bit to 38.8 pounds per square foot, but the ridiculous bomb provisions refused to go away. By that time the armament had grown to a level seen in the Curtiss XP-46 and the P-43A; in fact, some versions of the XP-47 had nearly twice the armament aboard the Lancer! Objectivity requires at least a peek at some other contemporary foreign lightweight fighter specifications for comparative purposes.

Bloch MB-700	700 hp.	Gr.Wt. 3858 lbs.	Wing Span: 19 ft. 7.6 in.
C.A.O 200	860 hp.	Gr.Wt. 5511 lbs.	Wing Span: 31 ft. 2 in.
Caudron C.714	450 hp.	Gr.Wt. 3858 lbs.	Wing Span: 29 ft. 5 in.
Caudron CR.770	800 hp.	Gr.Wt. 4740 lbs.	Wing Span: 28 ft. 6 in.
Dewoitine D.551	1100 hp.	Gr.Wt. 4740 lbs.	Wing Span: 30 ft. 7 in.
Potez 230	670 hp.	Gr.Wt. 3968 lbs.	Wing Span: 28 ft. 8 in.
Roussel 30	690 hp.	Gr.Wt. 3891 lbs.	Wing Span: 25 ft. 5 in.

The rare Roussel had the look of a half-scale Republic P-43. Certain of these fighters, such as the Dewoitine D.551, were not viewed in France as "lightweight" fighters at all but as their standard pursuit aircraft. Most of these types, including the

Sometime before autogiro magnate Walter Kellett joined Seversky Aircraft Corp. to interject some production and management expertise, Alex Kartveli's engineers designed the AP-10 (military designation XP-47). Powered by an Allison engine – like its competition, the Curtiss XP-46 – the product turned out to be a 4600-pound midget that must have made General H.H. Arnold happy. It went through many design creations before military planners were jolted into reality by Nazi successes in the spring of 1940. Alex "Sasha" Kartveli (c.) and C. Hart Miller, Dir. of Military Contracts (l.), were summoned to Wright Field early in June. Advised that the XP-47 and XP-47A pursuits were canceled and that the P-44 type was inadequate for the job, they were given new specifications of what was wanted – on a "crash" program. The ultimate response is pictured behind Miller, Kartveli and Maj. Russell Keillor, AC Plant Representative. (Author's Coll.)

***There can be little doubt that the Republic XP-47B would turn out to be the most evolutionary – certainly not revolutionary – of all successful fighters in WWII. In virtually every respect it was a growth-factor extension of the Seversky AP-4 of early 1939. In power and size the AP-4 was a two thirds-scale version of the ultimate design in an era when such scale models were often used to test concepts before final commitments were made. Even more to the point, the propeller, powerplant, production cockpit area, supercharger installation and even the landing gear were mere extensions of the AP-4J/P-44 mockup. The wing planform and even the original Clark YH airfoil used from 1930-31 – with an improvement originating with the Seversky AP-9 (S3 sharpened leading edge) – were unchanged. By contrast, the XP-38 that was laid out four years earlier was wildly imaginative.* (Mfr. 2408)**

terribly outmoded Morane-Saulnier MS.460, a fabric-covered simple contemporary of Seversky's P-35 in age and performance, were very short on range, protection, power and reliability. Most of these types never even attained production status.

Evidently nobody had spoken to General Arnold about the vast differences in requirements for most defense/chasse aircraft in Continental Europe as opposed to what was needed in America. Distances and missions were entirely different, something which the general (not being an engineer or fighter pilot) had totally failed to consider. The relatively new French light fighters were not moving down production lines in 1939. But even the best *Chasse*-types were totally outclassed by Messerschmitt Bf 109Es. Scattered reports concerning effectiveness of Dewoitine D.520s against American fighters in North Africa during TORCH in North Africa can be dismissed easily. They only faced P-40s and some Grumman Wildcats for a matter of days at the most. That is far afield from months of combat against Messerschmitt Me 109Fs. The D.520 fighters were certainly the best French defensive airplanes in production when France was invaded, but they might not have fared well against Republic EP1s. They surely would have been outclassed by high-flying P-43s.

At the conclusion of the Wright Field meetings, Model Spec. 88B was issued to reflect the XP-47 changes. Somewhat amaz-

***In contrast to the P-44 cockpit and published for the first time, the prototype XP-47B office looked like this, complete with centered test controller panel. Gone are any provisions for cowl-mounted guns. In 1942, test pilot Fillmore Gilmer had to bail out of this (reworked) cockpit at more than 400 miles per hour when his rudder and elevators burned away. This picture was taken two days before the XP-47B made its initial flight from Farmingdale on May 6, 1941.* (Mfr.No. 2411/F. Strnad)**

ingly, the contract price and original delivery dates held over from 4900-pound XP-47/XP-47A airplanes remained unaltered. The unarmed XP-47A was scheduled to be delivered within nine months (appropriate?) of Contract AC-13817 issuance. The Assistant Secretary of War signed the papers on 17 January 1940.

During the days when the "Roosevelt Recession" was rapidly fading from view, artist Frank Tinsley provided the illustrations (and probably the story lines) for a magazine, as I recall (without confidence), named *Bill Barnes Air Adventures*. Fighter aircraft named "Snorter" and "Scarlet Stormer" had all the wildest racing planes and foreign aircraft features rolled into one, including retractable floats. Alex Seversky had not even been so bold as to attempt such a feature on his fighters. Nevertheless, he and Kartveli had submitted serious proposals for construction of the Hughes Flying Boat of its day. They were far ahead of Howard and his associate, Bill Kaiser. The Seversky "Super Clipper" was a 120-passenger, "8-engine Lockheed P-38" with retractable floats! Were they serious? It followed that Martin and probably Boeing and Sikorsky had their own competitive proposals (and that was not in any way connected with the Boeing 314 derivative of the XB-15 bomber, circa 1934). As was their habit in those days, Pan American Airways had invited the proposals and probably breathed a sigh of relief as French, British, Dutch and Belgian forces were crushed by the *Wehrmacht* and the *Luftwaffe*. After deriving some publicity from the scheme, they could back out gracefully. Pan Am probably never had any intention of buying quantities of such monsters. It was also the fashion of the time for Air Corps artists at Wright Field to release designs presaging the combat aircraft of some future time.

When Seversky was no longer a factor in what emanated from the design room in Farmingdale, the Kartveli design team came out with their own Model AP-12 which was certainly low on drag coefficients and high on armaments. It did *not* feature a retractable float, although Blackburn Aircraft in Great Britain gave that theme a try with an experimental patrol aircraft. Despite the issuance of an inboard profile drawing which did everything but flash the "Untrue" sign, a few people have claimed the fairly large fighter was to be powered by the ubiquitous Allison engine. As any engineer would ask, why try to put contrarotation propellers on a V-1710 engine? Nobody needed to do that, and such frivolities in pre-war days were inviting defeat. If the AP-10 lightweight fighter was barely happy with the V-1710, the big AP-12 was going to have trouble getting off the ground. And why put the engine in a canister? Well, that was most likely the closest impression of a Wright Tornado 35-cylinder XR-1800 engine anyone is likely to see. More disturbing, there is no real sign of supercharging. Whether it was Kartveli, Pappas, or some other engineer who had the concept, the empennage would have made many pilots and flight controls people ecstatic. It was near optimum for the period and at least some of that feature showed up in two other Republic aircraft, the XP-69 and the P-84 jet fighter.

Quite frankly, it is fairly obvious the XP-69 (two ordered, but canceled because the engine failed to materialize) was the direct outgrowth of the AP-12. Republic's designation for the big escort fighter was AP-18, actually designed around the Wright XR-2160-3 Tornado 42-cylinder engine. Sadly, that non-engine was responsible for defeating the XP-69, paving the way for the less viable conception known as the Fisher XP-75. The AP-18 was a much better design with heavy armament, and it would have been far more productive from a manpower and facility standpoint to revamp it around the R-2800 aircooled engine or a pair of Allison V-1710s coupled at the propeller shaft by something like Menasco's Unitwin drive. Then, unlike the V-3420 Allison situation in the Fisher XP-75, the AP-18 could at least cruise on one V-1710. If Northrop managed the aircooled R-2800 in their XP-56, a combination of Republic and United Aircraft propulsion men surely had the talent to solve any installation problems in the XP-69. It would be safe to assume aerodynamicist Costas "Gus" Pappas had much to do with the AP-12 and AP-18. And there can be no doubt the Wright Tornado had put its hex on yet another airplane.

Strange thinking seems to have prevailed within the Navy and War Departments relative to prototype aircraft. Secondary projects warranted construction of two prototypes (as with the XP-47 and XP-47A and the Curtiss XP-46 and XP-46A) while only one XP-47B and one Grumman XP-50 were ordered. Those in charge seemed to have ignored the loss of the XP-38, Boeing's 299X (XB-17) and the North American NA-40X early in those programs. The sole prototype XP-47B, serial AC40-3051, came within a whisker of loss on its first flight, and the Grumman XP-50 was lost after but a few flights. Lt. Ben Kelsey's advanced armament concepts of 1936 were ignored in his absence (on other assignments) on nearly every pursuit development for about 4 years. European events in springtime 1940 awakened those "Sleeping Toms" to the fact that Kelsey had been more than 100 percent right. The new Thunderbolt was required to have six (optional eight) machine guns (Spitfire style), but .50-cal. bullets had far greater hitting power than any rifle-caliber bullet. Only four of the P-47 airplanes had those automotive-type access doors. (Author's Coll.)

Rapidly changing military requirements were now approaching full gallop. Suddenly the need for self-sealing fuel tanks and armor protection for the pilot, not to mention ever heavier armament, became critical items. Less than 30 days after the signature of the Assistant Secretary of War was inked on Contract AC-13817, Materiel Division was pressing Republic for those very changes. But hold your horses! Design gross weight

Sometime late in 1941, the first production P-47B Thunderbolt was ready for test flight. Like virtually every combat type of aircraft in the new Army Air Forces, it was cloaked in drab camouflage broken only by the star and meatball insignia. At the extreme right, in the background, a reminder of the Depression still existed with AMERICAN still emblazoned on that landmark smokestack. That was a remnant of the American Airplane and Engine Co. which had been facility owner since the dark days of 1931 until Hitler disrupted world peace. Centered in the background is the new Republic Bldg. 17, gleaming in peacetime paint. This P-47B, AC41-5895, was one of two production airplanes fitted with the car-door-type-access canopy. Proving that all P-47Bs did not have a tall, forward-raked HF antenna mast, this one features a mast like the one fitted to the XP-47B. Rudder featured upper mass balance weight. (Author's Coll.)

was projected to rise to nothing short of 6400 pounds. Who was in charge of this exercise in futility? One trip to the research records files would have revealed Republic's XP-47 had reached a higher weight than the much older Bell XP-39. Design gross weight of the latter type with the same engine included a long extension shaft required to drive the propeller and a heavy-weight cannon. Arnold's dream of a lightweight fighter was now bumping against a gross weight exceeding that of the RAF's Spitfire IAs. Somebody had at least abandoned the archaic light bomb requirement and it was finally gone for good. Once again the wing area increased, this time to 170 sq.ft.. Of course there would be an attendant schedule slippage, this time in the form of a 60-day increase.

Despite the pressures, Kartveli's designers completed the drawing changes for slightly under $10,000, an amazing feat by any standard.

Precisely neat, pilots and Bell P-39Ds of the 31PG(F) line up for inspection at Selfridge Field in 1941. Those men were unaware that war would descend on them so soon. Missing from the P-39s were the 37-mm cannons and wing machine guns. Component failure rates in the Carolina War Maneuvers were at unacceptable levels for P-39s and Douglas A-20s. Bell cockpit doors habitually came off in flight. (AFM)

As a major element in the corporate reorganization of 1940, Hart Miller was appointed Director of Military Contracts, a newly created position. He had barely found his desk when Materiel Division issued Contract Change Order No.1. It was formally approved on June 1st. The company was scrambling to define the XP-47's configuration, design and build the first P-44, manufacture and deliver the first of thirteen new-design YP-43 service test airplanes to the Air Corps. And, of course, Republic had to build and deliver 120 of the bread-and-butter EP1s to Sweden and get moving on yet another order for 2PA-204A light bombers for the *Flygvapnet*.

With the company having to expand rapidly from a work force of less than 200 in the first quarter of 1939 to at least ten times that number in a year, largely via the recall route, it did a very good job of delivering the first Swedish EP1-106 two weeks before Christmas Day, 1939. All of the fifteen fighters covered by the initial Swedish contract had been delivered by 15 January 1940 and fabrication of the next group of 45 moved along nicely with all 60 fighters delivered by June 2nd. Production of the first two dozen light bombers (Model 2PA-204) was behind schedule as the initial unit was delivered on 2 May, a day after all 24 were to have been handed over to the Swedish representatives. Production continued without any interruption on the additional 28 bombers as final assembly of a second lot of sixty EP1s continued on the parallel line in Building 5. Dade Brothers had a circus tent erected near Hangar 1 on the east side of the flying field to expedite the dismantling of completed EP1s and 2PAs because the hangar was too small to accommodate the workload. All 2PAs on the first contract and nearly half of the bombers on the second contract had been completed and sold off to the Swedish representatives when the U.S. Government requisitioned every airplane not yet shipped. All 60 fighters on the first two contracts had been exported along with two 2PAs. About two-thirds of the 60 EP1s on the third contract had been completed and delivered into Swedish hands at Farmingdale (really a technicality) and a large number had been nicely crated. The requisitioning occurred on 24 October 1940

Sad to say, Robert Woods was not the equal of Reginald Mitchell or Clarence Johnson. When he designed the Bell P-39 in lockstep with Larry Bell, he obviously had the right idea but lacked finesse in technique. The landing gear and armament requirements (specified by Lt. Kelsey) were just right – true state-of-the-art for 1936-38. Blame deletion of the turbosupercharger on NACA and a new project officer. Blame Woods for using an obsolete wing and failing to lengthen the aft fuselage or not using a ventral fin, both suggested by Kelsey after making the first XP-39 flight. Canopy style was excellent, execution bad. Doors often came off in flight. This P-39D from the 31st Pursuit(F) Group was participating in the 1941 summer war games in the Carolinas. **(AFM)**

As America was perched on the brink of conflict in the late months of 1941, a typical 311-mph upgraded Curtiss P-36C looked like this. A contract for 210 of the P-36As was awarded as a result of the 1937 "Pursuit Competition," but political influence and cost per unit far outweighed performance factors. It was the identical thinking that resulted in a huge Douglas B-18 bomber contract. That company's XB-18 had been outperformed by Boeing's Model 299 (the original B-17) by a significant margin. As the P-36 contract was signed, Great Britain already had the Hurricane and Spitfire in mass production. **(Author's Coll.)**

as a result of the Embargo Act. It is quite probable the EP1s already crated were the ones shipped to Luzon in the Philippines in December 1940. Not necessarily well suited to the role of advanced trainer, most 2PAs were redesignated AT-12 and found their way to training centers in the far southwest. If their service capabilities as trainer were unappreciated, it did not interfere with their acceptance by many senior officers as flying staff cars and rapid transports.

All sixty of the EP1s were promptly reidentified as Republic P-35As, and at least 45 of them must have been shipped to the Philippines where they were promptly assigned to the 17PS in the 4th Composite Group, a motley collection of Boeing P-26s, Curtiss P-40s and a few Lockheed C-40s. The group had Curtiss O-52s and Douglas O-46s, high-wing observation airplanes of dubious capabilities in late 1941. Douglas OA-4s and Grumman OA-9s also served.

DAWN OF THE REAL P-47 THUNDERBOLT...

It was the beginning of a new decade, one of hundreds or thousands that have marched on irresistibly at a well-regulated rate. Based on what had been experienced in the second and fourth decades of the century, we knew it would be exciting but not necessarily to our liking. Probably the most daunting question: Will we actually be here at mid century? Millions would not have that opportunity.

So-called "major" factors had once been something to generate excitement. By 1940 they were coming so thick and fast, all had become major or minor, depending on a person's interests. Some of the most fateful decisions were to be made in the spring of 1940. Struggles at Republic centered around major attempts to get the XP-47 firmly ensconced in the pursuit plane picture at a time when the Curtiss XP-46 appeared to be progressing toward procurement in a lively manner, if someone tended to believe reports (or were they rumors?). At that same time, in an unconcealed investigation of the entire fighter (pur-

Realistically and definitely sadly, it must be emphasized that the prototype XP-40 flew for the first time more than three years after the 8-gun Hurricane made its initial flight. Yet this 1942 picture shows that most available P-40s, as with this no-suffix-letter P-40, packed only two machine guns and lacked nearly every necessity for combat worthiness. Alexander de Seversky, at the time, tried to alert the American public to such dangers through his book "Victory Through Airpower." Totally misinformed "experts" (e.g., scribe David Brown, writing in PIC magazine in 1942) sought to smear the Major. Their "facts" would have you believe P-40s could clobber Mitsubishi Zeros easily. **(USAAF)**

Shown in company with his Republic P-35A, probably at Nichols Field in the Philippines as of July 1941, Second Lieut. John Geer was lucky to have survived his fighter days to move on to bombers rather than perish in combat. He relates that his final flights in the P-35As that remained flyable were only for reconnaissance missions. John survived the war to fly B-47s before retirement as a colonel. He felt far more comfortable flying bombers. (John Geer)

suit) program, the Emmons Board – named for its chairman, M/Gen. Delos C. Emmons, Chief of the GHQ Air Force (then the striking force of the Army Air Corps) – was delving into every facet of the experimental fighter program. On 19 June 1940, the Board issued a critical report which somehow received endorsement from M/Gen. Henry H. Arnold, Chief of the Air Corps. Here was the very man who seemed to be pushing hard for the Allison V-1710 engine and the Hyper engine program, not to mention the foundering lightweight fighter projects. His blessing was valuable because Arnold was a man with "clout" in the halls of Congress and the War Department.

One key target of the Emmons Board report was the Allison engine because too much of America's pursuit (fighter) airplane program hinged on "one type liquid-cooled engine."[1] If it was not evident to everyone, the real purpose of the board was to come up with a "no holds barred" recommendation about which fighters should be produced in quantity for combat operations. Finally someone had the guts to question earlier decisions supporting procurement of only one main type of engine powering every Army Air Corps production fighter to the virtual exclusion of all other engines. The board strongly recommended expedited development of fighters powered by aircooled radial engines and, if possible, having other sources to develop and produce liquid-cooled engines. Incredible! Finally – close to five years after the Rolls-Royce Merlin was certified to produce 1000 horsepower while at least three U.S. firms struggled to develop 1000 hp engines – some of the best brains in the Army Air Corps realized they had all eggs in but one basket. If the idea to have a British engine manufactured in this country was not on the agenda of the general's committee, it must have been suggested by somebody within days.[2]

Essentially anticipating the findings of the Emmons group, the Experimental Aircraft Division of powerful Materiel Division requested the appearance of Republic Aviation's top engineers and decision makers at Wright Field to discuss new fighter requirements and revision of existing programs. Of course "Sasha" Kartveli and C. Hart Miller responded by heading for Ohio on the night train. Capt. Marshall "Mish" Roth (M/Gen., USAF Ret.) was at the conference as representative for Major L.W. Craigie (L/Gen., USAF Ret.), who was at that time the assistant to Col. Howard Bogert. Roth recalls Capt. Benjamin Kelsey (B/Gen. USAF Ret.) was away on a mission centered in London, England, to evaluate the performance of all combat aircraft actively participating in the war.

"I can't recall after all these years," said Gen. Roth during an early 1970s interview, "exactly who attended the conference and my records have been destroyed." He did remember, however, "there was a great deal of discussion about the fighter concepts that came out of that meeting. Many of the old timers were appalled at the idea of a fighter that would gross out at six tons," (the weight of a Martin B-10 bomber).

Along the way, Kartveli had been asked to upgrade the P-44 design with a higher-powered version of the R-2800 engine, namely the -11 model which had only recently passed its tests for a military rating of 2000 horsepower. Fighter aircraft requirements were changing so rapidly it was becoming clearly evident early introduction of leakproof fuel tanks, protective armor plate and heavier armament in the existing P-44 airframe was infeasible because of limitations in the wing and fuselage structures, not to mention lack of space and increased wing loadings. With events in Europe dictating the ground (and air) rules, Kartveli and Miller could hardly argue against the need to meet the new requirements or for the idea all could be introduced within the envelope of the existing interceptor. Without dissent they agreed to the specific new requirements and the impossibility of accomplishing the job without a much larger airplane. Those parameters could not be met without a design gross weight of at least 11,600 pounds. Materiel Division's specifications now called for a speed of 400 mph at 25,000 feet[3] on

---

[1] This was a very obvious reference to the Allison V-1710 engine.

[2] Back in 1936-37 when Lt. Ben Kelsey was seeking a cannon to be part of his new fighter specifications, he actively pursued sources overseas because nothing was available in America. He was evidently the only advocate of such procurement which eventually resulted in the production of Hispano-Suiza 20-mm cannons in this country. But, let's face it; he did not have huge, powerful U.S. manufacturers producing a competitive product. If the Emmons Report was the trigger that led to Packard manufacturing Merlin engines, General Emmons deserves our undying gratitude.

[3] Your attention is drawn to the competition and contract of 1939, just ONE year earlier, in which the designated winner of a so-called High-Altitude fighter contest

Fourth Composite Group Republic P-35As, including 17PS commander, Captain "Buzz" Wagner's No. 17 at right, had heavier armament than the one flown by John Geer. But it was to little avail. Maintenance had been neglected to such an extent that the P-35As could not be regarded as combat weapons, except in desperation. (S. Hudek Coll.)

Military Power rating and an armament of not less than six .50-cal. machine guns. There was no thought at all of putting any machine guns in the fuselage; all *shall* be in the wings, and the preference was for eight guns (alternate armament). The archaic light bomb concept had already been trashed for the XP-47 "lightweight" fighter.

On one hand, these requirements seemed radical to some of the newer officers who had joined the expanding Experimental Aircraft Division at Wright Field, mainly because it is psychologically difficult for people to accept change. On the other hand, anyone on the cutting edge of technology – especially where life or death warfare is involved – is forced to accept change or lose the battle. It would appear that people at the

had a critical altitude rating on the engine of less than 15,000 feet. Full armament in the winning P-40 design was two machine guns mounted on the cowling, and one of those could be a .30-cal. gun. It would come as no surprise to learn some brave soul had told General Emmons, "This is a hell of a way to run a railroad." It appears Emmons had been pushed as far as he was going to go.

Pilots manned their new P-43s, ready to participate in the 1941 Carolina War Games at what was obviously a newly constructed airfield. These First Pursuit Group Lancers from Selfridge Field shared the maneuvers base with Lockheed P-38Ds in this rare photo. Crosses, ranging from green to red, were painted over the new fuselage national insignias. The newly devised tailwheel installation had been suggested by Lt. Mark Bradley, eventually assigned as Project Officer on the P-47 program. (Frank Strnad Coll.)

conference had forgotten (momentarily, perhaps) that the Lockheed YP-38 service test airplanes were nearing completion at Burbank. In fact, the first YP-38 would be flying within four months. The developed Lightning had a similar maximum gross weight and firepower (established as criteria by Lt. Kelsey four years previously); yet it still did not feature armor plate or self-sealing fuel tanks as of that date. Designer C.L. Johnson had confirmed all such improvements could rather easily be accepted within the envelope and performance parameters of the P-38 design. In Kelly Johnson's situation, the only real problem could be in artificial power restrictions (something that did not become a serious problem until the P-38G model was in production).

Pratt & Whitney's success in producing a new high-horsepower radial engine as the decade of the sluggish '30s ended spawned several advanced-design-concept aircraft. Martin, for example, broke completely with design tradition in producing the sleek B-26, seen here at the Middle River plant. Two big R-2800 Double Wasp engines, swinging two of the first new constant-speed 4-blade propellers, were enclosed in tight-fitting cowlings when the Marauder lifted off on its first flight on 25 November 1940. Surprisingly, the rapidly expanding Air Corps had ordered a huge quantity of the new B-26 medium bombers "off the drawing board." The initial aircraft assembled was a production B-26, breaking a precedent. When that first Marauder left the ground, there were only about a dozen aircraft in the Air Corps that were faster. **(S.I.)**

A LEGEND MUST END ...

More than a half century after the events occurred, almost every author who tells a story of Republic Aviation's historical events persists in stating Vice-President and Chief Engineer Alexander Kartveli designed the XP-47B Thunderbolt and proposed it to the Army Air Corps. In a word, that is not what happened at all as has been ventured in the immediately preceding paragraphs. Only the rarest of successful military aircraft have been "invented" by any designer or, as they say, "were made from scratch." It would be nice to believe the outstanding C.L. "Kelly" Johnson had designed the Lockheed XP-38 from his own ideas and pure concepts. It just does not happen that way. Johnson responded to a Request for Proposal (RFP) and a Circular Proposal under the immediate direction of his Chief Engineer, Hall H. Hibbard. It was Lt. Benjamin Kelsey who set up a specification document which strayed afar from conventional wisdom. But, even at that, Ben had been directed to have a design competition and he did not, *per se*, design the XP-38 or the (Bell) XP-39. But he did unlock the barn and show the designers working for various companies how to get in and what they were supposed to do. The best twin-engine interceptor designs came from Lockheed and Vultee, and the Johnson proposal had a clear advantage...even though some at Wright Field doubted it could achieve its stated goals. That it did in fact exceed those guarantees made the airplane a "quantum leap forward" as Ben Kelsey was wont to say.

Anyone who has ever been "on the inside" in engineering management or in Preliminary Design departments knows it just does not work in the way writers have enjoyed glamorizing for decades. In the same niche, test pilots do not climb into the cockpit and immediately try to see what the absolute top speed is at some impossible altitude. It makes for very palatable legends. But it just ain't so.

There is, in the case of the Thunderbolt, a lengthy paper trail from 1930 until the meeting in 1940 clearly revealing just how *evolutionary* the P-47 fighter was. No fighter anywhere in World War II was more evolutionary than the Thunderbolt. Does anyone even have a thought that Indianapolis race cars of note have sprung from a purist idea in any year? Rarely. The British Cooper rear-engine car was one of the few trendsetters. Although Lockheed's P-38 and Messerschmitt's Me 262 made the greatest leaps forward in their particular eras, even they were partially evolutionary.

As we shall show, Kartveli and Miller arrived in Dayton with the XP-47 lightweight fighter and the powerful P-44 interceptor branded into their minds. When General Emmons' staff officers revealed how the "game" had changed in a matter of weeks or, perhaps, months because of events elsewhere in the world, they backed their discussions with a set of new ground rules – essentially, pages of new *specifications*. In lockstep with nearly every invitee who attended that conference, neither Kartveli nor Miller had any idea what type of aircraft was going to be required when they walked through the doorway.

No XP-47 or P-44 in any shape, form or manner was going to meet those specifications. So, given the powerplant available

and what it had to haul around at 400+ mph at 25,000 feet, what was the shortest, fastest route to follow? Kartveli settled on the classic 300 square feet of wing area needed to accommodate the given weight and expanded what was an excellent design to do the job. Why reinvent the wheel and perhaps wind up with an XP-75??? Did he or anyone else have the nerve to commit to the preliminary design of the AP-12? Not unless there was some great desire to emulate those investors on the pavement of Wall Street, circa 1929 and 1930.

When Alexander Kartveli, young Hart Miller and Capt. Marshall Roth boarded a train (of course they had at least one compartment) to head in the direction of New York, there were written agreements that all work on the Allison-powered XP-47 and XP-47A would cease. A formal Change Order would follow to amend the contract. Furthermore, all development work on the P-44 would be ended within days, and work was to be initiated immediately and with priority to design and manufacture an XP-47B powered by a Pratt & Whitney R-2800-11 aircooled radial engine[4] with a guaranteed Military Power rating of 2000 hp at 2700 rpm and a critical altitude rating at 25,000 feet.

As the train sped toward the East Coast, the three men talked of what was in hand. They were being given the latest version of the Double Wasp engine, they already had a wing of proven design and if it had increased in area once it had done it a half dozen times, and they had a first-rate cockpit and canopy that was hardly going to warrant a great deal of design effort. It was good enough for the P-43 and the new P-44; nobody had even brought up the subject of a cockpit, so why try to fix it? All wing area changes would occur inboard, so they would gain needed space within the fuselage by moving the wheels outboard. With eight large machine guns and about 4000 rounds of ammunition required, there would be zero chance of putting fuel or oil in the wing. It seemed more than unlikely they could improve much on the S-3 airfoil. (If they knew anything about the new Hawker Tornado and Typhoon or the Westland Whirlwind, they should have been ecstatic at what they already had in hand.) An improved G.E. turbosupercharger had been promised and increased heat rejection would be simple to calculate. The biggest problems would be flight controls, weight & balance, fuel capacity and attendant oil capacity plus a redesigned landing gear. Their best aerodynamicist, Gus Pappas would be right in his element.

On May 5, 1940, the prototype XP-47B was rolled out of Hangar 1 and parked not far from the low chain link fence alongside the 2-lane roadway on the east side of Republic Airport. Nearby was a new AT-12 about ready for delivery to the Air Corps. The only marking on the pursuit plane was the standard U.S. ARMY painted on the lower surface of the wingspan. The fabric-covered rudder was coated in Army camouflage paint. The rest of the aircraft was all polished Alclad metal. By the time any ordinary photographs were taken in flight, a 1941 camouflage scheme had been applied and the red meatball was painted out on all national insignia. Before this airplane was lost in a crash in 1942, it had attained a speed of 412 mph at 25,800 feet. (Republic)

What about the stories insisting Kartveli had designed the P-47 on an envelope with General Arnold in attendance? Pure fiction. Here are the quoted words from a responsible Air Materiel officer who was sitting directly across from Republic's Chief Engineer in the roomette on the train. "Kartveli sketched up the XP-47B design on (board) the train," General Roth recalled vividly. "It was drawn on an envelope at that time; not at Wright Field and not at Farmingdale. I know because I was there."

At a different time and place, Hart Miller told me all of the preliminary work, up to September 1940, was accomplished without a contract or formal coverage. As newly appointed Director of Military Contracts, he was really not worried and expediency paid off. By June 12th the Model Specification was revised to reflect Kartveli's promises, identified as Spec. No. 88C. Procurement action had been initiated on July 10th under Authority to Purchase rules, leading to Change Order No. 2 dated 15 August 1940 on Contract AC-13817.

The fixed price of $486,760 awarded in connection with the lightweight XP-47 and XP-47A was augmented by $63,752 to take care of cancellation costs and to cover the single XP-47B equipped with a P&W R-2800-11 engine, three sets of blueprints,

[4] There is cause for considerable confusion about the precise engine installed in the prototype XP-47B. A report written by civil service Project Engineer Paul B. Smith states the R-2800-11 engine was specified, confirmed by an Air Corps technical report dated 1943. On the other hand, Pratt & Whitney's documentation lists the -17 and -35 engines as being installed in the XP-47B, and the Materiel Division Board of Inspection & Survey report confirms a -17 engine was actually in the XP-47B as of 14 April 1941. Furthermore, Change Order No.4 to the contract called for substitution of the -21 engine for the -11 (but that was not an on-site inspection comment, only a routine contract change). Most books and articles have reported the XP-47B was powered by a -35 engine on its first flight. The only report that could be totally relied on for accuracy would be Flight Test's airplane and engine log. It has not been located.

OPPOSITE: Almost totally ignored in history – certainly unknown to most – North American's XB-28 was very nearly a clone of the Martin B-26 in size, shape and engines. However, the XB-28 was turbosupercharged, had a fully pressurized cabin and featured remotely-controlled gun turrets like those used on the later Douglas A-26 and Boeing B-29. With nearly 2000 horsepower available from each Double Wasp coupled with great high-altitude capability, it was as fast as the Luftwaffe's Focke-Wulf Fw 190A fighter. Apparently the success of medium-altitude B-25 and B-26 types in the same medium bomber category, coupled with the surging demand from every direction for P&W's R-2800 to power successful new types in many classifications led to cancellation of the XB-28 project. In retrospect, it seems illogical to have committed so much effort to many downright unsuccessful, dead-end experimental and production programs as exemplified by Vultee's A-31/A-35 Vengeance and Curtiss's C-76A, various XP-60s and the XP-62 (3 years) while quickly abandoning the B-28. (Mfr./File)

Although the threat of West Coast invasion had not evaporated, at least until the Battle of Midway was history, and as some of the best brainpower in the USAAF labored to establish the logistics realities needed to successfully transfer entire groups of fighters (never before attempted), bombers and transports by air to Europe, the first flight formation pictures of P-47 Thunderbolts were released. Lowery Brabham leads in (Y)P-47B 41-5902, Joe Parker follows in (R)P-47B 41-5930 and Victor Pixey brings up the rear in the XP-47B. The 170 production P-47Bs were not deemed suitable for overseas combat and had soon gained the "R" prefix placing them in a Restricted category for operations. (Mfr. via Author)

stress analysis report, technical manuals, a 5-foot wind tunnel model and a spin-test model. That Change Order was definitized on September 6th. Scrapping all the earlier work on the Allison-powered P-47s was estimated to cost $60,000 including an Air Corps request to complete the full-scale wind tunnel model and mockup and transport it to NACA at the Langley Memorial Aero Laboratory in Virginia. NACA was to investigate engine cooling requirements, which it did, and pass the data on to "other manufacturers" (no doubt meaning Curtiss and North American Aviation).

No mockup of the XP-47B was requested or constructed, but a cockpit mockup was fabricated and subsequently inspected by Project Officers Major Laurence W. Craigie and Capt. Marshall Roth on 30 and 31 October. They reported their finding to M/Gen. Oliver P. Echols, Chief of the Materiel Division. A slightly mystifying situation existed in connection with the cockpit mockup. There is every indication that the entire production cockpit component (including windshield, canopy, structure, skin and everything down to, and including, the floor but not the control panel or consoles) for all razorback P-47s duplicated that employed on all P-43s. But the prototype and three other P-47Bs had a very different (non-sliding) canopy system with a left side door that was a cross between the Bell P-39 door and a much smaller door system used on Hawker's Tornado. For emergency exit, the right side door was essentially the same except it did not include any portion of the canopy top. Again, contrary to all previously published information by others, the four airplanes using this unusual canopy were the XP-47B, the first two production RP-47Bs (R meaning Restricted Use) and the last RP-47B which was completed as the XP-47E with a pressurized cabin. The mystifying parts: Who suggested the change and enjoyed enough influence to get Republic to try it on the initial group of fighters?; who canceled the idea and use?; and why didn't they stay with a canopy that seemed to work perfectly well?[5]

Work moved along rapidly and the real XP-47B was available for the full-bore engineering inspection conducted by a dozen engineers from the various laboratories at Wright Field. Their findings were reported to Mr. Paul B. Smith (Project Engineer), Major Ben Kelsey (who, as of that date, was with the Production Engineering Section) and Major Roth (P-47 Project Officer). The report stated an R-2800-17 engine was installed at the time the inspection was conducted. A handprinted notation to that effect also appears on the P&W Model Designation & Characteristics Report. Incredibly, considering the company history, the designers failed to include that omnipresent baggage compartment and door, a long-standing military requirement and fixture on all American military aircraft – since the beginning of the Army Air Service it would seem – so the inspection team specified it must be included prior to acceptance. That must have given Major Kelsey a real laugh. Would it still have to accommodate a golf bag and clubs at the very minimum? Somebody must have had the good sense to say, "ENOUGH!" A close look at the P-47D cutaway drawing is unlikely to reveal such a compartment.

Total fuel capacity (100 octane fuel specified) was measured at 298 gallons instead of the required 315 gallons. Fuel was not all carried in a single tank, the bulk being contained in the large tank immediately in front of the cockpit. As expected, there was no tankage in the wing at all.

[5] The question was never asked when I interviewed Gen. Craigie in the 1970s, and Gen. Roth did not recall it at all.

Good weather smiled on the XP-47B when it was parked outside of Hangar I during the first week of May 1941. Although the fighter was ready to go almost exactly a month ahead of schedule, it seemed to lack even minimal markings. In that same period, several Lockheed YP-38s had been completed out in Burbank, California, and each carried full Air Corps tail stripes and "meatball" style national insignia in the traditional manner (soon to be lost forever). By contrast, the people at Farmingdale seemed to be at a loss about the markings patterns to be applied. Although the bright aluminum skin remained in its natural state, the fabric-covered rudder was painted Army olive-drab. All other flight control surfaces were painted silver and the legend U.S. ARMY was painted in black letters across the lower wing surface in the same manner as it had been applied to YP-43s a few months earlier. There was not a sign of the four national insignia emblems on any wing panel.

During the daylight hours it was rarely more than 100 yards from the public highway running along the eastern boundary of the sod airfield, with only a 4-foot chain link fence separating the road shoulder from the hangar and that field. At one time the XP-47B was tethered to the concrete apron, the tail held down by a wide strap for high-power runup testing. The engine installed was equipped with a torquemeter.

Lowery Brabham, Chief Test Pilot under George Burrell, Manager of Flight Operations (having remained with the company after Seversky departed) brought the XP-47B to life on 6 May 1941 as he prepared to take off from Republic Airport for the first time. He completed the runup on the paved surface at the northeast corner apron with the half-million square feet of factory Bldg. 17 looming off to his right, still about a month away from completion, having been erected in record time during the winter months. The month of April had been a rather wet one, so the sod airfield, still without the least vestige of a paved runway, was in a somewhat questionable condition for heavy airplanes.

Of course Capt. Russell Keillor, Plant Representative for the absolutely new Army Air Forces, was an observer and Capt. Roth, as Project Officer was on hand to witness the first flight. When retired M/Gen. Roth was interviewed in the early 1970s he did not have any documents to refer to, all having been lost much earlier in time. But here are his comments as he remembered the operation:

"The initial gross weight – to the best of my recollection – was pretty close to 14,000 pounds, or some 2300 to 2400 pounds over the design gross weight of 11,600 pounds." (*In contrast, according to Lowery Brabham, the actual gross weight observed at Hangar 1 during weighing operations was "closer to 12,500 pounds." He did not say how much closer, unfortunately. One thing is certain. The weigh-in is a mandatory function before any aircraft is ever flown for the first time, so a record did exist somewhere.*) Weight control engineers were a luxury in those days. Lockheed had encountered exactly the same problem with the XP-38 when it was prepared for testing at March Field, California, at the beginning of 1939.

Gen. Roth continued his story: "The turbo installation was too heavy, as was the Curtiss prop, and there must have been 1000 pounds of extra beef in the landing gear." Just as Major Seversky was prone to comment in those contentious days, Kartveli was not an aircraft designer; he was educated as an electrical and structures engineer in France. Mish Roth confirmed that when he said, "Sasha Kartveli's strong philosophy about a rugged, solid aircraft had afflicted some of the engineers too deeply...and they went overboard. (Probably no 'tame' stressman would ever locate that fine line between strength vs. weight in the same place as a combat pilot would want it.)

The rapidly balding Brabham was not too concerned with all of this as he prepared to lift the biggest, heaviest of all single-engine, single-seat fighters into the blue for the very first time. He must have absolutely chortled every time he thought about his decision to terminate his military career and join Seversky when visions of that contract (AC-15850) for 773 of the P-47Bs and 54 smaller P-43s was signed less than 30 days after the "go ahead" was approved for creation of the XP-47B. Thoughts of the long-ago day when he and fellow pilot Hub Zemke had landed their P-6Es on this very airfield and met the Major must have at least flashed through his mind on May 6th.[6] This was a remarkable contrast to conditions prevailing just two years earlier.

The camouflage paint job used on the XP-47B (trailing) had evidently been rejected for production airplanes, while the (Y)P-47B leading featured the standardized scheme. A flat-plate armorglass windshield – similar to the type used on later bubble-canopy versions – had been installed experimentally on the airplane being flown by Brabham. Generally unknown until now, four Thunderbolts featured side-opening access doors: the XP-47B, No.s 1 and 2 P-47Bs, and the XP-47E. (S.I.)

In addition to the major construction project (a complete new factory) underway immediately across Conklin Street from

[6] As it ultimately worked out, only 170 of the P-47Bs were delivered to the AAF, with one additional B having been reworked into the pressurized cabin XP-47E. The remaining 602 airplanes were produced as P-47Cs (various change blocks were involved). All C models had design improvements, including a QEC (quick engine change) mount which resulted in lengthening the forward fuselage by 8 inches. A secondary benefit was virtually automatic compensation for tail heaviness. In the event, flight characteristics improved, producing another unexpected bonus.

old factory Bldg. 5, one of the most significant changes during the Republic years had taken place in top management. Former airline and aircraft manufacturing executive Ralph S. Damon was brought in to replace Wallace Kellett when he was elevated to Chairman of the Board. As president, Damon was a miracle worker, turning the company into a major builder of aircraft anywhere around the world. It is quite probable he adopted the highly regarded Lockheed System, a pioneer in so-called moving production lines and subassembly manufacture required to mass produce P-38 Lightnings. Strongly endorsed by the Materiel Division of the AAC (AAF in 1941) and the Office of Production Management (OPM) sometime in 1940, it became the model for many others. Republic and Ford Motor Co. (at Willow Run) were two of the prime examples of this concept in aircraft production. The Willow Run plant was not limited by real estate constraints and it benefitted from some of the better automotive production ideas. In addition it was designed to accommodate the much larger B-24 bomber from the start; therefore, it was (like Consolidated Fort Worth) a rather lengthy structure featuring some of the finest jigs and fixtures ever seen in any aircraft plant.

Mr. Damon, Kartveli, Hart Miller and Captains Roth and Keillor (both soon to be promoted) were in the group on hand to witness the flight of this impressively large fighter. Brabham lined up on what was soon to become concrete runway 4-22 (No.1). The prototype was backed into a corner almost immediately north of Hangar 1 allowing a bit over 3900 feet on turf before he was at Broad Hollow Road, at that time little more than a 2-lane country road.

Brab, as he was known, described the event during one of several interviews we had in later years. "Our field at Farmingdale had just been extended and there were no paved runways. Recent heavy rains had created a sea of soft spots which made landing there too risky, but a careful route had been selected for the first takeoff. Permission had been granted for the first landing to be made at nearby Mitchel Field, a long-established Air Corps field with newly extended (paved) runways."

The Double Wasp responded nicely to full throttle. Brabham had evidently held the brakes until he was showing 52 inches of manifold pressure. By the time he was at the intersection of runway 1-19 and immediately opposite the southeast corner of new Bldg. 17, which was to be finished in the following month, the tail was already up, Piper Cub style. He must have been off the ground in 2500 feet.

No airplane anywhere near as powerful had ever thundered out of Republic Airport. As it climbed into the crisp spring air, there was something the pilot did not know. A considerable amount of oil had accumulated on the exhaust manifold during a prolonged checkout period and additional oil coated the long stainless steel exhaust ducts leading back to the General Electric Type C-1 turbosupercharger. Without strong airflow around the cockpit during the checkout operations,there had been no indication of any problem. Although the R-2800 had an integral supercharger providing a certain amount of boost, much of the boost effect was lost in the ducting and intercooling system. Therefore, the turbo was employed for takeoff to overcome the losses, resulting in a bad combination of factors. That was not a good thing for a prototype first flight.

Here is what Brabham had to say about that important and closely watched event:

"Hot gasses for turbo operation came from each side of the engine and passed through a tailpipe on each side of the belly of the airplane. Shielding the hot tailpipes from the inside of the airplane were stainless steel shrouds. In order to control the amount of exhaust gases necessary to give the desired boost, there was a wastegate on each side of the engine aft of the cowl flaps, to 'spill' unneeded exhaust gases overboard or direct all exhaust to the supercharger.

"The waste gates were controlled by a governor which at low altitude was controlled by maximum manifold pressure and at high altitude by turbine rpm," he continued. "On the first

***General public awareness of the Republic P-47 Thunderbolt probably dates from the initial appearance on the cover and within the pages of a major national magazine in the winter of 1941-42. Pictured in a Christmas-like atmosphere at Farmingdale, the sole XP-47B appears to be ready for yet another test flight. The red meatball insignia virtually confirms the picture was taken just prior to the attack on Pearl Harbor.* (Life via T-Bolt Pilots Assoc.)**

No words or other views could be more effective in defining the importance of an engine in an airplane's success or failure. Even with an early series Rolls-Royce Merlin engine, beautifully sculpted P-51A aircraft should have been a match for any series Me 109 or Fw 190. And Lockheed P-38s could have been far more effective if powered with R-R Merlins. Bureaucratic decisions blocked Lockheed from even conducting tests, but millions were squandered on the G.M.-Fisher XP-75 travesty. (Author's Coll.)

flight, as soon as the waste gates closed and the oil on the pipes got hot, the pressure in the cockpit began to drop. The XP-47B had a one-of-a-kind canopy which open-ed to the side and could not be opened in flight. [Actually, two P-47Bs and the XP-47E had it too.] When the cockpit began to fill with smoke from the oil, my first reaction was to open a small sliding panel on the door [left] side of the canopy. This only made matters worse as it pulled more smoke in.

"Of course you immediately think of getting out as there must be fire somewhere. But you don't see any. Maybe it's best to wait 'til you see flames. But maybe she'll explode before you see any flames. And Long Island is pretty densely populated [in 1941?]. It would be necessary to go out over the ocean or Long Island Sound to get out. All of this races through your mind," he said.

"Thoughts of going back to Farmingdale were vetoed by the knowledge there were no paved runways and recent rains had made the field soft in many places. "Landing there was too risky."

With a lot of flight test hours under his belt, Brabham did not panic and actually began to make some observations about the flying qualities of the prototype while he was engulfed in smoke. He remembered his thoughts of abandoning the ship continued. "I made a very careful examination and evaluation of the flying qualities of the airplane. After all," he opined, "this was to be the number one fighter of the war, and if it was lost on that first flight with nothing known about it the program would have been set back disastrously. In the 15 or 20 minutes that the ship was airborne, I believe that I could feel its potential and later stated the opinion that the XP was to be one of the greatest fighters ever. Sticking with the aircraft, the first landing was made at Mitchel Field without any difficulty."

The XP-47B remained in a hangar at Mitchel for several weeks while alterations were made prior to the second flight. It was never delivered to Wright Field, deviating from what had been standard operating procedure. Viewed as a production prototype, it was assigned to Republic on Bailment Contract AC-27839 for the purpose of aiding in the solution of production problems. Accordingly, no official performance tests were to be conducted until the production program stabilized. However, prior to the unfortunate loss of the airplane in 1942, a preliminary high speed run had been made at an altitude of 25,800 feet. While pulling some 1960 horsepower, verified through interpretation of the (nose) torquemeter data, the airplane had attained a true airspeed

Every flyable Republic P-47 flight test airplane is depicted in this undated photograph, and the brand-new manufacturing building is shown in the background. The huge new factory, with Bldg. 17 being the key portion of sub-assembly and final assembly, was close to completion when the XP-47B made its first takeoff. The unpaved runways, after any rainstorm, could not safely be used to land a P-47. When this photograph was taken, all runways for a first-class airport had been completed along with thousands of square feet of taxiways and ramp areas. Several test Thunderbolts were painted zinc-chromate yellow overall for easy air-to-air recognition. The camouflaged Thunderbolts are shown equipped with experimental 205-gallon drop tanks. (Mfr./F. Strnad)

(TAS) of 412 mph. That speed equaled the corrected speed attained by the seven-ton, overweight XP-38 in the first quarter of 1939 on what was essentially the same total horsepower at approximately the same altitude. By one strange coincidence, the XP-38 was lost in a crash landing a few hundred feet from the Mitchel Field runway just east of Farmingdale when the Allison engines failed to respond to the throttles. Pilot Ben Kelsey was completing a record-setting speed dash from March Field, Calif. in seconds over 7 hours.

Of course it was unlikely anybody at Republic knew anything about England's Hawker Tornado, the RAF's newest fighter. That medium-altitude single-seater was about 15 mph slower than the XP-47B according to recorded information. Hawker's Typhoon, a near clone of the Tornado except for a different engine, was essentially a 400 mph airplane, but surely not at 26,000 feet.

Suddenly, and rather quietly – at least for the next 6 months while mass production got underway in new Building 17 – Damon and Kartveli had performed the minor miracle of helping to put American aviation back into the position of being a world contender. How could it be that two aircraft manufacturers literally 'hanging on the ropes" in 1938-39, neither of which had

Following the demise of George Burrell in the P-47B, Lowery Brabham was appointed Chief Test Pilot and Operations Mgr. early in 1942. Newly promoted vice-president Hart Miller (on wing of YP-47B) discusses the test program with Brabham. That flat-plate windshield with built-in mirror was not adopted as a standard item prior to the changeover to the bubble canopy configuration. **(C. Hart Miller)**

Seen at the controls of the semi-official YP-47B is Military Contracts Director C. Hart Miller who was involved in many early test flights. This airplane was serving as the service test "mule" for Republic, and may well have been the P-47 (AC41-5899) that suffered loss of the empennage on March 26. Tail number cannot be seen. **(Mfr./Author's Coll.)**

built more than about 100 airplanes in their history, were to put America back at the top?

Trying to trace the career moves of the more senior officers in Materiel Division or any part of it in a period corresponding to the Battle of Britain and, subsequently, the years immediately following December 7 at Pearl Harbor is difficult. In fact, without individual service records or diaries it is nearly impossible. There was layer upon layer of expansion, especially at Wright Field because it was the center of all aircraft development for the Army Air Corps and then the Army Air Forces when it came into being in the spring of 1941. Laurence Craigie became chief of all aircraft projects in the Experimental Aircraft Division. By that time Materiel had become a command. "Mish" Roth was Craigie's assistant in his assignment as Fighter Projects Officer. In a few short years the annual production of all military aircraft went from a mere 1141 in 1936 to 2141 in 1939, an uninspiring performance. With war raging in Europe in 1940, production of the military types rose to 6019, then zoomed to 96,318 in 1944, a performance never again to be equaled by any nation. A permanent landmark.

Such expansion was difficult enough to manage, but there was a major multiplier in technological progress. One overview example: Several manufacturers worldwide were producing fabric-covered biplanes in 1939; by 1944 there were several producing jet fighters capable of top speeds in excess of 550 miles per hour.

Major Roth said he had made one flight in the XP-47B prototype. In fact, he said he was the very first Air Corps pilot to fly it. "I was really amazed at the rate of roll which could be achieved with this very, very large piece of machinery. Its response to the controls was excellent," according to his recollections. (Other airplanes that he remembered flying included the Republic XP-47F and Focke-Wulf Fw 190A in 1943 and the Republic XP-47N in 1944.)

When the XP-47B accomplished the feat of flying at a TAS of 412 mph, it was powered by an R-2800-35 engine equipped with a torquemeter nose case. For various reasons, it had never been possible to develop more than 1960 horsepower output from that particular engine. However, shortly after production P-47B airplanes became available for experimental development work at Farmingdale and at Wright Field, the situation began to change. One of the initial P-47Bs from production revealed just what performance the prototype might have exhibited had it been equipped with the more definitive R-2800-21 engine.

Thunderbolt production work began in new Bldg. 17 before final construction was completed. Workers were either recalled or hired anew for training. Grading of the airfield did not start that summer. Paving work was yet to show much in the way of progress when the first production airplane was moved onto the airfield on or about Thanksgiving Day, 1941. Following a rather standardized military airport plan using a conventional A layout, but featuring a slanting crossbar, Republic eventually had three major paved runways and a reasonable amount of apron area. A small AAF installation was constructed at the south end of the field between the two key runways in 1942. There were two hangars for defensive P-47s and taxiways providing quick access to those two runways. The new cross runway forced a major rerouting of the old road on the east side of the airport. Cleverly named New Highway, it became a busy thoroughfare. A major 1940-41 expansion of the Ranger Aircraft Engine Co. facilities engulfed the original building at the southeast corner of the airport, completed even before construction started on Republic's tremendous expansion. A large segment of old Conklin Street disappeared under the Ranger plant, forcing traffic to permanently detour around the new engine factory.

In the meantime, there were some rather strange happenings. The Swedish *Flygvapnet*, "robbed" of its entire second-buy

***When production specialist Ralph Damon took over as president of Republic Aviation before the XP-47B made its initial flight, it must have been a stroke of genius on the part of someone in power. His management techniques were exactly what the corporation needed at a critical time. Neat rows of P-47Bs are shown in Bldg. 17 of the rapidly expanding Farmingdale facility soon after the Pearl Harbor episode. The fighter was producible; the team was hot; and the WPB approved production at no less than three major plants...Curtiss Buffalo included (a sad blot on the entire affair). Rising from obscurity, Republic was the equal of its Long Island associate, Grumman Aircraft, and absolutely made the New York firms of Curtiss and Brewster look like infamous quislings, even if they were only inept.* (Republic)**

quota of sixty EP1-106 fighters and the fifty 2PA-204 light bombers, turned to Italy and purchased a similar quantity of Reggiane Re 2000 semi-copycat versions of the P-35. Apparently nobody is going to locate data proving the Italians illegally copied the P-35 design in some unknown manner. The improved aerodynamics closely match the Seversky AP-2, lending credence to rumors they bought AP-2 plans and/or the slightly damaged AP-2 well before the Neutrality Law was written. Nobody can duplicate a design so accurately without having one example at hand or a set of drawings. Considering Italy's close alliance with Germany, its part in North African aggression, and the massive military support provided for Spain's Gen. Francisco Franco, the latter idea makes sense. If the Major had dealt openly with Reggiane, Seversky Aircraft would have disappeared with hardly a trace. Pure speculation, of course, and "it never happened." Of course it didn't.

Back in the southern states, a massive war game exercise was put in motion as the "Battle of the Carolinas." A fairly impressive contingent of Republic P-43s, decorated with large green crosses was engaged in that "battle." Was that training exercise more important than the defense of our key Western Pacific bases in the Philippine Islands? The War Department had previously made a judgment leading to seizure of 110 Republic airplanes already sold to Sweden, a neutral nation. Those airplanes were required, according to the Embargo Act, to fulfill a critical need for defensive aircraft. Redesignated P-35As, most were shipped westward to Nichols Field on Luzon, P.I., circa January 1941. If the need was so critical, why didn't the newer P-43s follow closely behind to substantially improve the defensive position? The assorted Republic P-35As of the 17PS, sometimes equipped with .50-cal. wing guns, were assigned to the 4th Composite Group. Young Second Lieutenant John Geer, just

recently having left cadet status, was a member of that squadron. By his own admission, John was not terribly anxious to be a "pursuit" pilot.[7] He really preferred being assigned to bombers.

[7] Retired Col. John Geer, having flown P-35As and other outclassed aircraft against Japanese opposition, was fortunate to be included in a group departing Corregidor for Australia at the last minute, early in 1942. Soon transferring to bombers, he eventually piloted Boeing B-47s in later days. Many years later he was in the author's propulsion systems technical manuals unit working on the Lockheed L-1011 program.

ABOVE: Less than 5 miles away from Farmingdale, Grumman's Bethpage plant turned out some experimental F6F Hellcats about a year after the contemporary XP-47B first flew. This version was also powered by Pratt & Whitney's spirited R-2800 Double Wasp (in this case, a new -27). It was originally supposed to have a Wright R-2600-16 engine with the experimental Birmann turbosupercharger, but in final form with the P&W engine, it was the XF6F-3. As with most U.S. fighters, the spinner eventually disappeared. (Americans developed the radial engine; why were the British, Germans and the Japanese successful in using spinners on radial engines while they rarely worked on U.S. powerplants?) (Grumman)

BELOW OPPOSITE AND RIGHT: SCOOP! On March 26, 1942, just three years and a month after Lt. Ben Kelsey crashed his Lockheed XP-38 on a golf course near Mitchel Field, N.Y., the tail assembly on the fifth production P-47B suffered a major structural failure only about 3 miles from the Kelsey crash site. Totally out of control, like Ralph Virden's YP-38 some four months earlier, the Thunderbolt fell into a sand trap on Salisbury Golf Course. The Chief Test Pilot and Operations Manager at Republic, George Burrell, leapt from the falling fighter at about 200 feet altitude. The P-47 broke its back on impact, and a fire erupted but was quickly under control. Burrell died on the way to the hospital. His parachute did not have time to deploy. (Ed Boss, Jr.)

Just about seven months after the XP-47B was airborne for the first time and less than a fortnight after a welcome "Turkey Day" in fateful 1941, the importance of that 1940 P-47 project definition meeting in Dayton began to sink in. General Emmons must have had a vision. His decision to draw the line and retreat no further in the face of the General Motors-Allison juggernaut took considerable fortitude and an impressive understanding of the world situation. After all he was close by when General Andrews voiced his unwelcome opinions about American airpower. In fact, Emmons was the immediate beneficiary of the punitive action, but he could have become the victim. For the few who understood the special implications of Thanksgiving Day 1941, the appearance of the first production P-47B on the furrowed sod of Republic Airport was a real blessing. Looking exactly like a camouflaged XP-47B, it was in almost every sense a duplicate. The black numbers on the tail read 15895, but were nearly undiscernible against the olive drab paint scheme. That airplane and its twin, serial numbered AC41-5896, were a bit different from subsequent P-47Bs in having cockpit canopies which duplicated the XP-47B design. They also were easy to distinguish from all other Bs on sight because the radio antenna mast was short and straight up. As the first production airplane sat alone and forlorn on a gray, cold weekend, it was at least backed up by the massive new assembly building. Yet it created a stark, lonely landscape compared to what the same viewpoint would reveal in just a year or two. The clock was ticking in a relentless manner and a deadly Japanese action with a world-shaking purpose was already in motion. Perhaps the appearance of the lonely Thunderbolt – probably not yet then its official, formal name – was meant to be an omen. The timing would soon be seen as a remarkable occurrence, actually uncanny.

The chain of command in the new Army Air Forces in 1941 still required some untangling if you take even an abbreviated glance at the structure. L/Gen. Delos C. Emmons, as commanding general of the Combat Command, was subordinate to the new Deputy Chief of Staff (Air) and Chief of the Army Air Forces, M/Gen. H.H. Arnold. Another major general, M/Gen. George H. Brett was also subordinate to Arnold, although he was officially Chief of the Air Corps. By that time it was purely a training and operations command which also featured M/Gen. Oliver P. Echols as head of the Materiel Command. As if that was not confusing, within a matter of weeks there was to be a very non-military lieutenant general in the midst of it all when OPM chieftain William Knudsen became L/Gen. Knudsen.

Sunday, December 7, 1941, certainly got Lt. John Geer's attention. Whether or not he got to fly on that day, the 8th on that side of the International Date Line, is unknown. But as the days sped by, he did fly some reconnaissance missions in his P-35A which evidently still had Swedish-type instruments aboard. His few flight sorties in a combat situation were essentially for reconnaissance purposes. By then they were down to a handful of flyable P-35As and not much else.

John did outlast his airplanes though, and that was something to be thankful for. With such a daunting start, fatal to so many, it is remarkable that he survived to carry on in a later "Cold War."

On the afternoon of 26 March 1942, the Thunderbolt program took a very hard jolt, providing what must have been a long, sleepless night for Sasha Kartveli, Ralph Damon, Hart Miller and a host of others. The jolt came from an event totally unknown until now to a vast array of former P-47 pilots, history buffs and about anybody with an interest in World War II. Although local press coverage was nearly as extensive as Lt. Benjamin Kelsey's 1939 crash in the same area received, it was barely mentioned in aviation magazines appearing a couple of months later. Good reasons existed for quieting all reports of the crash event because it involved the newest of America's fighting airplanes. Like the Kelsey/Lockheed XP-38 crash event almost exactly three years earlier, the scene of this particular crash was the local Salisbury golf course in Westbury, close enough to Mitchel Field to have their fire trucks arrive at the scene as quickly as the local Westbury and Carle Place units. It was only natural for military police to take charge immediately, and in a wartime environment it was certainly unlikely they would be

Just weeks after the first production P-47B rolled off the Bldg. 17 floor, it arrived at Wright Field and was assigned to the Service Test team (most likely operating out of Patterson Field). After just a few weeks of flight testing, the B model was deemed unsuited for combat operations. Part of the problem stemmed from difficulties in effecting an engine change. The redesign resulted in a lengthened nose that first appeared on C models; an extra benefit was improved maneuverability. The Burrell crash was caused by incorrect stress analysis computations, resulting in a major structural weakness that ended in complete separation of the empennage from the airplane. (WF Photo)

challenged. Although some exceptionally good photographs were taken at the crash scene even as firemen fought the blaze for 15 minutes, very few were published in newspapers. (They appear here for the first time in any book.)

Coming in a period when Republic Aviation was involved in a "battle" to overcome very serious production delays, it was a terribly unwelcome happening. Setbacks had been encountered as company employees struggled to meet initial delivery schedules. Ralph Damon was brought into the picture sometime in May 1941 in an effort to overcome the problems being encountered. Rapid expansion within the AAF following the Japanese attack on Pearl Harbor and the Philippine Islands had assigned Major Mark E. Bradley as P-47 Project Officer when Marshall Roth's responsibilities expanded quickly. Bradley, as a prime example, was quickly promoted to the rank of lieutenant colonel. He had been SPO for the Curtiss P-40 program at Buffalo, N.Y., for many months and had possibly retained that responsibility in addition to the P-47 program.[8]

Republic Flight Operations Manager George W.Burrell, the man involved in the Seversky EP1-68 and 2PA-BX sales demonstration tour of Europe in 1938-39, had elected to fly the fifth P-47B (AC41-5899, one of the semi-official YP-47Bs according to a Service Engineering document) on March 26th. He was attempting to determine the cause of a reported anomaly, wishing to get the airplane "sold off" to the Army. Flying above a 10-10ths cloud cover, he found it. In a moderate flight-load maneuver, the entire empennage broke away from the fighter. It is probable the old bugaboo Q-factor was encountered as the result of a sharp maneuver at low altitude. (In this event, the main cause was a very serious miscalculation on the part of stress engineers, a perfect example of Murphy's Law that could have had far more serious consequences but for Burrell's test flight that day.) As the mortally wounded P-47 fluttered earthward without direction, Burrell cut the switches and struggled to bail out. Although the newspapers claimed he was attempting to avoid a crash into a housing project (an impossibility under the circumstances), the aircraft's gyrations probably pinned him in the seat. He finally was able to exit at about 200 feet altitude. His parachute did not deploy completely and he fell to earth about 300 feet from the sand trap into which the Thunderbolt fell. When the fighter struck the ground, on a slight ridge, the airplane's back was broken. Fuel lines ruptured and the volatile gasoline ignited. Foamite spread by the Mitchel Field crews was most likely the major factor in controlling the fire so quickly. The 42-year-old Burrell died in the ambulance taking him to Mitchel Field, the medics totally unaware of his status as a civilian test pilot.

According to random commentary, Alex Kartveli himself immediately conducted the stress analysis investigation and discovered the flawed calculations. The entire empennage came as an assembly from the sub-contractor, but the failure location was in the fuselage at the attachment points. The twisted tailcone provided much needed evidence. (In this same month Major Seversky wrote some damning letters about Kartveli's engineering, stating without equivocation that Sasha Kartveli was not the least bit innovative in design, had no part in the overall design of the AP-4 which led directly to the XP-47B, but was an excellent man with details and problem solving. This all had to do with his lawsuit action against Republic. The corporation and the Major settled out of court.)

Incredibly, just five weeks later on May 1st, J.P.S. "Joe" Parker was flying the next P-47B off the line, AC41-5900, engaging in some necessary dive testing, the event in which the fabric covering used on the elevator and rudder surfaces ballooned and shredded. He was flying out over the Atlantic Ocean off Oak Beach when he was forced to "BOOK' (bailed out OK). Obviously somebody in Flight Controls engineering failed to study the Lockheed P-38 design. Years earlier, C.L. "Kelly" Johnson and structures specialist James Gerschler, in Burbank, Calif., had employed metal covering on all P-38 control surfaces from the time of XP-38 prototype construction. The complete empennage of the P-47 was quickly redesigned in conjunction with the aft fuselage structural correction. It should be noted that the mass balance weight near the top of the rudder post disappeared in the process. The miracle is that at least one of the two test pilots lived to report and one of the airplanes crashed without additional damage to the aft fuselage. The loss of two of the planes over the ocean along with their pilots in the disastrous early months of 1942 could have had the most serious consequences imaginable for the entire Thunderbolt program. Fate decreed otherwise.

Lowery Brabham was promoted to Flight Operations Manager at this time, to be joined by Carl Bellinger. Republic was able to complete its first P-47B delivery in March 1942, and the

[8] Retired General Mark Bradley, Hart Miller nor anyone else had ever made the slightest reference to this particular crash over a period of a quarter century. No book or magazine article appearing since WWII had even mentioned the event. Remarkably, the spectacular pictures were encountered as a researcher, Ed Boss Jr., brought forth a long-ignored selection of photographs during the author's first visit to the Cradle of Aviation Museum. The special connection to the 1939 Lockheed XP-38 crash near Mitchel Field, N.Y., and the missing empennage triggered an alarm.

production of Republic Lancer P-43s had been continued to maintain a stable work force. The last of thirteen service test YP-43s went to the AAC slightly ahead of schedule in September 1940. When it was necessary to cancel the 80-plane P-44 contract simultaneously with the Allison P-47/P-47A development to initiate work on the XP-47B, Hart Miller had to get back to Materiel Division later to calculate the impact on factory activity. Some fairly serious schedule errors were entered into the Current Company Contracts book corrected up to late 1940. But errors of a year or more were undetected until now. The last YP-43 was delivered in April 1941 with most going to Selfridge Field. However, Contract AC-13380 called for nearly 80 of the new P-44s by that time. What the 54 Lancers designated as P-43s were supposed to replace when change orders to the production P-47B contract was issued is unclear. The flawed schedule shows all 54 as being delivered before the first YP-43 was produced. They had entered the incorrect years on the entire schedule.

When agreements were reached to substitute eighty P-43As for the like number of P-44s, all of those Lancers received the serial numbers initially allocated to the P-44 Warriors. It created a situation where P-43As would have FY1940 serial numbers while the P-43s would have later FY numbers. The Y prefixes were eliminated, but most Lancers were soon redesignated as RPs anyway, indicating Restricted use. Then, to add to the confusion, but to keep the lines operating as P-47B schedules slipped, 125 of the P-43A-1 versions were ordered by using Lend-Lease funds for China. Then, two RP-43As were modified for photographic duties and designated RP-43C and RP-43D.

The following tabulation may help to understand what actually took place:

YP-43 13 ordered	R-1830-35 engines	Redesignated RP-43
P-43 54 ordered	R-1830-35/-47 eng.	Redesignated RP-43
P-43A 80 ordered	R-1830-49 engines	Redesignated RP-43A
P-43A-1* 125 ord.	R-1830-57 engines	China Lend-Lease(51)
P-43B150 reworked	R-1830-49 engines	Tac-Recon (cameras)
RP-43C 2reworked	R-1830-49 engines	Photo-Recon.
P-43D (As reqd)	R-1830-35/47/49/57	Rev. camera instal.
P-43E	P-43A-1 reworked to P-43D standard	

*Equipped with 4 x .50-cal mgs and self-sealing tanks.

The final P-43A-1 was completed and delivered in March 1942, allowing everybody to concentrate on P-47B production. Four P-43A-1 aircraft and four P-43Ds were transferred to the Royal Australian Air Force for tactical reconnaissance work in China.

One of the initial batch of YP/P-47Bs was assigned the task of obtaining much-needed performance information for at least a dozen different reasons. Of six airplanes routinely assigned to Flight Operations, one was assigned the performance task and was possibly flown by popular Joseph Parker, generally known as "Joe."

The fighter was equipped with an R-2800-21 engine, about 150 pounds lighter than the 11 engine and fitted with a different carburetor. All speed figures derived from the tests were TAS...True Airspeed numbers. The best number reported was 429 mph at 27,800 feet on 2000 horsepower. Down at 5000 feet, top speed was 352 mph, while at 34,000 feet it was clocking 412 mph with only 1845 hp available. If that seems low , just compare it with the power ratings furnished for a medium-level fighter such as the Grumman F6F-3. The non-turbo -10 engine used in that Navy fighter produced 2000 hp at 1000 feet and only 1550 hp at 21,500 feet. The -8 series engines used in many of the Corsair series were not one iota better. In the much later Voughts the engines produced 1550 hp at 26,000 feet, but contemporary P-47s were getting 1700 hp up there. It is a simple matter to say Grumman Hellcats and Vought Corsairs would not have been very effective over Germany in combat with Messerschmitts and Focke-Wulfs. In battle, the gap between 26,000 feet and 34,000 feet is an eternity. (All figures mentioned here are based on density altitude.)

Against some of the early P-38 series, P-47B climb performance was inferior, but on the other hand Allison-powered P-51s and Bell P-39s of the same time in space were hopelessly outclassed. Figures reveal the P-47B climbed to 15,000 feet in 6.7 minutes and to 20,000 feet in 9.75 minutes. Lockheed P-38Fs were getting up to that altitude almost 4 minutes quicker. At 35,000 feet, rate-of-climb figures on the P-47B deteriorated to 450 fpm with 305 mph indicated on 1365 horsepower. Remember that climb rate because one or two pilots from the 56FG in New England made some claims about performance figures believed to this day. That story is worth telling (see Chap.9) when some of the best combat pilots and group commanders of WWII make a little clucking sound, clear their throats and quietly refute the statements as "impossible".

But, "The time has come," said the walrus...to tell some heartbreaking tales, some encouraging facts coupled with some harsh ones, and get those single-engine babies flown (Oops) floated to the United Kingdom. Damn those torpedoes; full speed ahead.

With Wright Field number 31 painted on its cowling, another RP-47B cruises serenely over the Ohio landscape about midyear 1942. By that time, the small turtledeck windows on each side had been deleted in production. Control surfaces were still fabric covered and the rudder mass balance shows clearly. (WF Photo)

CHAPTER 9

AN INAUSPICIOUS BEGINNING

There were – and are of course – two things in aviation you could always count on...a learning curve and Murphy's Law. Well, almost always.

Test pilot Filmore "Fil" Gilmer was very good at his job, but he was not quite perfect. Just like his boss, Lowery Brabham, he had a very busy schedule trying to find out just how good his

airplanes could be or just how bad they might be. It was mid summer or as we came to know several years later, the Summer of '42. Going to the beach that year was not always good, especially if you were on an island named Guadalcanal way out in the Central Pacific Ocean area. Fil had flown Grumman F4F Wildcats in the U.S. Navy, among other types. And he probably was extremely familiar with the importance of that airplane's capabilities in the battles then raging over an island known to only a handful of Americans as the year dawned.

It was frequently Fil who drew the the assignment to fly the XP-47B, initially flown in 1941 by "Brab" Brabham. Since that time the airplane had been treated to a coating of near-standard camouflage paint, partly to prepare it for a publicity photo spread and cover display in America's favorite, *LIFE* magazine, on sale Friday, February 2, 1942. It was priced at 10¢, probably the best dime's worth you could find anywhere. The magazine writers did not divulge a secret. When the pictures were taken, the total number of flyable P-47s could be counted on the fingers of one hand. Almost exactly six months later, Fil Gilmer sauntered casually out of the Hangar 1 office to fly the XP-47B again. Unlike Brab (in 1941), he would be departing from a paved runway on a rather routine mission. The empennage problems of March and May had been partially corrected on the XP, but it still had fabric-covered control surfaces. High-speed dives were frowned on with that old timer.

The Navy-trained Fil Gilmer might well have been a doctor like his father. After attending Kentucky Military Institute, he completed three years of pre-med training at the University of Kentucky. (Perhaps he knew Hart Miller early on.) Deciding that the medical profession was not entirely his idea of a real life, Gilmer opted for a career as a naval aviator. Eventually he was sent to Opa Locka, Florida, and to Pensacola for flight training. Captain (later, Admiral) "Bull" Halsey, then Pensacola Naval Air Station commandant, pinned the gold wings to Fil's tunic upon graduation. Later, as a carrier-based bomber pilot, he served with VB-6 (Bombing Squadron Six) aboard the old carrier *Enterprise* before transferring to a battleship scouting squadron. He piloted Curtiss SOC-3 float-mounted biplanes, participating in dozens of those old gunpowder-charge catapult launches. He had spent some time flying Wildcat carrier-based fighters when his normal tour of duty was up in March 1942. Somehow he was put on inactive duty and joined Republic Aviation in the same month, just in time to encounter the tailplane problems. Under the leadership of Lowery Brabham, Fil soon became one of Republic's prime test pilots in Experimental Flight Test, in very good company with pilots like Mike Ritchie, Ken Jernstedt and Carl Bellinger, and, of course, the fatherly Joe Parker.

It was August 8, so Fil was thinking about going to the beach on Sunday, unless there was some pressing need to fly. Old AC40-3051 had enjoyed the company of several different Pratt & Whitney Double Wasp engines in its seventeen months on the Experimental Flight roster. The R-2800 was already warmed up by the flight line crew chief, so as soon as the 18 cylinders came to life, Fil made the short trip to the head of the runway. He ran it up to 2700 rpm and 53 inches manifold pressure, released the brakes and headed off in a southerly direction. The turbosupercharger regulator had been adjusted, permitting the pilot to obtain more manifold pressure than was required in attaining 2000 horsepower. It was needed to compensate for variable temperatures encountered as the long climb progressed. Weather conditions in August and failure of the engine to produce 2000 horsepower mandated the regulator setting changes. (Gilmer had set the boost control at a point where he expected to obtain 53 in. Hg. as part of his cockpit check.) He started to retract the landing gear, but when it was about halfway up the manifold pressure indication had moved past the redline limit.

Reacting quickly to the boost pressure, he released the landing gear control handle, taking the time necessary to reset his boost control. When the landing gear handle was released, the gear retraction process was interrupted[1], but not for more

A very relaxed Alexander "Sasha" Kartveli pictured sometime in the 1940s, probably early in WWII. On corporation documents he was listed as a vice-president and Chief Engineer back in June 1940. E.W. Walker was project engineer on the XP-47B program. Alexander de Seversky was no longer even on the B.O.D.. (S. Hudek Coll.)

OPPOSITE: Operating from First Air Force HQ at Mitchel Field, NY, on assignment to protect Republic, Grumman and factories in southern New England, six of many early model Republic P-47Bs were heavily involved in transitional training of pilots destined to operate out of British bases. The No.1 airplane in this formation was piloted by none other than newly promoted Lt.Col. Hubert "Hub" Zemke, just after he was assigned to command the new 56th Fighter Group. The No.26 fighter was being flown by Lt. Robert S. Johnson, soon to make his mark in history. The group would gain fame as one of the top scoring fighter outfits of the USAAF in WWII. (Hart Miller)

[1] This comment appeared in the official crash report, prepared by Gilmer and repeated in the company newspaper. My editor, Jeff Ethell - who, at times, has been known to fly a very slick P-47D - wanted to know if that was peculiar to the XP-47B. Nobody had raised the question in the early '70s when I first wrote that story. Nobody in the Thunderbolt Pilots Association asked the question at the 1979 convention in Los Angeles. Lowery Brabham, Bellinger and Gilmer are no longer with us, and they did most of the flying in the XP-47B. Who else would know?

While press releases from Republic and the news media touted the tremendous performance of the P-47, tactician Hub Zemke says that those early "interceptors" (and that is what they were intended to be) were "lead sleds." Republic strove to improve the breed and was soon producing great quantities of Thunderbolts in two plants hundreds of miles apart. At the same time, Curtiss dabbled and dithered around with P-47G production as if it was all a game. A significant revision to the P-47C expedited engine changes, at the same time improving overall maneuverability. Structural design errors in P-47B tail assembly attachment were rectified hastily, and all control surfaces were quickly metallized. This P-47C-1-RE (AC41-6086) was one of several undergoing test at Wright and Patterson Fields. One of three sent to Eglin Field in October 1942 for "combat evaluation" recorded a critical altitude TAS of 427 mph, but time to 20,000 feet was a dismal 14 minutes. At that time, Lockheed P-38F/G models were doing it in about 6 minutes, but level speed was under 400 mph; however, P-38G combat range was far greater. **(Air Materiel Command)**

than 10 to 20 seconds. Unfortunately, the tailwheel was – in accordance with that famous Murphy's Law – stopped directly aft of the stainless steel exhaust shroud partially enclosing the supercharger turbine wheel. It is a safe bet to conclude nobody had ever anticipated such a possibility...there was that hard rubber tire in the path of what amounts to a blowtorch flame. The test pilot continued his steep climb at military power, without the slightest indication that anything was amiss. Somewhere between 10,000 and 12,000 feet altitude, the XP-47B became a bit tail heavy. Fil found it necessary to keep firm pressure on the stick to maintain the 160 mile-per-hour climb rate. Airspeed was dropping conspicuously then, so he applied more nose-down stick force.

There was absolutely no response!

He applied more stick force, but speed continued to decrease. Alarmed, Fil slammed the stick full forward, but his reward was the smell of burning rubber and not one iota of change in the attitude of the fighter.

"Oh, God...No elevator control," swept through his mind. He reacted by rolling in full nose trim, simultaneously retarding throttle and boost controls. As the nose dropped below the horizon line, he rolled the trim control back. The nose was continuing a relentless drop.

He reacted by quickly moving the throttle and boost control handles forward, but his dilemma only became worse with the nose pitching down. As the XP-47B seemed, to him, to go past the vertical, his conscious thought was, "Get rid of the canopy." He pulled the release, ejecting the car-door style canopy – a 2-piece component peculiar to the XP-47B and three of the earliest P-47 production proofing airplanes – almost explosively. But for the security of his seat belt, he would have been sucked out of the cockpit in trail with the canopy. His eyes snapped to the airspeed indicator almost automatically. In later conversation he would recall "That airspeed was registering somewhere between 420 and 430 mph, and it was moving very rapidly."

Probably in a reflex reaction almost immediately after he cleared the cockpit, he grabbed the ripcord handle and pulled. *Wham!* "That was a wrong move." Subjected to a 400 mile-an-hour impact, one cloth panel ripped from the parachute canopy and other holes gave a clue to more rips.

Republic Aviation's extremely competent president, Ralph Damon, is shown discussing current situations with popular test pilot Joe "Swede" Parker as he prepared to take off in the P-47B/C (#170) development aircraft at Farmingdale. As the top official at Republic, Damon was the key to production attainments that totally eluded Curtiss Airplane Division. (This photograph had been autographed to flight line specialist, "Pop" Provo by Parker and Damon, reflecting their high regard for Provo's expertise.) **(Author's Coll.)**

"Damn! That hurt!" Fil cursed himself. By that time the fighter prototype was in a vertical dive, leaving a trail of smoke and ash behind. (The episode is colorfully illustrated by Bob Boyd within these pages.) Army Col. Tex Hulser who, fortunately, was cruising on Long Island Sound and saw the Thunderbolt plunge into the water kept an eye on the descending parachute and headed in that direction. When Gilmer went in, Hulser pulled him to safety. Although the survivor suffered from bruises and harness burns, he had no serious injuries. No attempt was ever made to recover the XP-47B or any major portions of it. Gilmer returned to the job of flying experimental aircraft the next day. After all, there was a war on.

Under the impetus of the war emergency, Republic Aviation – which had only come into being in October 1939 to avoid bankruptcy of Seversky Aircraft Corp. – enjoyed fantastic growth. In just three years, virtually everything that can be seen in this photograph had been constructed. Odd patterns seen on the airfield were a wasted attempt to camouflage the installation. An enemy launching an attack would have used many landmarks on Long Island as IPs for target headings. With dozens of completed fighters on the field, deliveries were feeding the growth of M/Gen. William Kepner's VIII FC groups in the autumn of 1943. (Mfr. 651)

Republic Aviation delivered the first true production P-47B (c/n 6) in April 1942, but that was a technicality because the airplane remained at Farmingdale for empennage retrofit and a continuation of flight testing. Because of the two serious component failures, a lot of "midnight oil" was burned at Republic and at Budd Manufacturing, fabricator of the empennage assemblies. Budd was a specialty firm with considerable experience in stainless steel fabrication. They continued as a major subcontractor to Republic during the war.

The XP-47B/P-47B Thunderbolt fighters set the pattern for the majority of those Farmingdale and Evansville airplanes yet to come from the production lines: (Deviations from the XP-47B pattern are indicated in boldface type.)

Single-engine, single-seat interceptor fighter. Pratt & Whitney R-2800-11 Double Wasp engine. Although the -11 engine, normally used in North American XB-28, was actually in the XP-47B during the Board inspection, a -17 engine was evidently installed for the first flight. R-2800-17, R-2800-21, R-2800-35.

Type C-1 General Electric Turbosupercharger.

Curtiss Electric 4-blade propeller.............. 12 ft. 2 in. diameter
Wing airfoil section... Republic S3
Wing span..41 ft./**40 ft. 9-5/16 in.**
Length (overall).......................................35 ft./**35 ft. 4-3/16 in.**
Wing area..300 sq.ft.
Wing dihedral (top surface)..................................... 4 degrees
Design Weights: Empty...............................8655 lbs./**9189 lbs.**
Gross..........................11,600 lbs./**12,245 lbs.**
Design gross................12,000 lbs./**12,500 lbs.**
Power loading..(normal gross wt.).......................5.80 lbs./bhp
Wing loading (normal gross weight).................38.67 lbs./sq.ft.
Wing loading (actual)...................................... 41.67 lbs./sq.ft.
Power ratings: Takeoff....................2000 hp/2700 rpm/52 in.
Military.............. 2000 hp/2700 rpm/25,000 ft.
Normal....................... 1550 hp/2550 rpm/S.L.
Normal.............. 1625 hp/2550 rpm/25,000 ft.
Internal fuel capacity..(design)............305 gal./**298 gal.(Act.)**

Armament: Six Type M-2 Browning .50-cal. machine guns, each with 500 rounds of ammunition. Alternately specified, two additional M-2 machine guns.

Performance............ w/R-2800-35 engine...412 mph/25,800 ft.

(No other performance figures obtained.)

As stated previously, the first pre-production (YP-47B) airplane was completed at Thanksgiving time, 1941. Only the first two pre-production YP-47Bs were manufactured with the XP-47B non-sliding cockpit enclosures (AC41-5895 and -5896). The 171st airplane in the P-47B series (AC41-6065) was completed as the XP-47E, employing a variation of the XP-47B canopy. The main feature of the E model was a pressurized cabin; it appeared to be much easier to seal a door-like canopy than a sliding unit. (The "YP" service test designation was evidently semi-official only.)

For unexplained reasons, the first P-47B (c/n 6) slated for delivery to the AAF appears all alone on some charts. The serial number AC41-5900 indicates it was the airplane being flown by Joe Parker on, of all days, April Fool's Day, less than a month after c/n 6 was completed. Because the Thunderbolt AC41-5900 was singled out from all other P-47Bs and was obviously exceeding the flight limitations applied to the type after Burrell's c/n 5 airplane lost its empennage, a new, "beefed up" aft fuselage tail support structure was, most likely, in place. We must presume Parker was beginning a series of dive and pullout tests to ensure integrity of the structure. Nobody ever suspected a major design anomaly in using fabric-covered control surfaces. Parker was the man for the obviously dangerous testing job because everybody from Ralph Damon to stock clerks respected his abilities as a tester. If this procedure seems strange, in mid 1942, people were often compelled to test for such serious design flaws in a flight vehicle because the contract did not specify delivery of a static test airplane. We can possibly conclude the one "skeleton airplane" ordered for delivery by 13 March 1941 (sic; tilt) was actually a stress test airframe. (The "1941" error permeated the entire Contract AC-15850 delivery schedule for months before discovery. One skeleton airplane was really intended for March 1942 delivery.) It's a fair guess to presume the first P-47B touching down at Wright Field, Ohio, was AC41-5905. Good photographs taken of a B model fighter show a small numeral 5 painted in white on the camouflaged airplane. Most fighters of that model were eventually redesignated RP-47B, condemning them to a life of restricted operation, sometimes as targets for weapons testing.

Air Service Command (ASC) had established a modification center concept intended to avoid assembly line slowdowns every time some design change had to be incorporated. It was a good scheme. Manufacturers could go on at a steady rate building airplanes to a given design. Once the aircraft were turned over to the military, they were flown or towed to mod centers and then treated more or less like major overhaul candidates. (At late stages of the war, every P-38J or L converted to a photographic reconnaissance version was the product of a modification center.) The War Production Board was single minded in its desire to increase all aircraft production rates to unheard of levels and with the least possible delay. Almost concurrently with the rollout of the first actual production P-47B, a decision was made to construct a second-source manufacturing facility at Evansville, Indiana, in a far southwest corner location of that state. Although the area had always been primarily farming country with only a few manufacturing plants, the Board was convinced it could be made to work. One strong point was the availability of high capacity railroad service and a major 4-lane highway headed in the direction of Chicago. A modification center being built at the same location made the idea logistically attractive.

In the next move, the WPB authorized start of construction in a farm field north of the town, adjacent to the Evansville Municipal Airport. The factory site actually fronted on that 4-lane highway. In one of the finer home front episodes of the war, construction was begun (in April) on a factory almost immediately after winter officially ended and while the ground was still thawing. A large factory sprang from that hard ground and was in the process of building the new fighter by September. Farmers and milkmaids had to be trained at breakneck speed during the summer months and hundreds, perhaps thousands, of logistics problems had to be solved. The first two proofing P-47Ds in a team of four were produced in September according to official Department of Commerce statistical studies, but Republic Aviation figures show the first one was not done until 23 October 1942(sic).[2] Significantly, those first airplanes out the door had Farmingdale written all over them. It had fallen upon the New Yorkers to step up subassembly production in support of their western neighbors. Therefore, the task of getting the factory on line in a hurry was simplified. In fact, the first four fuselage assemblies for the initial Evansville (RA) airplanes arrived by train. The wings were manufactured by another nearby subcontractor and mated to those fuselages. On a parallel path, uninspired Curtiss Airplane Division employees in Buffalo seemed to be walking on eggshells. (In later years, a former Curtiss Chief Engineer and successor to Donovan Berlin, explained how "top management" absolutely demanded that the WPB allow them to build "what they knew best." Essentially simple-design, decade-old H75/H81 Curtiss Hawks. Top-rung executives were

[2] Tremendous growth, tragedy and wartime catastrophes certainly generated numerous record-keeping mistakes in 1942. Department of Commerce records indicate the first two P-47D airplanes assembled at Evansville were delivered in September, but the Farmingdale records show the first one delivered on October 23rd. However, same day newspaper accounts report the affair on 19 September. A current affair report can hardly be in error. Newspaper masthead and dateline information is almost never in error. Other data in the New York records add to the confusion, not the least of which is stating the second batch (110 airplanes) all bore RE suffixes, which would indicate they all came from Farmingdale. Factual reporting indicates the P-47D s/n AC42-22250, bearing the name "Hoosier Spirit" and incomplete national insignia, was test flown and handed over to the USAAF Plant Representative on 19 September. However, that part was a special unexplained situation.

Horrendous problems associated with the new liquid-cooled engines forced the RAF and Hawker to seek a modicum of success in switching to an aircooled radial, the Bristol Centaurus CE.4S. It cranked out a smashing 2210 horsepower. But the initial Centaurus Tornado was done "quick and dirty" at best. Even with its thick wing and ill-conceived cowling, it managed to attain a speed of 402 miles per hour. The revised version (shown) tended to solve many overheating problems, swinging a 4-blade propeller in place of the 3-blade type used initially. Identified as HG641, this British approach to the P-47 concept boasted of a dozen .303 cal.machine guns buried in the thick wing section. Rearward vision from all Tornado prototypes was terrible. This rework of the first version made its initial appearance in November 1942. (Author's Coll.)

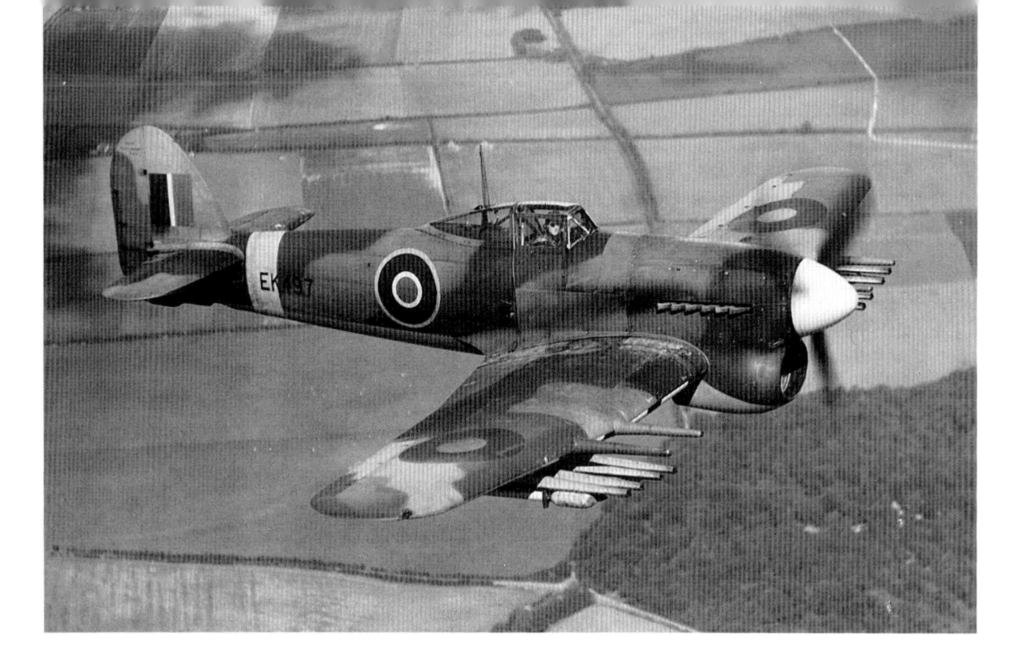

In its initial form, the prototype Typhoon (P5212) was a near duplicate of the first Tornado, but it featured a Napier Sabre engine and the radiator was located immediately behind the propeller. It flew for the first time on 24 February 1940, but a second prototype (P5216) became airborne for the first time on May 3rd in 1941, only three days before the XP-47B made its first flight. Designer Sydney Camm did not use state-of-the-art technology, and was a firm believer in thick-section wings. This production version, the Typhoon 1B, shows that it was an important service type but it never was able to be considered a dogfighter on a par with Focke-Wulfs and Messerschmitts. (Author's Coll.)

more interested in deriving a high return for the least difficult work. The C.E. to whom we refer was the much respected Walter Tydon, the man primarily respnsible for creation of the fastest of all Warhawks, the P-40Q.)

With Evansville factory construction still in progress, the first plane out the door, "Hoosier Spirit," (a P-47D) was completed, test flown and accepted by the USAAF in a one-day sprint effort on September 19th. There was a flaw in its appearance: it had a "kluge" paint job with the national insignia star missing from both sides of the fuselage. The main incentive for all the activity that day stemmed from the nearly daunting numbers of high-level AAF and civilian executives on hand for the delivery event. For the company, Ralph Damon was in town to lead the ceremonial activities in a wartime atmosphere of dedication and intense patriotism. He had traveled to Evansville in company with other key company executives Hart Miller and Mundy Peale. Representing the AAF was B/Gen. A. W. Vanaman from Wright Field (described in newspapers as commander of Air Materiel Division, but unlikely since M/Gen. Echols was – we believe – still wearing that hat). He had flown in aboard his converted pre-war Douglas B-23 bomber, by that time cloaked in wartime camouflage paint.

Mr. Damon opened ceremonies with a short speech. Among his comments, "It is interesting to note that the first delivery date set for these planes was June 1943...and that nine months' time has been saved." Following his speech and introductions of VIPs, a flight demonstration featured the first Evansville-produced P-47D. Piloted by Evansville Chief Test Pilot Victor F. Pixey, the fighter taxied out precisely at 11:30 a.m. with two Farmingdale P-47Ds at the wingtips. Colonels Russell Keillor and Mark "Moe" Bradley were at the controls of the visiting aircraft. (Bradley had replaced Marshall Roth as SPO – Special Projects Officer – when Bill Craigie and Roth were moved up the ladder to increasingly important positions. By then, Craigie was at least a brigadier general.) The three enthusiastic pilots proceeded to put on an exciting 15-minute program to demonstrate the attributes of the newest series Thunderbolts.

Some months earlier, the initial group of first-line P-47s was activated in the New England-New York defense area of the First Air Force. It also turned out the be the only active fighter group ever committed to operations with the P-47B/RP-47B version of the Thunderbolt.[3] That fighter organization was a unit eventually to gain fame in Europe: the 56th Fighter Group.[4] Initiated by orders issued on 14 January 1941, the 56th Pursuit Group was then composed of three squadrons,, i.e. the 61PS (for Pursuit Sqdn.), 62PS and 63PS. Initially based at Savannah Army Air Base, Georgia, they moved to Charlotte Army Air Base, N.C., in the summer of 1941 as part of the U.S. Army Air Forces (from June). In October the group redeployed to Myrtle Beach, S.C.,

[3] It has been positively confirmed by nobody less than Col. Hubert "Hub" Zemke, Col. Cass Hough and C. Hart Miller that not a single P-47B ever went overseas as an operational fighter. Any report to the contrary is in error. All B models eventually were classified as RP-47Bs and placed on Restricted operations as trainers or test vehicles of some sort after Zemke took the 56FG overseas to the ETO, Eighth Air Force.

[4] The 56th Fighter Group was assigned initially to defend (if necessary) Republic Aviation, New York City and industry in the surrounding area and in New England. It was the first and only P-47B Thunderbolt group assigned to a combat role, especially in the Zone of the Interior, it ultimately played a major part in the buildup of VIII Fighter Command, Eighth Air Force, in the ETO, and its combat record in the war was superior.

While most Hawker Typhoons with Napier Sabre engines were fitted with 3-blade propellers of large diameter, this late series fighter featured a 4-blade DeHavilland Hydromatic (license-built) propeller and a bubble canopy. The early series Typhoons and the failed Tornado type featured car door canopy units like the Bell P-39 Airacobras. (Author's Coll.)

and at least partially re-equipped with Bell P-400 and P-39D Airacobra fighters. The 56PG was fully involved in the "Battle of the Carolinas", especially on 24 October when the defensive forces intercepted an "enemy" attacking force, their primary commitment. Bell Airacobra maintenance was not optimum. Brake problems and the loss of cockpit doors in flight could hardly be looked upon with favor. Incidentally, those P-400s were from the massive French-British order, becoming part of the USAAF when the RAF refused to accept the airplanes for failure to meet specifications. One squadron had been sent into combat service, soon proving conclusively to the British, at least, the type was unable to fight effectively against the *Luftwaffe* Messerschmitts and Focke-Wulfs in Europe. (Their own Westland Whirlwinds and Boulton-Paul Defiants suffered the same fate.)

Dipl. Ing. Kurt Tank, a first-rate pilot, freely admitted that he was heavily influenced in his Fw 190 fighter design concept by the 1935 Howard Hughes H-1 (sometimes called Model A) record-breaking racer. To see the prototype Fw 190 V1 taxiing toward you (on film) would be convincing proof that Tank was an honest man. Even the very short original wing was deja vu, given memories of the World Landplane Speed Record version of the Hughes airplane. Upon deletion of the ducted spinner, the Fw 190 V1 was a clone of the H-1. The rare Fw 190 V5 shown here still had the original small-wing configuration. Observe that the Germans were using paddle-blade propellers in 1940! **(William Green Coll.)**

When the shock and excitement of the December 7th attack on Pearl Harbor and the Philippines wore off, the group deployed as follows:

Group HQ and 61PS(F) to Charleston Municipal Airport, S.C.

62PS(F) to Wilmington, N.C.

63PS(F) remained at Myrtle Beach, S.C.

Evidently the Curtiss P-40s had been written off and the squadrons were largely equipped with Bell P-39s and Curtiss P-36s, all being flown to the brink during every waking hour.

On the group's anniversary in January, the squadrons moved from the sunny south to the wintry north, becoming part of the First Air Force, I Fighter Command. Headquarters was established at the permanent base of Mitchel Field on Long Island, N.Y., 56th Group HQ was established in Teaneck Armory, N.J., and the squadrons were dispersed as follows:

61PS(F) "commandeered" Bridgeport Municipal Airport, Connecticut.

62PS(F) moved into Bendix Airport, N.J.

63PS(F) was lucky enough to move into new semi-permanent quarters at Republic Airport. With new paved runways and taxiways, it was in great shape, and two new first echelon hangars were built with basic facilities at the south end of the field. While on a special assignment for group commander Col, David Graves, Lt. Hubert Zemke (who's then current function was to fly a desk at the 56th headquarters) went across the field to Republic Aviation and got a closeup introduction to at least a few of the Thunderbolts out on the flight line. One of those had to be the P-47B/C being flown heavily, usually by Joe Parker, as one of the test airplanes flaunting either a yellow cowling and empennage or a full coat of paint matching zinc-chromate primer. That B/C airplane may have had the new 8-inch section installed, *primarily* to accommodate the QEC (Quick Engine Change) package which would greatly reduce time involved in changing engines. Here was, with some certainty, the precursor of what the AAF considered the first combat-worthy P-47s, to be identified as P-47Cs. Presumably all P-47Cs were produced with the QEC package and the attendant increase in overall length of the fuselage, but that was an incorrect assumption. In fact, the first group of P-47C-RE fighters, some fifty-eight in number, were essentially improved Bs without benefit of the forward fuselage structural changes.

By June 1942, the first of 37 airplanes built as P-47Bs on company Project No.89, were being allocated to the 56FG. At the same time, group command passed to Col. John C. Crosswaithe. Only thirteen of the officers were assigned to the new 56FG

The Luftwaffe did not hesitate to convert Fw 190As into fighter-bombers as early as the summer of 1943, with the production models being identified as Fw 190G-3s. (This pair, apparently being flown by II/SG 10 pilots over Rumania in 1944, is typical of a type being dispatched against Eighth AF bomber assaults in the skies over France and Germany in the Regensburg-Schweinfurt days.) By injecting 96-octane fuel – instead of water – into one air intake, maximum sea level speed was raised to 356 mph in clean condition. Yes, that's 356 mph on 1870 hp., not overly impressive. Best speed at critical altitude was no better than 410 mph. Incidentally, there are serious doubts about the fabulous numbers of Allied airplanes shot down by Luftwaffe pilots because of credit allocation methods. **(Author's Coll.)**

Headquarters, the others being scattered to the 90FS and the 80FG, the latter transferring to Farmingdale's Republic Airport from Selfridge Field, Michigan. In the shakeup turmoil, HQ and the 63FS moved to Bridgeport Municipal Airport to join the 61FS, while the 62FS relocated to Bradley Field, Windsor Locks, Connecticut. (During the first half of 1942, with the Army Air Forces having gathered itself into an operating entity, the combat units were in transition. First there were the Pursuit Squadrons, then Pursuit (Fighter) Squadrons and finally, Fighter Squadrons at varying times during the period.)

In July 1942, Lt. Zemke suddenly was promoted to Major Zemke, and this was in the same general period when 2nd Lt. Cass Hough was jumped to captain and then major within weeks.[5] Following a spate of other key assignments, Maj. Zemke moved to command the 56FG on 16 September 1942. He had commanded the 80FG for a brief period at Farmingdale, so he was well known around Republic Aviation, most likely having something to do with a publicity event involving himself. Several P-47Bs were "prettied up" as best they could be and Lt.Col. Hub Zemke (!) led a modest formation in flight, posing for camera work by an aerial photographer, most likely Hans Groenhoff or Rudy Arnold. Many black & white photos and numerous Kodachrome color transparencies froze those scenes for those of us not fortunate enough to be there.[6]

Colonel Zemke was a no nonsense pilot who quickly learned how anyone rising from lieutenant to lieutenant colonel in about

Early in 1943 as Republic P-47s began to arrive at U.K. ports in marvelous wood crates from Dade Brothers in New York, or fully assembled on decks or below decks of tankers and freighters, all assembly and checkout work eventually was taken over for a time by the British Reassembly Division. This Lockheed Aircraft Service organization operated mainly at Speke Aerodrome adjoining Liverpool and at Renfrew, Scotland. Here, some P-47C-5-REs are seen at Speke in company with a Ventura, Hudson and Bermuda. (Lockheed BRD)

---

[5] In January or February, civilian OPM (Office of Production Management) chief William Knudsen suddenly became Lt./Gen. Knudsen, commanding a mix of military and civilian personnel in the new War Production Board. My father, a Principal Engineering Consultant for OPM, moved from civilian to Major Al Bodie in one day as well. He was to report to M/Gen. Oliver P. Echols at Wright Field as an advisor. The wave of a phenomenal expansion program had come ashore at last and thousands of new leaders were created overnight.

[6] The brilliant color photograph on the dust jacket of this book somehow survived in original form for more than a half century while most others from 1938-42 only survived as copies of copies. The shock of acquiring this item can hardly be described. It must be recalled that Kodachrome did not even come into widespread use until sometime in 1939. Few (if any) cameras had color-corrected lenses, and purchasing film in 4x5 size emptied the coffers in a hurry. The film could be processed only at Kodak in Rochester, N.Y.. All color photographs appearing in this book are wonderful survivors from the "birth of an industry" era, having every right to be viewed with the same awe we have accorded Matthew Brady photos from the Civil War. Neither the movie industry nor magazines having the stature of *LIFE* magazine were prepared to record the worldwide war engulfing us in the same year the process became reliable for color stability. Therefore, on the homefront, war was almost universally in black and white. The dauntless young men who went to extremes to obtain color photographs in war zones and then weather the rituals needed for processing deserve more than a salute.

Soon after Col. Ben Kelsey and Maj. Cass Hough flew to the U.K. in Operation BOLERO piloting Lockheed P-38Fs (in the "Fuddy Duddy" flight) as escorts for B/Gen. Frank "Monk" Hunter's B-17E, the new VIIIth Fighter Command general asked Kelsey and Hough to set up an Air Technical Section at Bovingdon close by his HQ at "AJAX." Higher command sent Kelsey off on many other P-38 assignments, so deputy Hough was functional C.O. until Jan. 1944 when Kelsey returned. The fourth Thunderbolt (a P-47C-2-RE) to reach the United Kingdom was AC41-6182, shown upon arrival at Bovingdon in January 1943. With no provisions at all for external fuel tankage, combat radius of action hardly exceeded that of contemporary Spitfires. For one entire year, only two P-38H (few J models) groups were able to escort bombers into Germany. Until the "universal" wing and belly tank provisions were available, P-47 ops beyond the Rhine were virtually impossible. Cass Hough was the key to range extension. (VIII FC)

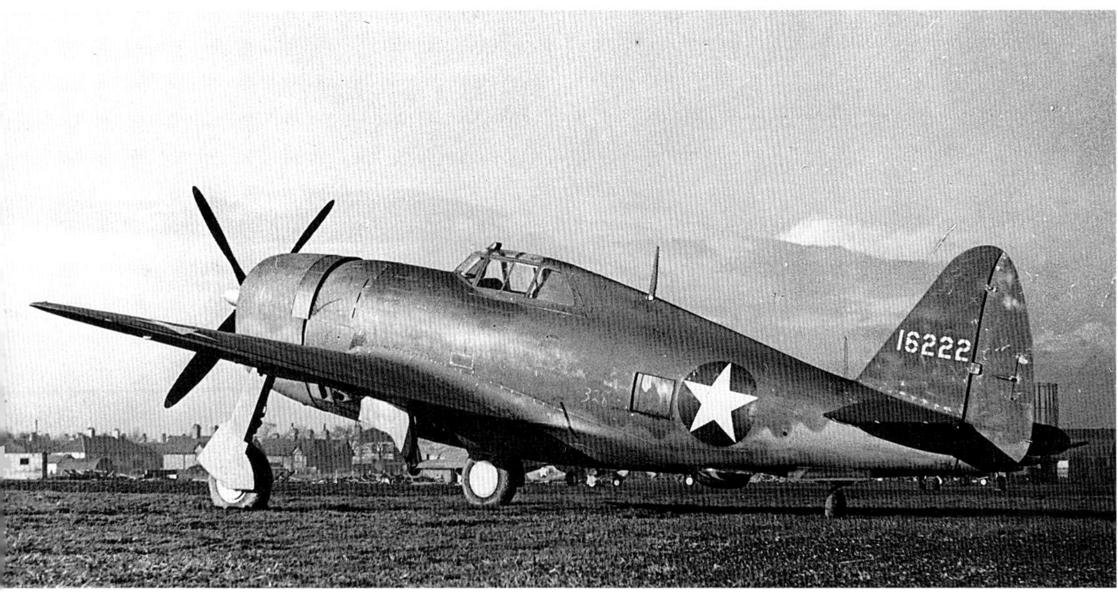

Early model Thunderbolts (T-bolts, Jugs) arriving in the U.K. via the sea lanes were generally processed at the British Reassembly Division facility at Speke (Liverpool) where this P-47C-2-RE was pictured shortly after New Year's Day, 1943. (No P-47Bs were sent overseas for any reason.) Teething troubles in the ETO included a need to switch from HF to VHF radio for compatibility with the RAF and serious range limitations. Cass Hough, by then a "light" colonel, warned Gen. Hunter in mid year that without external fuel capabilities, T-bolts would never perform their POINTBLANK role. Hunter told him to "get it done" fast. Hough and Capt. Bob Shafer and the Technical Section did just that in weeks. Fourth Fighter Group pilots were unhappy with big, short-range P-47s and, like the 56th and 78th groups, tried the big Republic-supplied 205-gallon (unpresssurized) belly tank. It was a major flop. Next step was to use P-39 type 75-gal. metal tanks for missions first flown on 28 July 1943 according to Col. Hub Zemke. **(Lockheed BRD)**

six months in command positions must change his standard lifestyle fairly quickly and not have a very thin skin. At the same time he was attempting to build a proficient fighter group, he was also very much involved in doing what the First Pursuit Group had been doing with the YP-38s and P-38Ds in the previous year – performing all the functions normally handled in peacetime by a service test unit working out of Wright Field. But he lacked experienced pilots who had managed to live many years by being canny and using common sense. The 56FG was populated with hotshot, "I'm king of the world" fliers whose attitudes carried through into overzealous piloting. Naturally there were many bent airplanes and dead pilots accumulating in a fairly short period of time.

Which brings us to one rather typical situation laced with much fantasy lingering on to this day. It was bad enough in its original form, but inflammables added by no less personages than Sasha Kartveli and Republic's director of public relations at the time ensured its survival. Two young pilots in the 56th performed impossible feats one day up there over New England while flying a pair of those RP-47Bs, and the story is worth telling if only to show why some flying stories are close kin to fish stories. Because Kartveli did not even try to convince them of the impossibility of obtaining such performance from an RP-47B, and with Costas Pappas not at the scene, it has become impossible to convince those pilots it could not occur in the manner described. Physics alone prove the feats to be far beyond the Thunderbolt's capabilities. We will open by quoting a Republic press release dated 1 December 1942.

It began like this: "*New aerodynamic and speed frontiers were established November 15 at an East Coast Air Base when two Army lieutenants, in Republic P-47 Thunderbolts, dived these powerful fighter planes at 725 miles an hour, and probably became the first human beings to hurtle through space at a speed greater than that of sound.*[7] More than twelve miles per minute." (Author's bold emphasis.)

This source document may be the only original copy still existing, although we can speculate on the chances of Col. Harold "Bunny" Comstock having a copy which is probably in a better state of preservation. But the key point to be made is the value of reporting it exactly as Republic Aviation executives surely approved of it in 1942. There had to be other factors affecting claims like those described in the press release, and the most important are presented here. The two new USAAF pilots involved were flying RP-47Bs from the batches of 37 or 126 production Bs. As of November 1942, all of those fighters were operating under official flight safety restriction orders because few, if any of them, had been retrofitted to incorporate aft fuselage structural improvement for empennage attachment. None had the redesigned empennages with metal skins replacing the fabric-covered control surfaces. Those very serious, unpublicized accidents which took Burrell's life and nearly did the same to Joe Parker had the most serious implications. Both accidents, restricted status of the aircraft involved and knowledge that both Army pilots had been air cadets in 1941 have to be considered in evaluating their joint report. In addition, the Air Force and Republic were both in need of good news for a change. We now return to the news quotations.

Lieutenant Harold Comstock, 22, of Fresno, California, and Lieutenant Roger Dyar, 22, of Lowell, Ohio, did not reach that unexplored aerodynamic horizon where the airflow theoretically breaks up, splashes against the plane's surfaces, and may cause the craft to lose its essential flying characteristics and flop around like a drunken pigeon. But they did attain a speed where the air pounded against the vertical, or "straightened out" control surfaces with such tremendous force that their joy sticks were frozen solid. When the stick won't respond to a young man's muscular tug while the young man is plummeting earthward at substantially more than 1000 feet every second— well, as Lieutenant Dyar puts it, it results in "that old unsatisfactory condition."

[7] Although many fliers over a half century, well into recent years, have claimed with much intensity (and even anger) they personally flew faster than the speed of sound, one factor remains as a deterrent to such stories. The best aerodynamicists, test pilots and physicists have proven without any doubt this fact: ***No airplane anywhere in the world in World War II attained a speed in excess of the speed of sound.*** With or without wings attached. C. L. Johnson was the most learned person in the Allied world about the subject of compressibility and the speed of sound as early as 1939 or 1940. The first recorded encounter with the phenomenon known as compressibility of air by an airplane was probably the YP-38 being flown by Major Signa Gilkey doing service test work out of Patterson Field, Ohio. Fortunately he was able to recover from the ordeal, but others were less fortunate. Suffice to say, while Kartveli and Pappas generally contend that P-47s could fly up to Mach 0.92, Kelly Johnson said their instrumentation failed to compensate for errors caused by the static pressure sensing devices, creating serious instrument errors. Limiting Mach No. on the P-38 was approximately M=0.68. It is conceded the P-47 may have been able to attain M=0.83, but even that is exceptionally fast for the time. More to the point, the RP-47Bs had structural limitations on them, primarily in the empennage attach structure and in the fabric-covered control surfaces which had already caused the loss of a new P-47B.

With this 56FG P-47C-2-RE Thunderbolt displaying the newly required I.D. markings dictated by the staff at AJAX, a large media event was generated on March 10 to introduce the P-47 as the backbone of VIII Fighter Command (FC). While 4FG T-bolts went off on a sweep (for emphasis) as their initial combat mission, 56FG airplanes were available for still and movie camera coverage at King's Cliffe. The fighter shown is believed to be Lt. Col. Hub Zemke's mount assigned to the 62FS (coded Z64, AC41-6209), but another pilot evidently occupied the cockpit while the still obscure commander made sure that the media event was a success and that the press was duly impressed. (VIII FC HQ)

The hyperbolism continues. *Fortunately, the Thunderbolt which Alexander Kartveli designed to stand terrific stress, also is designed for split-second maneuverability. Lieutenants Comstock and Dyar were at the plane's best fighting altitude,*[8] *and when physical laws prevented use of the stick, both pilots resorted to the crank which controls the elevator trim tabs. "When I rolled back on the tabs, the plane shuddered as though it had been hit by a truck. Frankly, I wondered whether the tail section was still there, but the ship was as well knit as the Siamese twins," Comstock declared.*

Although the speed of sound is 736 miles per hour at sea level, it decreases as the density of the air decreases. (At approx. 32,000 feet, the speed of sound is 670 mph. Auth.)

When the pilots took off on that bright Sunday, neither had any intention of becoming a candidate for the title of world's fastest human being. Each was to do horizontal speed runs at 35,000, 30,000, and 25,000 feet. When they completed their work at 35,000 or thereabouts, they dived merely to reach more quickly their next level of operations.

[8] In the context of a 1993 telecon, Harold Comstock told me they had climbed to an altitude of 48,000 feet. When asked for verification of that figure, he was emphatic about it. We know for certain that contemporary, high-flying P-38F airplanes flown to their maximum capabilities by VIII FC's Air Technical Section at Bovingdon by Lt. Col. Cass Hough (the Deputy Commanding Officer) were absolutely at their limit at less than 43,000 feet, and oxygen system capabilities were overstressed. Lightnings had very high-lift airfoils; the P-47's S-3 airfoil was more in the compromise area. Col. Hub Zemke had commented during a telephone interview session: "My experience with the P-47B was that it was near its absolute limit and on the ragged edge at 32,500 feet."

Led by a newly arrived P-47D-1-RE (AC42-7870), a flight of T-bolts consisted mainly of P-47Cs in this 62FS formation. Markings reflect a certain fluid status, but the 56FG had not seen fit to use the large national insignia on all four wing panels. Generally unrecognized has been the fact that while D models had the increased number of cowl flaps, many early Ds did not feature the deeper belly line believed to have been inaugurated with the "Dog" airplanes. It seems that some early P-47Ds were retrofitted with those features that allowed for installation of standard bombs and drop tanks on the centerline. LM ★R is a typical unmodified T-bolt with the smooth keel line. (Roger Freeman)

Breaking left on takeoff, a P-47C-2-RE reveals the original smooth keel line of the Kartveli design. "Sasha" was adamant in his stand to resist marring that unflawed belly line with the rework needed for carrying centerline external stores and in his opposition to tank/bomb pylons under the wings. That 8-inch segment inserted in the fuselage immediately forward of the windshield was hardly noticeable, but it did improve maneuverability and expedited powerplant changes. (USAAF)

Lieutenant Comstock started his downward course at 36,500 feet. His eyes fixed on the instrument board, he saw the airspeed indicator leap as the 2000 horsepower engined Thunderbolt began its vertical plunge. At 26,000 he tried to pull back on the stick and in the one second that it took him to travel more than the next 1000 feet, he realized that he must use the elevator trim tab.

Lieutenant Dyar, while diving, was able to joggle (sic) his stick so that he had slight control over the ailerons, and wobbled his wings as he descended. (Most of two paragraphs here which pertain to psychologists' contentions about reactions to speed, and the pilots' personal feelings during the dives have been eliminated as irrelevant to the main theme.)

While both of their speeds are listed officially[9] *at 725, it is quite likely that Lieutenant Comstock exceeded that figure and touched that mystical realm where the airflow spatters and sound waves begin to act on control surfaces with unpredictable results.*

Second Lieutenant Comstock's Form 5 showed 189 hours in a P-47 at the conclusion of that flight. Lieutenant Dyar's Form 5 time in a P-47 amounted to 265 hours. By the time Harold Comstock had been credited as an ace with five aerial victories, he was a major. Roger Dyar was killed in action while flying with the 63FS.

When Lt.Col. Cass Hough's 1943 report of his Thunderbolt dive test was interpreted and reported in a news release from Headquarters, VIII FC (AJAX), the interpretation recorded Hough's P-47C-2-RE had reached a speed of 760 miles per hour at 27,000 feet. He had written there was a sudden jump in the speed from 385 mph, so it is necessary to recall Kelly Johnson's statement about the air speed indicator problems caused by the static pressure effects at high altitude. Although nobody had raised an eyebrow about supersonic speeds and the speed of sound in December 1942 in connection with the Comstock episode, a wall of flak seemed to go up in connection with Hough's technical report. An identical situation occurred when his comparable dive in a P-38F from 41,200 feet on 27 September 1942 at Bovingdon received unwanted worldwide attention from news reports. In that particular case, it was unfortunate for all concerned when Lockheed's vice-president of Engineering, Hall Hibbard (Kelly Johnson's boss), released the story to newspapers with extreme exuberance. Also unfortunately, Kelly Johnson, Chief Test Pilot Milo Burcham or test pilot Tony LeVier did not openly criticize the Army Air Forces and Lockheed releases as being totally inaccurate. Privately, Burcham ridiculed it. Why not? After all, in 1942 the most knowledgeable people in the Allied world on the subject of compressibility encountered by aircraft were Kelly Johnson, Burcham and LeVier. The status of German knowledge about the subject at that time is unknown. Neither Kartveli nor Pappas at Republic had any real experience with Dr. Mach's Number, and it would be impossible to substantiate the Comstock-Dyar reports.

(Just for comparative purposes, Comstock had logged 189 hours in P-47s. At about the same time, Lt.Col. Hough had 525 hours on an assortment of first-line and experimental fighters plus an amazing 4,971 total time on non-military aircraft, or a sum total flight time of nearly 5,500 hours. His engineering education and experience far exceeded that of Comstock, Burcham or LeVier.)

Claims regarding supersonic, transonic or close brushes with the speed of sound have generated smoke and fury for a half century, at least. It is time to put lie to the claims, once and for all. When the situation reared up again in 1988, some of the

[9] By the Federation Aeronautique Inter. (FAI), perhaps? In the pilots' service records? National Aeronautics Association?

Col. Hubert Zemke has confirmed that the first fighters his group received, after the 4FG was supplied, were P-47C-2-RE and -5-RE models. This P-47C-5-RE and virtually all others for the 56FG were assembled and checked out by his 1st Echelon men who had to show the 33rd Service Group "how it was done," Markings were not up to requirements established by AJAX in that large stars were not yet painted on four "corners" of the wing and no yellow surround was on the fuselage stars. No internal plumbing or bomb shackles were available for carrying external fuel tanks or bombs. For interceptor missions, Zemke was totally unimpressed with P-47C climb speed and level flight acceleration. (USAAF)

most experienced and knowledgeable engineers and test pilots were contacted and the hard-core lines of their statements are presented here. These are direct quotes:

No WWII propeller-driven aircraft even came close to reaching Mach l.0 in a dive, due to the astronomical drag rise of the aircraft propeller shortly after exceeding approximately Mach 0.83... Herbert Fisher, former Curtiss-Wright Chief Test Pilot and a PhD Aeronautical Sciences

...no reciprocating engine airplane broke the sound barrier. I personally did the original terminal velocity dives on the 'Jug' and starting at between 35 and 40 thousand feet, the maximum Mach number achieved was .82 or .83... Lowery L. Brabham, former Republic Aviation Director of Flight Operations and Chief Test Pilot.

When Col. Arman Peterson's 78FG arrived in England at Goxhill they were a Lockheed P-38 group, but their aircraft and all pilots below flight leader status were soon requisitioned for Operation TORCH duty in the MTO. Moving to Duxford with newly arrived P-47Cs in April 1943, Col. Peterson soon had them flying combat missions. The experienced Maj. Harry Dayhuff is pictured with his P-47C-2-RE bearing the name "Mackie." Code letters applied were MX★ Z, and that large star under the port wing indicates that the "4 corner" national insignia scheme was in effect. Dayhuff went on to command the 4FG. (Col. Harry Dayhuff)

Anyone who ever reached Mach One in a WWII propeller aircraft ain't here to tell about it. Tony LeVier, Lockheed Aircraft Corp. former Director of Flight Operations and Chief Test Pilot.

Re: the max. speed on the P-47 and P-51. Bob Hoover and I did some work at Wright Field in 1946 and '47 to see just how fast we could get the aircraft up to. Both were instrumented and the max. Mach No. for each, straight down, wide open from as high as we could go was as follows: P-47 Mach 0.805, P-51 (Mach) 0.81. General Charles "Chuck" Yeager, former fighter pilot and test pilot.

And in ultimate conclusion, a final word from the late, great Kelly Johnson. *I am surprised to hear that some WWII fighter pilots are still maintaining that they dove their airplanes through Mach 1 in the war. When Colonel Cass Hough reported his P-38*

incident, I checked into the matter immediately and found the solution. Pilots were indicating very fast dives, particularly at high altitudes with various aircraft at the time including the P-38, P-47, P-51 and Spitfire. The reason they saw these false data on the cockpit instruments was very simple. It had to do with the fact that the static system was generally hooked to a pitot tube and had a substantial delay in it reading in a dive, giving false values of the airspeed indicator, altimeter, rate of climb indicator, and the lines themselves could not vent fast enough to measure true static pressure in a screaming dive. The thing that happened then was the same as applying suction on the static side of the airspeed indicator....

Do I hear any rebuttal from the opposition???

The 56FG was alerted for overseas duty on Thanksgiving Day 1942. The group, pioneering in the use of a fighter that had not even gone through a normal service test program and was virtually a production airplane from the outset of design, suffered a very high accident rate. But it is well to recall from those early days of helter skelter Army Air Forces expansion how high accident rates were with Martin B-26s, Consolidated B-24s, Curtiss AT-9s and Lockheed P-38s, just to name a few. Overexuberance was not the least of the causes for fatalities.

In the wee hours of the morning on 6 January 1943, the members of the 56FG boarded HMS Queen Elizabeth, berthed at Pier 90 on the North River, following a cold, wearying trip from Camp Kilmer. Lt.Col. Zemke and his staff had been led to believe their RP-47Bs would be shipped to England for training and operations, but that was a less-than-accurate prediction. Instead, the B models were assigned to a different group in the Z.I. for training. Actually, that was a bonus. Beginning late in the month, Republic P-47C-2-REs and P-47C-5-REs would start to trickle over to their first assigned base at Kings Cliffe from the British Reassembly Depot at Speke Aerodrome on the outskirts of Liverpool.

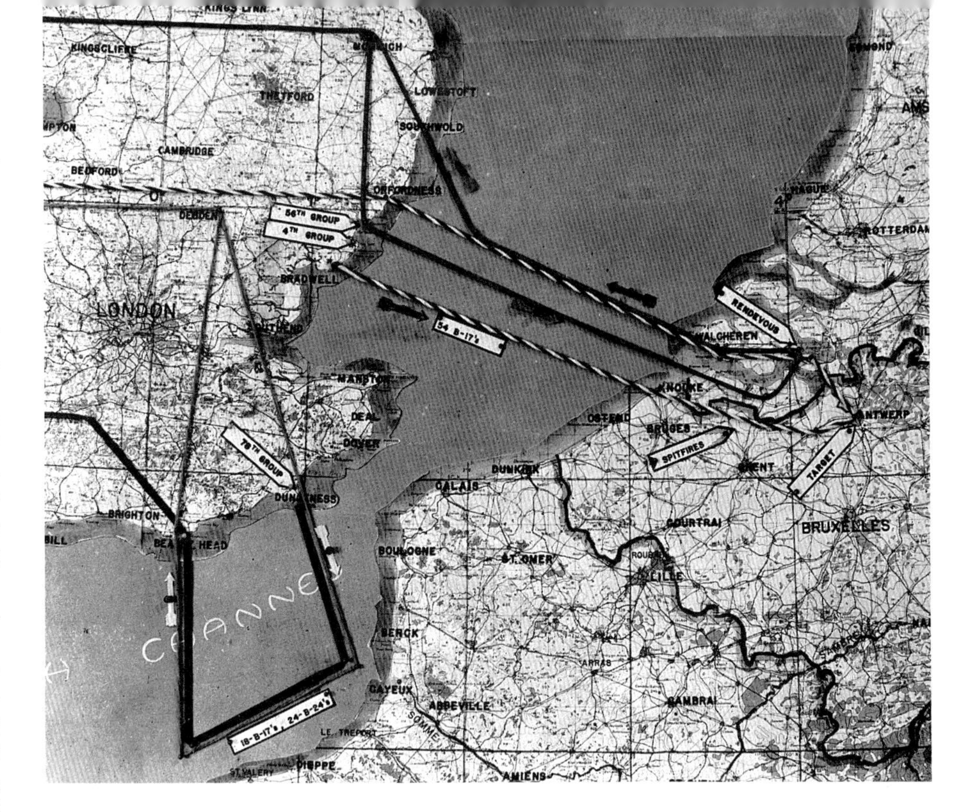

What we have here is identified as the route map of the first of many P-47 Thunderbolt bomber escort missions over the Continent. This was a withdrawal Ramrod flown by 4FG and 56FG aircraft when 54 Boeing B-17s of the 1st Bomb Wing attacked former Ford and General Motors plants in the Antwerp, Belgium, area on 4 May 1943. At the same time, the 78FG flew a Circus with a mixture of diversionary B-17s and B-24s aimed at Abbeville. A force of RAF Spitfires provided bomber escort approaching the target area. German Air Force (GAF) aircraft attacked the formations over Festung Europa and even out to mid channel. In the excitement, Walker Mahurin misjudged his high approach speed as his flight attacked four Fw 190s that had ganged up on a Spitfire. Although no e.a. could be claimed, they did scatter the Focke-Wulfs. All part of the learning curve. (56FG)

Before attempting to even touch on the massive operation being established in the European Theater of Operations (ETO) in Great Britain, a few other topics of interest occurring in the U.S. deserve recognition.

FLYING THE SEVERSKY P-35 IN 1942

Although thousands of AAF pilots flew Republic P-47s, Lockheed P-38s and North American P-51s during and after World War II, those who experienced exciting and extremely trying flight situations probably number only in the hundreds. Realistically, pilots in that group who are willing (or able) to elaborate on their flying experiences is relatively small. In contrast to the thousands of P-47 Thunderbolts that rolled from the two Republic factories, only about seven dozen P-35 type single-seaters were ever

Just as Lt. Robert Johnson returned to base in a shot-up, basket-case P-47C aircraft, Lt. Justus D. Foster of the same 56FG brought his P-47C fighter back to base with extensive damage from cannon fire. (Actually it looks more like an 88-mm shell exploded on contact.) With the tire sustaining serious damage, the ensuing ground loop upon landing ended in a noseover. (USAAF)

manufactured by that mass producer's progenitor, Seversky Aircraft. Furthermore, the last of those P-35s rolled out well before the decade of the 1930s became history. With all that taken into consideration, what were the possibilities of finding any pilot who had logged many hours of flight time in P-35s, had experienced several exciting moments in the cockpit, remembered a great deal about the type and, nearly of equal significance, was willing to relate the story?

Well, a certain highly qualified colonel by the name of John B. Stirling, having hung up his flight suit after flying his 101st combat mission over Vietnam in a McDonnell RF-101 on May 1, 1967 at the age of 47, fit the bill. He was, at the time, commander of the 20th Tactical Reconnaissance Squadron operating from a base in Thailand. The good Colonel Stirling had been chasing clouds in airplanes for a long time, dating back to pre-Pearl Harbor days when he was training Royal Canadian Air Force student pilots in North American Harvards. Early in 1942 he transferred from the RCAF into the USAAF as a 2nd Lieutenant. Quickly assigned to Craig Field in Alabama, he found himself in the role of an instructor in AT-6s, a home-grown version of those Canadian Harvards. Yearning for more excitement in the air, even at Craig, he had observed several Curtiss P-36s and Seversky P-35s parked at Base Operations which were under their jurisdiction. It was also obvious most of the senior instructors and training supervisors (captains) gravitated to the Curtiss P-36s, shying away from the Severskys.

But let Colonel Stirling relate his experiences with the P-35, a pursuit type which made up at least one quarter of the entire first-line Army Air Corps defensive force in the last half of the 1930s. With the outbreak of war in Europe in 1939, the Swedish Government found itself with an abundance of obsolete aircraft for defense of the country and a tiny production capacity. As pointed out in an earlier chapter, they bought 120 more powerful and better armed Republic EP1-106s. The Army Air Force, under Presidential authority, had commandeered the final sixty aircraft on the Swedish contract, redesignating them Republic P-35As, and most of these were serving in the Philippines when the Japanese attacked on December 7.

By mid year 1943, after having received a production go-ahead in early 1942, Curtiss Buffalo was only able to say that it had built half of the 40 examples of the P-47G-1-CUs (not quite up to Republic's P-47D-1 version). The initial 20 airplanes constructed as P-47Gs were equal to Farmingdale's C models. And Curtiss-Wright did not even have to build a factory. Starting at nearly the same date, Republic constructed a brand new factory and mod center in the Indiana farm fields near Evansville and had flown their 115th airplane – a combat-worthy P-47D-1 just about 100 days after this Curtiss airplane was photographed. Virtually all of the Evansville employees were farm people, while Curtiss had an experienced industrial base to tap. When the WPB finally found out how poor the Curtiss effort was (probably via Truman Commission reports – in the newspapers), they canceled 4220 airplanes. Buffalo had turned out no more than 354 T-bolts by March 1944; the built-from-scratch Evansville plant had turned out 1141 useful P-47Ds by the end of 1943! An XP-40Q (AC42-9987) converted from a P-40K receives mechanics' attention in background. **(Curtiss)**

I made contact with Captain John Hester, who was the base operations officer and OIC of the P-35s and P-36s. The captain authorized me to fly the P-35 type, and he personally supervised my first flight. He made it a requirement that I study the appropriate Technical Order, complete a brief and simple questionnaire, and finally I had to undergo a lengthy verbal interrogation regarding the technical aspects of the aircraft.

Within a short time I was enjoying flight in the P-35, completing perhaps a dozen flights. This bird was really very touchy, especially on takeoff and landing, and it demanded more attention than any aircraft that I had ever flown before. The big Twin Wasp geared engine and three-blade propeller was a real contrast to the AT-6 since it produced 950 horsepower. As I remember in that hot Alabama summer of 1942, the powerplant also generated a great amount of hot air, a fair amount of which flowed into the cockpit.

The landing gear system was relatively primitive and somewhat of a bother to operate. (In fact it was appropriate to call it primeval. Seversky P-35s were only the second operational type in the pursuit category to feature a retractable landing gear. Worse, it was a giant step backward from the system used on Consolidated's P-30/PB-2A and was actually only semi-retractable. – Auth.) *According to the T.O., the*

One of the more interesting photographs to come out of the homefront sources in WWII was this night firing scene with an early P-47D model blasting away with all eight .50-cal. machine guns. **(Author's Coll.)**

landing gear could be retracted and extended in one of two modes: electrically or manually.

In practice, it was necessary to use both modes. After takeoff, the gear was retracted electrically; then the gear mode handle was set to the MANUAL position. Next, a small crank below the gearbox was rotated a few times to ensure that the long worm-screw mechanism had the gear positively up. To extend the landing gear, the reverse procedure was even more important to ensure that the worm-screw had extended the gear fully and it was locked down. Because of this relatively complicated procedure, the standard fighter break and the tight landing pattern was essentially impractical.

One day, after turning downwind to land at Craig Field, a brake part failed and the rudder pedal on one side was flapping uselessly. This was a real predicament, and it could prove disastrous if the fighter swerved off the runway and crashed, possibly into the large number of aircraft parked on the ramp. I made a long, straight-in approach followed by a gentle, tail-high touchdown. The Seversky rolled right down the center of the runway, continuing straight off the end and up onto the Craig Field golf course, eventually turning 90 degrees before stopping on the green. (Lt. Stirling was most fortunate to have landed with the tail high because P-35s were noted for a ground looping tendency that would occur abruptly, especially while in a three-point attitude. Auth.)

The P-35 design had some bad features. The most dangerous, I believe, was the use of integral fuel cells in the wing section. (Before the Battle of Britain, most Seversky/Republic designs used the "wet wing" concept to maximize fuel capacity and range. They were prone to seep, weep and sometimes, but rarely, leak. There was no way in which they could sustain battle damage without serious problems arising.) *These tanks were prone to leak, causing belly landings to be dangerous because of the fire hazard. The P-35 was also prone to go over on its back, and for this reason the pilot's seat could be released to fall backward, allowing the pilot to escape into the oversized baggage compartment in the aft fuselage. From there he could exit through the large baggage hatch located on the port side of the fuselage.*

To the best of my knowledge, the Seversky P-35s flew for the last time on July 4, 1942, under the following circumstances: Lt. John "Pappy" Herbst and I engaged in an aerial dogfight that day. We were both flying P-35s in the vicinity of Selma, Alabama, starting our engagement somewhere around 10,000 feet. We ultimately found ourselves trying to get on each other's tail right down "on the deck." In our intense enthusiasm, we practically flew down the center of the main street of that town at high speed. When we landed, we fully expected to be disciplined severely. We really didn't care because we both wanted to be shipped out of Craig Field to get into the war. However, while we were dogfighting, some major in a P-35 experienced a total power failure, forcing him to attempt a "dead stick" landing on an auxiliary training field located 10 or 15 miles from Craig.

That landing gear actuating system or low altitude probably affected the major's judgment because the P-35 touched down with the landing gear retracted! For unknown reasons the wing tanks ruptured, the fuel ignited and the pilot was burned to death in the cockpit. Because of the holiday, no firefighting equipment that normally would have been present at the auxiliary field was at the crash scene. Why didn't the pilot immediately leap from the cockpit? Why didn't he release the seat and escape through the aft escape hatch? Later on we heard that not only was the canopy still closed, it was evidently jammed. In those days, it was virtually impossible to break out through the canopy.

As a result of that July 4 fatality, all P-35s in the AAF inventory were grounded, relegated to Class 26 to be employed in the training of mechanics. Herbst and I remained undisciplined for our dogfight at Selma, and we were probably the last pilots to fly Seversky P-35s anywhere. (For the record, the Swedish *Flygvapnet* continued to operate their Republic EP1-106 versions for several years. All USAAF Republic P-35As overseas were either lost in combat or destroyed on the ground by Japanese attackers before the fall of Bataan or Corregidor. A few P-35As remaining in the United States were virtually indistinguishable from the Severskys, and quickly joined them in Class 26. – Auth).

Capt. John "Pappy" Herbst became a 15-victory ace in the Fourteenth Air Force, having been promoted to colonel by war's end. He was killed in a P-80 during an airshow maneuver in postwar days. That accommodating, but strict, Craig Field operations officer, Capt. John Hester, eventually became a major general. While he was commander of the Seven-

Securely mounted on a ramp at the then raw Eglin Field, Florida, late in 1942, the third production RP-47B (R for Restricted operation) was tested by Air Proving Ground Command to determine the exact vulnerability of the supercharger system to machine gun and cannon fire. The First Priority tests were conducted in November and December, with machine gun bullets and 20-mm cannon shells fired into the fighter, especially into the turbine rotor while it was spinning at operational speeds. Point blank mounting of the cannon for one series is shown. (APGC)

Evidently being flown at Farmingdale, NY, an early P-47D-2-RE exhibits a generally overlooked feature of many Thunderbolts that were produced in the latter months of 1942. That all-important keel modification which permitted installation of a shackle unit to carry a 500-lb. bomb or a 75-gallon drop tank was not yet installed. The deeper belly line seen on most P-47s produced in the war years was yet to be incorporated. Many T-bolts delivered in this form were modified by use of retrofit kits later on. This fighter appears to have developed an oil cooler leak that needed attention. **(Republic)**

teenth Air Force in Europe in 1965, General Hester died from injuries sustained on the fifth and final parachute jump during a training course session.

IN ANTICIPATION OF COMBAT

Republic Aviation completed the first of fifty-eight P-47C-REs (AC41-6066) on 14 September 1942. By early October the much improved P-47C-1-REs began to come off the assembly lines in Building 17, the new factory assembly building at Farmingdale. The earlier group of C models were basically B models with strengthened aft fuselage structure for empennage attachment, metal-covered flight control surfaces and a revised fin/rudder unit without the exposed mass balance weight seen on all earlier versions of the P-47. A revised oxygen system with additional capacity was installed and there were radio equipment changes. However, the QEC package accommodation was not part of the production change, so the fuselage length did not change. The revisions to the rudder increased overall length of the airplane by an inch. Col. Zemke had not found the RP-47B model to his liking, but with the 8-inch fuselage extension in the firewall area, revised primarily to accommodate the QEC package, a rather significant improvement in overall maneuverability was noticeable, perhaps even conspicuous. With the standard Curtiss propeller in place, overall length had grown to 35 ft. 5-3/16 inches. Fuel capacity was 305 gallons, giving a trained pilot a maximum cruise range of about 835 miles, while normal combat range was no better than 550 miles.

Availability of the P-47C-1-RE, ostensibly the first really "combat worthy" Thunderbolt model, resulted in dispatch of three to Eglin Field, Florida, the newly created Air Proving Ground Command. On orders from Wright Field, they were flown directly to Eglin. B/Gen. Muir Fairchild, then Director of Military Requirements, had issued a blanket order instructing the APGC to determine the tactical suitability of every new fighter arriving at the base. His tone carried an extreme sense of urgency. By that time there was wide-ranging awareness about the unsuitable performance of the Bell P-39D and P-400 fighters in the South Pacific arena. General Fairchild most certainly knew 1942 plans to include P-39 Airacobras in Operation BOLERO had been reversed, with only Lockheed P-38Fs being involved in the ultimate operation of flying fighters across the North Atlantic Ocean. In fact, an alternate plan to ferry P-39s, more than likely aboard the USS Ranger, to a point where they could then fly off to England had been abandoned.

Testing of the P-47C-1-REs began on October 27th and was concluded by Thanksgiving week. This program was set up to participate in a test program involving a Lockheed P-38F, Bell P-39D-1, P-40F and a P-51 (no suffix). Part of the program required mock air-to-air combat. All flight testing was conducted by the Tactical Combat Section, with Special Project Officer Captain L. B. Meng in charge. Every aircraft assigned was cloaked in standard camouflage finish, with no special changes permitted, especially for the speed trials. One or more of the Thunderbolts exhibited the best high speed performance – at all altitudes – in the entire group of AAF fighters. (That had to be a shocker to every Allison devotee or Merlin advocate. Even Hub Zemke had misgivings about flying any big blunt ended mass, expecting seriously compromised speed performance.) How could Republic's admittedly "enormous" radial-engine fighter ram its way to the forefront in speed? Possibly more shocking, those P-47Cs were faster than any fighter in the world in production trim at rated power. At an altitude of 30,000 feet, using calibrated instruments, one of the Jugs was able to demonstrate a top speed of 427 miles per hour.

However, in the rate-of-climb category, the Thunderbolt was a bitter disappointment...it could not generate a rate any better than 3000 feet per minute. Turning to the XP-38 of the late

From the first day that Hough and Shafer designed the paper drop tank system until the first flight date was only 16 days in May 1943. All routine channels were bypassed, evidently with President Roosevelt's personal blessing. According to Hough's own Form 5, the tank demonstration drops were apparently made at Burtonwood Air Depot (near Liverpool), witnessed by many high ranking military and civilian personnel on June 26, that being the only day on which he flew an early P-47 (a C model) at B.A.D. in 1943. The special mount on the C keel and the P-47D cowl that had been retrofitted are noteworthy. That's Cass in the cockpit. Within 30 days of the concept, he said operational versions were being delivered. The ultimate success of that project is reflected in the thousands used on P-47s and P-51s. **(Cass Hough)**

Two very, very familiar faces (and one body) belonged to none other than Bob Hope and his ever-present singer, Frances Langford, during a visit to the 78FG base at Duxford on 3 July 1943. The P-47D was the mount (normally) of Capt. Robert E. Eby. Hope and Langford seemingly popped up everywhere that military or naval personnel created a base of operations. (Jack Oberhansly)

1930s and recalling it was designed to reach 20,000 feet in six minutes, which it accomplished on 1030 hp per engine, P-47C evaluation pilots were not pleased to find it took 14 minutes to reach that altitude. The mock combat phase showed the Thunderbolt to better advantage in rate-of-roll, dive acceleration and, a bit surprisingly, in level flight acceleration – but only after the inline-engine types began to overheat and were forced to open coolant shutters. Turning radius of the P-47C was greater than any of the competition, with the P-38F doing unexpectedly well in that department. In fighting to tighten the turns, the Jugs always tended to "mush" or, when attempting to turn with any of the others they demonstrated high-speed stall characteristics. Top-scoring combat pilots ranked rate-of-turn at the top of their "want" lists.

Despite Col. Zemke's expressed disappointment with a T-bolt's acceleration in level flight, it is surprising to learn there was never more than a 50-yard gap among the adversaries – decidedly slower than any of the fighters being evaluated, the P-40F was not included. Col. Zemke was absolutely right, however, because only the P-38F (of the competing trio) was ever capable of beating the Messerschmitts and Focke-Wulfs in the ETO and MTO. (In time the P-47C was a match for the Germans, but lack of good range remained a handicap.) The 50-yard handicap faded rapidly as the Allison-powered types began to overheat. (Lockheed P-38Hs would have possibly encountered a problem in that category if the pilots chose to use all available higher power. The leading-edge intercoolers would be overwhelmed with rapid heat rise.) From a level flight attitude and at the end of diving maneuvers, the P-47s were about equal to the P-40F in zoom ability. In contrast, the P-39D, P-38F and P-51 could outperform the Jug every time. Close-in fighting was, rather surprisingly, one of the best events for the P-47Cs at Eglin Field, primarily because of the very excellent rate of roll. Pilots found they could roll into a reverse turn and break off combat virtually at will, a great advantage in combat.

In view of the P-47C's deficiencies, the Tactical Combat Section pilots universally *recommended avoiding dogfighting with any type of first-line enemy fighter likely to be encountered in a combat zone*. That sort of commentary would grab Zemke's attention in a hurry. He would quickly remind one and all: "You fight with the tools you have, and the arena is most unlikely to be

OPPOSITE: There can be little doubt about the final destination of most P-47s after Operation POINTBLANK was given new impetus in June 1943 with the ultimate aim of preparing for the launch of OVERLORD in about one year. That key destination, of course, was the British Isles, and the greatest concentration of the Eighth Air Force men and machines was no great distance from London. It was only natural that everyone would gravitate to that massive city sooner or later. A main focal point was certainly Piccadilly Circus with its entertainment, garrish signs, missing street lights and a preponderance of double-decker buses and London cabs. (Author's Coll.)

When Lt. Robert Johnson's badly mauled P-47C-2-RE was moved to the boneyard at Kings Cliffe since it could not join in the transfer to Horsham St. Faith, he flew a borrowed P-47C-5-RE like the fighter shown here. This particular 62FS Jug featured a nearly perfect likeness of cartoon character, "Li'l Abner" and bore the name "Torchy." It was assigned to Capt. Eugene O'Neill, leader of Red Flight. (AFM)

one of your own choice." One terribly surprising omission from the Eglin Field trials was Combat Radius of Action, with and without external tanks. The P-38F would have run off and left the pack, but then it would have been an interesting tussle among the also-rans. Only the P-47C and P-51 in that group lacked any sort of provisions for external tankage. (The P-38F was fully qualified and equipped to carry two 165-gallon drop tanks, and test pilot Milo Burcham had demonstrated the capability to take off and fly safely with two 310-gallon drop tanks. As a reminder, on 27 August 1942, he piloted a P-38F-5-LO equipped with those tanks over a partially corrected distance of *3167 miles* in *13.67 hours* – nonstop – and that was without the benefit of being able to jettison empty tanks. Question: Did anyone in a P-51D circa 1945 ever even match that? Any P-38L-5-LO could have *exceeded* the recorded distance.)

Unfortunately, when the results were submitted to General Fairchild, his remarks eventually were absorbed into the labyrinth of "impossible to locate" files.

Responding with little or no enthusiasm, or was it merely understanding(?), to pleas from the small number of pilots who

With some rising stars in the "formation" at AAF Sta. 365, famed WWI ace Eddie Rickenbacker and another ace from the "Great War" of a quarter century past, B/Gen. Frank O'D. Hunter, visited the new home of the 56FG on 29 July 1943. Col. Hubert Zemke, talking vigorously with Rickenbacker is followed by his flying executive, Dave Schilling. The sartorially correct "Monk" Hunter strides along at the left of the group. A Beech C-45 may be seen in the background. Rickenbacker addressed most members of the fighter group later in the day, even as a mission was launched. Ten days earlier, Zemke had been awarded the group's first DFC. (56FG)

had become familiar with the bird, Alex Kartveli assigned somebody the task of providing a given quantity of external fuel. He may have been responding to a request for ferrying operations, or somebody from Wright Field may have sent photos or drawings of a bulbous belly tank provided for use on Bell P-39Ds. When that fighter type was originally set to participate in Operation BOLERO, a massive "cow's udder" affair was created for the Airacobras. In the event, a near duplicate of the P-39D container was created by Republic with a (gasoline) capacity of 205 gallons. No attempt was made to pressurize the tank for high-altitude operation, leading to the assumption the tank was *never* intended for combat operations. At least for some unknown time period, no attempt was made to evaluate drop separation from the aircraft in empty or partially filled conditions. Provisions for a 4-point attachment and supply hookup of the drag-producing glandular protuberance were provided on P-47C-2 and C-5-RE airplanes destined for overseas delivery to the ETO.

Some flight testing was certainly accomplished in October 1942. One early photograph of seven flight test airplanes on the line in front of Hangar No.1 at Republic Airport (refer to Chapter 8) shows two camouflaged P-47C-1-REs equipped with the 205-gallon tanks. (The first five test aircraft in the lineup were painted yellow to ward off unwanted aerial encounters.) Only two of the five test aircraft were B models, including "8-Ball." Rather surprisingly, a P-47C-1-RE (AC41-6149) marked as #265 in the lineup, appears to be carrying a 110-gallon metal teardrop external tank. We can only presume ASC would not release that tank for production until it was thoroughly tested under simulated combat conditions. Processing through the chain at ASC usually took months, even in the face of desperate combat need.[10]

If Dade Brothers constituted the entire shipping organization responsible for transporting Thunderbolts overseas from the East Coast, at least, they were certainly busier than ever before. Although it remained routine to ship QEC packages overseas in crates, and many other component assemblies went in that manner, it was no longer appropriate to crate complete airplanes. Cocooning methodologies were invented by somebody, and used almost universally during the war. One very complex effort involved loading the P-47s in the holds of freighters, on the decks of freighters and tankers (the latter known to ride low in the water, compounding the problems of corrosion prevention). Possibly as many as 450 of the C-2 and C-5 versions of the Thunderbolts were shipped to Liverpool, England, or to Scotland from the Port of New York for processing through one

[10] In the absence of any related documentation, photo interpretation of one single obscure picture can produce an appropriate summary account of what was actually occurring in the relevant time frame. This account could not have been generated without the single battered photograph rescued from trash by long-time Republic employee Frank Strnad.

Although the Republic-designed 205-gallon unpressurized belly tank shown in this rare photo did extend the range of 4FG T-bolts in some early 1943 combat missions (carrying only about 100 gallons of usable gasoline), drag and altitude limitations were unacceptable. Capt. "Mike" Sobanski's P-47D-1-RE, marked QP★F, is shown taxiing out with his wingman, probably on the July 28 mission to Emmerlich, Germany, as trials with those "pregnancy" tanks were carried out. "Caught out," some Luftwaffe outfits sustained unexpected losses at the hands of the 4FG. The 56FG used 75-gal. P-39/P-40 tanks that day. (Sobanski via C.A.M.)

of two British Reassembly Depots (BRD) operated by Lockheed Overseas Corporation (LOC).

Lt. Col. Cass Hough, commanding VIII FC Air Technical Section at Bovingdon, flew over to BRD at Speke to check out on a newly arrived and processed P-47C-2-RE, most likely the fourth such airplane to arrive in the ETO, AC41-6182. It is truly unfortunate pilots rarely entered the complete designation or the a/c serial number on their Individual Flight Record (formally known as Form 5). Hough first lifted off in the fighter on 23 January 1943 and was most certainly one of the first to fly a P-47 in England. After putting eight hours on it over a period of a few days, he flew to home base on 29 January, then hopped over to Duxford the same day to show off the new fighter to members of the 78FG, possibly touching base with Col. Arman Peterson or Major Harry Dayhuff. If the 56FG or 4FG actually picked up the first of their allocated P-47Cs on January 24th, Hough was certainly at Speke and probably even had some discussions with the fighter pilots. As for the 4FG, he would have been aware of the situation if they had already picked up a couple of P-47Cs assigned to them. They would have flown back to Debden the very day the Jugs were turned over to them.

Although the original members of the 78FG had arrived in England a couple of months ahead of the 56FG personnel, they had trained diligently on Lockheed P-38s, then worked up on them in England from early December. When all P-38F/G Lightnings were commandeered, along with most of the pilots, for Operation TORCH, the small cadre of key flying and ground personnel remaining in England were soon told of the planned switch to Thunderbolts. To be truthful, the 78FG expressed no joy at the idea of rebuilding with new pilots and unfamiliar, unproven aircraft. What had been a very cohesive fighting group was suddenly a handful of dissociated pilots and ground crew members.

As for the 4FG, most pilots and ground crew were, to say the least, appalled by the blunt-lined massiveness of the Thunderbolts and in comparison with their elite, elegant Supermarine Spitfires. If the T-bolts had enjoyed some significant range advantage and high rate-of-climb performance or acceleration over the Spits, the changeover would have been less onerous. As things were, the P-47Cs could not even reach as far as Antwerp in a *Rodeo* let alone as part of a *Circus* (refer to Glossary). Any prolonged air-to-air combat heightened the potential for splashing in the Channel. Add to that an inherent human opposition to change and you encounter a serious problem. The problem was really never cured for at least half the group's pilots until the Thunderbolts gave way to the Mustangs.

Why did the 56FG rather wholeheartedly embrace the P-47 while the 4FG never warmed to the big fighter? The answer most likely centered on the group and squadron commanders. It is not so much a matter of what goes up on the chalkboard in black and white, but more of an undercurrent spread by osmosis. Oh, Hub Zemke was not really enthralled with the Thunderbolt from day one, but he showed no fear, hatred or open contempt for the type. As he was wont to say, "You fight with what they give you."

The powerful Thunderbolts had barely appeared on the flight lines at Speke, Burtonwood, Langford Lodge or Renfrew when orders were issued to the Air Fighting Development Unit (AFDU) at RAF Station, Wittering, to begin tactical trials and evaluate the results for the Air Technical Section's director at Bovingdon. Obviously the director, none other than Lt.Col. Hough, had asked General Hunter to arrange for such an order.

Nothing had been done anywhere to evaluate night flying and night fighting performance of the P-47C. Therefore it was a relatively high priority item on the agenda. Flying characteristics of the type were thought to be good for a fighter aircraft, but there were some serious shortcomings. The landing light was far too bright, but that could be rectified with ease. However, when

Clean-cut Capt. Winslow "Mike" Sobanski was one of the oncoming stars of the 4FG in 1943 while flying P-47s in the 336th and then the 334th Fighter Squadrons. He and Duane Beeson (334FS) seemed to do better in the P-47s than many other group members. Sobanski is shown about to embark on a mission in the spring of 1943. In a 1944 Mustang mission, the aggressive pilot was KIA. (AFM)

the supercharger waste gates were open, the flare of the exhaust compromised the pilot's night vision. Testing was accomplished on a moonless, clear, starlit night, slightly darker than most "fighter nights." Observed from another aircraft, the exhaust flames could be seen for distances up to 250 yards from directly ahead or in any beam angle. With closure of the waste gates for supercharging, a red glow was visible at the supercharger hood, and exhaust flames shot out sporadically. Although not visible to an aircraft under attack from below and aft, supercharger exhaust presented an illuminated bullseye for a defensive gunner when the T-bolt broke off its attack, exposing the underside of the fighter. Attacks from above would be hazardous to the attacker because of that supercharger glow/flame. The AFDU report recommended removing the P-47 type from any consideration for use as a night fighter.

In daylight operations, USAAF pilots assigned to the project were pitted against an experienced pilot flying a Focke-Wulf Fw 190A-3 which had been captured and brought to first-rate flying/fighting condition. Gross weight of the Fw 190 was about equal to that of a Republic P-43 Lancer at 8300 pounds, while the P-47C-2-RE grossed 13,140 pounds. Battle tactics included attack and evasion by each aircraft at altitudes of 26,000 feet and 6000 feet. Although the high-altitude combat evaluation would have been the most valuable, that had to be abandoned when the BMW radial engine in the captured fighter began to cut out badly at high power settings. Maintenance people were unable to correct the problem. Based on the mock combat that did take place at 26,000 feet, the P-47C was determined to be at a distinct disadvantage, both in the attack or evasion modes. Pilots quickly found they were unable to perform any maneuvers which would allow them to evade except to enter a very steep dive angle. Obvious to all, somebody was going to somehow develop effective tactics for combatting the *Luftwaffe*'s prime fighter aircraft. And very quickly. Whenever an inexperienced P-47 pilot turned into an attack being attempted on a line slightly off center, the Focke-Wulf pilot could get off some blasts right at the beginning. Of course that was predicated on the effective range at which evasion tactics began. Even at the end of three full turns, the Fw 190 was still positioned to chalk up additional hits. When any climbing turn was attempted by a P-47 pilot, it proved "fatal" in terms of mock combat. Even with the AFDU pilot having a disadvantage of not being highly experienced in Fw 190 combat tactics and lacking the newest and best operating equipment, the Jug did not come off to advantage.

Whenever the P-47 pilot was pressing the attack, he could not gain the upper hand if the Fw 190 turned to evade. If the enemy aircraft climbed into a turn, results were even slanted more toward the German type. Any failure to quickly break off the chase could immediately result in the AAF pilot coming under aggressive attack. Only when the Focke-Wulf attempted escape by diving away did the Thunderbolt gain an advantage. A dominant position could be maintained during the dive with some ease, permitting some effective gunnery to take place. Surprisingly, the high-altitude Republic product could turn with the German aircraft at 6000 feet or less, at least if he did not attempt to zoom with the enemy.

Although the date cannot be confirmed, a visit by the Duke of Windsor to Farmingdale in company with B/Gen. Frank Hunter must have occurred upon the general's reassignment to the Z.I. after he was replaced by M/Gen. William Kepner as head of VIII Fighter Command. Bareheaded man is test pilot Parker Dupouy, there to point out salient features of the P-47D. **(Lowery Brabham)**

Realistically, the two combatant aircraft should be viewed in a direct time-sensitive light. They were far more comparable than most people realize. Even the German high command recognized their Fw 190A-2 series fighters were the first fully combat-worthy versions, with anything earlier being more or less comparable to the Republic P-47B and P-47C/P-47C-1 models. In their Fw 190A-3, they had a fully qualified type powered by a BMW 801D-2 engine rated at 1700 hp for takeoff and emergency power at sea level with 1450 horsepower available at about 18,700 feet. Gross weight of the fighter was (officially) 8770 pounds. For comparison, the P-47C-2 and C-5 models had 2000 hp available for takeoff and at 25,000 feet with the turbo-supercharger. Normal power was 1625 hp at 6500 feet and at 25,000 feet. Normal gross weight could be as high as 13,500 pounds. Accounting for the lack of advantage on the part of the Fw 190A-3 at 6000 feet, its maximum speed in that realm was only about 330 mph while P-47Cs were regularly clocking approximately 352 miles per hour there. At 20,000 feet, TAS for the Fw 190 was just under 385 mph, with the P-47C able to hold its own , but the German type could not maintain the power-required output for more than one minute (in override boost). The big Double Wasp could easily maintain high power for 5 to 10 minutes (military power), even before water injection was an added feature. Those early P-47Cs had a very slight maximum range advantage of perhaps 50 miles, but half of that was eaten up before they got to the enemy. The Focke-Wulf Fw 190 had almost exactly two-thirds the wing area of any P-47 fighting in the ETO during the war. Another shocker – information fortunately not available to anybody except possibly group commanders (doubtful) – fewer than 250 of the Jugs were delivered to combat groups in England by springtime 1943, while more than 2000 Fw 190A-2 and A-3 versions had been delivered to the *Luftwaffe* by the time the 56FG landed in the ETO. What did the P-47 type have as a co-equal backup? Nothing at all. But the backup for Focke-Wulfs was at least a co-equal in the form of late-model Messerschmitt Me 109s. In excess of 40 percent of all German first-line single-engine fighter production was represented by Fw 190As as of 1 January 1943.

Why were the Nazis capable of designing and building a directly competitive radial-engine fighter which weighed in at some 5000 pounds less than an early series Thunderbolt? Why were their much lighter fighters able to outclimb our turbosupercharged P-47s at nearly any altitude up to perhaps 25,000 to 30,000 feet, and perform those feats with mechanical supercharging? There is no single answer, of course, but a young Lt. Benjamin Kelsey saw it rather plainly in the 1930s. Army Chief of Staff Gen. Malin Craig's budget control effectively blocked development of multi-speed, multi-stage mechanical superchargers. Then an unwarranted commitment to development of inline liquid-cooled engines stemming from high-level love affairs with small frontal area engines merely compounded the situation. Perhaps the part played by Lt.Col. Edwin Page, almost completely unknown to the general public in the decades of the '20s and '30s, was far greater than anyone can recall. Despite the tremendous successes of the Wright Whirlwind and Pratt & Whitney Wasp in the 1920s and early 1930s, the Army Air Corps hierarchy turned their backs on aircooled radial engines for fighters, at least, by 1935.

The questions relating to mechanical multistage, multispeed supercharging were pressed in an interview with the late General Kelsey in 1973. "The Army Chief of Staff, Malin Craig, blocked every request for funds, and nobody of rank would fight for such development," he responded. Over in Germany, the first independent aircraft radial engine development – as opposed to license-built engines – initiated by BMW resulted in the 801 series. It featured two things largely ignored in the United States by the Navy Bureau of Aeronautics, the NACA and by the Army Air Corps, namely a single-stage, two-speed mechanical supercharger and direct fuel injection. The deep effects of Isolationism, the Great Depression, and industrial lobbying cannot be ignored. Even after a half century, the evidence lies in the results. American executive and congressional leadership in the 1920s decade cannot be looked upon with tremendous respect or warrant admiration for performance.

In spite of bad seeds sown in past decades, America survived and became the true Arsenal of Democracy. There was no shortage of ingenious Americans who could compensate for past errors in judgment or selfish interests. Those men were being tested many places on earth, and the Arsenal was going to prove its real worth.

Col. Hub Zemke[11] has expressed a strong view that one of the more significant conceptual planning mistakes made by the Army Air Forces was based on the belief bombardment forces, light, medium or heavy, could rely almost entirely on self protection. The theory of fewer than twenty fighter groups (actually, fifteen was tops) in the Eighth Air Force being needed to support as many as forty-three heavy bombardment groups plus a smaller number of medium and light bombardment groups was badly flawed. The colonel contended the totals should have been absolutely reversed, or even three fighter groups to one bomb group would have been more logical. While AAF and RAF bombing did wreak havoc on Germany, much of the damage only proved to be superficial. German production of aircraft actually surged forward in 1944, and production of every other weapon of war or supporting equipment may have been just as unimpeded by the bombing. Wounded, yes; destroyed, no. Yet once Allied fighters gained absolute control of the air, or nearly so, the tide surged against the Nazi war machine!

Consider each heavy bomber had to be manned by up to ten or eleven men while each fighter only involved one man in the air, it seems clear 20 to 25 groups of heavy bombers should have been supported by 40 to 50 groups of fighters, with great emphasis on range extension. Early mastery of the air over *Festung Europa* by short-range and long-range fighters could have been achieved with lesser loss of men and greater concentration on production of fighter aircraft – but not the ones proven to be inferior to the enemies' defensive fighters.

Most certainly the final results of the war prove we did many things correctly, though proof also exists that better ways can be even more productive. What more proof could there be than the Republic P-47? The all-important decisions of the Emmons Board to scrap the Curtiss P-46, Republic's P-44 and Allison P-47 programs at a time when the War Production Board did not yet exist have never been fully recognized. General Emmons, representing the element that would do the fighting and General Oliver Echols, representing the side that would supply the fighting equipment, deserve our undying gratitude.

Some time in February, 1943, Lt. Col. Hough took off in his new P-47C-2-RE, determined to find the cause of at least two accidents in which Thunderbolts had dived straight into the ground, purportedly from high altitudes. At least one of the accidents was fatal. (Unfortunately for history, neither the actual dive date nor the aircraft serial number were included anywhere in the reporting document.) Otherwise, details included were as plentiful as those included by the pilot in his Lockheed P-38F dive report submitted late in the year, 1942.

Colonel Hough flew to an altitude of 38,000 feet (indicated), established a speed routine of 195 mph IAS and then half rolled into a vertical dive. He managed a successful recovery in the vicinity of 12,000 feet. He submitted the report to AJAX, General Frank Hunter's VIII FC headquarters.

In February-March 1943, Hough traveled by Douglas C-54 to the Zone of the Interior, primarily to visit Republic Aviation at Farmingdale and Lockheed Aircraft Corporation in Burbank. At that time he more than likely had been exhorted by his boss, B/Gen. "Monk" Hunter, to return to America. The General was no stranger to the value of publicity. He wanted Hough to furnish data to Lockheed about his compressibility dive in the P-38F. The subsequent dive in a P-47C was to be discussed with Republic. During the course of that trip Lockheed's Hall Hibbard issued the news release to the *Los Angeles Times* and other news media about Hough's P-38F accomplishment. He was not aware an even flashier news release telling of Hough's 780 mph dive in a P-47 was soon to hit the press. It was also going to be a surprise to the colonel.

[11] It was not just planning being flawed either. The outspoken Col. Benjamin Kelsey arrived in the ETO in mid 1942 as a full colonel, but after a lifetime of outstanding contributions to airpower, he was lucky to retire as a brigadier general. Col. Hubert Zemke arrived in the ETO less than six months later in command of the 56th Fighter Group. He fought until finally being downed after a truly inspired career as a combat group commander. After the war, his military career as a tactician continued unblemished. Zemke's status with a knowledgeable aviation-minded public rivals that of Charles Lindbergh. Upon retirement he remained a colonel. By contrast, B/Gen. Frank O'Driscoll Hunter arrived in the ETO to *lead, command and inspire* VIII FC. Less than a year later he was back in the Z.I. amid great bitterness, promptly receiving a promotion to major general. The record is not lost to posterity.

Numerous Republic P-47D-5-REs are depicted below decks in what must have been an escort carrier, or possibly the USS Ranger, upon arrival in Belfast, N.I., on 11 August 1943. By the time they had arrived, that Type 2 national insignia was definitely obsolete, but the new D-5 series Jugs were welcomed with open arms by VIII FC. Later photographs of fighters in transit or in open storage upon arrival confirmed that Republic Aviation had been instructed to delete the fuselage insignia markings. All airplanes in this photo were part of a shipment called Project UGLY. (Author's Coll.)

LEFT: A bird with 17-league boots was formally conceived in the terrible financial environment of 1933, with Boeing receiving the contract for a prototype two years hence. Designated XB-15, this aerial monster weighed in at more than 35 tons when it made its first takeoff on October 15, 1937. Unsupercharged and underpowered, it served the Air Transport Command during WWII (redesignated XC-105) by flying heavy cargo between Miami Army Air Field and the Panama Canal Zone. Smaller B-17s performed the general bomber requirement more effectively in battle. (Rudy Arnold/S.I.)

OPPOSITE TOP: From a retrogressive Seversky SEV-2XP hopeful for a 1935 pursuit plane contract – through at least two "ugly duckling" formats culminating in the contract-winning SEV-1XP design of 1936 – Major de Seversky took every measure necessary to ensure victory in the Air Corps pursuit competition. Half a century later, this charismatic P-35 stood as a monument to the tenacity of the Russian expatriot. Embellished with accurate markings of the 94th Pursuit Squadron commander in the late '30s, the restored Seversky P-35 mirrors a pride of workmanship that was prevalent in the Great Depression years. (AFM via D. Menard)

BELOW: Skillfully photographed with the new Kodachrome film in 1939 when the Seversky P-35 was still a first-line pursuit airplane for the Army Air Corps, this example was from the Selfridge Field 1st Pursuit Group Hq. & Hq. Squadron contingent. All but a handful of P-35s were stationed at Selfridge upon delivery. That gracefully sculpted wing worked wonderfully on all Seversky and Republic airplanes with only minor changes introduced from 1933 to 1945. (Rudy Arnold)

LEFT: Logically we can entitle this Paint Your Wagon *(with apologies to the playwright). For the 1939 National Air Races, pilot-owner Frank Fuller, Jr., had his Seversky SEV-S2 racer painted a metallic bluish-green with yellow accents. Fate decreed that it would be the last of the annual racing events for six long years and the last appearance of any racing Severskys.* (Hans Groenhoff/S.I.)

BELOW: In the latter half of the 1930s decade, Curtiss P-36A and C pursuit airplanes shared America's air defense role with the similarly powered Seversky P-35. This P-36A, assigned to Wright Field's Materiel Division, was obsolescent in 1941 when both types were being replaced by Bell P-39s, Curtiss P-40s and Republic P-43s, none of which could ever be regarded as suitable for effective air combat in the ETO. Most U.S. fighter aircraft paid a penalty, being designed from the outset to defend North America with its great distances between coastlines. (Hans Groenhoff/S.I.)

It's showtime! As war clouds gathered in Europe late in the 1930s decade, America sought to "avoid foreign entanglements." The impact of massive orders from nations opposing Naziism was yet to affect U.S. industry, and the Army Air Corps was still limited to fewer than 5000 aircraft of all types. Curtiss P-36Cs from the so-called Headquarters Air Force were swathed in colorful coats of camouflage paint, essentially for general public consumption at air shows. (S.I.)

With a great need for haste, the AAF issued specifications for a fast bomber in January 1939. Upon the outbreak of war in Europe, Glenn L. Martin Co. received a contract for 1100 of their radical B-26 (Model 179) bombers just days later. Powered by two of the same Pratt and Whitney R-2800 Double Wasp engines of the type soon to be specified for the Republic P-44 and ultimately in 1940 for the XP-47B Thunderbolt, the B-26 Marauder was the first 300-mph medium bomber. One of the five earliest B-26s off the line is shown, possibly in November 1940 when the maiden flight was made, at Martin's Middle River, Md., factory – also newly built. (Martin)

Flaunting its Seversky P-35 heritage, the first of 50 Republic AT-12 advanced trainers is shown in flight over Long Island, New York, with Chief Test Pilot Lowery Brabham at the controls. All but two of the 2PA-204A Guardsman dive bombers that had been sold to Sweden were seized by the U.S. Government in a "crisis" action in October 1940. The AT-12s retained their pre-war style markings for some considerable time, were generally regarded as optimum "hack" machines for use by the more senior officers. (Rudy Arnold/S.I.)

ABOVE: Early in the winter of 1941, service test examples and pre-production models of the Bell Airacobra were being delivered as P-39C versions (Bell Model 13s) in the cold surroundings of Buffalo, N.Y.. With Capt. Ben Kelsey totally involved in P-38 and other activities, the turbosuperchargers were omitted from all P-39s. Here was the NACA, devoted entirely to aerodynamic improvement, dictating tactics. By the summer of 1942, all P-39 useage planning had to be deleted from the BOLERO program, the RAF had rejected the type as an operational fighter, and even the AAF had soon relegated the newest models to ground support and reconnaissance missions in the MTO. (Hans Groenhoff/S.I.)

Following the 1939 pursuit plane competition at Wright Field, the losing builder of pursuits – soon to undergo a major reorganization and be renamed Republic Aviation Corp. – received a contract for thirteen yet unseen YP-43s. They were inspired by the private venture Seversky AP-4. Another losing entry, the Seversky XP-41, was soon assigned to a test program at NACA, Langley Field, Va., where this extremely rare color photo was taken for a magazine story, circa 1940. (CAM/Ed Boss)

Although the Curtiss XP-40 "failed the exam" in the 1939 pursuit competition – only the Seversky AP-4 could perform well above 20,000 feet – the Curtiss Airplane Division received the nod and a contract for 524 pursuit planes. The XP-40 was, in most respects, a 1935 airframe with a 1,030 hp Allison V-1710-19 engine up front instead of the 1,200 hp of the radial R-1830 of the first-line Curtiss P-36C. With the same wing, 170 less horsepower and weighing 750 pounds more, the customer and constructor would have us believe it was 31 mph faster than the parent P-36C. For the P-40 production model shown, the weight increased another 340 pounds but the speed increased another 15 mph!(?) and it had a miserly two 0.50- caliber machine guns. (Hans Groenhoff/S.I.)

Led by a gleaming Republic YP-43 development of the private venture AP-4, a covey of Army Air Corps pursuit airplanes all swung Curtiss Electric propellers as they cruised the sky for the benefit of a topnotch aviation cameraman. In the stack were a Curtiss P-40, a Bell P-39C and a Lockheed YP-38, all flown from the Wright-Patterson AAF complex in the summer of 1941. The tubby YP-43 Lancer showed some of the Grumman and Brewster influence (cross pollination?) of job-hopping engineers living on Long Island. (Rudy Arnold/S.I.)

ABOVE: Designed at least three years after the Lockheed XP-38 was committed to paper in a completely different world atmosphere, the Republic XP-47B was the most evolutionary of the ten best fighters of WWII. Nearly every feature had appeared earlier in Seversky Aircraft design layouts dating back to that company's "Executive" and AP-4 pursuit proposals of 1938 and far earlier with respect to the wing. The de Seversky-Gregor-Kartveli brand marks were everywhere on the XP-47B, finally having paid off after years . It is improbable that any Thunderbolt ever looked better than the prototype did in this pastoral Farmingdale scene in May 1941. (Rudy Arnold/S.I.)

Constructed in parallel with a brand new factory and administration building that would add about 1,000,000 square feet of floor space to the 220,000 that existed, the Republic XP-47B was ready for its first flight on May 6, 1941. Lowery Brabham, hired the previous summer as a test pilot, departed Farmingdale's sod runway 4-22 and planned to land at Mitchel Field because it had paved runways. After rains had soaked the Republic field, Brabham did not wish to risk a nose over on landing the heavy fighter on soft turf. (S.I.)

If you subscribe to the adage that "If it looks right, it probably is right," the Bell Airacobra likely shattered your concept. Robert Woods' design showed great promise in 1937, but a combination of errors canceled out the potential. The British Purchasing Commission took over a massive French order immediately after Dunkirk in 1940. Of 675 Airacobra Mk.Is on order, fewer than 80 were accepted for RAF service, and early Rodeos in October 1941 revealed severe deficiencies. As a result, deliveries to the RAF were terminated and the remaining Airacobras were taken out of operation less than 3 months after the first mission was launched. One weaponless example, AH621, is shown in flight over Western New York in 1941. (S.I.)

In 1939, with Lt. Ben Kelsey totally involved in the Lockheed P-38 program, the Curtiss P-40 won out over the Seversky AP-4 (and XP-41) in spite of inferior altitude performance. A major factor: the Battle of Britain was still to come and reveal the need for best performance above 15,000 feet. Responding to another ill-defined/ill-conceived Army Circular Proposal, Curtiss produced two XP-46 (Design CP39-13) pursuits. Powered – again – by the Allison V-1710, the XP-46 weighed as much as the older XP-40 and performance was less than ½ percent better. Republic responded with the AP-10, one-third lighter but powered by the same engine. It was subsequently known as the XP-47/XP-47A. Eventually the XP-47 was only 1000 pounds lighter than the XP-46 and could have passed as a fraternal twin. The Emmons Board, reacting to new information from England, quickly cancelled the XP-47/-47A. Curtiss's XP-46A is shown at Buffalo in the winter of early 1941. (Warren Thompson)

A rather colorless trio of Grumman F4F-3 navy fighters from Fighter Squadron 5 (VF-5), basically assigned to the USS RANGER, display the immediate pre-war camouflage paint scheme adopted for Navy and Marine aircraft. After Pearl Harbor the Bureau of Aeronautics was forced to adopt some vastly revised color schemes to meet the challenge of global warfare conditions, but there was no way in which colors could be changed with any immediacy as battles raged for mere survival. (Hans Groenhoff/S.I.)

BELOW: In a radical departure from every existing concept of a carrier-based fighter anywhere on earth, Grumman Aircraft proposed a twin-engine type. It was ordered by BuAer as the XF5F-1 on June 30, 1938, at a time when brand-new single-engine biplane fighters were still being manufactured (until nearly a year later). Named Skyrocket, the new interceptor was first flown on April 1, 1940, some 14 months after the faster, better looking Lockheed XP-38 performed that feat. But the XF5F-1 did fly about 60 days before the Vought XF4U-1 made it. Despite being the most powerful of the trio at 2400 horsepower, the Skyrocket was the slowest at 383 mph top speed. (Rudy Arnold/S.I.)

History shows that the most important role played by Republic P-43 Lancers was keeping the old Seversky factory operating while a new plant was constructed, providing time for the engineers to produce the new XP-47B, and maintaining a cadre of experienced workers. The Republic attempt to supply a viable fighter by developing the P-44 was doomed to failure by range limitations. Total combat duration was only 2 hours, fuel load hardly greater than that of the Lancer and the R-2800 having a healthy thirst. As in the XP-47B, there were no provisions at all for external tankage. Few, if any, P-43s were involved in West Coast defense in December 1941.This example was photographed at San Angelo, Texas, late in that year. (Col. Fred Bamberger)

ABOVE: Men of destiny were at the controls of this flock of RP-47Bs of the 56th Pursuit Group (Fighter) early in 1942. Assigned to protect Republic Aviation at Farmingdale and the New England defense plants, squadrons and flights were dispersed to various fields to provide the needed defense against any kind of sneak attack. Leading the formation on this publicity mission was Lt. Col. Hubert "Hub" Zemke, newly assigned group commander, in the colorful No. 1 airplane while Lt. Robt. Johnson piloted No. 26. (S.I.)

Republic RP-47B fighters of the 56PG(F) are shown in simulated patrol activity over Long Island, N.Y., during the early post-Pearl Harbor days. This group was probably the only operational group to fly the P-47B model. Differences in photographs of the same flight must be attributed to film used, poor preservation or varied photographic exposures. (Frank Strnad Coll.)

RIGHT: Lt. Earl Hayward and his new Republic P-47B were the subjects of this excellent 1942 color photograph taken at a site reported to be Sikorsky Field, Bridgeport, CT, with a WWI type hangar and a camouflaged Douglas B-18 in the background. Both the war and the AAF's very large red-nose Republic RP-47B were extremely new and loomed most impressively over the young 56PG(F) pilot who soon would be posted to the ETO. Hayward bore a striking resemblance – at least with his helmet on – to Francis (Gabby) Gabreski, then in the RAF. (Rudy Arnold/S.I.)

ABOVE: A top-rung Vought-Sikorsky test pilot – believed to be Boone T. Guyton – is shown with the very first production Vought F4U-1 Corsair on the day of its first flight, 25 June 1942. Power had been increased to 2000 hp by installation of an R-2800-8 engine. Top speed was stated as 415 mph., just about as fast as the XP-47B and the earlier XP-38, but at a lower critical altitude of course. Nobody had a carrier-based aircraft that was as fast, even on the drawing boards. It was probably somewhat tougher than the Thunderbolt or Lightning fighters. True, American fighters were very big, but what foreign high-performance fighters were such prodigious weight carriers (read that as long range and assault weapons carriers)? (S.I.)

With its stars brilliantly displayed, a brand new Lockheed P-38D climbed into the blue just weeks after the Pearl Harbor attack. If the "combat ready" D model was really not suitable for the job on December 7, the same was certainly true of the P-39Ds, P-40Bs and Cs or P-43s. However, the Lockheed P-38E model proved to be suitable for the job then, and hundreds of effective P-38Fs were in the pipeline during the next 90 days. No P-51s could be relied upon for defense action for most of 1942, and it was mid-year before early-series P-47Bs were on station. (Hans Groenhoff/S.I.)

ABOVE: Only one single photograph of this airplane, referred to as the "Double Twister," was ever taken according to Republic top executive/ pilot C. Hart Miller. This early application of contra-rotating propellers to a P-47B was accomplished so discreetly that its installation and single test flight was unknown to all but a handful of people at Farmingdale. Many "insiders" including chief tester Lowery Brabham insisted that no such P-47 existed, but others even recall working on it. A simple oversight (not neutralizing fin offset) before the only flight trial nearly cost Miller his life. (C. Hart Miller)

Two of the earliest P-47D-REs, shown in flight off the coast of Long Island, were externally identical to C models. There was no bulged keel line, no provisions for any external fuel tanks on these P-47D-REs which should be logically regarded as production "proving" airplanes from the brand new Evansville, Indiana, plant. (USAAF)

Inasmuch as the P-47 Thunderbolt specifications called for a fighter that would perform in the interceptor role, confirmed by the airplane configuration, production P-47B and C models had no provisions at all for what the AAF called "range extension." None of the B types were sent to a combat theater, and not many C models went to the ETO. None were committed to the SWPA, Alaska or the MTO. Most were assigned to conversion training units or groups scheduled to receive advanced models upon arrival overseas. A P-47C-5-RE (AC4-6558) is shown taking off in 1943. (USAAF)

Considering the intensive defense put up by the enemy in protecting the Romana-Americana, Concordia, Vega, etc. oil refineries around Ploesti during the low-level attack called Operation TIDALWAVE on August 1, 1943, it is miraculous that all four Consolidated B-24Ds shown here survived the raid. "Joisey Bounce" and three others from the 93BG (8AF) based at Hardwick were evidently on a routine practice mission. Republic P-47Cs and unmodified early Ds of May-June 1943 lacked the range to escort ETO bombers more than 75 miles over the Continent from the coast. Every P-38F and G had been stripped from the U.K. for North African duty in connection with Operation TORCH.

Largely unheralded Major LeRoy A. Schreiber, at minimum a double ace, flew this Republic P-47D-2-RA (LM★T) in scoring heavily against the best the Luftwaffe could field in 1943. It is important to know that many P-47Cs and early Ds were retrofitted to carry drop tanks, with some undetermined number receiving the complete "bulged belly" treatment that accommodated a centerline external drop tank with as much as 150 gallons of fuel. The 108-gal. version was most common. Retrofit modifications could be made at depot, subdepot and even group-level facilities. This picture was taken at Halesworth, Suffolk (Sta. 365). (AFM)

Another photo of the Evansville-built P-47D-4-RA (AC42-22794) reveals erosion of the red cowl paint (a temporary addition?) and several white dots on the camouflage paint. Those were inspection stickers applied as the fighter approached factory "sell off." (Frank Strnad Coll.)

Beginning with a rather definitive P-47D-4-RA production configuration, the razorback Thunderbolt was beginning to have more than interceptor range when equipped with a drop tank. But it still lacked a "universal" wing with bomb, drop tank and rocket launching installations. Luftwaffe pilots soon learned that you might dive and split-S away from a P-38H or J, "but don't try it" against a P-47D. Derided in the British journals for their "huge" size, Lightnings and Thunderbolts were merely on the cutting edge of the future. (Republic)

A large white oval with a 3-digit number on the cowling generally identified all Air Proving Ground Command (APGC) aircraft operating out of Orlando, Florida's Pine Castle AAF. This Republic P-47D-4-RA was flown in mock combat against a Lockheed P-38H and a North American P-51A Mustang to evaluate its combat capability. Evidently most mock combat activities assigned to Eglin Field actually took place at Pine Castle AAF. (USAF)

A 55FS Lightning pilot at Wittering, in this case Lt. Arthur N. Rowley, faced the camera of a fellow pilot while perusing a Republic P-47D-5-RE assigned to the 352FG out of Bodney. The visitor on that October 1943 day was typical of the main fighter type then being flown by VIII FC in the ETO. The new commanding general, B/Gen. William E. Kepner, had a mandate to make his command the protector of Eighth Air Force bombers then targeting German industry as part of POINTBLANK. (USAAF/AFM)

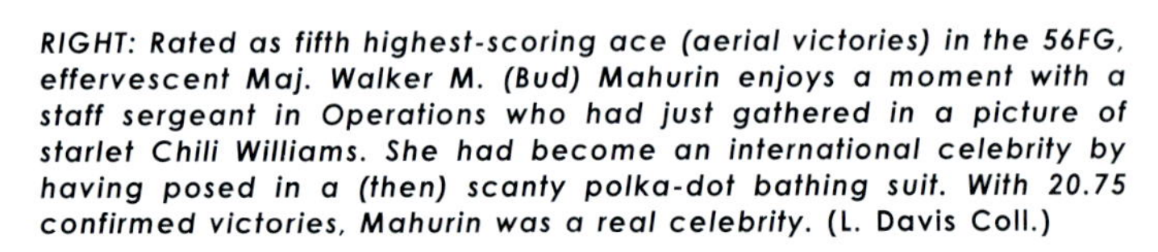

RIGHT: Rated as fifth highest-scoring ace (aerial victories) in the 56FG, effervescent Maj. Walker M. (Bud) Mahurin enjoys a moment with a staff sergeant in Operations who had just gathered in a picture of starlet Chili Williams. She had become an international celebrity by having posed in a (then) scanty polka-dot bathing suit. With 20.75 confirmed victories, Mahurin was a real celebrity. (L. Davis Coll.)

In their eagerness to demonstrate the capabilities of whatever type fighter they were flying and their own newly gained prowess, aggressive new fighter pilots often would "attack" other breeds flying in stateside skies, and even those of the same breed. Republic test airplanes, carrying sensitive recording equipment on carefully choreographed flights, usually had sections (or the entire airplane) painted in zinc-chromate yellow for high visibility. Combative "aggressors" usually got the message. The pilot of this one (No.5) is probably the local AAF Plant Representative or one of his staff members.(J. Campbell Archive)

During the rather short tour of duty in OTUs, there was little time for niceties and no individual pilots to "personalize" their favorite aircraft. Such was the destiny of virtually all training aircraft. A Republic P-47D was serving in an OTU but, assigned to the 514FBS, it displayed a bright blue cowl leading edge for quick identification. Congaree Army Air Field, SC, served the 406FG as the aircrews trained for the job that lay ahead with the Ninth Air Force in the ETO. (Stan Wyglendowski via J. Ethell)

BELOW: Black, white and red "petal" paint flashes and "E" number prefixes identify various training squadron airplanes on this Training Command base at Walterboro, S.C., in the autumn of 1944. Documents indicate that OTUs had relatively high accident rates, especially mid-air collisions. (C. Jaslow via J. Campbell)

BELOW: Discovery! Having photographed airplanes for more than a half century, your chronicler's life was lit up by acquiring wartime era Thunderbolt color photographs of superlative quality from every theater of operations. It's beyond expectations. This Republic P-47C-5-RE, marked VM★A, belonged to the 551st Fighter Training Squadron of the 495FTG that called Atcham (Sta. 342) home in 1944. Old 16530 may have passed its prime, but it surely was not a WW (War Weary) as it was fully armed and flown with pleasure by more than a couple of pilots in the ETO. (Robt. Astrella via J. Ethell)

The Eighth Air Force began to permit applications of a great array of the most colorful markings on fighters, at least on a widespread basis, sometime in the summer of 1943. This 353FG P-47D-6-RE, coded YJ★H (called H bar), reveals that new or waxed camouflage paint could appear extremely dark. The yellow/black checkerboard cowl and rudder trim tab paint scheme became high-visibility markings. So-called "nose art" as seen here was really much closer to the center of gravity. (R. Astrella via J. Ethell)

Proving that the excellent North American Mustang airframe design of 1940 had failed to excel as a fighter as early as 1941 only because it had been saddled with a low-altitude Allison engine, the type did a Dr. Jekyll and Mr. Hyde act upon installation of the Packard-built Merlin V-1650 supercharged engine. That same fate should have fallen to the Lockheed P-38, or at least half of the production airplanes. (NAA)

OPPOSITE: Could anyone attending the June 1940 specification and contract definition meeting that created the XP-47B design concept ever have foreseen that production versions might carry more fuel externally than the prototype could carry (all internally) at maximum design gross weight? Perish the thought. And it could be accepted as an absolute fact that designer Kartveli would have been the least likely to have that foresight. The very thought would have been repugnant. But just four years later P-47Ds routinely carried 375 gallons of fuel in three drop tanks. Typically, a P-47D-20-RE, named "Belle of Belmont," was flying with the 56FG while carrying two 150-gallon drop tanks, as in this scene as it prepared to join others on a Ramrod over Festung Europa. (Mark Brown/USAFA via J. Ethell)

BELOW: Very early in the "game" in the UK, it became evident that gunners aboard the "big brothers" drew a bead on anything with a round engine configuration, thinking Focke-Wulf. Those Luftwaffe Focke-Wulf Fw 190s were far too deadly, begetting snap decisions. As of February 1943, all combat P-47s bore wide white cowl bands, similar tail stripes on the vertical and horizontal surfaces plus oversize national insignia on all four outer wing surfaces (top and bottom) for instantaneous recognition. Early block P-47Ds, like Aldwin M. Jucheim's HL★J(J bar), carried only the 108-gallon belly drop tank developed by Col. Cass Hough's Technical Operations Section at Bovingdon. Stores pylons for wing installation had not yet been developed by Republic. (USAAF/AFM)

A very dark finish on this ETO-based P-47D-21-RE, is probably the product of multiple wax coatings applied, while the lack of any unit markings indicates that the fighter has just exited a depot after a major repair or overhaul cycle. Date and location are unknown except that it was in the United Kingdom. (Robt. Astrella via J. Ethell)

BELOW: Never have WWs (War Wearies) looked better, unless you were a downed airman floating in the cold North Sea or English Channel. This Republic P-47D-5-RE ,coded 5F★S, and those in the background, belonged to Detachment B of the 65FW at Boxted. The non-squadron, in the summer of 1944, was the only 8AF air-sea rescue unit extant during that busy year. (Mark Brown/USAFA via J. Ethell)

Cosmetic surgery was yet but a promise for this 5ERS P-47D when it was seen at Boxted. Although propeller and cowl paint looks fresh, the bulk of this WW marked 5F★A really had that "weary" appearance. Evidently still doing the channel patrol, its armament had been reduced to four guns. That Malcolm hood (the canopy) may have been installed when the airplane was still flying combat missions. (Mark Brown/USAFA via J. Ethell)

BELOW: Airplanes in an active combat theater were subjected to an accelerated aging process, but a little TLC (Tender Loving Care) could do wonders for a War Weary (WW). One Republic P-47D-1-RE (5F★Z) assigned to the little publicized Detachment B, later redesignated 5th Emergency Rescue Sqdn. (ERS), evidently had been processed through depot overhaul (Fourth Echelon), coming out with a new lease on life. Paint removal, a new late-model engine and the latest in propellers were backed up with a full complement of guns. (Robt. Astrella via J. Ethell)

ABOVE: Slow timing the P&W R-2800-59 B-series engine over the sparsely populated eastern portion of Long Island, a brand new P-47D-22-RE razorback (AC42-25716) seemed to revel in its bright new skin, sans camouflage. Deletion of the O.D. paint saved up to 70 pounds of weight and reduced drag. This model T-bolt was fitted with a paddle blade propeller. (Rudy Arnold/S.I.)

BELOW: Those ever-present Marston pierced-steel mats show unexpectedly through the turf at an East Anglia airfield southwest of Cambridge. The site was AAF Station 122 at Steeple Morden; the fighter was a P-47D-22-RE of the 355FG. Providing proof that any aircraft might differ from the specific norm, this aircraft was still fitted with a Curtiss "toothpick-blade" propeller instead of one of the newer paddle-blade types. (A. Cal Sloan via Ethell)

Two Republic P-47D-23-RAs, having just returned from a combat mission, feature some unique Eighth Air Force markings. Of particular interest are the large (oversize) national insignia applied to "all four corners" of the wing and the very obvious 78FG checkerboard cowl markings. These Evansville-built ultimate razorback Thunderbolts show off the newly installed Curtiss paddle-blade Type 836 high-activity propellers with a 13-foot diameter. (Mark Brown/USAFA via J. Ethell)

A bit of reverse English showed up at Boxted in late summer, 1944. The newest Thunderbolts, essentially bubble-canopied P-47D-25-REs, appeared with the violent contrast of brilliant red cowl rings and customized camouflage schemes. At the same time, razorback P-47Ds – which would routinely be accepted in any O.D. paint coatings – were likely to have molted their layers of paint. These mission-bound 63FS Thunderbolts, with a bubble-canopy D-25 leading and at least a pair of unpainted razorbacks in the mix, show the unexpected contrast. (Mark Brown/USAFA via J. Ethell)

Massed Thunderbolts with a P-47D-25-RE (UN★B) in the foreground prepare to depart Boxted, England, in the late summer of 1944. If not for the war, it was a scene of pastoral pleasantry. But the thunderous roar of many powerful 18-cylinder engines and whirling 13-foot propellers shattered the air at AAF Station 150 in County Essex. Colorful P-47Ds taxi out for takeoff on what appears to be a Ramrod. External tankage evidently amounted to 216 gallons for each Thunderbolt of the 63FS at start up. Weather at the time of takeoff can only be described as delightful. (Mark Brown/USAFA via J. Ethell)

Following stentorian Russian-language protestations (per Sasha Kartveli) that the aerodynamic lines of the P-47 were "destroyed" by adoption of the 360-degree-vision bubble canopy, new P-47D-25-REs delivered were actually a smidgen faster than their immediate razorback siblings, the P-47D-23-RAs. Nearly as significant as the improved vision for combat resulting was an accompanying increase in internal fuel capacity (from 305 to 370 U.S. gallons). Although adequate range with drop tanks already had been achieved for Ramrods well into Nazi airspace, General Doolittle's full commitment to the P-51 with Merlin power early in 1944 resulted in a major revamping of ETO missions for P-47s and P-38s. T-bolt AC42-26429 leads this newly manufactured group over Long Island Sound. (C.A.M.)

LEFT: Like the Ford emblem on a car, the visages of Francis "Gabby" Gabreski and Hubert "Hub" Zemke are almost instantly associated with the Republic Thunderbolt. This wartime photo of Gabreski illustrates why the young man was virtually a recognition symbol associated with the P-47. And his boss, Zemke, epitomized what it takes to be a good fighter pilot, with aggressiveness leading the way. Both men were multiple aces. "Gabby" flew 153 combat missions in WWII before that trait bit him. Unharmed in a low-level (strafing) encounter with terra firma, he became a P.O.W., robbing him of the opportunity to run up an even more impressive score. (S.I.)

ABOVE: Home, home on the range. The ubiquitous drop tank. As Col. Cass Hough had warned B/Gen. Frank O'D. Hunter in early spring 1943, P-47 range extension would be critical if the Thunderbolt was ever going to perform in the escort role to Deutsches Reich targets. Wright Field officialdom was evidently acting as if on a pre-war basis, assuming that they were aware of the desperate need for bomber escort in the Eighth Air Force. However, the Regensburg and Schweinfurt raids were soon to reveal that they should have been proceeding at flank speed. Col. Hough's organization at Bovingdon came up with operational drop tanks in weeks after Hunter told Hough to "get it done." Piles of range extenders – glue-impregnated paper (and, ultimately, metal) tanks – were successfully utilized as the planes were modified, first by retrofit and in then current production P-47Ds. (AFM)

BELOW: Two of the 61FS armorers assigned to Maj. Francis Gabreski's P-47D-25-RE when he commanded the squadron for a second time were Sgt. John A. Koval (l.) and Sgt. Joe DiFranze (r.). They are shown at work filling the ammo trays. Work of the armorers was critical to a fighter pilot because jammed weapons could easily bring an abrupt end to a promising career. (Alain Pelletier via J. Ethell)

BELOW: Captain Walker "Bud" Mahurin, a popular and successful fighter ace in the 56FG of the Eighth Air Force was pictured (as part of a media event) pointing to a distant destination in Festung Europa that was the day's target. Until the best bubble canopied P-47Ds arrived on the scene in the ETO, escort missions were usually limited to penetration and withdrawal escort assignments. The avid observer looks even more youthful than Mahurin. (AFM)

ABOVE: More colorful than most early post-war racing airplanes, Col. David Schilling's P-47D (coded LM★S), in September 1944, was highlighted by one of cartoonist Al Capp's favorite cartoon characters, "Hairless Joe." In airshow weather, the multiple ace is pictured about to take off on a Rodeo or Rhubarb over France or the Low Countries. His mount carries a centerline 150-gallon drop tank. (Mark Brown/USAFA via J. Ethell)

BELOW: Another side to the coin. As good weather held in the U.K. in the late summer days of 1944, the new C.O. of the 56FG – Col. David Schilling – had his P-47D-25-RE armed with two sets of 4.5-inch rocket tube clusters in preparation for one of those frequent Rhubarbs as the OVERLORD troops were breaking out of the beachhead. Rockets could be an effective weapon against enemy armor and railroad locomotives. (Mark Brown/USAFA via J. Ethell)

Did any warplane ever look better? Many of us had forgotten that WWII was not carried on in just monotones of gray and black and white – just like Bogart in Casablanca. In this 1944 photo, the Jug looks more like a float in the Rose Parade. Truly, there was beauty in this beast, the P-47D-25-RE in which Colonel Schilling aggressively led the 56FG – "Zemke's Wolfpack" – against an enemy that then came to full realization that it had its back to the wall. (Robt. Astrella via J. Ethell)

Parked on a very heavily used servicing pad, a 56th Fighter Group P-47D-28-RA is shown undergoing routine post-mission squadron-level (1st echelon) maintenance. All eight .50-cal. machine guns were removed from LM★A for cleaning or replacement. Drop tanks, ammunition, oil, fuel, oxygen, fluid for the water-injection system and the guns will all be taken aboard as part of the turn-around process. AAF Sta. 150 was home to this 62FS fighter. (Mark Brown/ USAFA via J. Ethell)

LEFT: Several well-known aces were among the 56th Fighter Group pilots being briefed for a mission. While no details were provided with the picture release, it was certainly a typical mission briefing that occurred up to several times a week as the tempo of war increased and weather conditions were cooperative.(AFM)

BELOW: Ol' Fireball Roberts would have found the yellow-nose, yellow-tail YP-47M much to his liking. It was a hotrod. Here, the second of two prototypes carries a 110-gallon belly tank as it prepares to depart Republic Airport on a test flight in the winter of 1944-45. Mike Ritchie was usually at the controls, although Chief Test Pilot Lowery Brabham spent many hours scooting around the New York-New England skies in the prototypes at astonishing power settings. He claimed to have pulled more than one hp per cubic inch of displacement rather frequently. Anyone care to dispute the Chief Test Pilot's words? In this picture, a military pilot is at the controls. (C.A.M. via Ed Boss)

Surrounded by lamentable conditions, a P-47D-27-RE belonging to the 404FG (9AF) awaits a target assignment. Carrying three 500-pound bombs and a complement of rockets mounted on new zero-length launchers, this Thunderbolt is loaded for bear. It carried the logo, "I'll Get It." (C.A.M. via Ed Boss)

When the Ninth Air Force came out of mothballs as 1943 drew to an end, the mission mandate was to carry out tactical warfare in support of Allied invasion forces until the mission was accomplished. While that force began operations with North American P-51Bs, the majority of its fighter-bombers were soon to be Thunderbolts and Lightnings as Doolittle's strategic Eighth AF/VIII FC concentrated on the P-51 Mustang for long-range escort. Fighters of the 514FS, 406FG, spent most of their WWII combat days in France and Germany after little more than 90 days of basing in the U.K. This P-47D-26-RA taxiing on the ubiquitous Marston mats could have been at any one of nine or more bases it operated from on the Continent. (Stan Wyglendowski via J. Ethell)

With brilliant sunlight glistening on the semi-elliptical wings that distinguished the Thunderbolts, a factory-fresh P-47D-27-RE out of Farmingdale was pictured over a fairly solid overcast before being shipped to a war zone. The New York (Long Island) plant produced 611 of the D-27 version. (Ed Boss Coll.)

In marked contrast to the environment at Boxted, home of the 56FG, this P-47D-27-RE assigned to an unidentified 9AF group (possibly the 404FG) was operating from a former Luftwaffe base in Europe. The conditions seemed evocative of a trash dump as "Bull IIIrd" sits quietly awaiting another mission call. The pilot was identified as Lt. James A. Mullins, but needs confirmation. (C.A.M. via Ed Boss)

Cruising serenely on the way to a Nazi target, part of a 381BG formation (pictured late in 1943) seemed remote from war over fleecy clouds. If that central B-17G-70-B0, VE★N, completed its mission and returned safely, it surely was with the aid of an VIII Fighter Command escort during penetration and withdrawal phases. M/Gen. William Kepner, having replaced B/Gen. Hunter after un-escorted missions against German Vaterland targets decimated Bomber Command formations, brought in P-38 units that had overtrained in the Z.I. despite the critical need in the ETO. (S.I.)

If the Lockheed P-38s deserved any criticism for Allison engine failures in the ETO – essentially absent elsewhere – the North American P-51 and P-51A fighters warranted cancellation. And that is exactly what occurred! Incapable of taking on Messerschmitt Me 109Fs and Gs or Focke-Wulf Fw 190As over Festung Europa because of V-1710 deficiencies (lack of power, a low-level critical altitude), those early P-51s hardly fared better than Bell P-39s and (later) P-63s. In the formation of Orlando-based APGC fighters that include a Lockheed P-38H-5-LO, a Republic P-47D-4-RA and a P-51A, notice the absence of a Bell P-39 or Curtiss P-40. (Hans Groenhoff/S.I.)

There can be no doubt that after M/Gen. William Kepner replaced B/Gen. Frank Hunter as commander of VIII FC (in the autumn of 1943) it was quickly transformed into a far superior fighting machine. All available Lockheed P-38Hs were expedited into long-range escort duty for the "heavies" and P-47 range extension became a priority effort. General Kepner flew 24 various missions while in the ETO, with 16 of them being in this 2nd Bomb Division control aircraft, a P-47D-25-RE (AC42-26637). Provisions for two .50-cal. machine guns had been deleted from "Kokomo" well before the date of this photograph, 25 March 1945. Kepner's role in fitting bomb/tank pylons on P-47s was significant. (Robt. Astrella via J. Ethell)

The very colorful "Angie" was a P 47D-27-RE assigned to the 512FS of the 406FG commanded by Col. Anthony V. Grossetta. Other squadrons in the group were the 513FS and 514FS. Immediately prior to D-Day, the group was carried as part of XIX Tactical Air Command, 303rd Fighter Wing. At that time the group was based at one of the Advanced Landing Grounds in Kent, close to the English Channel. This photograph was taken months later, of course, at an airfield on the Continent. In the last mission of the war for the group, nineteen P-47s attacked a large cargo ship at Lubeck, Germany, firing rockets that set the ship on fire. That mission was on 3 May 1945. The squadron code for "Angie" appears to be L3★U. (S. Wyglendowski via J. Ethell)

Operating from an advanced airfield on the Continent, one of several from which close-support Rhubarbs were launched against targets in France and Belgium in support of ground forces, many P-47D-28-RAs like this 514FS fighter totally disrupted German transport and communications. They attacked everything from couriers on motorcycles to lengthy trains and key bridges with more than a little success. Destruction of an enemy tank was as important – perhaps even more – as burning a Messerschmitt. (Stan Wyglendowski via J. Ethell)

When the first Thunderbolts arrived in the ETO, there were no provisions for mounting either belly or wing drop tanks. Therefore, the big powerful airplane had some speed margin over the smaller Spitfire, marginally more range, but inferior climb, acceleration and maneuverability. By the time this P-47D-28-RE version was being allocated to squadrons – in time to support the great invasion of Continental Europe – it could easily tote three 108-gallon drop tanks like these. With prodigious fuel capacity, it could take a direct route to Berlin and furnish target penetration protection or protective cover on the way out for the "heavies." Thunderbolts could, by then, match or exceed Luftwaffe defensive fighters in combat capability deep inside Festung Europa, something no Spitfire group could ever accomplish. (Robt. Astrella via J. Ethell)

Paddle-blade propellers and more powerful versions of the Double Wasp engine provided P-47s with greatly improved climb and lifting capabilities together with better performance at fighting altitudes. This pair of P-47D-28-REs, probably just in from the Service Command depot at Warton, could each carry 450 extra gallons of fuel to feed the voracious appetites of those powerful engines in combat. (Robt. Astrella via J. Ethell)

BELOW: Cruising sedately over southern Bavarian mountains of Germany, Lt. George McWilliams's P-47D-30-RE, bearing the markings D3★V and assigned to the 397FS of the 368FG, was photographed from an accompanying Thunderbolt. It was one of those rare P-47s that carried the petaloid style cowling markings. The date was sometime in 1946. (Arthur O. Houston via J. Ethell)

The red corolla petal marking used on the cowling of this 395FS Republic Thunderbolt assigned to the Ninth Air Force's 368FG was rarely seen on combat arena P-47s. Less blatant versions seemed to be more commonplace on stateside trainer P-47s. In this photo taken at Straubing, Germany, in 1945, a 373FG Thunderbolt may be seen in the background. It was identified as a visitor from Illesheim. (Arthur Houston via J. Ethell)

Another F-47D assigned to the 527th Fighter-Bomber Squadron out of Neubiberg – probably the other wingman on the same flight – is seen high over a snow-laden mountain. (Logical question: Why did it take nearly a half century for quantities of WWII color photos to appear? In the post war era, individuals merely forgot that they had the material and no market existed.) (Col. Robt. W. Casey)

Republic Thunderbolts were as likely to be seen carrying Lockheed-type 165-gallon drop tanks (often referred to as 150-gallon tanks) as any other type. Operating out of Neubiberg, Germany, in 1948, an F-47D-30-RA of the 527FBS, 86FBG, shows off the postwar European theater markings. In combat operations during the war, the outfit won DUCs in Italy and in Germany. (Col. Robt. W. Casey)

Pictured from the aft seat position in an AT-6, which was an uncommon bird in Italy, a Twelfth Air Force razorback P-47D of the 57FG, 64FS taxies in the rain at Grosseto, Italy, in the late months of 1944. That red tail was not an authorized component of the squadron marking scheme, especially in that it seems to have been sprayed right over the fighter's call numbers. (Col. Fred Bamberger)

On what appears to be sand-based taxiways amid a sea of mud, a 57FG (Twelfth Air Force) pair of P-47Ds moves past a North American AT-6 Texan toward possible gridlock on the runway. The scene was Grosseto, Italy, late in 1944. At most times, the airfield was a busy parking lot for fighter-bombers and light bombers (Douglas A-20, etc.), many of which may be seen in the background. It had to be a photographer's heaven. (Col. Fred Bamberger)

Seen through the windshield of a North American B-25 Mitchell at Pomigliano, Italy, a flying signboard was a favorite in the 79FG. It was a P-47D bearing the profile of a blue horse, the name "The Trojan Warhorse" and the code X65 as a member of the 86FS. All was combined with the colorful tail markings and the "wild" squadron emblem on the cowl that depicted a caricatured Comanche on the warpath. As the preceding aircraft ran up their engines, the B-25 was bouncing; therefore, the picture is not critically sharp. (Col. Fred Bamberger)

LEFT AND LEFT BELOW: Rarely – if ever before – seen in full color, several P-47Ds of the 79FG in Italy were captured on film by the co-pilot of a North American B-25J flying out of Grosseto to Cesenatico in 1945. Unfortunately the lack of proper communications blocked chances of getting the flight to move forward. Result: the focal plane shutter in the camera picked up the black blades on the B-25's propeller as intrusive blurs. The spectacular tail markings on 79FG Thunderbolts varied greatly in application. Capt. Fred Bamberger, assigned to a major intelligence job, was a pre-war photographer in New York. Nearing age 80, he still wields a mean camera and pen. (Col. Fred Bamberger)

As a Douglas C-47 drifts in to a squishy landing at Pomigliano, the pilot of "The Trojan Warhorse" champs at the bit awaiting a signal to takeoff. Air traffic was extremely heavy in 1944. Thunderbolt X65 carries a 75-gallon centerline droptank. (Col. Fred Bamberger)

Half a century after the first Republic Thunderbolts were committed to combat in Europe, a near perfect example has been identified as the only flyable P-47 still on active charge with any nation's air arm. True, the actual P-47D-30-RE (AC44-20339) flown in combat by Ens. Frederico G. Santos of the Forca Aerea Brasileira (FAB) fell to enemy guns near Spilimbergo, Italy, on April 23, 1945. However, the example pictured, technically a P-47D-40-RA, is painted to accurately represent the airplane flown by Santos in the MTO during WWII. (Dan Hagedorn Coll.)

Return to Bataan. Lugging a pair of 500-pound bombs on a short-range close support mission in the Philippine Islands in 1945, a P-47D-30-RA of the Fuerza Aerea Expedicionaria Mexicana's 201 Escuadron was beautifully and uncommonly photographed in color. The FAEM squadron operated as a functional part of the Fifth Fighter Command's 58FG in the Western Pacific in mid summer 1945. Operating with a mixture of borrowed AAF and their own Lend-Lease P-47Ds, the 201 Esc. flew about three dozen close support combat missions in the retaking of the Philippines. (NARS/Dan Hagedorn)

ABOVE: Against a magnificent background of cumulus clouds at Nanking, China, a P-47D-30-RE (CAF 4202) formerly operated by the USAAF 81FG awaits its Chinese Air Force pilot. Somewhat battered 110-gallon drop tanks are to be seen on the wing mounts. In the background, a former 1st Air Commando Group (10AF) Thunderbolt is seen to be clearly marked with the CAF 11th FG insignia. (McKay via L. Davis)

BELOW: Instead of the typical stacks of drop tanks in the background – so common in the ETO at U.K. bases – there were rows of fuel drums behind a 14AF Republic P-47D-30-RA someplace in the CBI Theater of Operations. Markings of the 81FG adorn this Thunderbolt from Evansville, parked on an excellent crushed rock ramp. Very little machinery was used in construction. As the Indiana factory, close by the Ohio River and fronting on US41, continued to produce P-47Ds, the Farmingdale facility began to turn out large numbers of N models and a priority series of P-47Ms. (McKay via L. Davis)

OPPOSITE: The tiger shows his stripes in the Philippines. Cruising serenely over Northern Luzon during the return from a strafing/bombing mission in 1945, a 58FG Republic P-47D-23-RA displays all variety of stripes. Permission to apply pre-war Air Corps tail stripes had to emanate from a top command official who had the fondest of memories about those very popular stateside markings. (USAAF)

RIGHT AND BELOW: "Carefully maintained and operated by wealthy owner," might have been an appropriate placard for this virtually new-condition P-47D-30-RA (AC44-32734) that was passed along to the Chinese Air Force 11FG by the American government. We seriously doubt that the "Lend" portion of Lend-Lease was used to any widespread extent, but the "Lease" portion might have meant involvement of land and sweat investment. The USAAF national insignias were crudely obliterated, then overlaid with the Chinese AF emblems. Black and white tail striping covered the AAF 81FG tail marking on the rudder of this T-bolt parked at Nanking. (McKay via L. Davis)

ABOVE: Far from Farmingdale and Disney's Burbank studios, Donald Duck brightens the cowl of an AAF Republic P-47D bearing the name "My Better Half." Neatly awaiting action in the wings – pun not intended – off stage at Hangchow, China, are numerous new North American P-51Ds of the Fourteenth Air Force. Chiang Kai-shek's pilots inherited a very large number of second-hand T-bolts after arrival of the replacement aircraft. (McKay via L. Davis Coll.)

Little has been written (or illustrated) about Generalissimo Chiang Kai-shek's Chinese Air Force in World War II except that they flew many of the Republic P-43A Lancers, and that the general was completely behind the Flying Tigers participation. Late in the war in the Far East, with arrival of North American P-51Ds in the 14AF as replacement aircraft, Republic P-47Ds were transferred to the Chinese 11FG. A CAF P-47D-23-RA (former AC42-27730) assigned to the 11FG is shown taxiing on a gravel perimeter strip. (McKay via L. Davis)

Never previously seen in print, a factory photograph shows three new USAAF P-47D-25-REs flying in line abreast formation and being joined by a similar type in the markings of the Forca Aerea Brasileira and one destined for the Royal Air Force of Great Britain as a Thunderbolt II. Although numerous D-25 series fighters were delivered to the Brazilian military, the airplane pictured was one of the later D-27 versions. It soon shed the Brazilian markings and went to an AAF group in July 1945. (C.A.M.)

The odds against finding two or more monochromatic photographs of different P-47 Thunderbolts in WWII days wherein the last three digits of the tail call numbers are identical must be astronomical. The prospect of finding even a single color transparency circa 1944 – such as P-47D-28-RA AC42-29222 pictured here – must be incalculable. Yet, at least <u>three</u> such airplanes are reproduced in color in this publication. With more than 15,600 Thunderbolts manufactured in two plants, it is unlikely that even ten had triple-digit endings to tail numbers. (C. Jaslow/ Campbell Arch.)

ABOVE: Apparently this "Brazilian" P-47D-27-RE, AC42-26777, was only painted in Forca Aerea Brasileira colors for publicity purposes. That accompanying Thunderbolt II was authentic, however, going to the RAF as programmed. Once again there is a legitimate AC number ending in triple digits. (Frank Strnad Coll.)

RIGHT: High over a major airfield near Hartford, CT, in the winter of 1944, a Training Command P-47D-28-RA is seen joining a formation. The training of Thunderbolt pilots at OTUs was then about to be reduced considerably, even if the XP-47N became a success – something that the AAF hierarchy viewed with skepticism. (C. Jaslow/Campbell Arch.)

All but a few Curtiss-built P-47G-CU Thunderbolts were assigned to training groups as the Curtiss Airplane Division continued in its determination to "shoot itself in the foot," so to speak. That huge division only delivered six P-47Gs in all of 1942. Incredibly, the trainer-fighter shown here was the sixth P-47G built (AC42-24926). As of July 1943, this equivalent of a Republic P-47C was based at Thomasville, Georgia. In the entire year of 1943, with the need for fighters at crisis levels, Curtiss could produce no more than another insignificant 271 of the P-47Gs! Production wobbled to a halt in March 1944 with a sickly output of 77 airplanes in the entire quarter. Republic at Farmingdale had single-handedly turned out exactly twice that number in January 1943. That was fourteen months earlier and merely a single month's production! (Joe Sherwood via C.A.M.)

While the WPB was still placing massive orders for Bell P-63E/F King Cobra models as late as 1945, contemporary P-47M/N, P-51D and Hawker Tempest II fighters were anywhere from 40 to 80 miles per hour faster, higher flying and – above all – had proven themselves to be superior aerial combat aircraft in the toughest proving ground of all, the airspace over Festung Europa. This Bell P-63A-10-BE was photographed at Portland AAF, Oregon, being one of a small quantity – out of 1725 built – that stayed with USAAF units. Most P-63As went to Russia via Lend-Lease arrangements. Only a handful of E and F models were constructed before cancellation orders were issued. (James Kunkle)

The wide stance of a P-47N-1-RE is well illustrated by this wartime photograph taken on Tinian Island in 1945. Routine maintenance was being performed by the crew chief in the heat of the day. A total lack of squadron and personal markings indicates that this long-range Thunderbolt was a new arrival, straight out of production lines in New York. (R. Stoffer/Campbell Arch.)

Former Ninth Air Force Lockheed P-38 and North American P-51D pilot Lt. James Kunkle was recycled back to "Uncle Sugar" at Portland, Oregon, after wintering in Europe as part of the Occupation Forces in Germany. In due time, he had seen most of the Mustangs transferred to the Royal Swedish Air Force (Flygvapnet) to replace the Republic EP1-106 and similar Reggiane Re.2000 types that had survived the war years. A Republic P-47N-25-RE seemed certain to become his next mount in carrying the war to Japan. (As it turned out, the Japanese surrendured. Kunkle was soon to make the transition to a brand new Republic P-84B jet fighter.) (James Kunkle)

Sparsely marked with the emblem of the First Air Force, a very neat and clean P-47N appears to have been visiting a naval air station circa 1946. Independent air force status was still to come, and so were the skinflint days of Louis A. Johnson as Secretary of Defense. One scarcely known sidelight of the Korean War (sorry, Police Action) was that while P-47s were evidently not considered for use there, North American Aviation responded to a USAF request (circa 1951) for conversion of P-51s by installation of a powerful radial engine, most likely the P&W R-2800. It never progressed beyond the design proposal stage. (D. Menard/L. Davis)

German Luftwaffe hangars, British RAF Hawker Tempest II fighters in camouflage paint, bright metal North American P-51Ds of the 59FS, 33FG constitute the "peaceful" setting at Neubiberg, Germany, after cessation of hostilities (1946). Centaurus-powered Tempest IIs were late arrival equivalents of ground-strafing P-47Ds in size and speed, but not high-altitude performance or combat range. Successor to the low-altitude Typhoons, the prototype Tempest II first took to the air 27 months after the XP-47B's initial flight. All Tempests featured a wing planform and chord/thickness ratio closely emulating that of the P-47. Only 452 of the radial-engined fighters were completed as of VJ-Day. These Tempests were operated by No. 26 Squadron out of Gutersloh, Germany. (James Kunkle)

"Happy Warrior," apparently a Republic P-47D-27-RE (unconfirmed), became virtually anonymous with the obliteration of national insignia and squadron markings. Previously flown as a part of the 9th Air Force, it was obviously destined for transfer to one of the air forces of an ally, most likely in France or Italy. A badly damaged former Luftwaffe base appears to be the site at which this Thunderbolt received its separation papers from the USAAF. In the distant right backdrop a large factory or office building apparently was untargeted or it remained undamaged for unknown reasons. Bombardiers did not have many perfect days to equal the one pictured in winter months. (Fagan via Menard)

A decade of Messerschmitt Bf/Me 109 production came to an end in April 1945 under incessant pounding by Allied bombers of Operation POINTBLANK. German aircraft production had actually accelerated to approximately 4400 airplanes per month in August 1944, then plunged steadily to zero. Most GAF production centered on fighters in the last years, represented at the end by this Me 109K interceptor being viewed by a U.S. Army soldier. The first Messerschmitt Bf 109 V1 had flown at Augsburg-Haunstetten, Germany ten years earlier when that G.I. was a child. (S.I.)

Encounters with camouflaged P-47D (or N) Thunderbolts in postwar days were as uncommon as seeing a boy with green hair. That was a factual situation even at Oscoda, Michigan, when the major film "Fighter Squadron" was being filmed. However, at least one such fighter was seen in that very configuration as shown here. Unfortunately, application of the paint obliterated the tail number and overspray shows up on the fuselage national insignia, but this was definitely an official USAAF airplane in active service. (D. Menard/L. Davis)

Having served their masters dutifully, hundreds, even thousands of our "Little Friends" had this typical reward levied upon them in the immediate backwash of WWII. The basic reasons involved greed on the part of industrial powerbrokers who wanted assurances that surplus commodities could, in no way, impact sales of new production items. Some destructiveness involved, of course, the inevitable stupidity on the part of certain policymakers. Crushingly intermingled in this tiny view of a great heap at but one site in faraway Ondal, India, shortly after VJ-Day, are tons of P-47s and P-51s. (L. Davis Coll.)

Nearly a decade after the P-47Ds with bubble canopies started to come off production lines, F-47D-30-RAs (and D-40s) were being delivered to air forces like the Fuerza Aerea Dominicana. The colorfully marked Jugs are shown at San Isidro in 1954, having joined the FAD in June 1953. (Dan Hagedorn Coll.)

Even in the uninspiring post-war, tight-budget days when the USAAF gained status as an independent service, colorful markings did not entirely disappear. This red-cowl F-47D, equipped with two 165-gallon P-38 drop tanks, was as colorful as most Eighth Air Force fighters were during the last 12 months of air war in the ETO. Unfortunately, data furnished relative to the organization and location proved to be inaccurate. (D. Menard/L.Davis)

CHAPTER 10

THE BATTLE IS JOINED

AUTHOR'S PRONOUNCEMENT:

An idealistic view of a history book project would be to cover the entire subject selected in total scope and substantial material detail. But nobody really has the time, stamina or financial resources to attain such an optimized goal. In our society, nobody can set out to locate the fountain of youth any more than they can expect to finish the ultimate history book I have described.

As established early on in this document, the goal is to definitively describe the origins, development, testing, capabilities and performance of the Republic P-47 Thunderbolt up to and including its final configurations. This tome was never intended to do more than outline the wartime combat record of the fighter, people associated with the airplanes, organizations which were involved and day-to-day operations, a logical goal. To delve into the wartime combat performance of the Thunderbolt in just one group or squadron in one theater of operations would certainly result in creation of a complete book defining that specific subject. Where a specific type of aircraft has been covered in microscopic detail, any notion of providing comparable coverage in numerous theaters of operation for up to twenty dozen squadrons would be – to put it bluntly – lunacy.

The only logical approach is to provide a more generalized picture of the tasks, geographic location, enemy strength and the overall accomplishments. Some

***Scattered about a former permanent RAF base in England, several relatively new P-47D-2-REs display the latest correct markings that prevailed in the VIII FC as of 1 July 1943. Included was the star and bar with a red surround. The large format (55-in. dia.) national insignia on the wing underside was applied at an assembly depot. Thunderbolt in the foreground does not feature a bulged belly line while the same model at left definitely features that important revision.* (Ray Bowers)**

ABOVE: Mark this date: September 1941. In spite of commonplace protestations to the effect there was an Air Corps/AAF directive precluding use of external drop tanks on pursuit aircraft, Curtiss P-40Es were everywhere with 52-gallon drop tanks as seen here in September, and sometimes with 75-gallon tanks. If some emergency existed, it was certainly kept from the American public, and much of the military. Obviously nobody was silly enough to enforce any obsolete paperwork; in fact, somebody was writing contracts for production of the drop tanks. The Emmons Board, part of the general's Combat Command, completed its report about what was needed in the way of fighters (based on air war reports from Europe during the Battle of Britain.) M/Gen. Emmons, wishing to avoid unnecessary arguments, obtained the written endorsement from Air Corps Chief, Gen. Arnold before going to Air Materiel Division with requirements. But, with all the high-powered people in Combat Command and Materiel, there was a serious failure to communicate or understand. In determining the "shape" of the required P-47B fighter, everybody seemed to forget the big gas eater up front was going on a fighter with no space for fuel. Why have eight heavy weapons and 4000 rounds of ammunition if you can't get to the "playing field" chosen by the enemy? Almost exactly two years after this P-40E was pictured with an external drop tank, the AAF was "jury rigging" P-47Cs and Ds to carry the 75-gallon metal tanks already available. (C-W Corp.)

of the more outstanding examples of combat can then be described adequately and comprehensively. I have elected to embrace the proven adage, a picture is worth a thousand words...but with each picture adequately supported on a foundation of physical and functional wordage.

This is my approach chosen for Chapter 10 and following chapters with the specific exception of Chapter 11.

Substantial evidence exists to support the reasoning behind the decision. Just to give a generalized concept of the scope of war operations, a high-level decision was made sometime around VJ-Day to properly commit to paper the most accurate, yet concise, history possible about the Army Air Forces in World War II . The USAF Historical Division assigned to a staff of historians, researchers and editorial personnel the well defined task of abstracting thousands of documents and memoirs to provide a factual narrative that would prove to be readable by persons other than dedicated scholars. In its basic form written by famous authors and historians, it eventually (over a period of some years) became a seven volume production. Each volume is packed with close to 1000 pages of precisely edited text. Pictorial and art coverage in any single volume is, when compared to this P-47 Thunderbolt book,

As a crew chief pulls (pushes?) the prop through on the P&W Double Wasp engine to clear the lower cylinders prior to starting, pilots of the 353FG at Metfield await the order to fly. Seated on the jeep at left is Jack Terzian with Walter Beckham immediately behind him. In this case, the P-47D carries a 75-gallon metal drop tank. (Jack Terzian)

Even as Operation BOLERO efforts reached a crescendo, this Bell P-39F, seen on 12 June 1942 at Lockheed Air Terminal, shows off its normal equipment – a 75-gal. metal drop tank. The white paint blotting out the red meatball in the insignia shows the airplane was in the configuration before the Pearl Harbor attack. Any perceptive person always knows every Mitsubishi A6M Zero attacking on December 7, 1941 carried a drop tank of about 70 gallons capacity. It was no fluke; it was logical planning. But eighteen months later, the Eighth Air Force (run by dozens of generals) was still sending heavy bombers and mediums over Fortress Europe with escorting fighters on a short leash. When T-bolts finally were equipped with a motley selection of tanks and Col. Cass Hough's team raced to provide a useful pressed-paper centerline tank, there were no provisions at all for this powerful fighter to carry ANYTHING underwing. Alex Kartveli was opposed to such "ornaments" being hung on his sleek fighter. But who was the customer and who was responsible for military tactics? It wasn't Sasha. As will be shown, B/Gen. Hunter was treating the war like a sabbatical. Generals Arnold and Barney Giles were about to send him back to the 1930's. Long overdue. (Lockheed Aircraft.)

minimal at best. The two key "chief" editors (really editorial directors) of the seven volumes – Wesley F. Craven from Princeton University and James L. Cate from the University of Chicago – worked with people having nearly unlimited access to military and civil records. To support the effort, massive funding was available, especially in comparison with private standards.

There was no place for detailed analysis of any specific airplane type, and it was not even attempted. But they soon found that six or seven thousand pages would prove to be inadequate, even with minimal attempts to provide coverage for any specific combat episode. An additional volume was prepared for The Army Air Forces in World War II series, entitled Combat Chronology 1941-1945, authored by K.C. Carter and Robert Mueller. Although each significant day in the various active Air Forces was described in an extremely abbreviated style, the volume contains just barely under 1000 pages. Add to appropriate coverage two more volumes detailing the configuration of Combat Squadrons of the Air Force - World War II (authored by Maurer Maurer) and Air Forces Combat Units of World War II and you have nearly a library. Publication was by the University of Chicago Press or by the USAF History Division through the Government Printing Office.

The value of those books – Priceless.

Could those works be completed with any hope of great accuracy and depth today? I have serious doubts about it, no matter how much financial support might be authorized. But that may be one man's opinion, or it could be a universal opinion for those well versed on historical research and the publications business.

At age 71, I am firmly convinced it would be essentially impossible for any author born after 1945 to research, write, illustrate and publish any in-depth history of the P-47 Thunderbolt, necessarily including evolution of its Seversky predecessors, which would be anywhere near as comprehensive, accurate and well illustrated as this volume. Such authors lack the definitive resources, access to authoritative personnel and the personal contact with the decade and a half during which all took place. Simple examples: Could they have even the most remote idea of what Hollywood Boulevard was like on Friday night in 1941-42? Or what it was like to drive on Hollywood Way in Burbank during a shift change? Or what it's like to be caught in a blackout in Los Angeles with anti-aircraft guns popping off in the distance. Most of the key personnel involved in design, development and test in that fifteen-year period have soloed to another realm. Republic Aviation Corporation records have

The demands placed on aircraft in combat naturally resulted in a massive dose of Unsatisfactory Reports (URs) to flow through HQ back to the ZI (Zone of the Interior, i.e. United States). Many of those URs led to experimental projects and subsequent test programs. Result: testing was carried out at many locations other than Wright Field, Ohio, and Eglin Field, Florida. Combat evaluation was routinely carried out at Pine Castle AAF/Orlando AAF, Fla., where this P-47D-4-RA out of Evansville's plant went against P-38s, P-39s, P-40s and P-51s. Air Proving Ground Command (APGC) airplanes carried distinctive numbers on their cowlings for positive and quick identification. (APGC)

The 82FS of the 78FG is shown assembling for a Ramrod mission in which 75-gallon metal belly tanks were being used for (probably) the first time. Believed to have been taken on 7 August 1943, the Duxford-based P-47Ds (and a few Cs) came under the command of a new C.O., Major Jack (Jake) Oberhansly. This mission was to be a withdrawal escort for the bombers which were met at Heinsburg, Germany. Lt. Col. Harry Dayhuff had just been appointed group executive officer and deputy to new group commander, James J. Stone, Jr., succeeding Lt. Col. Melvin McNickle who had become a POW. One peculiarity of this group was in the eight abreast takeoff formations. (Jake Oberhansly)

been destroyed, scattered, lost and generally dispersed into private collections, making them all but untouchable. Military records were destroyed by the thousands in a St. Louis fire; photographs formerly preserved at the Photographic Records Center in Arlington, Virginia, were transferred to a different command at Norton AFB, California, where many treated them with indifference, carelessness and even annoyance. Budget restrictions blocked preservation of nitrate films and many other items at crucial times, condemning the materials to destruction. With Norton AFB doomed to closure, nobody can accurately predict the future of the archives. Like old soldiers, much of it has just faded away. What insider was likely to protest?

This chapter and Chapters 12 through 19 in this book rely heavily on large numbers of appropriate, choice photographs documented with detail- and fact-filled captions. The story lines carry through the various theaters of war in support of relevant photographic coverage, coming to a conclusion with – hopefully – something less than a quarter-million words. Anything more will be overkill. Although the film title, "The Longest Day," gives the impression of substantial coverage of the invasion of Festung Europa, critical analysis would reveal coverage was limited to scraps or fragments of key overall happenings. And so it has to be with this book.

For those wishing to expand their detail knowledge even further by learning about individual squadron, group, theater or specific battle action, a select list of recommended reading (see Appendix F) is provided. Of course it is not extremely comprehensive, and that is by design (although this writer has most certainly not read every good book written about the war). Many books cannot be recommended because research depth is lacking. Some of the books are long out of print, but may be located with only minor difficulties. The Air Force Museum Research Division, Library of Congress, and used-book

A gathering of Thunderbolts at Meeks Field, Iceland, on 9 August 1943 – about a year after the BOLERO P-38Fs passed through – signaled the pioneering attempt to deliver P-47s by air to the ETO. Air Transport Command's 2nd Ferrying Group was to fly ten of the P-47D-5-RE from Delaware to the U.K. for VIII FC. Lockheed 165-gallon P-38 drop tanks were jury-rigged for the trip because Republic Aviation had not yet devised pylon mounts. One aircraft groundlooped enroute, sustaining some damage. Barry Goldwater, later to become a U.S. Senator, flew the No.3 airplane, "Peggy G". He confirmed that the Jug could be made to fly for 7 hours on one fillup, adding "my backside was never so sore." (AAF)

dealers are certainly the most obvious search points. But be aware, more than 250 individual P-47 Thunderbolt squadrons served in WWII (see Appendix G). And, yet, P-47 fighters were not participants in every theater of operation. They did not enter combat in the Aleutian Islands campaign, and little has been divulged about the role played by P-47s shipped to Russia.

Therefore, as President Harry Truman was inclined to say, The Book Stops Here! The scope and manner of presentation had to change. Otherwise, the manuscript would have continued on forever, never to be read by anyone. Or, it may be more contemporary to say, "Watch for Thunderbolt II, III and IV...even more." It could prove to be yet one more longest day.

ABOVE: Unfortunately for all concerned, this was Capt. Winslow "Mike" Sobanski's final T-bolt, a P-47D-10-RE. With the national insignia located exactly as the fighter left Farmingdale, 334FS code letters QP★F had to be added in this unorthodox manner. Sobanski, scheduled for promotion, was KIA in mid 1944. The 4FG promoted him, posthumously, to rank of major. (W. Sobanski via C.A.M.)

* *

By the beginning of the second half of 1942, Operation BOLERO was a "done deal" and literally hundreds of Lockheed P-38 fighters, Boeing B-17E bombers and Douglas C-47 transports had flown to Great Britain in an amazing, pioneering operation. Mission success was beyond expectations. Earlier, on 25 June, General Spaatz had set up his Eighth Air Force headquarters at Bushy Park (Code name WIDEWING) just outside London and plans were proceeding to prepare the way for Operation ROUNDUP, the projected invasion of France scheduled for early 1943. It proved to be an overly ambitious plan, about to be delayed further into spring 1943 when the decisions were made to sever numerous bomber and fighter groups from the Eighth Air Force. They had been committed to Operation TORCH, the invasion of North Africa. (Although three P-38 groups were operating as functionaries of the Eighth Air Force, they were actually already operating within the framework of the Twelfth Air Force. It really mattered little; they were going to the MTO over any and all objections.)

The only operational fighter and bomber groups in the ETO in the summer of 1942 were the 31FG flying Spitfire Mk.IXs for high altitude escort of 97BG Boeing B-17Es and some low-level RAF Douglas Bostons flown by AAF pilots. Lockheed P-38Fs were beginning to arrive over the BOLERO transatlantic route, but no Bell P-39s were operational. On 28 July, B/Gen. Frank Hunter arrived in England aboard his B-17E after having been escorted across the

ABOVE: After sustaining destructive shellfire, like this P-47D-11-RE of the 63FS which was a victim of a direct hit by an 88-mm shell, scores of Jugs managed to return to base. More than a few Thunderbolt pilots complained to their crew chiefs that they had a rough engine, only to learn that one or more cylinders were missing from the R-2800 engine. This pilot was not identified, but should have received accolades. (56FG)

Obviously not giving an immediate post-mission description of aerial combat – the gun muzzles are still plugged – one 335FG fighter pilot tells of his exploit as his compatriots observe the techniques certain to make them aces. The pilot of "Jani" had chalked up an aerial engagement and was really illustrating it for the publicists. He was probably describing action of 12 August 1943. (Jack Terzian, 353FG)

In a picture probably taken in the rough late months of 1943 at Halesworth, according to Col. Zemke, 56FG pilots (plus a few others) are given a prebriefing about a tough mission. (Personnel, by row, initial, name and squadron.) First: S. Burke 63rd, M. Quirk 62nd, C. Reeder Jr. 62nd; Second: J. Patton 61st KIA, J. Brown 63rd, W. Janson 63rd, Not IDd; Third: First 3 not IDd, B. Smith 63rd, R. Westfall 63rd, A. Cavallo 63rd; Fourth: J. Powers Jr. 61st, R. Johnson 61st, W. Aggers 61st, J. Carter 61st; Fifth: F. Klibbe 61st, S. Hamilton Jr. 61st, D. Thompson Jr. 33rd Service Gp.; No IDs in last two rows. Several of these men were or became aces in the 56th Fighter Group. (56FG)

BELOW: Air war seemed to agree with Capt. Walker "Bud" Mahurin. Like the white hat guy with a grin, the 56th Group's Mahurin always seemed to be in good spirits. After having shot down two e.a. in just minutes during a single mission on 17 August 1943, he received congratulations from many squadron mates. Until that day, Mahurin had been depressed about his combat performance. His final WWII combat record was 20.75 aerial victories. (AFM)

Sometime in late summer of 1943, Thunderbolts of the 84FS, 78FG, wait for the flare signal that would send them off on what was probably their first mission using pressurized fuel tanks. On July 30 they had flown the first belly tank mission, but most likely it was with 75-gal. metal tanks that were unpressurized. At least they were able to ease a bit over Germany to Haldern. The group claimed 16 victories that day. The P-47D-6-RE in the foreground is Maj. Eugene Roberts' "Spokane Chief," WZ★Z. (AAF)

Another pair of external fuel tanks that underwent development at the VIII FC Air Technical Section were a flat-section pressurized paper tank ("A dismal failure" according to Hough) and this flat-section metal tank ready for test around September 1943. His early series P-47D was still not equipped with external stores pylons and it has the "toothpick" propeller blades. Carrying a nominal 150 gallons of fuel, tanks like this were ultimately made of near identical upper and lower halves with mating flanges. They were widely used in 1944-45, but only on P-47s. The 38-year-old (1943) colonel was pretty ancient by fighter pilot standards, but he flew numerous missions (mostly in P-38s), one of the earliest P-59A jets (at Bovingdon) and one of the first four YP-80As upon their arrival in England. Interestingly, he flew a P-47M and a P-47N at Bovingdon! In fact, on a mission in the M, he had to abort after reaching the Continent (ignition). On 2 December 1944, at age 40, Hough flew the YP-80A for 7 hours 25 minutes, including shooting four landings. (Cass Hough)

North Atlantic to Ayr, Scotland, by none other than the "Four Fuddy Duddies," better known as Colonels Ben Kelsey, James Briggs and John Gerhart and Major Cass Hough in P-38Fs.[1] Their flight, identified (appropriately) as "Lightheart Red," had been joined on the last leg by Lt. Don Starbuck of the 94FS. Hunter established his VIII FC HQ at Bushey Hall (not to be confused with Bushy Park), giving it the code name AJAX. (On 23 Feb., B/Gen. Ira Eaker had assumed command of VIII BC in England, later establishing his Headquarters at High Wycombe, assigning the code name PINETREE.)

Now consider this. The RAF had certainly advised USAAF top commands of the inept (see Webster) performance of their Bell Airacobras (redesignated Bell P-400 when taken over by the AAF after British refusal to accept further deliveries). Despite the warnings, the 31FG was scheduled to receive P-39Fs and P-400s. At roughly the same time, the campaign against Major Alexander Seversky, sponsored by vengeful factions within the War and Navy Departments, was under way in the press. The most strident attacks in magazines and newspapers stated Bell Airacobras

LEFT: Historical records about this event stated that the pilot had his instruments shot out, but the unscathed lieutenant found his way back to Halesworth. In landing, the P-47D assigned to the 56FG veered off the runway into soft dirt and the port wheel sank in. The fighter cartwheeled, ending on its back with fuel spilling from the 108-gallon belly tank. Evidently the supercharger ignited the fuel, but fast action by a firefighting crew averted disaster. With the canopy buried in soft dirt, the pilot would have been unable to escape. In the aftermath, he was extracted unharmed. (56FG)

---

[1] The youngest of these men, and it wasn't Cass Hough, was probably about twice the age of most fighter and bomber pilots then assigned to the quite new Eighth AF. Individually or collectively, these four men - self-depreciatingly known as the "Four Fuddy Duddies" - had more genius packed into them than the vast majority of mortal men. And despite their ages, every one of them was as brave as the most aggressive fighter pilot. They all had proven it over and over. Not only that, they were among the nicest people you ever met.

Col. Hub Zemke described Lt. Glen D. Schiltz, Jr. (correct spelling), as a "very young, tenacious pilot." On 17 August 1943, Schiltz (in the 63FS) scored the first triple kill of enemy aircraft (e.a.) for the 56FG, and Lt. Walker "Bud" Mahurin did well on that same day. Schiltz was credited with eight aerial victories during his tour of duty in combat. Having survived the rigors of WWII in the ETO, he was KIA in Korea. (AFM)

and Curtiss P-40s – both previous targets of Seversky's pen – were more than a match for Japanese Zeros in the Pacific Theater. But concurrently, the first AAF Bell P-400s to operate at Guadalcanal with the First Marine Air Wing were unable to cope with Japanese bombers at higher altitudes, let alone battling Zeros. Was self delusion rampant within the commanding ranks, or were they simply remembering what had happened to Gen. Billy Mitchell and to Gen. Frank Andrews?

RAF Bomber Command also took a very dim view of Eighth Air Force daylight precision bombing plan against Germany. However, when a dozen B-17Es, escorted by Spitfires, made the first VIII BC attack against railway marshalling yards at Rouen-Sotteville on the Continent, 17 August 1942, it was deemed a success. Just two days later the 8AF launched 22 bombers on a successful diversionary raid, but that formation of Flying Fortresses only managed to drop an unimpressive 34 tons of bombs. The big event of the following day was issuance of a joint RAF/USAAF "Directive on Day Bomber Operations Involving Fighter Cooperation," aimed at the coordinated day and night bombing effort. Within 24 hours a "warning shot" was fired across their bow. It came in the form of a 12-plane bomber mission dispatched to attack Rotterdam's shipyards. A lack of proper coordination caused the Spitfire escort to miss the rendezvous, and the B-17s were promptly attacked by a formation of 25 Messerschmitt and Focke-Wulf fighters. The mission against a shoreline target was very promptly aborted. What did they think was going to happen when the destination was well within German borders? Or was it "Mirror, mirror on the wall, who's the smartest of them all?"

There seems to have been one saving grace, almost universally omitted from the equation. For a number of reasons, the *Luftwaffe* failed to establish an aggressive interception campaign against the daylight bomber attacks, with but one or two startling exceptions, for nearly a year. Those exceptions both occurred at Bremen. Mission No. 52 was carried out by 106 bombers, and the German defense brought down 16 of them, the greatest number of bombers lost on any mission to that point and a very high percentage loss. Until June, loss rates returned more or less to previous levels, but on the 13th (Mission No. 63 to Bremen for attacks on U-boat yards) the *Luftwaffe* was again very aggressive. Although VIII BC had launched a force of 152 aircraft against Bremen, only 102 unescorted bombers engaged in the attack. Once again sixteen of the bombers failed to return to bases. A 15½ percent loss ratio couldn't be sustained for any length of time.

Surely the Germans learned some lessons from the Bremen experience and their own defense posture. For the bomber groups flying deep into Germany on 17 August 1943, it was pure hell. The initial heavy bomber wave was to hit industrial and aviation targets (primarily the Messerschmitt plant) in far-distant Regensburg, then turn south to reach bases in North Africa. It was a move which at least partially upset German interception plans. A force of 146 bombers was launched, but only 127

It is quite probable that this P-47D-2-RE (AC42-8001, LH★V) was one of the first Dog model Thunderbolts to carry the Hough-Shafer developed pressurized, paper, 108-gallon drop tank as shown here. Flown by a 350FS pilot, the airplane exhibits the Type 3 national insignia with red surround, only authorized during a 2½ month period in the summer of 1943. (Author's Coll.)

Appearing to be too young to vote (probable) or get a driver's license (possible), Lt. Vic Byers certainly had no trouble chasing Messerschmitts in his P-47D. He was a member of the 353FG known as the Slybird outfit. (William Tanner)

managed to reach the target area. Of the striking force, 24 – nearly one in five – failed to reach North African bases. Last minute weather-related operational changes resulted in a much later takeoff and assembly time for the second wave – Schweinfurt raiders – so the coordinated attack plan for the day had to be revamped. That bomb wing fought off sustained, furious and innovative attacks most of the way to the target. After what was declared a successful attack, they were then forced to fight their way back across Germany without any accompanying long-range fighters. In the process, of 230 bombers launched only 183 managed to reach the target area. Losses due to flak and fighter attacks were, in final analysis, 36 Flying Fortresses. Total VIII Bomber Command heavy bombers lost in attacking the two distant targets was 60 airplanes with at least 600 men aboard. The loss ratios at both targets were at an intolerable 19 percent, and they were not sustainable.

Losses would have been substantially greater but for the determination of pilots flying VIII FC Thunderbolts, equipped only a few days earlier with 205-gallon "udder" drop tanks and, in some cases, 75-gallon pressurized metal teardrop tanks. Employed by the 56FG for the first time on the 12th, the "cow udder" tanks were only marginally useful. Those pressed paper tanks had a limited life cycle before they began to leak seriously, they were unpressurized and separation difficulties were difficult to overcome. Typically for the period, on the 17th the 56FG provided penetration escort for the heavies, the extra fuel adding some 30 minutes to their round trip flying time. As the '47s broke for home, the GAF interception coordinators launched their aircraft to attack the bombers. The American

The saga of Capt. Robert S. Johnson and his historic brush with death on 26 June 1943 has been told and retold many times. Attacked by Fw 190s while on a Ramrod to the Paris area, his P-47C-2-RE (AC42-71235) named "Half Pint" sustained massive damage almost instantly. Unable to open or jettison the jammed canopy, he managed to extinguish a fire but had a serious hydraulic leak. Attacked repeatedly by a single 190 as he staggered toward England, the P-47 struggled on. Johnson even managed to land the pile of junk decently, groundlooping to a stop without brakes. Replacing a borrowed P-47C-5-RE on 27 August 1943, this P-47D-5-RE, AC42-8461, was coded HV★P and was named "Lucky." Johnson rated it his best T-bolt; he scored 20 victories in it, including one over 206-plane Luftwaffe ace Hans Phillips. "Lucky," with a different pilot at the controls, was lost in a mid-air collision over Holland on 22 March 1944. (C.A.M/Ed Boss)

Originally trained to fly Lockheed P-38s in the ETO in 1942, the 78FG soon lost all airplanes and most personnel to the North African Allied Air Force in a replacement mode. Senior aviators remained with the Eighth Air Force and soon received Republic P-47Cs to initiate operations under rapidly changing tactical planning. While the new Thunderbolts probably met the requirements of the 1940 change order and Republic Spec. #88C, they were ill-suited to perform the job that evolved in the summer of 1943. In effect, the "bomber people" had misled themselves and the "fighter people" about what would be needed to carry out a daylight bomber offensive. Operation POINTBLANK, the Combined Bomber Offensive against Germany was revised in June 1943, putting great new demands on the VIII FC. By that time, Maj. Jake Oberhansly was flying a P-47D-1-RE in the 82FS as MX★X, appropriately named "Iron Ass." (Jake Oberhansly)

pilots, having no desire to wind up with bone dry tanks before reaching the English coastline, had no alternative. In the late afternoon, refueled and re-armed, they again rose to escort the Schweinfurt bombers out of Germany during their withdrawl from the target. In this event the 56th was capable of penetrating 50 or 60 miles deeper, across the border and into Germany. Taken by surprise, the *Luftwaffe* lost many single-engine and twin-engine interceptors to the AAF pilots.

The August 17 experiences of the 4FG and the 78FG were quite similar, but by that time both groups were pretty far removed from the udder-like tanks. The 4FG airplanes had been equipped with them for a mission on 28 July with approximately the same unimpressive

A couple of eagles and aces are shown enjoying the comfortable, if not luxurious, commander's suite of offices at Debden in 1943. They are Col. Chesley Peterson (l.), then C.O. of the 4FG (and at age 23 arguably the youngest AAF colonel) and his deputy, Lt. Col. Oscar Coen. Both had scored aerial victories as Eagle Squadron members, and both flew Spitfires and Thunderbolts after transfer to the USAAF. (London Press)

With his crew chief in attendance, Lt. Col. Donald J. M. Blakeslee is shown in his P-47D, evidently preparing for an early morning departure from Debden. By the end of 1943, nearly all missions were flown with 108-gal. paper drop tanks on the keel shackles. Concurrently with Doolittle's arrival as 8AF commander, Blakeslee took command of the 4FG. Having led the new 354FG in their P-51B Mustangs on familiarization sweeps in December, he promptly lobbied intensively for re-equipping with P-51s, with success. Although virtually all "Big Week" missions were Ramrods flown by P-47s, conversion of VIII AF groups to Mustangs was completed by September 14. The famed 56FG did not convert.

results. Not any happier to have utilized them on 30 July, the 78FG managed to abandon them by 12 August when they flew their first mission with 75-gallon pressurized teardrop external tanks. Those tanks had been around for a rather long time, and it is somewhat disturbing to learn they actually gave an increased flight duration of 20 to 40 minutes over what was gained with the big, cumbersome 205-gallon monstrosities.

On that disastrous August day, one long year after the Eighth Air Force arrived in some strength in Great Britain and several months after Operation ROUNDUP was supposed to be launched, there were NO long-range escort fighters in the ETO. By long-range, we mean aircraft with a combat radius of action at least equal to the distance to target. Fighters capable of escorting the bombers over deep-penetration targets in Germany were not in inventory. Not in England. Agreements reached with the British about fighter escorts for bombers proved to be hollow because the RAF could field nothing with the necessary range and performance. The AAF was stretching beyond its limits to supply long-range P-38 fighters for General George Kenney's Fifth Air Force and to the Fifteenth Air Force in the MTO. The War Production Board was – can remiss really be strong enough? – remiss, if nothing more, in failing to establish a second production source early in 1942 (or even in 1941 when the vast order for Lightning Is was still a fact) for P-38 Lightning production. Lightnings were big, complex fighters and a second source was sorely needed. Where in the world were some available Lightnings with range capabilities to match the needs? Right here in the Z.I., with fretting pilots at the controls wondering if they would ever see the order of battle! They were overtrained and over here, not over there. We refer to the 20FG and the 55FG, both anxious to go, but seemingly unwanted.

Colonel Hough went to General Hunter in his AJAX office with the admonition about P-47s never being an escort factor unless they got some viable range extension. Did "Monk" Hunter rise to the need and make the short trip to WIDEWING to demand some allocation of Lockheed P-38s? Did he write a letter to Gen. Spaatz or send a teletype to Gen. Arnold? Not on your tintype, and that was some weeks before mid August, exactly the timespan Hough needed to develop those paper drop tanks which were so useful. Just as M/Gen. Brett had been the wrong man for the job as commander of the Fifth Air Force and was quickly replaced, B/Gen. Hunter had neither the required aggressiveness nor attentiveness for the VIII FC assignment. But Hunter was also victimized by his counterpart, Gen. Eaker. The bomber CG was so firmly entrenched with his belief in the defensive capabilities of the heavily armed B-17 and B-24 aircraft and their combat formations on paper, the realities escaped him. All pilots possessed different capabilities and bomber crews were not all functional at the same level. On the other hand, Eaker was painfully aware of Hunter's foibles and had written him off as ineffective. His big failure: not making it an issue.

For nearly two decades, the fighter had been looked upon by the Army as a limited role interceptor or, worse, a showpiece. As a result, Arnold and Eaker viewed the heavy bomber as the route to a strong and efficient air power. However, the goals were surely sabotaged by forced procurement of "bombers"

Former 71 Eagle Squadron pilot Lt. Duane W. Beeson from Boise, Idaho, flew with the 334FS to become a multiple ace in P-47Ds flying out of Debden. He is shown with his gear as he enters the cockpit of P-47D-1-RE (AC42-7890) bearing the I.D. letters QP★B. Debden (Sta. 356) was a long-term home to the 4FG, the only group to fly Spitfires, Thunderbolts and Mustangs with USAAF markings. By war's end, Major Beeson had 19 confirmed air victories over Luftwaffe aircraft plus 4¾ aircraft destroyed on the ground. (AFM)

ABOVE: When Walker Mahurin had scored 11 of his 20.75 victories in the air, he was pictured on the wing of his P-47D-5-RE "Spirit of Atlantic City, N.J." in the winter of 1943-44. Only one 108-gallon drop tank could be carried on early D models, and in this case it was a metal version. Once wing mount pylons were installed outboard of the landing gear, three tanks could be carried with relative ease, adding 216 gallons of useful fuel. (AFM)

like the Douglas B-18. Nobody on earth would ever have viewed B-18s as anything even closely akin to a Fortress in the Sky. As the war proved to all, it was more likely to be seen as a lead sled.

Sharing the bomber mentality with Ira Eaker, General Arnold could see no valid reason in early 1942 for not awarding the VIII FC plum to his longtime friend, Frank O'Driscoll Hunter. After all, "Monk" had been a fighter ace in "The War to End All Wars." He had been involved with pursuit airplanes throughout the poverty-level existence of the 1920s and even Depression-strapped budgets of the 1930s produced some improved results and status. Analysis of VIII FC stateside origins in the early months of 1942 reveals Col. Ben Kelsey and Capt. Cass Hough were the two key people in leading Hunter away from the Bell P-39F Airacobra as a weapon of choice for the ETO job. At least they were aware, if nobody else was, the Bell was a short-range, medium-level interceptor unable to meet the criteria emanating from Battle of Britain evaluations. The Airacobras aided Kelsey and Hough greatly by almost constantly failing to meet the challenges arising from preparations for Operation BOLERO, proving their point. The handful of Republic P-47Bs existing at the time were not going to provide the right combination, and neither would the P-43, already out of production.

Lockheed P-38Es and Fs were already meeting the challenges of the Alaskan-Aleutian campaign, and Kelsey and Hough both knew the Lightning was the best choice for the job at hand. When the fighters took up stations in England, who could have anticipated the switch to Operation TORCH and transfer of all Lightnings to North Africa? Perhaps that cut the heart out of Hunter for he seemed to go out of focus. Without the only weapons capable of carrying out what had to be the defined mission – bomber escort – he was perceived to be adrift. Instead of fighting for the life of his VIII FC, he chose to make the best of life in Merry England. Gen. Spaatz, with several enormous jobs laid on his back, obviously relied on his subordinate generals to carry out their own tasks and relieve him of any additional responsibilities. How was he to know Hunter had surrounded himself with a staff which others could see as being almost totally incompetent, even disinterested.

It took the Schweinfurt-Regensburg Raid, Mission No. 84 on 17 August, to trip the alarm wire, set the sirens screaming and threaten the very existence of the jobs held by Arnold and Spaatz's Eighth Air Force. You can bet the British equivalents of "I told you so" were whispered or shouted throughout the British Isles. But well before that fateful day, Gen. Arnold's ire was being raised by the old WWI ace.

Suddenly, Arnold de-

A very tired trio at Debden, following a tough 1943 mission, consisted of (l.-r.) Lt. Duane Beeson (top scoring 4FG ace in Thunderbolts), Col. Jesse Auton from VIIIFC and Maj. Henry Mills, a 334FS ace. As a general, Auton later commanded the 65th Fighter Wing. Beeson was credited with 19 aerial victories before becoming a POW on 5 April 1944, joining others in Stalag Luft IV in due time. (4FG)

ABOVE: Tough as a Great White shark, Thunderbolts regularly shook off the effects of near disasters, usually flying again in due time. Test pilot Carl Bellinger managed to separate a couple of cylinders from the case during a detonation test run on this P-47D-2-RE, with the engine promptly gushing a flood of oil. With a sick engine and opaque windshield, Carl managed to find Republic Airport. With flaps extended, he did not (or could not) extend the landing gear, so a belly landing was made on the sod. T-bolts featured a very strong keel. (Carl Bellinger)

cided to fire Hunter, but it must be presumed M/Gen. Barney Giles (scheduled to become a Lieutenant General and Chief of Air Staff to Arnold within days) suggested they allow Hunter to hang himself. Just make him give a synopsis of his command actions during his tenure and what actions he had taken to provide VIII BC with protective escorts over the targets on their missions.

To explain that in abridged terms, VIII FC B/Gen. Frank Hunter had come under fire from several high-ranking officers some six months before Doolittle displaced (a really accurate description) L/Gen. Ira

LEFT AND ABOVE: Republic test pilot Ken Jernstedt, former AVG fighter pilot, had a bad day while testing one of the new R-2800-C series engines along the coast of New England at 32,400 feet. As indicated by the hole torn in the top of the cowl, one complete cylinder and piston assembly departed from the crankcase. Fire erupted, and Jernstedt began a descent in the general direction of Long Island. Losing altitude, but with the fire having dissipated, he realized there was little chance of reaching home base. Having spotted an airbase near Bridgeport, Conn., he made an emergency approach, but on initial touchdown the propeller and reduction gearbox broke loose. Flailing the cowling and wingtip as it gyrated away from the razorback P-47D, it proceeded to mangle itself and destroy the gearcase. Jernstedt guided the T-bolt glider to a safe halt. (Carl Bellinger)

Photographed on the return leg of yet another mission, Capt. Fred LeFebre's P-47D-1-RE (YJ★L) bore the personal "Chief Wahoo" marking. The "Chief" was yet another popular newspaper cartoon character of the era. This 353FG Jug illustrates the manner in which the fighters were marked for every mission flown in the early days of ETO operations. Most high-scoring units had to abandon the practice to merely record air and ground victories. **(Col. Fred LeFebre)**

Eaker. Secret documents reveal great anger on the part of General Henry Arnold, triggered by well-founded accusations against Hunter and, far more damaging, by actual happenings. The decorated ace from WWI warranted a widespread reputation for "pleasure loving" life, and he was not on a 24-hour-a-day crisis communication basis with his field commanders. In fact, he distanced himself from those with the greatest knowledge of what faced fighter pilots on every combat mission. Evidently General Giles, the man who had been largely responsible for establishing the tremendous pilot and crew training programs out of the totally inadequate late 1930s Air Corps training concepts, had conducted an intensive investigation relating to charges against Hunter. At that time, Giles was AC/AS Operations and Requirements. Let us present some relevant factual material from daily routines, classified correspondence, and personal contacts with first-party personnel.

a. While two fighter groups with long-range P-38 equipment and experience trained and overtrained in the Z.I., just as the mission of VIII Bomber Command was building to a crescendo, Gen. Hunter made no attempt at all to demand transfer of those groups to the ETO. He had every right to insist on reassignment of trained groups to replace P-38s lost to the North African campaign launched by Operation TORCH. In direct contrast and half a globe away, Gen. George Kenney beseeched, threatened and cajoled AAF Headquarters, on a weekly basis, to send every P-38 which could be made available to his Fifth Air Force.

b. The fighter groups remaining under Hunter's command in the ETO, in the critical quarter before arrival of "interceptor range" P-47Cs and early Ds, flew Spitfires and Bell P-39s. Nobody was about to challenge the *Luftwaffe* 50 miles inside *Festung Europa* with P-39F Airacobras and even the well-honed RAF was not properly equipped to go that far to protect "Flying Fortresses." The question arises: Why didn't Hunter demand

Although "Hairless Joe," an Al Capp character from his "Li'l Abner" comic strip, was selected by Maj. Dave Schilling to decorate the nose of his P-47D-1-RE (and later mounts), it might have been better to have used "Earthquake McGoon." Free-spirited Schilling used up a few fighters during the war, but this T-bolt was surely reparable following a landing gear malfunction. Records indicate that many Jugs sustained midair fires in that first year of ETO operations for various reasons. **(Ray Bowers)**

Another Tankless Thask, Ollie. (Pun intended) Assembled in Finger Four flights, P-47Ds of the 350 and 352FS, 353rd Fighter Group, head off on a training sweep in the summer of 1943, prior to their initial combat mission of 11 August. At that time, Lt. Col. Joseph A. Morris was group commander, only to go MIA on August 16. Lt. Col. Loren G. McCollom was immediately thrown into the breach. Like the 355FG, the 353rd scored very heavily in the ground strafing role. **(William Tanner)**

a new definition of the role for VIII FC? Was it to be an unneeded interceptor organization or was it going to provide escort over the Continent? Historical documents seem to indicate an aversion by everyone to discuss the question. Hunter ignored it.

c.Early combat-worthy P-47s were in a state of basic readiness for at least several months before General Hunter was confronted with a giant-size problem. Faced with an urgent demand by his trusted aide, Lt.Col. Cass Hough, he gave the verbal, impatient response to develop range-extending drop tanks for the P-47s. Up to the moment Hunter spoke, he could not have even conceived of the serious problem rising like a tornado. According to Colonel Hough, he told Hunter "the P-47 is never going to do its job without more range," and "that big old bathtub tank won't do the job. So he said to me,'Figure out a way to do it.'" It is also a fact, from the direct source, Colonel Zemke made two real attempts to impress upon Hunter what serious problems had to be solved and soon. The response: "Put it in writing." Zemke made no more attempts to deal directly with the general. No direct knowledge about attempts by the 4th Fighter Group commander, Col. Edward W. Anderson, or the 78th Fighter Group commander, Col. Arman Peterson are known. Unfortunately for all, Hunter did not want to be faced with problems, his staffers knew it and they were not about to make waves.

d. Curtiss P-40Es and Fs, and Bell P-39Ds and Fs, were equipped with tear-shaped drop tanks in the Z.I., in the Aleutians, in Iceland and England in 1941 and 1942. If nobody in the entire United States had reason to recognize the range limitations forced on P-47 combat groups as they were dispatched to the ETO and SWPA, alarm bells should have jangled in AJAX beginning in January or February 1943. At Fifth Air Force HQ, General Kenney certainly recognized the P-47's shortcomings in a matter of minutes. These short-range interceptors had absolutely no provisions for carrying external stores, especially external drop tanks. With 75-gallon tanks widely available, why couldn't anyone install at least one (up to three) on P-47Cs??? Evidence exists revealing the flight test status of 110-gallon teardrop tanks on at least one P-47 fuselage mount at Farmingdale late in 1942 or early in 1943. Everyone still tends to say, "The AAF was opposed to installing external tanks on aircraft," an attitude extending back as far as the 1920s. Such commentary is pure tommyrot. Virtually every P-40, from the D series on, was equipped to carry a 52- or 75-gallon drop tank. Bell P-39Ds all had the provisions for carrying the tank, and usually did carry one. For the projected participation in Operation BOLERO, strenuous efforts were being made to install one of the "udder" style tanks on each participating P-39 fighter. It may have been the prototype of the similar P-47 "drop tank."

(To paraphrase some comments made by General Hap Arnold after the war was won, when someone asked him about the handicaps faced by P-47 fighter groups when assigned to VIII Fighter Command in the ETO, he was apparently more than anxious to bare his chest. He said the Planning Staff had done

Thunderbolt pilot Capt. Fred LeFebre flew his P-47D-1-RE quite successfully as a member of the 351FS operating out of Metfield (Sta. 366) and Raydon (Sta. 157). The first three 353FG C.O.s went down on the Continent; Lt. Cols. Joseph Morris and Loren McCollom in 1943, Col. Glenn Duncan in 1944. In those active days, top 353FG and 8AF ace (at the time) Walter Beckham became a POW. First VIII AF C.O. to go down was Armand Peterson of the 78FG. **(Jack Terzian)**

When Loren McCollom was still commander of the 353FG at Metfield, Lt. Col. Glenn E. Duncan gathered his valued crew together for a group portrait with his P-47D-5-RE. Duncan succeeded McCollom when the former 4FG pilot was shot down on 25 Nov. 1943, only to meet a similar fate the following July. Duncan managed to evade capture, worked with the Dutch underground, and eventually returned to command the group again. All of his later P-47s carried the name "Dove of Peace," although his 19 aerial victories and 7+ strafing targets hardly allowed him to make such a claim. (AAF)

Even in black and white, the misguided attempt to camouflage Republic Airport failed dismally in wartime. Hangars at the right, as viewed looking almost due north from immediately above defense squadron airplanes parked on an inactive east-west runway, were virtually uncamouflaged but for paint. Furthermore, with hundreds of P-47s awaiting delivery at any given time and no real attempts being made to hide the Ranger engine factory (top right), only somebody with 20-400 vision could have missed the facility. The compass rose and taxiway leading to it might have been a good aiming point for any bombing aircraft. That strip, with gun butt to the left, aims right at Bldg. 17 (the main assembly plant) and the old Seversky Bldg. 5 immediately north of it. Photographed on 9 October 1943, almost exactly two and a half years after the XP-47B first left the sod runway (starting at upper right), the huge facility is in stark contrast to the peacetime days. (Mfr.651)

Some small British manufacturers and Lt. Col. Cass Hough's unit at Bovingdon deserve the undying gratitude of hundreds of P-47 Thunderbolt pilots, not to mention thousands of men in VIII Bomber Command. In a matter of just a few weeks, the Technical Operations outfit designed, developed and pushed production of pressed paper & glue pressurized drop tanks for P-47 range extension. Hough tested them; then flew on sweeps as far as the Rhine River on proving flights. Ultimately the same design was done in thin steel. Painted tanks in foreground are the metal type; those painted silver (right background) were paper tanks. Drag counts were not high, and they worked. Later, hundreds of the P-51Ds used them just as successfully. Air Materiel Command's verdict: "Rejected. They will not work." Heroic deeds come in many forms. The mass-produced 108-gallon tanks were not small. (56FG)

a hell of a poor job in planning for any daylight mission escort by fighters.)

General Arnold acted out his postwar comments during the war, not with too many words – especially for the news media – but by direct, if camouflaged, action. He had initially drafted a TWX (or cable) for transmittal to M/Gen. Ira Eaker at WIDEWING in mid 1943 after General Barney Giles, on behalf of others, had made numerous serious charges against Hunter. Dissuaded from sending it, Arnold sent a letter (with enclosures) to Eaker. Confronted with the material, Hunter was ordered to draft a supportable response to each charge. According to those who should know, many of his rebuttals were distortions of the facts.

Within days, Hunter was out and on his way to the Z.I., where Gen. Arnold eventually pinned a second star on him and gave him an irrelevant assignment as CG of the First Air Force at Mitchel Field, New York, (effective 17 September 1943) for the remainder of the war. It could not have been easy for the Chief to lash out at his old friend, a WWI ace, but Hunter had run a "9 to 5 operation" and enjoyed an apartment in London while thousands died giving their all. Even worse for Arnold, he had to ultimately remove Eaker from a job not well done. In retrospect, it was accomplished with such finesse, most of the public and much of the military had no idea that it was severance with pay in the "politics mode." They simply created the Fifteenth Air Force. M/Gen. Doolittle was actually given a not-too-subtle pat on the back and the assignment of getting Operation POINTBLANK (the Combined Bomber Offensive) on track to support OVERLORD, the planned invasion of Europe.

With little fanfare, M/Gen. William Kepner took the helm to run VIII Fighter Command. At last the right man for the job was at the controls. Almost within a matter of days, the 20FG and the 55FG, long-range P-38 fighter groups were scrambling to head for the ETO. They were to receive the newest, but not yet the best, Lightnings with which to escort the bombers into the bowels of Europe. In 1943, the best available were P-38Hs, not the longest ranging nor most powerful of Lightnings. But, you had to fight with what they gave you, as Col. Zemke stated with such candor. Defending bombers over targets such as Regensburg was well within their capabilities but, alas, there were only two groups even close to being combat ready much before Thanks-

According to some sources, this photograph was taken following the 26 November 1943 raids on the Continent in which no fewer than 633 Eighth AF bombers were launched on the largest operation to that date. However, Col. Zemke has stated that this 56FG picture was taken not long after they settled in at Halesworth on 8 July. The aggressive, successful fighter pilots were (l.-r.) Cook, Morrill, Bryant, Truluck, Mahurin, Comstock, Schilling, Gabreski, Craig, Stewart, Klibbe, Brown, O'Neill, Petty, Valenta and Carcione. (Author's Coll.)

giving Day, with Lt. Col. Frank James's 55FG carrying the main load. On the other hand, Merlin-powered Mustangs were hardly more than a mirage before New Year's Day, 1944, and groups of P-47s capable of carrying more than a single 75- or 108-gallon centerline drop tank were certain to remain a massive minority in the ETO and MTO for the remainder of 1943. With the appearance of underwing pylons in the autumn months, the T-bolts were given a new dimension. Bovingdon's 108-gallon pressed-paper tanks generally supplanted the single 75-gallon metal tank, essentially doubling the external fuel tankage without a great increase in the overall drag. The astonishing success of Operation BOLERO, wherein large numbers of Lightning fighters made the transatlantic crossing with the aid of drop tanks (development of which was attributable to Col. Benjamin Kelsey's gutsy "beyond the scope of official authorization" actions), should have generated major repercussions. If there was any priority project underway which exceeded the drive to produce P-47 fighters, it certainly had to be something very special. Why the mission of VIII FC was not clearly defined immediately after Regensburg, and why the use of underwing drop tanks did not become an attendant priority will never be known.

It could be envisioned by only a few on New Year's Day, but the Thunderbolt was soon to move away from the role of escort fighter to "flying artillery." That was akin to being born again, but it would hardly make Alex Kartveli grin broadly. He would grin and bear it. In the meantime, after Schweinfurt-Regensburg Round 1, tactics and the nature of the P-47s themselves were to change overnight. Gone was the "interceptor" limitation which had been such an onerous constraint of the Jug. Escort duties increased tremendously in the final quarter of 1943, and enemy aircraft fell to their guns in great numbers. However successful

With a flick of the stick (Bics came later), Capt. Michael Quirk rolls his P-47D-11-RE to display the famed Seversky-Gregor wing planform that was as recognizable as the twin fuselages of the P-38. This 56FG, 62FS airplane is of special interest in that it carries a yellow rudder, depot-relocated national insignia on the fuselage and the "four corner" wing markings. No attempt was made to enlarge the starboard wing insignia, but the added emblems were of the 55-inch dia. Type 4 style. VIII FC had granted the 56FG permission to identify squadrons in the group with colors painted on the rudder of each fighter. The airplane was obviously well used. **(Kevin Brown)**

Five of the 56th Fighter Group's top aces are seen in December 1943 with their new C.O., Col. Robert B. Landry (r.), who had replaced Col. Hub Zemke (returned to the Z.I.). The aces (l.-r.) included Gabreski, Walter Cook, R. Johnson, Schilling and Mahurin. Landry, obviously, was old Air Corps. (56FG/91495AC)

they were, it could have been far better. North American's P-51A, a full five years newer than the German Messerschmitt Bf 109E, was doomed by total inability to match Me 109F or G performance. Installing the Rolls-Royce Merlin was a significant alteration, and it worked wonders. Thunderbolt range-extension modifications, accomplished within weeks to create the efficient P-47N at a later date, were already committed to engineering drawings sometime in 1943. The disasters and near disasters associated with daylight precision bombing of Germany's industrial, transport and oil targets over a six-month period after the Fourth of July 1943 could have been attenuated by intensive range-extension efforts. If the task had been defined properly, work could have been initiated by Thanksgiving 1942. Alas, there were just too many roadblocks.

The Lightning as savior was never going to happen. Insufficient training programs had been put in place; just two groups, even three or four, were never going to constitute a real fighter force in a theater such as the ETO. Nothing less than a half dozen would suffice.In the face of odds, Gen. Kepner made the best possible try. An early second-source production facility instead of some of the more useless and disgraceful projects, plus at least a 50-percent allocation of P-38s with Merlin power, could have worked wonders. Instead of becoming involved with the lunacy of the XP-75/P-75A program, a mating of Rolls-Royce Griffons to the Lightning airframe would surely have created a wondrous airplane every bit as rapidly as North American was able to develop the P-51D. None of that is a pipe dream; all of the elements were in hand or available. As early as 1942, Lockheed – with cooperation, instead of dismissal, on the part of the WPB – could have developed a 450-mph escort fighter with a 1200-mile combat radius of action using about 85 percent of existing P-38 components. With P&W R-2800 turbosupercharged engines and deletion of Prestone cooling, a 4000 hp high-altitude fighter could have been in service by January 1944 (or earlier).

The main reason for pointing to these concepts is to expose the really bad decisions made in the heat of war when far simpler and more logical decisions should have been in the forefront. Many politically motivated decisions affecting production can be every bit as devastating as the Battle of Kasserine Pass or the Battle of the Bulge. Surely there was adequate brainpower, no matter the work load, to clearly see what should have been done.

Three of the top-scoring P-47 fighter aces in the 4FG – although at the time of this picture they had probably just transitioned to P-51Bs – were (l.-r.) Lt. Duane Beeson, Capt. Don Gentile and Maj. James A. Goodson. Beeson had enjoyed the greatest air-to-air victory success flying the Jug. On April 13th, Gentile was making a low pass at the Debden airfield when his Mustang's prop hit the ground more than 20 times. He pulled up to avoid the 336FS area but soon crashed and broke the back of his newest fighter. Major Goodson, flying his new P-51D on a Rhubarb against an airfield at Neubrandenburg on 20 June, was struck by ground flak and became a POW. He had only recently become C.O. of the 336FS. (AFM)

Lt. L. Avakian's P-47D bore an interesting assortment of markings in the "optional" category, albeit significant and very important to the pilot. The diamond-checkered cowling (353FG) carried the name "Mole" and that was backed up by "Lucky" painted on the fuselage side panel. "Dolores" was painted at the wing-fuselage juncture. Some 42 mission emblems were added behind the cowl flaps. Avakian evidently scored victories over two e.a., followed by the destruction of a locomotive. Other markings reveal that four targets had been bombed by the P-47 pilot. (Ray Bowers)

Republic Aviation's ability to rise from virtual obscurity and surge forward with one of the greatest production demonstrations of the war should have been a beacon to guide those with less talent. Where Curtiss, Brewster and G.M.-Fisher failed on numerous projects, Republic maintained its focus and rose to the task at hand.

The crescendo of air-to-air combat became more intense as summer weather faded into autumn, then winter. The struggle to provide even moderate-range escort for the HBs taxed the best abilities of commanders at squadron and group levels. Generally unrecognized is the total unavailability of P-47D-16 and D-23 types in the combat arenas until 1944. Tooth-

LEFT : Having lost his brother to the war, Lt. John T. Godfrey went on to score 18 aerial victories sweetened by 12.6 additional e.a. destroyed in strafing attacks. Flying "Reggie's Reply" (VH★P) in the 336FS out of Debden, he did a fine job of avenging his brother's death. By early August 1944 he was a major and C.O. of the squadron, only to be shot down in a P-51.(AFM)

BELOW: Some thirty-two recognizable Republic P-47Ds of the 353FG were caught by a camera as they climbed out on a short-range mission (no tanks), a rather unique photo. Often seen as companions to the 56FG airplanes, it is not surprising to see them flying in near optimum "Finger Four" formation. Col. Hub Zemke was an early proponent of the RAF-type combat formation, and it soon supplanted anything that might have been taught stateside. (Jack Terzian)

LEFT: Nestled in close to the wingtip of a Consolidated B-24 (certainly not a B-17 with those cowlings), Lt. Charles Reed's P-47D was photographed on (probably) 20 December 1943 while returning from a Ramrod to Bremen, Germany. The flaps on "Princess Pat" appear to be about one-third down, providing a clue that airspeed must have been around 150 mph. Reed was in the 63FS and was credited with at least four confirmed e.a. destroyed. (AFM)

LEFT: Often overlooked in evaluating performance of various fighter aircraft in the ETO, at least, was the fact that in 1943 the "playing field" was anything but level. When it was pretty well leveled by mid 1944 by range-extension tanks, Continental airfield locations and more fully developed P-38s, P-47 and P-51 fighters, the balance of power became more evident. But well before that happened, adversarial "round engine" fighters were put into war game situations to determine weaknesses and strengths of the respective opponents. A P-47D razorback and a Focke-Wulf Fw 190A "buzz" an unidentified 8AF fighter station, almost certainly before May 1944, after having engaged in a bit of non-fatal dogfighting. (Ray Bowers)

BELOW: Evidently returning from a successful mission, with all external stores expended, a flight of 353FG Thunderbolts sweeps across the flying field at AAF Sta. 157 (Raydon, Suffolk), not far from Ipswich near the east coast of England. Pilots who were extremely critical of the Jug's size and heft when the fighters first reached the combat arenas – especially those who had flown Spits at any time – were not especially objective about the job to be done. The Supermarines were never going to be escorting the heavies to Berlin, Schweinfurt or Regensburg, and it took more than a little effort to accomplish such a feat in the Jug. With no escort fighters, POINTBLANK would have been on hold pending arrival of B-29s and B-32s, and they would have been subjected to attacks by Me 262s and Me 163s. And so it goes. (AAF 69631AC)

pick propellers, extremely limited external stores mounting, and other attributes handicapped the fighter groups in the ETO (and elsewhere). Eighth Air Force A-2 (Intelligence) reports revealed actual significant increases in German fighter aircraft production despite increasingly intense bombing attacks in the Combined Bomber Offensive. But the best P-47 types being supplied to VIII FC were P-47D-5s through D-11s, and few (if any) of those models had a two-droptank capability. The victories scored seemed to indicate there was some factor(s) favoring the rising tide of Thunderbolt squadrons in action. Possibly the major factors were development of fighter tactics and pilot training. M/Gen. Kepner's interest in his groups and the ability to work well with the commanders certainly played a significant part in successes. His performance was in stark contrast to that of Hunter and his staff.

One serious handicap never overcome was the ratio of fighter groups to bomber groups. To quickly gain control of the air, the ratio should have been reversed, but that was no more likely to happen than to see VIII BC move to night area bombing. Far more efficient P-47D-21 and D-23 aircraft were coming, but they did not even go into production until January 1944.

It cannot be denied. The battle had been joined in 1943.

Revealing best P-47 razorback characteristics – as revered by Sasha Kartveli – and camouflage, markings and details, a production P-47D-11-RE gets a magneto check prior to moving out for takeoff. The forward tip of a metal, unpressurized 75-gallon drop tank of the standard type used on P-39s and P-40s is visible in this wartime factory site view. Overseas, Hough's Air Technical Section promptly adapted the 108-gal. tank pressurizing system to the smaller unit. (RAC H-1069)

RIGHT: As the tremendous, unpublicized (but impossible to hide) buildup for OVERLORD proceeded night and day, seven days a week, Republic test pilot Carl Bellinger was massaging the controls of profusion of Jugs. A masterful crew of experimental test pilots, headed by Lowery Brabham, twisted the tails of everything new and untried. (C.A.M./Ed Boss)

ABOVE: When Lt. George Perpente's P-47D-15-RE "Fran" sprouted three of the glue and paper 108-gallon drop tanks (date unknown, unfortunately), the Thunderbolt was finally approaching true maturity. Still unable to fly escort over targets as far away as Berlin, or close to that combat radius in formation, nevertheless the P-47s were in a position to jolt the Luftwaffe beyond expectations. The extra 324 gallons of "Kickapoo Joy Juice" (sorry, Al Capp) nearly doubled the range. Lt. Perpente flew with the 351FS of the 353FG based at Metfield. (AAF 68920AC)

OPPOSITE PAGE: Five P-47Ds from the 353FG joined up with a B-17F of the "Bloody Hundredth" (100 BG) with markings EP★E which was straggling badly in a quest to reach its base at Thorpe Abbots. It would appear that 100th and the 353rd, which had been on a withdrawal Ramrod, had encountered GAF fighters. The "Little Friends" and "Big Friends" relationship is well illustrated here. (Jack Terzian)

ABOVE: Amidst the furious rivalries generated in the ETO by several pilots seeking the status of top ace in VIII Fighter Command, Maj. Walter Beckham was the quiet, studious type. A cousin of Republic test pilot Lowery Brabham, the major ran up a leading score of 18 aerial victories in only 57 missions for the 353FG. Here he stands by his P-47D-5-RE that shows much wear and tear at a time when he was a captain. As C.O. of the 351FS he flew a slightly later razorback Jug named "Little Demon." After scoring his 18th victory in Feb. 1944, he joined other top-rated aces in a Stalag Luft. (AAF)

Cols. Ben Kelsey and Cass Hough were back together again when General Doolittle expanded the role of VIIIFC's Air Technical Section to the Eighth Air Force Maintenance and Tech Services. In actual fact, Deputy C.O. Hough was functional commander most of the time because Kelsey was usually on TDY. In 1940 when Hough was commissioned as a Service Pilot at Selfridge Field, he had already logged more than 4971 hours of flight time. His development of 108-gal. paper/metal pressurized drop tanks, built by Bowater-Lloyd and others, had historic significance. Cass's P-47D mount had a one-off canopy arrangement to accommodate test observations. Taken at Bovingdon, 21 Jan. 1944. (USAF)

BELOW: Lt. Col. Jack Oberhansly, Deputy C.O. of the 78FG, was pictured with his crew sometime in mid 1944, probably at Bassingbourn. A new P-47D had replaced his former mount, "Iron Ass," and it was not to get the same name, most likely in keeping with his higher status in the group. "Jake" finished up with the five victories shown, but in December he rejoined his long-time pal Col. Harry Dayhuff who had taken command of the 4th Fighter Group. It is obvious that Oberhansly and Dayhuff spent a very long time in the ETO combat arena. (Author's Coll.)

BELOW: Considering the fact that it had not yet left the factory, the 204th Curtiss Thunderbolt – a P-47G-15-CU – looks like a patchwork quilt. The people in Buffalo were still trying to figure out how to build the things, and 1943 was nearly off calendar. Unhappy slave laborers all over Continental Europe were building better fighters in greater numbers (by far). Like Brewster and some others, Curtiss-Wright was committing slow suicide. After a glance at Republic, the Truman Commission could have been applauded for serving hemlock to the C-W Corporation. (Curtiss Wright)

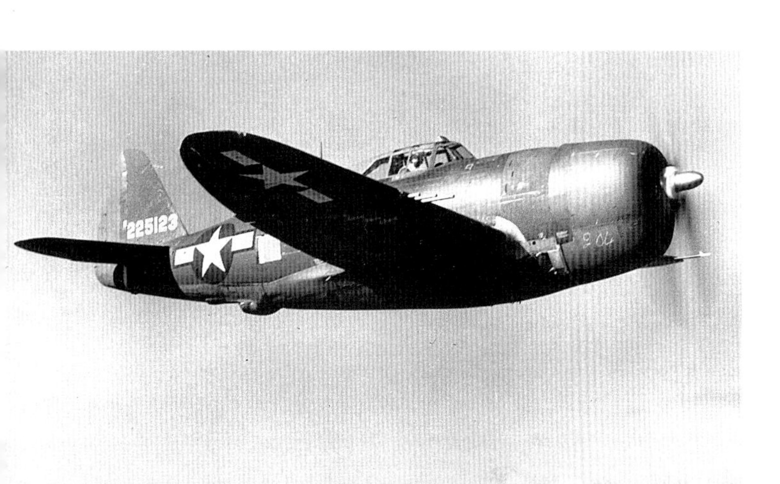

Captain Robert S. Johnson gives a friendly wave to some admirers as he sits in the P-47D-15 named "All Hell," his final mount in the 61FS. This photograph was taken after 13 April 1944, when he had scored his 25th victory. Rearview mirror and gunsight details are nicely revealed along with his updated victory symbols. It was time (22 April 1944) for this aggressive fighter ace to move over to the 62FS. Whatever the reason, Johnson never recorded any victories over aircraft on the ground. Gabreski and Bob Johnson turned out to be the highest scoring aces in the 56th Fighter Group. (USAF)

BELOW: Among aircraft called upon to engage in operational competition and aerial combat exercises at the Tactical Center, Orlando, Fla., in WWII were a P-38H, a P-51A and a P-47D-4-RA. Unfortunately, such actions often tended to "muddy the waters." Actual combat experience in the ETO, MTO, CBI and the SWPA was gained far earlier and more definitively. Allison-powered Mustangs were better than any P-40, but deployment was limited. Without the Merlin, it would have been just one more "forgotten" fighter, a "single-engine Westland Whirlwind" perhaps. (APGC)

CHAPTER 11

EXPERIENCES IN EXPERIMENTS

In the early days of World War II, public opinion and pressures from many venues forced the War Department to concentrate much of its expenditure of time, manpower, production capacity and funding on border defense. It was only natural for the AAF to issue circular proposals or even direct authority to develop effective interceptors. The Lockheed XP-38 and Bell XP-39 were, in their day, really atypical pursuit aircraft. Fighter Projects Officer Lt. Ben Kelsey was, in reality, looking for the best way to circumvent obsolete rules and regulations forced on the Army Air Corps by the United States Congress. Hence the "interceptor" appellation, created entirely as a subterfuge. Then, at a later time and place, Alexander Kartveli and company, seeking a home for an idea, reached out for an engine to make the P-43 Lancer something more than a cleaned up, supercharged version of the P-35A. It surely wasn't going to impress anyone if they took a page from the Curtiss book and installed an Allison in the Lancer. Fortunately for all concerned, the Navy Bureau of Aeronautics had commissioned Pratt & Whitney to develop a 2000 horsepower aircooled radial engine. It was evidently a success from the beginning.

With the Navy shying away from liquid-cooled engines because of disappointing experience with Curtiss, Packard and Wright engines in the 1920s, that branch of service devoted the bulk of their funding to efficient radial engines. The Air Corps needed good liquid-cooled inline engines to replace the Curtiss Conqueror line and to meet a strong demand for engines with small frontal areas. Contracts for development of so-called Hyper engines were issued in the 1930s and Wright Aeronautical offered their XR-2160 six-row aircooled Tornado as a potential 2000 horsepower unit. The AAF was interested in Wright's R-2600 twin-row Cyclone, but it was an engine of limited growth possibilities and it did not take well to exhaust gas

After Army Air Corps chieftans were apprised of the performance of Luftwaffe fighters (altitude, speed and maneuverability) when coupled with their equipment (armament, armor, self-sealing tanks), all lightweight fighter aspirations were put on hold. The performance of Me 109s and Fw 190s shocked even the British despite their faith in RAF Spitfire and Hurricane capabilities. Any hopes for Curtiss XP-46 and Republic AP-10/XP-47 lightweight fighter development were doomed. Seversky's abandoned AP-12 concept, obviously inspired by Wright Aero's (then) proposed 42-cylinder liquid-cooled radial Tornado engine, evidently led to the Republic AP-18, a fighter concept that deserved much more consideration. As the XP-69 came to 3/4-scale mockup status – unfortunately tied to the ever evasive Tornado engine – it can now be seen as a far superior, much earlier idea of what Fisher's (GMC) ridiculous XP-75 tried to be. The XP-69 was designed to gross approximately 10 tons, armed with two 37-mm cannons and four .50-cal. machine guns. If fitted with an Allison V-3420 or Wright R-3350 (both existing), it would have outstripped the P-75A on every score, and at least two years sooner. It is obvious that the P-84 jet fighter had its empennage initiated in this design. (Republic)

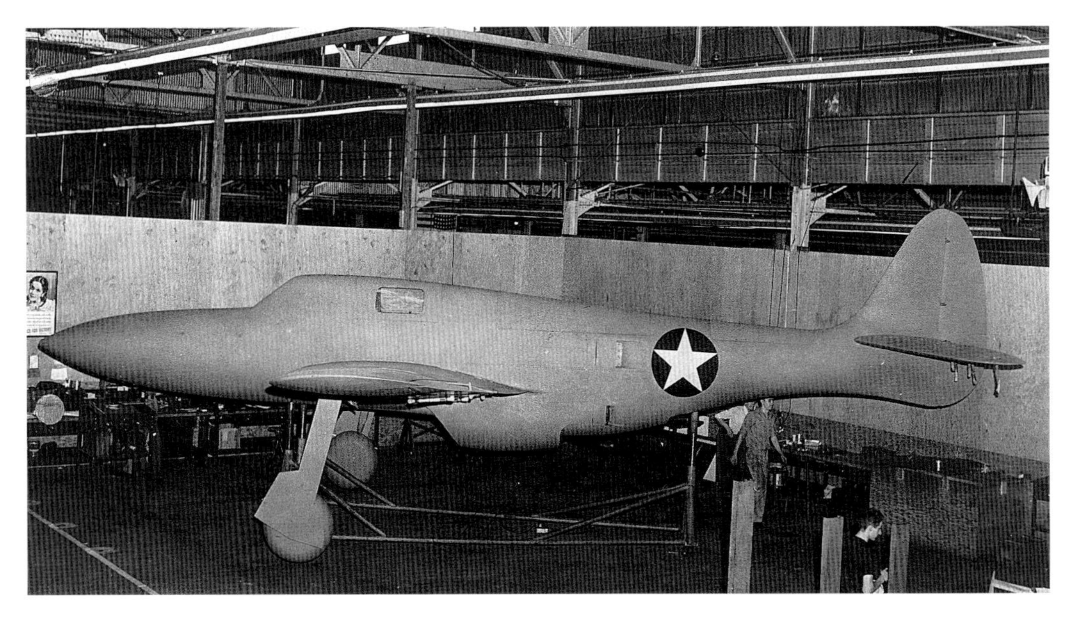

supercharging. It was limited to about 1600 hp for the duration. For several reasons, the big Wright R-3350 Duplex Cyclone was looked upon as *the* engine destined for the Very Heavy Bomber (VHB) programs. It had to be much larger and heavier with a 550 cubic inch displacement increase over the R-2800 Double Wasp and essentially the same horsepower.

WHEN WRIGHT IS WRONG

Exhibiting a great desire to expand their AP-12 concept to meet the new requirements, Republic offered the AAF a complete departure from their Thunderbolt program by constructing a large ¾-scale mockup of their Model AP-18 design. Photographic angles were not conducive to providing even a clue about the wing planform, but it was a mid-engine aircraft with rather conventional Thunderbolt-style turbosupercharging. Designers at Republic quickly realized the basic impracticality of their earlier AP-12 design. Foremost among implausible goals was the contrarotating propeller at that stage of development. Another was the tremendous deficiency in viable cooling air and aspiration air intakes. For a fighter aircraft, even a pure interceptor, the pilot's field of view was severely limited.[1] Kartvelian design characteristics were seen in every aspect of the AP-12 configuration. Sasha did not like anything breaking up the purity of aerodynamic lines. Contrary to previous speculation that the AP-12 was to be powered by an Allison without so much as a supercharger, every design feature would tie it to the five-row Wright X-1800 "corncob" radial engine. (Wright was seriously touting an output of about 370 hp per row of cylinders, or about 1850 hp for five rows. A bit later, a figure of 2000 hp out of six rows – 42 cylinders, 2160 cubic inches – was quoted.) If the X-1800 version was actually going to produce 1850 horsepower, that would have been just right for the AP-12.

Kartveli's AP-18, intended primarily for enemy bomber in-

The RAF high command received a rude shock during the Battle of Britain when they found that Luftwaffe Me 109s and Fw 190s could climb to higher altitudes, thus giving them the advantage of diving on defensive Spitfires and Hurricanes. That information was an unpleasant surprise to Army Air Corps GHQ Air Force and Materiel Division chieftans for they had only a handful of ATSC service test YP-38 and YP-43 fighters capable of reaching comparable altitudes. Even so, their big worry was Air Corps oxygen system capabilities. They quickly launched research into cabin pressurization for fighters with contracts for Lockheed XP-38A and XP-49 types and for Republic's XP-47E. The latter type, coated in chrome yellow paint, made its initial flight in September 1942, but any success with cabin pressurization was not achieved until July 1943. AAF Capt. George Colchagoff, project officer at Farmingdale, did most of the test flying. (Mfr. 7079)

[1] Aerodynamically the AP-12 was a perfect forecaster of what fuselage shape would ultimately be employed by Bell Aircraft to create the XS-1 supersonic rocket plane. Cockpit shape, general wing positioning, and fuselage shape continuity were about equal. The empennage shape for the AP-12 was possibly even an advance improvement because a ventral fin was included in the design.

Although some young P-47B pilots claimed to have reached levels of 48,000 feet, veteran civilian test pilots and experienced AAF pilots flatly scoffed at such attainments. Altitude performance of the XP-47E could hardly be described as scintillating. In fact, even when the Eclipse-Pioneer cabin supercharger managed to operate during a complete test flight, the airplane could hardly fly high enough to tax it. A major modification involved installation of a P&W R-2800-59 water-injection engine and one of the earliest paddle-blade H-S propellers. Repainted in standard AAF camouflage colors, the XP-47E was flown to Wright Field by Capt. Colchagoff where he was serving as the new project officer at ATSC's Fighter Branch. The war came to an end before cabin pressurization for fighters became fully operational. (ATSC)

ABOVE AND BELOW: *Focke-Wulf designer/executive Kurt Tank dismissed the laminar flow wing theory as inconsequential and impractical. In some ways he proved to be correct. North American Aviation was the first aircraft manufacturer of consequence in the U.S. to embrace the idea, designing their NA-73 Apache/Mustang fighter for the RAF around it. They had outstanding success. On the East Coast, production of P-47Bs had hardly begun when Chief Engineer Kartveli and Project Engineer Don Weed sought to modernize the Thunderbolt in 1942 by wedding it to a laminar flow wing. The 44th production P-47B was selected for modification in June. A shortage of skilled personnel, problems associated with bringing the Jug to high production rates, and other serious problems resulted in fabrication of an imperfect wing. As a result, their XP-47F prototype had performance below the level of production P-47Cs. After extensive company, AAF and NACA testing, it was destroyed in a crash not far from Langley Field, Virginia, in October 1943. ([illegible])*

terception at high altitudes, was predicated on knowledge of German and Japanese bomber development programs without solid intelligence about progress. The interceptor was required to be swift and have a superior rate of climb, heavily armed, and constructed with a pressurized cockpit for the pilot. With a requirement for an operational ceiling near 50,000 feet, that type of cabin was believed to be a necessity. The connection between the AP-18 and Republic's XP-47E might very well have been direct. Lockheed had been relatively successful with their pressurized cockpit development work for the XP-38A and the XP-49, but the Farmingdale firm was having less success with their efforts. Many pilots knew P-47Bs struggled to climb above 35,000 feet and even the latest C/D models were pretty near the limit at 38,000 feet. (Col. Hough's formal report on the compress-

ABOVE AND BELOW: The Army Air Corps had mounted fighters on ski gear since Curtiss PW-8s engaged in maneuvers in Michigan in the mid 1920s, and at least one Curtiss P-36 was mounted on skis before WWII. The Japanese attack on the Aleutians soon led to mounting a North American P-51A, a Bell P-63A, Lockheed P-38G and J types, a Curtiss P-40N and a Curtiss-built P-47G-1-CU on ski gear. Testing of the P-38s was accomplished at Ladd Field, Alaska, but the P-51A and P-47G were tested at Wright Field in the winter of 1943-44. All had retractable skis except the P-47, the latter being mounted on Federal skis attached by means of streamlined pylon struts. (ATSC)

ibility dive test in a newer, improved P-47C-2-RE revealed a total inability to climb past 38,000 feet.) Although oxygen systems had been inefficient at the end of 1941, fairly rapid improvements were made within a year. It was most unlikely any P-47 model was going to need cabin pressurization. Lockheed P-38s had been flown at 42,500 feet, and the XP-49 was *projected* to go higher with Continental engines.

In common with several contemporary designs, the AP-18 featured a mid-fuselage engine installation, a ventral radiator location based on NACA wind tunnel testing with the XP-47 full-scale model, and a long extension shaft driving contrarotating propellers. There was a direct progression from the earlier AP-12 concept, but far more practical. The pressurized cockpit was to be located well forward, in fact over the wing leading edge. Pilot vision, even rearward, would be at least as good as razorback P-47s. Evidence points to the use of a wing with straight leading and trailing edges, utilizing the new laminar-flow airfoil (scheduled for testing on the XP-47F). Presumably several Thunderbolt test programs were actually serving as "mule" programs for the AP-18. Incredibly, at least at the time, the design called for a normal gross weight of 18,655 pounds propelled by 2350 horsepower – at least according to the USAAF Model Designation - Army Aircraft book and other limited data sources. The aircraft was to have a wing span of 51 ft. 8 in. and an overall length of 51 ft. 6 inches. If the weight and dimensional figures were properly recorded, the proposed top speed of 450 mph had to be rather optimistic. But completely beyond belief was a specified overload gross weight of 26,165 pounds! This would have given the aircraft – an interceptor at that – a weight-to-power ratio of more than 11 pounds per horsepower, more than double the figure quoted for the XP-47B. What could they have been thinking?

A military designation of XP-69 was allocated to Republic for the AP-18 upon issuance of Spec. XC-622-10, and Contract AC-22238 was awarded for two airplanes, to be powered by the Wright R-2160-3 (Tornado) engine. In retrospect, the 2350 hp figure derived from another source was somewhat optimistic too. Even a post war figure rated the engine at no more than 2000 horsepower. (That same manufacturer had difficulty extracting 2200 hp from their aircooled R-3350 radial in 1944.) The contract for two XP-69s was eventually canceled, most likely because the Wright Tornado was more figment than fact. Even the well respected *Janes All the Worlds Aircraft* has never published one photograph of the Tornado engine, X-1800 or XR-2160. This scandalous, probably fraudulent, situation could only have remained cloaked in relative secrecy during the war to keep America's enemies or Congress from learning of it.

Air Materiel Command's (it grew) Armament Section people must have had a field day at one time during the war. Republic's XP-69 was to be armed with two 37-mm cannons and four .50-cal. machine guns. A contemporary imaginative design for an interceptor, the McDonnell XP-67 "Moonbat" out of St. Louis was about the same size at the Republic airplane, but it had two engines and about 2700 gross horsepower. Why any rational, aware group of engineers would have specified use of no fewer than six 37-mm cannons cannot be comprehended. That Browning weapon was proven to be essentially incompatible with airplanes during YP-38 trials in 1941. An American Armament

Wright Field's Experimental Section, evidently having paid attention to reports submitted by VIII FC combat pilots (and probably others), sought to increase Thunderbolt speed by reducing skin friction. A Republic P-47D-20-RE received a coverage of glossy camouflage paint and was then waxed after all joints were sealed and taped. For at least a couple of reasons, no national insignia or other markings were applied. (ATSC)

Corp. version was another fraudulent farce that could never have passed any serious procurement tests. Another questionable item relates to the XP-69 mockup created within a wooden enclosure in Bldg. 5, the original assembly building. After a full-scale mockup and critical wind tunnel model of the earlier AP-10/XP-47 was fabricated, why would Republic build a ¾-scale mockup of the AP-18? There was plenty of space available, and *if* an engine had become available it could not have been installed. Rapid, panic expansion at Wright Field was at least one of the culprits. No XP-69 was ever built, but one XP-47 was constructed and actually flew...far too late to be of any use at all.

LIKE A FISHER OUT OF WATER

Political pressures, inept management and other factors most certainly led to the creation of classified Project MX-317, the G.M.-Fisher XP-75 *Escort Fighter,* several months later. Donovan Berlin's semi-comic design was a truly composite aircraft in its original form. Why would any sane team of people in aircraft consider bolting a conglomeration of obsolete aircraft assemblies together? Some powerful, but inept, civil servant or overzealous AAF "patriot" must have had that brainstorm. Not enough? They then, very irrationally, wed that Rube Goldberg machine to a totally unproven Allison V-3420 engine and an absolutely troublesome, essentially unworkable dual-

rotation propeller system. The AAF would have been years ahead by commissioning some redesign activity on Republic's XP-69 to install two single Allisons or the V-3420 in a more logical, modern airframe. Illogical? Berlin and Fisher had to totally redesign the P-75 before receiving outrageous production contracts. What was the outcome? Mass-produced Lockheed P-38L-5-LOs, in the final year of the war, were about 35 miles per hour faster than a P-75A, had proven 3000-mile range capability, and were supercharged for high-altitude flight (P-75s had unproven 2-stage mechanical supercharging).

That "best of all" Lightning model was being delivered *en masse* by October 1944 from, at long last, dual production sources (hallelujah!). And, as if that were not enough to "deep six" the General Motors product from Cleveland, the Republic P-47N was created to perform the same function in a trice...while Fisher Aircraft Division of G.M. struggled mightily to produce SIX of their P-75A-1-GCs as of VJ-Day, 1945. Ultimately some 2944 were cancelled, and somebody had the good sense to pull the plug on plans for another 2000. In far less time, Republic built no fewer than 1667 of the long-ranging P-47Ns. And in the same time frame Lockheed delivered no fewer than 2520 of those outstanding P-38L-5-LO models, and the second production source managed a further 113 from Vultee's Nashville factory. Republic's P-47N and Lockheed's P-38L models demonstrated in tests and in combat the capability of outperforming the projections Fisher made for the P-75A. G.M.-Fisher (Cleveland) estimated the cost of two XP-75s (in 1942) would be $815,000.00. In three fiscal years, they built eight XP-75s (six of them almost 90% changed in design and with NO cost saving existing components) for a total actual cost of $9,110,021.56 and that includes a credit of nearly $650,000 for contract cancellation charges! Nobody went to prison. There has to be a touch of Socratic irony in all this.

PRESSURE AND MORE PRESSURE

One similar program, MX-88 initiated in April 1941, was worse than the XP-75 project. It proceeded almost unnoticed for five years. It was intended to produce a pressurized fighter with virtually the same capabilities demonstrated by Republic's XP-47E. It was one of those Curtiss (Buffalo) "'til eternity" projects, the XP-62, a caricature of the XP-47E. One obese airplane cost the taxpayers $2,903,637.06!

Materiel Division's early days fascination with the pressurized cabin or cockpit had been "cooking," so to speak, since a modified DeHavilland DH-4B biplane (known as an Engineering Div. USD-9A) from WWI staggered into the air in 1926. Adventurous Lt. John Macready was operating the airplane from within a pressurized tank built into the airframe of the Liberty-powered biplane. The interest really began to grow with the success of Lockheed's XC-35 pressurized-cabin version of the twin-engine Electra airliner in the mid 1930s. The Air Corps won the Robert Collier Trophy for that development in 1937. Later on, Lockheed received a contract change order to construct a pressurized cockpit version of their new P-38 fighter (66 ordered) for research and development purposes. Diverted from the production group, it was modified to become the one and only XP-38A. The airplane had limited success.

In April 1941 the Materiel Division asked Republic Aviation to investigate design parameters for pressurizing a P-47B cockpit. The AAF had a strong desire to obtain useful data about cabin pressurization for current and future use. Much was to be learned about sealing materials, pressure regulators, cabin supercharging, and windshield fogging. A contract change order, No. C.O.12, was approved on 16 October 1941 to initiate Classified Project MX-146. Capt. George Colchagoff, a brilliant engineer-pilot who later was involved in the X-1, X-2 and Dyna-soar programs, was appointed Project Officer on the experimental project under development at Farmingdale. The last

Pratt & Whitney produced the R-2800-57 C-series engine, identified by the split gearcase, for use in the P-47M and N, but development work was carried out with a -14W type which was nearly indistinguishable from the -57. Most interestingly, the P&W Model Designation & Characteristics chart lists an X-12 engine for dual-rotation prop drive. Obviously that was the engine fitted temporarily (for one flight) in the P-47B "Double Twister" that remained unknown to many key RAC executives. (UAC)

P-47B (AC41-6065) was taken from the production line, (via Letter of Intent) in mid September 1942. Evidently a major part of the modification work was completed on the assembly line because the major alteration was associated with the canopy installation, a modified version of the XP-47B non-sliding door-style canopy using special seals. Most of the work involved installing an Eclipse cabin supercharger, sealing every opening to the cockpit and reinforcing the structure.

Estimated cost for the experimental installation and basic testing work was $161,000 on a fixed-price contract. In most respects, the fighter configuration was closer to being a P-47C-RE than a B in the final group. The airplane was

OPPOSITE: Without "insider" assistance and attendant good luck, at least three rare versions of the Thunderbolt would have remained unknown forever. There was the "Double Twister" P-47B, one P-47D with that XP-47J powerplant and this undetectable "XP-47M" disclosed to me by test pilot Carl Bellinger. Probably without official sanction, a P-47C-5-RE "Mule" was fitted with the new 2800-horsepower (WEP) P&W R-2800-14W engine. Cloaked in chrome yellow paint, this old timer was flown extensively by Bellinger (in cockpit). According to chief tester Lowery Brabham, this unofficial XP-47M and the No. 1 YP-47M were frequently powered up to 3600 (!) hp. Fortunately for the AAF, when the V-1 Buzz Bombs were encountered in England, Republic engineers already had a solution. The new engine selected for the P-47N was fitted to D-30s already on the lines to create 130 rapid transit P-47Ms. Official testing was accomplished with two YP-47Ms in May 1944. (Carl Bellinger)

redesignated XP-47E for this experimental program, and for testing in the Farmingdale area it was coated with zinc-chromate colored paint. Republic pilots flew the airplane for the first time on 24 September. Although the secret XP-69 project was not mentioned in paperwork, it is pretty logical to assume the real benefits of the program would be applied to the more advanced aircraft. After all, Republic had no experience at all with pressurization of aircraft, and it is likely Wright Field's flight sciences laboratory was deeply involved in cabin design and probably furnished all Eclipse pressurization components. Specifications called for maintaining a cockpit altitude of 10,000 feet when the airplane was operating at a density altitude of 35,000 feet. At that altitude, the cockpit was to maintain a pressure differential of 6.65 pounds per square inch. Pressurization equipment was estimated to weigh 60 pounds, but structural changes and cabin sealing would likely add 220 pounds overall. In completed form, the airplane specifications were expected to meet these criteria:

It's a safe bet that 99 out of 100 people familiar with the Jug would identify this airplane as the XP-47J. Never before seen in print – any view, anywhere – until now, this P-47D-15-RE (AC42-75859) photographed on the approach to Runway 4-22 is really just an XP-47J lookalike. Having passed over the Ranger Eng'g. flight test hangar, it is just clearing a parking lot at the N.E. corner of Republic Airport. Tests were being conducted on the powerplant installation destined for an airplane that was officially the fastest AAF piston-engine fighter of WWII, the XP-47J (see story, pictures). Major differences in this powerplant when compared to other P-47s included a large spinner, cooling fan and a new CH-5 turbosupercharger with greater output. This test airplane has the standard 8-gun armament, turbo exhaust shroud and narrow tires. The experimental J model had 6-gun armament, fatter tires and a different supercharger installation. Cars ranged from Ford Model As to the rare Lincoln Zephyr convertible in the end stall. (Mfr. via Ed Boss, Jr.)

Design Gross Weight........12,900 pounds
Physical Dimensions.........As P-47C-RE
Powerplant.......................Pratt & Whitney R-2800-21 engine
Max. High Speed............429 mph, military power at 27,800 feet

***Designed specifically to overpower the Focke-Wulf Fw 190 series fighters in the ETO, the Republic XP-47J is shown soon after rolling out of the Experimental shop at Farmingdale on 16 November 1943. Featuring a more powerful P&W engine and a new GE high-performance supercharger, it was supposed to have Curtiss or Aeroproducts contrarotating propellers. That appears to be Lowery Brabham, Chief Test Pilot, in the cockpit.* (Carl Bellinger)**

Design Range..................550 mi. @335 mph, 10,000 ft.; 205 gal.
Normal Range..................800 mi. @280 mph, 10,000 ft.; 305 gal.

Serious problems manifested themselves very early in the test program, forcing a delay in conducting flight trials until July 1943. Upon receipt of what was regarded as a pre-production model of the cabin supercharger, testing resumed. Within a very short time, the unit failed at an altitude of 37,000 feet. However, this was considered part of the learning curve, adding to the limited knowledge of such systems in fighters. By December 1943 they had managed to complete the factory phase of system testing. The highest cabin pressure recorded was 11,000 feet with an airplane altitude of 35,000 feet. The company had hoped to utilize the XP-47E in connection with the Project MX-495 airplane (XP-69?), but that request was rejected after Republic had moved forward and added several changes to the system. As a result, the AAF would not reimburse the company for those changes.

Flown to Wright Field on 9 July 1944, the XP-47E had been forced to shed its yellow coating for a normal camouflage paint scheme applied at the factory. In the meantime the new Hamilton-Standard high-activity propeller blades became available. It is not known if the conversion was made as part of Republic's overzealous effort or if it took place at Wright Field. However, total project cost did not exceed the $160,000 estimate. Captain Colchagoff reported on performance of the cabin pressurization equipment, revealing unimpressive performance. He recommended discontinuance of the project, and especially voiced opposition to further consideration of the pressurization equipment supplied for that test program.

Possibly Republic learned some good lessons about the difficulties they would encounter in future programs. The beautiful XF-12 reconnaissance airplane had an optimum fuselage shape for cabin pressurization, serving as the only propeller-driven type built by Republic during the war to feature a successful cabin pressure system.

IF IT AIN'T BROKE, DON'T FIX IT!

Classified Project MX-116 was initiated in 1941 exactly two months before the XP-47B prototype left the ground. The Fighter Projects Office at Wright Field had evaluated new data about laminar flow airfoil developments and key men were well impressed with its potential. A modification contract, AC-19738, was awarded to Republic to obtain data about the relative performance of a P-47B airplane equipped with a conventional planform wing featuring the new airfoil. The AAF would furnish the aircraft (AC41-5938), the National Advisory Committee for Aeronautics (NACA) would provide a complete wing assembly and Republic would accomplish the modification. The company was not required to perform the full flight test program, but they did perform an evaluation program on two short-span, full chord wing sections to provide data about surface smoothness and to evaluate the effects of wing surface irregularities. When Republic's team mated the government-supplied wing to the airframe, creating the single XP-47F, they included the QEC engine mount package at the same time to bring the airplane close to P-47C-2-RE standards. An array of pseudo gun barrels were included on the leading edge to simulate an actual condition. Mating of the wing to the Thunderbolt was completed by 25 June 1942. The span was 42 feet, providing a wing area of 322 square feet.

Three key engineers (Eastman Jacobs, Harvey Allen and Ira Abbott) at NACA, Langley Field, were primarily responsible for development of the laminar flow airfoil. Jacobs was the primary developer of the overall shape, and he was heavily involved in development of manufacturing equipment needed to maintain an extremely smooth contour. In dealing with a few manufacturers who chose to become involved in the program, Jacobs maintained control over all design teams and wing parameters. For the P-47 program, at least, all moveable components of the wing remained as original except for compatibility with the straight trailing edge. Most of the basic wing structural design remained as before, although the changes in airfoil shape and required skin smoothness demanded some structural revision. The test wing was only designed to withstand 3-4g loading. No machine guns were installed, and the ammunition and gun access doors created a set of very serious problems.

In the cruise speed range, the goal was to gain a range increase of 20 to 30 percent. However, under practical manufacturing techniques, Republic Aviation's Experimental Deptartment encountered insurmountable manufacturing problems. No trials were conducted at high speeds because of the lower structural strength on the XP-47F combined with inherent P-47B flight limitations relating to the empennage attachment and the fabric-covered control surfaces.

Project engineer on the XP-47F/MX-116 project was Dr. Jacobs. The modified airplane was accepted by the AAF on 20 July 1942 with the first flight evidently made by Chief Test Pilot Lowery Brabham, who performed most of the pre-delivery testing. Additional testing was completed at Wright Field, then the XP-47F was flown to Langley Field's NACA Laboratories in the winter of 1942-43 for further evaluation. Performance figures essentially duplicated figures obtained with a standard P-47B featuring the normal S-3 airfoil. Actual costs of the project exceeded the $128,000 estimate by $55,347.00. With the laminar flow wing installed, the XP-47F could easily have been mistaken for any one of a few Curtiss P-60s with an R-2800 engine installed, but that was no cause for euphoria.

The XP-47F was flown to the Evansville, Indiana, modification center in March 1943, possibly (unconfirmed by records) for retrofit with a standard P-47 wing. Ultimately the airplane was wrecked at Hot Springs, Virginia, on 11 October 1943, killing the pilot, Capt. A.C. McAdams. The cause of that crash was never determined; could the empennage problem have taken a late bite?

At a somewhat later date, a Republic P-47D-20-RE tested at Wright Field was prepared with great care to extract the maximum high speed performance. All external markings were deleted and the fighter was given a special smooth surface finish. Although the airplane was photographed, nobody contacted in the group of people formerly assigned to Air Materiel Command could recall anything about it. Fighter pilots in various theaters of operations resorted to several different schemes, generally fairly successful, aimed at increasing speed. The primary approach was using fine steel wool to smooth the paint and then apply a highly buffed coat of wax. Numerous combat pilots, at least in the ETO where they were facing some of the foremost *Luftwaffe* pilots and latest model aircraft, elected to remove one machine gun from each wing. The weight reduction was beneficial unless the quantity of ammunition for the other guns was increased. Other pilots elected to remove some of the armor plate protection, but it is unlikely Capt. Robert S. Johnson ever considered that option after his early escape from almost certain destruction.

SKI LIFT

Materiel Division conducted numerous experiments with fighters mounted on ski gear. That was especially true in the early stages of the war when any final determination of Japanese Navy potential for carrying out continued assaults on the Aleutian Islands chain and Alaska seemed impossible. Fully or partially retractable ski gear was installed on a Lockheed P-38G-5-LO. A modified set of normal Federal skis was installed at Ladd Field, Alaska. Other types fitted with ski gear included a Curtiss P-40N-15-CU (also at Ladd Field). Although the P-38G suffered some damage when the landing gear collapsed during a landing in January 1944, general results had been encouraging. Additional testing proceeded after March 8 when one of the ten P-38J-1-LOs built for service testing was mounted on ski gear. Not long thereafter, any need for such devices on fighter aircraft vanished with the Japanese naval and aircraft threat soundly eradicated and Nippon's navy and army fighting for survival in the south.

At Wright-Patterson Army Air Base, a North American P-51A was mounted on retractable ski gear which seemed to work out just fine. A wind tunnel model of a Bell P-63A was fabricated with a ski gear setup looking much like the one installed on Lockheed's P-38G. No picture of the actual airplane was uncovered, and the report in which the model appeared did not give a hint

It was a cold winter day in 1943, and the XP-47J is seen being prepared for flight testing at Farmingdale. In contrast, a Ford Model T can be seen behind the company security man (wartime transport was at a premium). Republic claimed that its pilots had attained a verified speed of 493 mph at a pressure altitude of 33,350 feet and that higher performance could be obtained with a supercharger and propeller change. Impatient to beat the delivery date, the company made the changes at its own expense. (Mfr. 7061)

that the actual airplane was constructed. An early model Curtiss-built P-47G-1-CU Thunderbolt was, rather strangely, mounted on fixed landing gear struts attached to streamlined Federal skis. The large proportions of the tail ski would seem to indicate it was not retractable. Hopefully, the engineering included a fireproof enclosure for the attachment pivot. A canvas or leather covering could easily have resulted in a repeat of the 1942 XP-47B accident.

On 5 August 1944, test pilot Mike Ritchie flew the modified XP-47J equipped with an R-2800-14 engine and a new G.E. supercharger (Model CH-5) at a claimed speed of 505 mph at 34,450 feet over a company test course on Long Island. The run was only in one direction and was obviously not timed by FAI officials in wartime conditions. It was, however, the fastest recorded flight of any WWII piston-engined aircraft. After delivery to Wright Field, AAF test pilots failed to exceed a verified 500 mph speed. Republic vice-president C. Hart Miller is shown with the XP-47J on the last day of November 1944 when it was pictured at Farmingdale just days before being delivered to ATSC at Dayton. (C. Hart Miller)

DOUBLE YOUR DISPLEASURE

One Republic (Farmingdale) Thunderbolt proved to be so much of an enigma it was not even known to the chief powerplant engineer or the manager of flight operations. Perhaps anyone working at Republic's airport thought it was one of those struggling Buffalo-built Curtiss P-47Gs which became lost (or a runaway), or even more likely a version of the 3-engine Curtiss P-60 series fighters. *Well*, not really 3-engined. It just seems that way. Curtiss had floundered around with so many versions of an airplane in the P-60 series, they could hardly keep track. At various stages the same fuselage and wing may have had a Packard Merlin V-1650, an Allison V-1710 or a P&W R-2800-10. Why these disarmingly unlike (dissimilar) and unliked aircraft bore only slightly different numerical/alpha designations will never be known of course, but did it matter? While Curtiss struggled like a dinosaur in quicksand to manufacture P-47G Thunderbolts, their engineers produced a Double Wasp-powered fighter with new laminar flow wings, a contrarotating propeller plus size and weight equal to a P-47B. Guess what! The airplane was slower than a P-47B and could not fly nearly so high. And that brings us to the enigmatic Thunderbolt mentioned earlier. One bright day in 1942 a dirty-yellowish P-47B fired up near Hangar #1 and, in company with another mellow-yellow Thunderbolt, taxied a few yards to the runway. One was being flown by v-p Hart Miller and his wing man was AAF Plant Representative, Maj. Russell Keillor with a 35-mm camera hanging from a neck strap. Modest markings on the Miller airplane included a number 48 on the cowling. The number indicated it was serial number AC41-5942, the fourth airplane built in the final RP-47B production batch of 126 which probably had the structurally reinforced empennage support but fabric-covered control surfaces.

Now, this yellow P-47 with a difference was seen by nobody! What really made it different? (And invisible!) Well, it was the only P-47B that ever flew with a contrarotating propeller up front. (But the biggest surprise is yet to come. Although few people at Republic ever even acknowledged the P-47B "Double Twister," there is an amazing set of events exposed in the story of the XP-47J.) Miller's mount had an extremely large proboscis of Jimmy Durante proportions protruding from the cowling. That black bullet out there was at least two-thirds as long as the yellow cowling with six propeller blades churning the air. But nobody saw it, or so they said. How could Lowery Brabham not know about it? Was he absent from his office? When Chief Powerplant Engineer Ray Higginbotham was asked about it, he said he never heard of it – but he also stated he had been ill and his age may have been a factor in not recalling its existence.

Climbing in formation to 25,000 feet with this "Double Twister" trailing the "chase" airplane, Miller paused in this planned routine just long enough for Keillor to take a photograph. And, with a war on, it just happened to be Kodachrome film. (He either took only one picture or more likely the film just vanished as the years rolled by.) In performing a few basic maneuvers, Miller realized that with no torque to contend with, the Jug was constantly destabilized for some reason. All P-47s had a vertical stabilizer offset to compensate for torque effect, but nobody had thought to neutralize the vertical fin. Suddenly, the rudder went to the stops and locked. Under some circumstances, pilots did some improper thing and got into a deadly flat spin. Somehow Miller did manage to avoid that particular problem and recovered control.

The prime purpose for this test was to obtain evaluation data in connection with the proposed use of a dual-rotation propeller on the projected XP-47J. Other information has not come to light, but by the next day No. 48 did not have an engine or propeller. The double spinner and six-bladed propeller were returned to Aeroproducts, the manufacturer.[2]

Back in the final quarter of 1940 when M/Gen. Hap Arnold and Office of Production Management's first director William Knudsen visited Republic Aviation to inspect the facilities and managerial conditions, they departed with trepidation in their minds. The company was obviously devoid of any real mass production management specialists. Wallace Kellett's rhetoric did nothing to ease their fears. Republic was in desperate need of the best managerial personnel in the industry. They went directly to Undersecretary of War Patterson, explaining their

[2] Until work on this book began, a quarter century of research had convinced all that only two Thunderbolt fighters had ever flown with dual-rotation (or contrarotation) propellers. All researchers knew about the No.2 XP-72 being so equipped, but not one of them had ever encountered the P-47B "Double Twister." I learned about it directly from C. Hart Miller, who promptly gave me the only photographic print ever made. With the discovery of the P-47D equipped with a complete XP-47J powerplant also came the eventual information about both Aeroproducts and Curtiss six-blade dual-rotation propellers being installed on the P-47D. After supplying R-2800 engines for the Curtiss XP-60C and Republic's special P-47B and P-47D airplanes, P&W predicted a long delay in solving the gearbox problems. That and appearance of the XP-72 effectively "killed" the XP-47J program.

AAF Project Officer Capt. George D. Colchagoff accepted delivery of the one and only XP-47J on 8 December 1944 for transfer to Wright Field, Ohio. During the flight he ran into bad weather and had to divert to Toledo, but he managed the short trip to Dayton on the 9th after the storm had passed. Upon its arrival at Dayton it was photographed with its "Superbolt" nose art that had been applied by an RAC artist. (ATSC)

Among the many schemes put forth for towing troop gliders into combat zones was one tried out at Wright Field, employing a slightly modified P-47D-5-RE. With machine guns and the gunsight removed and a tow hitch installed aft of the tailwheel, it made at least one flight with a Chase XCG-14 in tow. If it had been proven feasible for a P-47 with full armament and at least one droptank to tow a loaded CG-4A glider to a combat drop zone, each massed group of gliders would have had their own protective fighter cover. (AMC)

ABOVE AND BELOW: The remote location of Muroc Army Air Base in the 1940s made it an ideal location for testing various schemes, no matter how "hairbrained" they seemed to be. One project pursued to completion at Muroc was a plan to tow salvageable P-47 (and other single-engine aircraft) from combat airstrips to rear echelon repair bases. With the propeller either removed or feathered, a flexible harness would be attached to the airplane. Some powerful aircraft like the B-17F seen here could either take off with the fighter in tow, or a Brodie-type snatch pickup rig would allow a multi-engine aircraft to fly past and literally snatch the fighter from the ground. Authorized by Project MX-584, air towing was demonstrated at Muroc AAB between October 19-28, 1944, using a tired RP-47B as the retrieved aircraft. A Boeing B-17F served as the towplane. (Air Materiel Command)

high-priority need for the very best. It would be necessary to literally hoist the pitifully organized minor league factory team off its collective butt and carve out a real production machine to occupy a half-million square feet of projected manufacturing and assembly floor space. Patterson immediately went to American Airlines and appropriated their operational vice-president, Ralph Damon, on a more or less temporary assignment. Mr. Kellett was boosted into a Board Chairmanship position, something not previously on the management list. Damon became president on May 1st just as the prototype XP-47B was completed, and who was going to protest? It was a good move. When Damon left Republic some 28 months later, the company had two well-oiled manufacturing plants operating at flank speed, grinding out about 450 fighters a month. Just about ten months after Damon took charge, he brought production expert Alfred Marchev aboard as his assistant. When the former production expert – at, of all places, Curtiss-Wright – departed from Farmingdale in September 1943 to return to American Airlines, his aide, Marchev, had a firm grip on the helm. (We won't think of asking why Damon was not moved to Curtiss at Buffalo. But there is no harm in wondering.)

Fred Marchev had come to America from Switzerland, but perhaps that was only a temporary stop on the route from Russia. Anyway, the team of Marchev and Kartveli were the two key men in charge at Republic. The legend about Alex Kartveli suffering pain when viewing the ultimate P-47N Thunderbolt is probably more fact than fiction. What picture did Sasha have on his office wall? Nothing other than the XP-47J, the designer's way of snubbing those repugnant things like bubble canopy cockpits and blunt wing tips, not to mention things under wing on which to hang bombs and external fuel tanks. Ornaments, he firmly believed, do not make for streamlined shapes. The temperamental Russian was just not pragmatic. Fighter pilots hated to fight the battle of the fuel gauge needle, and they would put up with a slight decay in speed in order to fly farther and pound the enemy with bombs.

TOO HOT TO HANDLE?

Not many weeks after Fred Marchev joined Damon as his assistant, all hell seemed to break loose. The war was going badly, to say the least, production of P-47Bs was not moving forward as planned, and then there were the devastating empennage problems. It is a certainty somebody was looking with a jaundiced eye at the manner in which P-47 gross weight figures were rising. And those people were also looking at the Focke-Wulf Fw 190A, wondering why our airplanes were so much heavier and yet had no performance edge.

Part of it goes back to an earlier comment about the lack of weight control engineers at Republic. One can sense Damon and Marchev had more than one intensive conversation with Kartveli and Hart Miller about such matters. By July 1942, alarm bells were jangling about the escalating weight of the T-bolt. A study was initiated for weight reduction by (a) deletion of equipment, (b) refining basic structure design and (c) by using more efficient methods to install accessories. It also seems they had borrowed a page or two from the Germans. After all, the AAF and the RAF were spending a lot of money and manpower working with captured weapons of war, so it seems logical to presume there was a method in their madness.

A program proposal was presented to the AAF in a conference convened early in November 1942; that led directly to a Materiel Command conference at Wright Field on 22 November. The subject was a new fighter carrying the designation XP-47J. Republic was given the objective of increasing performance of the Thunderbolt, primarily by means of the following changes:

a. Install an upgraded R-2800 engine utilizing water injection and fan cooling (already in use on the Focke-Wulfs and Messerschmitts).

b. Use a new and improved propeller, especially one featuring dual-rotation. (It also seems obvious why Curtiss, Hamilton-Standard and Aeroproducts had been directed to improve propeller performances.)[3]

c. Relocate an improved type, higher-altitude turbosupercharger toward the center of gravity as much as possible.

d. Revert to the original concept of six .50-cal. machine guns, with ammunition per gun limited to 267 rounds.

e. Delete the aft fuel tank.

[3] Somebody must have observed the very wide-blade propellers being used on the Fw 190A, Junkers Ju 88s and some of the Me 109s. American firms had made great strides in control hubs for the propellers, but they and the Powerplant Laboratory at Wright Field, headed up by Col. Page, seemed to have implicit faith in the "toothpick" propellers. Fan cooling of radial engines, according to the evidence, had not even been on the agendas of U.S. propeller manufacturers.

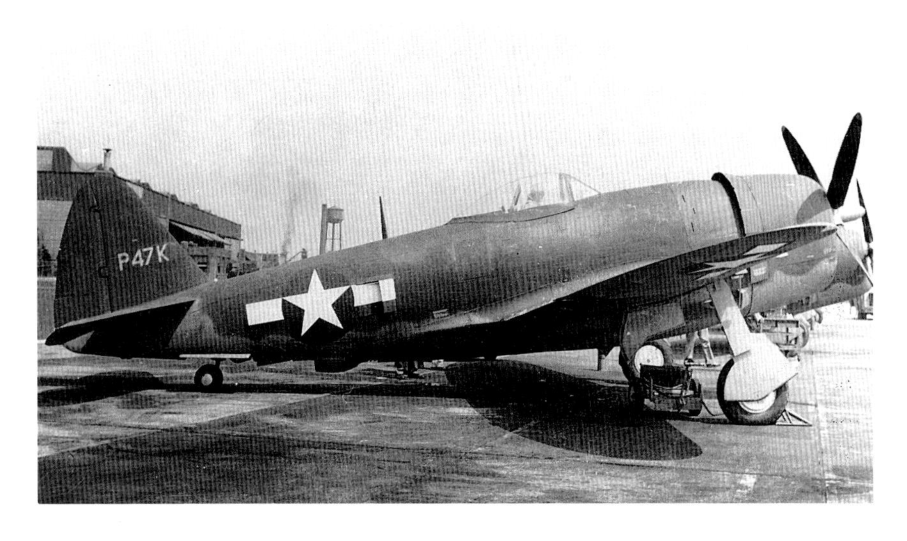

ABOVE AND BELOW: British successes in forming Perspex canopies for Supermarine Spitfires and Hawker Typhoons, and not forgetting the practical installation of Malcolm hoods on RAF Mustangs, led to AAF desires to put bubble canopies on several U.S. fighters. Curtiss engineer Walter Tydon was an early advocate of such canopies,and the second XP-40Q was probably the first AAF prototype to feature one. Republic was ordered to rework a P-47D-5-RE (AC42-8702), and this is the resulting conversion. Redesignated XP-47K, it was more than satisfactory. A single XP-47L, converted from a D-20, was similar but it featured increased internal fuel. When these and other changes were combined in one design, the P-47D-25-RE designation was adopted (for unknown reasons). There is evidence that the trial P-47N extended wing was first flown on the XP-47K. **(Mfr. G-84)**

f. Limit radio equipment to the VHF and a single SCR-274 radio.

Although the conference ended at that point, more changes were ordered on 31 November. A team of Republic personnel went back to Dayton on 5 January 1943, meeting with representatives from the Engineering and Production Divisions. By then the Materiel Command was throwing the door open to incorporating as many improvements as possible (including those in production tooling), and "changing the design of the airplane to obtain the maximum performance possible."One could almost detect a degree or so of panic. Company representatives threw a bucket of cold water into the conversations by pointing out the manufacturing department would find it necessary to change approximately 70 percent of the tooling.

In all seriousness, the AAF representatives concurred with the assessment by Republic, but they insisted "under no circumstances should the production rate of P-47s be permitted to fall off while changeover to the P-47J model was taking place." The response by company personnel is unprintable. The word "impossible" did have a prominent spot in the quotation. Part of the AAF "demand" was predicated on the maximum number of airplanes on order which could be converted (?) to P-47Js. Converted, with a seventy percent change in production tooling?

Cutting through all the mass of documentation, they initiated the program on 1 April. Very appropriate. The AAF was to get two XP-47Js plus a static test airplane, etc. for $1,027,650.00. After all that, the Army pulled in its horns and allocated only 30 percent of the quoted costs in its Letter of Intent, but formal Contract AC-39160 was approved on 18 June, upping the pot by $4500.

Amazingly, a 1/6-scale model was to be furnished at a price of only $1000.

The first of two XP-47Js was ready for initial flight testing sometime in July with the second airplane to be in that program by September. Specifications called for use of the Aeroproducts dual-rotation propeller. Mockup inspection took place during a six-day period late in April, and many revisions were requested. However, Republic was authorized to proceed with the first prototype in accordance with the original contract requirements to save time. The company wanted a torquemeter on the nose case of the R-2800-57 engine for use with the contrarotation propeller, but P&W was yet to design the installation and would not have it ready for another year or so. To the discomfort of Sasha Kartveli, Republic was instructed to employ the "free blown" bubble canopy design on the second XP-47J in order to improve all-around vision and egress from the airplane. The total price increased by about $210,000.00 to prepare the change. It was then the AAF decided it wanted two new fuselages, one for the flight test aircraft and the other for the static test program. They said to scrap the partially completed fuselage shells.

Recalling what the objectives were for this new aircraft, it seems amazing the Army wanted the turbosupercharger fully enclosed, providing some exhaust jet thrust. As if to spit into the wind, they wanted the deleted auxiliary fuel tank installed. The inspection committee wanted no fewer than 71 changes, which meant scrapping 30 percent of all drawings completed as of 17 July. As a result, Change Order No. 2 would cost another half million dollars. They also halted scrapping of the first static test fuselage, planning to use it for earlier tests of engine mounts,

ABOVE AND BELOW: As a result of some experience with 2-seat TP-40N trainers, Capt. George Colchagoff was appointed Project Officer for the conversion of two Curtiss-built P-47Gs into dual-control gunnery trainers. The Evansville plant performed the surgery, creating two TP-47G-16-CUs that had dual controls and full armament installation. The proposed instruction program never proceeded beyond the testing stage because of changing requirements. A young WASP pilot is shown in the front seat during a flight over Long Island, NY. (George Colchagoff)

etc., and then the fun really started.

A letter sent by Republic on 31 July contained performance calculations for P-47D-10, D-15, P-47J and P-72 airplanes. It said production could not start on either the P-47J or P-72 in less than ten months from contract initiation. And, if both types were wanted, there had to be a 9-month interval between programs. After all the fuss and feathers, Republic recommended scrapping the P-47J program except for completion of one flight test airplane in order to conduct the test program as planned. One feature of the XP-47J which had not even been mentioned in reports (at least key reports) was use of considerable spot welding of skins in the wing panels.

A normal 689 Board inspection of the completed XP-47J was held during the week preceding Thanksgiving week. The first flight took place on 26 November 1943 and, as might have been expected, a normal propeller was employed. After about 10 flight hours, metal chips were detected in the engine oil system. P&W had supplied an engine with a so-called "light-weight" crankshaft, but the replacement engine (installed in February 1944) was to feature a heavy crankshaft, water injection and other features to permit War Emergency Power operation (2800 horsepower). Not long afterward P&W notified all parties about some components in the R-2800-57 "C" engine which were not compatible with WEP ratings. Changing direction again, the AAF allocated funding for procurement of an R-2800-14 engine for April 1944 installation.

At that point, discoveries which should have been common knowledge to at least a brigade of key people at Republic and Materiel Engineering came to light. But rather mysteriously not one person had ever even referred to certain happenings, even in an offhand manner. First, a complete XP-47J powerplant – with a 4-bladed propeller and fan cooling – was flying around in a P-47D-15-RE (AC42-75859), a fact never before mentioned anywhere.[4] Yet everyone was surprised when Republic representatives announced they had been flying a P-47D (most likely the P-47D-15-RE) with a variety of propellers, including Curtiss and Aeroproducts dual-rotation units and regular and modified 4-blade single-rotation propellers. The information was discovered in an absolutely official document of great importance. Again, Chief Powerplant Engineer Ray Higginbotham was unaware of these tests (or had forgotten).[5] Testing of the dual-rotation propellers did not reveal any marked improvement over the single-rotation units. However, increases in performance were expected at the higher power ratings available in the XP-47J and with other projected propulsion improvements. At times it began to look like this was a spinoff of Lockheed's ill-fated five-different-engines XP-58 Chain Lightning program.

[4] To my great good fortune, the original negative (long ignored by everybody but Ed Boss Jr., as being just another XP-47J photograph, albeit one of high quality) was in mint condition and available. The rare bird fairly flew off the negative at me, begging to be recognized. It was, and is featured in this book.

[5] This was discussed with Ray Higginbotham several years ago, but he was in failing health at the time.

Pratt & Whitney threw some ice-cold water on plans in March when they informed the AAF of engineering difficulties relating to the reduction gearing for the XP-47J's R-2800-61 engine. United Aircraft could not even accurately guess when the problems might be resolved. (History will show such gearbox failures played a large part in failure of Northrop's B-35 bomber program which extended into the post-war era. In those years, any exceptional performance expected from gearboxes proved to be unmanageable. Aircraft totally defeated by such problems in later years included the Douglas XA2D-1 and North American's XA2J-1.) As a result, a large single-rotation 4-blade propeller with paddle blades was installed on the XP-47J.

What would appear to be an unlikely engine candidate for installation in a P-47 Thunderbolt, an inverted vee 16-cylinder liquid-cooled unit, was Chrysler Motors XI-2220. Initially planned for installation in the Curtiss XP-60C, the XIV-2220 (original and correct designation) engine – designed to produce 2500 horsepower – was not ready in the summer of 1942. The Curtiss installation was canceled in September. (Author's Coll.)

Republic submitted copies of formal flight test reports to the AAF Project Officer, claiming a maximum speed of 493 mph had been attained by the prototype at a pressure altitude of 33,350 feet. The key report stated this was accomplished at a verified power output of 2800 bhp during a flight on 11 July 1944. The XP-47J was, at the time, equipped with a G.E. CH-3 turbosupercharger. The manufacturer believed a better performance could be achieved by installing a CH-5 unit to increase the critical altitude of the engine. Permission was granted to change the supercharger and install a better propeller, in this case a 13 ft. diameter Curtiss assembly.

With the changes in equipment, company test pilot Mike Ritchie was at the controls when a speed run was made over a calibrated course on 4 August 1944. His test report dated 5 August, claimed "a speed of 505 miles per hour was attained at a density altitude of 34,450 feet with 2730 bhp."

Capt. George D. Colchagoff took delivery of the XP-47J (AC43-46952) at Farmingdale and ferried the airplane to Wright Field, Ohio, arriving there on 9 December 1944. (As the captain was approaching Toledo, Ohio, on the evening of the 8th, he learned about severe weather coming into the Wright Field zone. He decided to RON at Toledo rather than risk any problem with the experimental fighter. He made the short hop down to Dayton the next morning.) Some maintenance was performed at the base and official AAF testing began in January 1945. These initial tests involved maximum power trials at various altitudes up to the critical altitude in order to verify the Republic claims. Nearly all testing was completed when a serious failure occurred in the exhaust manifold system at an altitude of 36,000 feet. In the ensuing investigation, the AAF found the system to be inadequate to handle the greatly increased power (2800 bhp). When the company installed the larger capacity CH-5 supercharger, they did not change the constricted exhaust stack which was being used to take advantage of jet thrust. As a result, increased back pressure led to failure of the manifold system.

The Army report on the Ohio trials said they were unable to achieve maximum power, but a true airspeed of 484 mph was recorded at 25,350 feet on 2770 bhp. The P&W R-2800-57 engine had a Military power rating of 2100 bhp at 2800 rpm and 54 inches manifold pressure, limited to 5 minutes duration. War Emergency Power rating was 2800 bhp at the same rpm but boosted to 72 inches manifold pressure. Maximum fuel capacity of the airplane was 287 gallons, and design gross weight was 12,400 pounds. While the wing span was standard, overall length of the XP-47J was 36 ft. 3-3/16 inches.

The Official Performance Summary report lists a maximum speed of 507 mph, while Republic Test Report No.51 dated 27 January 1945 gives a maximum 2800 bhp speed of 502 mph. Col. George F. Smith signed the report for B/Gen. L. C. Craigie, Chief of the Engineering Division.

When the yellow brick aircraft engine building was completed at Highland Park, Michigan, progress on the V-16 engine permitted engineers to conduct preliminary 50-hour flight rating tests in the proper testing cell. For earlier low-power runs, the engine and propeller had been mounted on a 40-mm anti-aircraft gun carriage. (Author's Coll.)

EPILOGUE,

With the ability in hand to crank out 2800 brake horsepower at altitudes above 33,000 feet and the addition of 400 pounds of jet thrust thrown in, the XP-47J revealed just how remarkable was the wing on which all P-47s flew. (The comparable Hawker Tempest/Fury wing was not designed until at least a decade later.) Recalling the period in which it was designed, in the era of fabric-covered biplane fighters, only about three years after Lindbergh became the first to fly eastward across the Atlantic (nonstop at that), its performance was incredible. It housed up to four times as many heavy caliber machine guns as America's prewar pursuits could carry in the fuselage. It could support the widest track landing gear on any fighter in WWII. And it was apparently able to fly at jet fighter speeds without scaring pilots to death. How many pilots of Thunder Jugs had any idea of how slippery the wing was? And could they conceive the idea of the wing having been designed by two immigrant Russians, Alexander Sevérsky and Michael Gregor in 1930-31? It is terribly important to recall who was the first designer in history to put the turbosupercharger into the aft fuselage of a fighter, leading to the XP-47J's additional thrust. Just think what the Major could have accomplished with something other than a pauper's budget to work with. During WWII, the Government literally thrust money on any company with half an idea – as borne out by the Curtiss-run P-47G, P-60, XP-62, XF14C-1, C-76 Caravan and the Fisher (read it as General Motors) XP-75 or Wright Tornado programs. Alexander Seversky might well have been in a class with Kelly Johnson. The real facts bear out the validity of the arguments.

All who flew the P-47 Thunderbolt deserve to know how remarkable it was – and why.

Nearly a quarter century after the XP-47J was widely reported by official sources to have exceeded 500 mph in level flight at an altitude of 34,500 feet, test and race pilot Darryl Greenamyer was determined to exceed that speed. Piloting his highly modified Grumman F8F Bearcat navy fighter over a 3-kilometer straightline course at Edwards Air Force Test Center on 16 August 1989, he established a new official world speed record for piston engine powered aircraft. His average speed for four passes over the course required to be recognized by the FAI, the formal governing body for the world, was 483.041 miles per hour.

Way back on 26 April 1939, *Luftwaffe* pilot Captain Fritz Wendel had established a legitimate claim to the record speed with an average observed speed of 469.22 mph in a Messerschmitt Bf 209 V1 (reported for many years to have been a Bf 109R). The distorted information release was eventually proven to be very effective Nazi propaganda. Willy Messerschmitt's speedster had a very powerful but highly experimental limited-life engine and only about 115 square feet of wing area. (Quite interestingly, Republic Aviation's Alex Kartveli had submitted the AP-10 design for a lightweight fighter to the Air Corps within a year. It had approximately the same wing area and gross weight, but only half the power.)

On the day Greenamyer made his speed run, his highly modified P&W R-2800-34W Double Wasp was churning out more than 3300 horsepower using "doped" fuel. Although the F8F Bearcat airframe was about 25 years old, it had benefitted from masterful care and modification, featuring the latest modifications born of state-of-the-art engineering and nearly ideal flight conditions (100-degree day). Yet his Grumman, named "Conquest I" was well short of the 500 mph figure, and time continued to march on. Almost exactly another decade later, Steve Hinton piloted his extensively modified P-5l Mustang, appropriately

***As early as 1942, Chrysler had proposed installation of the new XI-2220 engine in a P-47B, but when they received a contract to convert two Thunderbolts in August 1943 the aircraft committed to the project were P-47D-15-RAs. Chrysler subcontracted the conversion to Republic's Modification Center at Evansville, Indiana. The first of two XP-47Hs, AC42-23297, is shown prior to its initial flight after the 689 Board Engineering Acceptance Inspection was conducted on 25-26 July 1945.* (Mfr. E-2510)**

named "Red Baron RB-51," to break the record on 14 August 1979 at a phenomenal speed of 499.04 mph. But he did not crack that 500 mph "barrier" and his record was to stand for another decade.

A different re-engined Grumman Bearcat, piloted by Lyle Shelton at Las Vegas, New Mexico on 21 August 1989, was ready for one more try at that elusive figure of 500. His engineers and technicians had managed to shoehorn an even larger (3350 cubic inches displacement) Wright R-3350 Duplex Cyclone inside the cowling of the F8F-2. Finally, with the benefit of 45 years of technology in hand and a ton of money expended, the seemingly impenetrable 500 mph barrier splintered. Shelton's No. 77 completed its four speed runs, flying at an altitude of about 300 feet, to set a new FAI-recognized record of 523.586 mph.

Swinging a 13-foot-diameter Curtiss propeller, the Chrysler engine was rated at 2500 horsepower for takeoff. The Detroit manufacturer had planned to use a new axial-flow supercharger, but it was not ready for flight testing. A General Electric CH-5 turbosupercharger was used in conjunction with the centrifugal blower mounted on the engine when the XP-47H left the Evansville runway on July 27. The pilot is reaching back to close the canopy as the main landing gear retracts. **(Mfr. B-961)**

It must be recalled, not as an excuse, Mike Ritchie was just performing his daily job on 5 August 1944 when he flew into history books. His XP-47J was not a stripped racer, not even polished. It carried six standard machine guns and all other normal equipment; the military did not sponsor it with assistance of any kind; and there was to be no glory in it because the war was raging all over the globe. But Republic management reported they had an even faster fighter in the beautifully cowled XP-72; their reward was XP-47J program cancellation. And the war ended before the XP-72s could be brought up to final standard where the USAAF was likely to give it the old college try.

TWO IN PURSUIT OF EXCELLENCE

Nobody can deny necessity is the mother of invention. A need arises, talk ensues and somebody usually has a workable solution. Or somebody has the ability to make an idea workable. One such need arose rather early on in the war, generated by the pure size and power of a single-seat fighting machine. More than one neophyte fighter pilot, faced with being sent aloft alone in the 2300 horsepower Lockheed P-38D or E, probably lost a meal. Rather fortunately, some Lockheed flight test engineers and pilots had observed a very strong need to make first-hand observations of conditions encountered by pilots in a vast variety of conditions. By removing radio equipment and doing minimal rework, they were able to gently squeeze men – especially men of smaller stature – into the area immediately behind the pilot where they had such a marvelous view of everything most of the cramping pain was forgotten. Record-breaking, world-girdling test pilot Jimmy Mattern saw great potential in saving student pilots from fatal errors by taking them on as little as one flight where they could observe the instructor pilot at work. He easily convinced Training Command General Barney Giles of benefits to be derived and was soon engaged in a well-supported program of "Piggy-Back" training. It was immediately successful in reducing the P-38 Lightning accident rate by 75 percent.

Curtiss was given a contract to modify P-40N fighters into 2-place gunnery training aircraft with accommodations about equal to an AT-6 Texan trainer. No true 2-seat versions of the P-38 Lightning were introduced in spite of a very definite need in the critical 1942-44 period. The demand for fighter and reconnaissance versions in combat zones was so great, no production aircraft could be spared for modification. (General Kenney, commanding the Fifth Air Force, lobbied almost incessantly for more P-38 shipments to the South Pacific area. In the MTO, Fifteenth Air Force groups were glad to get second-hand airplanes from the ETO.) For the gunnery training role, Piggy-Back P-38s worked rather well despite cramped conditions for the instructor.

Experience with the Curtiss Warhawks soon caused issuance of a document known as TI-1584 in December 1943, authorizing a study to determine the feasibility of converting some P-47 Thunderbolts for similar duties. A decision to modify at least two Curtiss-built P-47G airplanes came early, but a lengthy delay stemmed from a lack of suitable facilities to accomplish the required modifications. The Evansville (Indiana) Modification Center was ultimately selected to perform the necessary rework, beginning in May 1944. The major changes required extending the cockpit canopy forward over a new cockpit in

A Jug full of water (actually coolant liquid) is shown in a rare air-to-air portrait that reveals the need to install scoops to force air through the intercooler. Chrysler's preliminary flight test program came to an abrupt halt during the 27th flight with the No. 1 flight engine, having logged only 18 hours, when the propeller shaft fractured soon after takeoff. The skilled pilot managed to make a successful deadstick landing. He also managed to save the unleashed engine from a runaway explosion. **(Author's Coll.)**

front of the normal seating position. All essential flight controls were duplicated in the forward cockpit position, including navigation and engine instruments. Even the turbosupercharger was retained. It was, however, necessary to remove the upper half of the main fuel tank to provide the required space. As a result, internal fuel capacity was reduced to approximately 182 gallons. Two P-47G-15-CU razorback airplanes (serial nos. AC42-25266 and -25267) were selected for conversion to TP-47G-16-CU models.

Capt. George Colchagoff, being a very enthusiastic appointee, turned up as Project Officer on the two-seaters. It seems he had moved out of bombers into the Fighter Branch at Wright Field because of some very special talent he had for rocketry in prewar days. When that project requirement did not manifest itself immediately, he was assigned to P-47 projects in the rapidly expanding Air Technical Service Command. Additionally, his familiarity with the virtues seemingly inherent in two-seat conversions of first-line fighters dated back to transition from bomber operations to fighter testing. In January 1944, he joined a group of fighter pilots at a gunnery training base near St. Petersburg, Florida to sharpen the accuracy of the men assigned, many of whom had combat experience. One, a full colonel, was a very proficient gunnery expert, a hallmark of the true fighter pilot.

During a 2-week course in which many passes were made at a sleeve target towed by a P-47, only one person in the group believed himself to be in need of dual time in a Curtiss TP-40N with a highly qualified gunnery instructor. That was Colchagoff.

"I was the only guy – and I'm glad I did it for several reasons – who took advantage of the chance to fly in that TP-40N for an hour or two. The instructor chipped away at me from the back seat, and I was amazed at the improvement in my shooting ability," retired Colonel Colchagoff recalled. "This guy was breathing down my neck, giving second by second instruction. Even with no previous fighter experience, I got the second highest score in a group of about fifteen. That hot-shot colonel couldn't get over the fact that he only got a few more hits on the sleeve than I did. It was enough to make me a raving enthusiast for two-seat versions of every first-line fighter type."

Evansville specialists managed to get an almost complete duplicate cockpit in front of the regular position. It was just as roomy and had all necessary controls and indicators. Even to the normal gunsight.

"After we got the trainer developed, I was ordered to deliver it to the Training Command people at Mitchel Field, New York. By the time this came up, I had a lot of time in P-47s (from Project Officer assignments at Farmingdale, etc.) and this figured to be a routine flight, I thought. A civilian writer, Douglas Ingalls, who was living in Dayton and worked for the paper, asked me if he could come along. He was to be the first non-pilot civilian to fly in a P-47" (at least in the United States). "It was a beautiful day," he recalled, " and as I usually do when I have a non-rated copilot, I let Ingalls handle the controls for awhile. Just as I relaxed, there was a loud whirring that hit us and the nose dropped abruptly and turned left. I immediately realized I had not switched tanks and we had run dry. Ingalls was very upset, but I calmed his fears and told him it was quite routine."

But the day was not over, and Murphy's Law had yet to catch up with the P-47. "The second thing that happened that day was much more

Bundled up against the winter chill, RAC test pilot Mike Ritchie prepares to perform some high-power, extended duration or War Emergency Power flying at Farmingdale in midwinter conditions. By that time the production P-47M-1-REs delivered to the ETO were encountering some serious ignition breakdown problems at high power settings. The yellow-cowled #2 YP-47M is equipped with a 110-gallon belly drop tank, rarely seen stateside in the war years. **(Republic)**

Chief Test Pilot Lowery Brabham is seen, sans goggles, ready to fly the No. 1 long-range XP-47N with two M44 1000-pound bombs mounted on the wing pylons. Republic people outdid themselves in accomplishing a nearly impossible feat of converting the short-range P-47 into a long-range fighter. Army Air Force generals were scheduled to cancel all P-47 production, knowing that P-47D aircraft were not well suited to the impending Pacific Theater campaign against Japan. That job was to be assigned to P-51Ds and P-38Ls which had much greater range. The XP-47K fitted with the long wing proved that performance was not compromised. RAC engineers were able to cram about 111 gallons of additional fuel into each wing panel. The XP-47N grossed a bit over 10 tons. (Mfr. H3697)

serious. Arriving at Mitchel, I circled in the pattern and headed for a landing. As I approached the fence I noticed the airplane was going into a left skew. It was necessary to hold more and more rudder, and by then I had the stick well over. But the darn thing was trying to do a corkscrew on me. Suddenly I noticed, to my horror, that one flap was down and the other one had not moved at all. This is something, I had heard, which killed a lot of pilots. Near the stall, 100 feet above the ground in a 13,000 pound fighter, and completely out of shape. I grabbed the flap lever and the flap had come up about half way when I hit the ground. The tires developed a few flat spots, but otherwise it turned out OK. My passenger did not even seem to be aware that anything was amiss," he said, grinning impishly. Colchagoff remained in the service, eventually serving as Project Officer on programs like the Bell XP-77 fighter, the XS-I/X-I, the X-2 and other supersonic and space vehicles while routinely based at Wright-Patterson AFB.

After slightly more than 600 of the P-47C series had been manufactured, Evansville plant airplanes began to trickle from the production lines as D models to distinguish them from Farmingdale P-47s. However, ground rules changed quickly and the latest series airplanes from Long Island became D models too. It was just at the time when the block numbering system to identify changes likely to affect logistics and maintenance was adopted. The manufacturing plant identification letter system in the designation also came into widespread application. Fuselages for the first four Indiana-built fighters came from New York and received the D suffix letter while P-47C versions continued to stream out to the ETO. Evidently a few Cs were sent to Australia, primarily (it would seem) to set up maintenance facilities properly.

One situation which has led to confusion in aircraft identities was issuance of retrofit kits for installation of the "bulged belly" keel line normally associated with P-47Ds. To make matters more difficult, many of the first groups of D models were produced without the bulged keel because the assemblies were not yet in production. Test pilots at Farmingdale were flying at least two P-47s fitted with 205-gallon ferrying tanks during the first quarter of 1943, but it was never intended they be used in a combat situation for range extension. At the same time, a yellow test P-47 was observed on the Hangar 1 flight line with a larger version of the P-39/P-40 type 75-gallon teardrop tank. Tanks of that shape and size had been in use for some considerable length of time, especially in North Africa and in the Far East/SWPA areas. In the ETO at the same time all Lockheed P-38E and F models, equipped to carry two underwing drop tanks of at least 165-gallons capacity, were long gone to North Africa. Even in normal conditions, the P-38 Lightnings had greater range without external tanks than any fighter operating out of England. Lacking warnings from AJAX near London, how would Materiel Command or Republic really know the range problem was terribly serious.

With nothing but a couple of Spitfire groups and one P-39

Pratt & Whitney's R-2800C engines seemed to lose none of their famous reliability in spite of generating far greater power than earlier production versions. The R-2800-57 engine (as shown) was initially intended for the P-47J, was tried in the XP-47L and then was standardized on P-47Ms and Ns. It should be noted that the -14W version of the C-series engine was a USN model, not intended for turbosupercharging at sea level. Scintilla magneto ignition is fitted to this engine. Versions fitted with General Electric ignition were the -73 and -81, easily identified via the larger black magneto units. (UAC)

group, VIII FC was little more than a command marking time with little in the way of a serious mission requirement. Lt.Col, Cass Hough had returned to the Z.I. for a short period in the first quarter of 1943, but no storm warnings went up. Lt.Col. Hubert Zemke was commanding the only Thunderbolt group anywhere. His job was to pull a fractioned outfit together to serve as the main air defense interceptor group on the East Coast. Why would he be caught up in any range-extension campaign? When Cass Hough was back at Bovingdon and dealing with a multitude of technical projects, he was astonished by the reports relating to use of the "200-gallon" belly tank. It did not take long for him to confront B/Gen. Hunter with a crisis. Pointblank, he informed his chief of the limited range problem – gained first hand in a P-47C and an early series P-47D. If the Thunderbolt range (mainly, combat radius of action) could not be increased substantially, the airplanes would never play any significant role in the ETO. It was not a problem with which the general wished to deal in a period of relative calm. Lt.Col. Hough dealt with the problem, as we shall see in Chapter 12.

Within one month after it was completed and test flown by Brabham and others at the factory, the first XP-47N was ready for extensive testing by Materiel's Experimental Section, but all testing was delegated to the APGC at Eglin Field, Florida. Of all single-engine fighters built during WWII, none had a wider-track landing gear than the P-47N. Although a few Ns were flown in the ETO during the war years, it is doubtful that they did battle with any enemy aircraft. (APGC)

We really do not have any evidentiary clues about the origins of the bubble canopy movement in the USAAF, but the methodology came directly from British sources. Vision limitations with the Hawker Tornado and Typhoon prototypes were a serious flaw in the airplane design, and progress steps were taken to correct them. At the same time, the Malcolm hood was developed for use on Spitfires and RAF Mustangs. At one time Cass Hough had a special canopy on a razorback P-47D. It featured a smooth teardrop bubble in the side panel of a normal sliding canopy, a bubble large enough to accommodate the human head with a helmet. For the Spitfire and Whirlwind fighters, the British manufacturer of Perspex (the equivalent of Plexiglas) developed a method for creating free-blown teardrop canopies. When they appeared on some models of the Spitfire and other aircraft, they attracted the attention of Curtiss Chief Project Engineer Walter Tydon. He was so strongly impressed with what he observed, he managed to obtain a small contract to convert one P-40N, unofficially referred to as an "XP-40N." It improved pilot visibility almost beyond belief. That airplane also featured a fully curved windshield similar to the one used on the earliest NA-73 Mustang for Great Britain. Although a formal contract had been issued to build one XP-40Q featuring many of Tydon's drag reduction ideas, there were eventually three XP-40Q airplanes flying. In its final configuration with bubble canopy and a laminar-flow wing, it became the fastest of all P-40s built in WWII, being credited with attaining a speed of 422 mph on 1425 hp at critical altitude. It was also, by far, the best looking P-40 of all time.

Materiel Command issued a change order for creation of one prototype P-47 airplane, reworked to feature a pared down rear fuselage and a full blown or vacuum-formed bubble canopy. It was to be combined with a new windshield featuring a flat armor glass panel with improved optical quality. Chief Engineer Kartveli was livid with rage. His temper was a well-known feature of the man, but the customer prevailed. The 300th P-47D-5-RE (AC42-8702) razorback airplane was taken from the production line and committed to a $64,464.24 modification. Kartveli complained of a possible 30 mph reduction in speed and a serious

In a rather unusual view, the first XP-47N, c/n 4964, shows off its new wing planform in flight over the Ohio landscape near Wright Field. Despite nearly universal skepticism among command-level officials about Republic's ability to create a long-range, viable fighter aircraft without serious loss of performance in a matter of weeks, RAC personnel performed a virtual miracle. In fact, instead of canceling all P-47 production, the AAF eventually took delivery on 1816 production N models, with 1667 of those being manufactured at Farmingdale. (AAF)

***ABOVE AND BELOW:** The Eglin Field team was ready and waiting when that first XP-47N arrived, having quickly developed an extensive and intensive test schedule. By 24 August 1944 they had equipped the prototype with a pair of 330-gallon Lockheed P-38-style wing tanks...but these were made of wood. For unknown reasons, capacity had increased by 20 gallons per tank from the normal 310 gallons. Although tests were conducted with full tanks and there were no accidents, there was a strong recommendation against use of tanks anywhere near that capacity in normal operations. Loss of one tank on takeoff was likely to be fatal because of the sudden 1-ton asymmetrical leverage. Sway braces were installed, front and rear. (APGC)*

loss of directional control. Complete with camouflage paint and P47K painted on the fin, the Republic XP-47K flew for the first time in July 1943. If there was any reduction of speed at any altitude, it was insignificant. Any loss in directional control was evidently unrecognized until certain unfavorable characteristics became evident in combat. A very simple design change to add a dorsal fin worked well to permit field retrofit with little difficulty.

The AAF made immediate plans to introduce the K model into the production contracts, but some intermediate changes of another nature forced alteration of those plans. By change order, the final P-47D-20-RE (AC42-76614) was converted to become an XP-47L. Externally it was indistinguishable from the XP-47K, but internal fuel capacity was increased by 65 gallons to 370 gallons and this was combined with the bubble canopy change (the XP-47K did not have the increased fuel capacity, a change introduced at the P-47D-15-RE block series).

One cannot help but wonder why Major Seversky, back in the 1930s, did not see great potential in the formed Plexiglas canopy panel used only on one Seversky airplane, the AP-7 demonstrator. He was already part way to creating the slick bubble canopy. Rather strangely, it was the failed Westland Whirlwind – an advanced, beautiful design defeated almost entirely by engine choice and operating systems – which established the trend to bubble canopies.

THE WATER JUG

Nothing could seem more improbable than a P-47 Jug full of water. Well, at least part water, part Prestone. And it is quite logical to place the wartime project in the right context from the very beginning. There was never any serious thought of producing any P-47 Thunderbolt mated to a liquid-cooled engine. So it is logical to inquire about the manner in which two such Thunderbolts logged a fairly respectable number of hours in the air while flying out of the airport at Evansville, Indiana.

It all began with a visit to Washington, D.C. in June 1940 by representatives of Chrysler Corp. from offices in Detroit, Michigan. They wished to discuss an aircraft engine project with B/Gen. G.H. Brett, Chief of Materiel Division. They mentioned discussions with Mr. William Knudsen led to Chrysler's interest in developing a new aircraft engine. The men were seeking advice about what kind of powerplant would be of maximum usefulness to the Air Corps. General Brett suggested a 1700-1800 hp engine of the liquid-cooled type featuring fuel injection and supercharging. By 4 April 1941, design work was far enough advanced for the Materiel Division to issue a contract covering building of two 16-cylinder engines designated (then) as XIV-2220-1, supercharger development, and allied materials. Essentially, the engine was to consist of two 8-cylinder, 60-degree inverted engines in the 2000 hp class. After procurement of four additional complete engines, plus some 2-cylinder versions (typical test examples), was authorized, the Powerplant Laboratory people at Wright Field were critical of the expenditures by Chrysler. Col. Edwin Page was particularly critical, recommending against an award of an additional $9 million. However, that was ignored and the funds were allocated to continue the program. (Compared to the Wright Tornado engine project, details of Chrysler's big XI-2220 – as it was redesignated – were fully divulged.)

By 31 December 1943, many commands were of the opinion the engine would never be ready during the war; no airplane was available for the installation; the engine was exceptionally long; and it was not going to be more powerful than the Packard Merlin, Rolls-Royce Griffon and certain Allisons. Chrysler was given 6 more months to develop an engine with a takeoff and military rating of 2500 hp, and (incredibly) Col. J.M.

Gillespie, new Chief of the Powerplant Laboratory, reported the engine had completed its 50-hour test according to specifications. Nobody really wanted to get Republic Aviation involved, so Chrysler was asked to modify a P-47 under their engine contract. Work on the two axial-flow superchargers was reduced to design work only, so $850,000 became available for the modification of an aircraft.

Somehow or other, General K.B. Wolfe entered the picture in September 1944 and directed modification of two additional P-47s, but that was cut back to two airplanes (total) in October. A couple of P-47D-15-RA fighters were brought from the test center at Orlando, Florida, to Republic's Evansville Modification Center. Chrysler had concluded it would be impractical to attempt the modification work at their Detroit, Michigan, Ordnance Tank Repair Garage, so they subcontracted the work to the Modification Center. Both of the P-47Ds were converted to XP-47H configuration, each being powered by an XI-2220-11 engine utilizing a new General Electric CH-5 turbosupercharger. The XP-47H was a 13,896 pound aircraft with typical P-47 dimensions except for overall length of 38 ft. 41/8 inches. Serial numbers for the two airplanes were AC42-23297 and AC42-23398. A 13-foot-diameter Curtiss propeller was used.

Weighing in at 13,427 pounds (design gross weight was 14,010 lbs.), the first XP-47H flight test was conducted on 26 July 1945, immediately following the 689 Board engineering acceptance inspection. Using this single airplane, some 18 hours of flight time was accumulated in 26 flights. As the fighter was taking off on the 27th mission, the propeller shaft broke, forcing the pilot to make a hurried emergency landing with a really dead stick, or perhaps none at all. After that last flight, the No.1 airplane was delivered to Materiel Command on the 7th of November 1945, after the war had ended. Some reports have stated there were no more flights, but they were not correct. The one remaining flight engine was installed in the second XP-47H, with flight testing getting under way in September.

The airplane had some different operating characteristics with the Chrysler engine. A takeoff rating of 2500 horsepower was achieved at 3400 rpm and showing 71 in. Hg.(manifold pressure), and the figures applied to the military rating at 25,000 feet as well. The airplane must have produced an interesting sound. With a maximum recorded speed of 414 mph at 30,000 feet, it was hardly a phenomenal performer for 1946. It also had a cooling problem, exceeding the redline in Army Hot Day conditions. And with the cooling doors wide open, drag was so high they could not exceed 394 mph at military power settings.

So ended the saga of the saga of the XP-47Hs. The first conversion had cost $900,688 and the second airplane was completed for $270,753. Of course the development costs for the Chrysler engine are not included. In reality, the airplanes were nothing more than flight test beds for the development of engines. However, it did create some interesting looking airplanes.

MMMMMMMMS

With the fewest of exceptions, most people with any serious interest in the P-47 Thunderbolt believe the high-speed P-47M was developed as a response to the appearance of Adolf Hitler's first indiscriminate retaliatory weapon, the V-1 unguided – but basically preprogrammed – missile. That is really a false presumption.

Chief Test Pilot Herb Fisher (Curtiss-Wright Corp.) made numerous test dives in Thunderbolts equipped with a variety of supersonic propellers and reversible-pitch propellers. Highest Mach number attained with any late-model P-47 in postwar days was M=0.82, and Fisher stated that "it is almost impossible to obtain accurate data due to severe buffeting..." Especially notable is the fact that by then the Machmeter had been developed. They did not exist in WWII days. This P-47D-30-RA bailed to C-W was fitted with one of the straight thin-blade propellers. (Walter Boyne)

In truth, there are two major clues to prove the fallacy of that conception. First, there was the XP-47J project initiated in April 1943 for development of a high-performance version of the P-47 equipped with the new "C" series engine, long before the appearance of any V-1s. Second, and most important, was the early appearance of a very special P-47C-5-RE (at Farmingdale), subtly packing one of the very first "C" engines. It was an unofficial XP-47M "Mule" (AC41-6601) utilized to work out any bugs in the higher powered version of the basic R-2800. Test pilot Carl Bellinger was most frequently the pilot of this "wolf in sheep's clothing" airplane. He was very enthusiastic, even in later years, about the extra performance capability packed into the "sleeper." (The one and only photograph taken of the "XP-47M" appears in this book; it shows Carl Bellinger in the cockpit and the late-series cowling.) Bellinger also revealed it was on this same airplane the first wing set with 27-inch stub inserts was installed to test the design to be utilized on the XP-47N. Reference to supportive documentation served to confirm his statements about this "Mule."

Equipped with the -14W and, later, with the -57 engines in the R-2800C series, plus the larger diameter propeller, the "XP-47M" flew like greased lightning. Finished in zinc-chromate yellow-green paint, it also had a bright orange-yellow cowling and vertical/horizontal stabilizers. The rudder and elevators appear to have been camouflage-paint color, and the forward cowl ring is believed to match those surfaces. The color scheme was to discourage overzealous new fighter pilots from "bouncing" the fully instrumented experimental fighter while it was performing test programs.

And, of course, it would be folly to presume a much more powerful version of the R-2800 was developed in a crash program specifically to combat the V-1s and the newly encountered Messerschmitt Me 163 rocket-propelled fighter. Many hours of test cell running and perhaps hundreds of flight hours had been logged on "C" engines before they were fully qualified for any commitment to operational use in combat. It was really coincidence for R-2800-57 engines to be available at the very time V-1s began to hit London.

Bellinger stated the "XP-47M" was the finest Thunderbolt he ever flew. Some additional comments relating directly to the real P-47M came from Chief Test Pilot Lowery Brabham. He had this to say: "In order to provide extra power in short bursts for air combat and to assist in heavy-load takeoffs, a system of injecting water directly into the cylinders was perfected for the P-47D series [after several hundred had been built: Auth.]. This addition of water allowed exceptionally high manifold pressures to be used without the accompanying onset of pre-ignition detonation.[6] A war emergency rating of 2800 horsepower eventually evolved to benefit the P-47M under rigorous conditions of battle. This," he said,"was an unbelievably high horsepower for each cubic inch of displacement."

(While the reliability of the P&W Double Wasp was its main attribute and prodigious horsepower was produced, fairness demands we point to the Napier Sabre. The British engine was producing around 2200 horsepower as a military rating, not WEP, out of a mere 2240 cubic inches displacement with mechanical supercharging. However, it had serious vibration problems and was never able to develop a reputation for reliability anywhere near the equal of the R-2800.)

"In testing the new WEP rating," Brab continued, "we instrumented an engine completely, including detonation detectors which allowed us to determine just how high we could run the manifold pressure and carburetor air temperature without (experiencing) detonation. All horsepower was measured with a calibrated torquemeter."

Then, with quiet authority, Brab dropped this bombshell: "On the test engine – as installed in the first YP-47M (42-27385) – we repeatedly ran up to 3600 horsepower with no recorded problems!" That's 3600 horsepower out of 2800 cubic inches on Grade 100/130 gasoline in 1944-45. "After about 250 hours of very intensive testing, the engine and airframe were adjudged to be dangerously worn and were retired from the program. After the war ended, I received a call from a record-setting pilot named Bill Odom, involving a request for information about a P-47 he had bought from the War Assets Administration as surplus. He intended to compete in the 1947 Bendix Trophy "R" Race from Los Angeles to Cleveland," Brab said rather flatly.

"I asked him for the identification number, looked up its history and called him back to inform him that his airplane

[6] Pre-ignition detonation is a destructive enemy of an engine, a real engine breaker. At the very least it causes severe power loss.

Curtiss-Wright's top test pilot Herb Fisher looks up at the special thin-blade scimitar-shaped supersonic propeller, one of a series he was testing on a P-47D-30-RA (not an N as reported by Fisher) in post-WWII days. Development flying was done at Caldwell, N.J., with great care since the life cycle of the thin blades was extremely short. (Walter Boyne)

was the one we had so harshly used in testing. He was informed that it was not the best airplane for his purpose. Odom insisted that the aircraft and engine were performing flawlessly, and that he had no worries about it in the race. Modifications to his airplane, the No.1 YP-47M, included sealing the ammo bays to carry fuel (wet wing), installation of two 165-gallon P-38-type drop tanks and – dangerously – there was a fuel tank installed in the vicinity of the turbosupercharger, which sort of scared me. I believe it was fortunate several of these 'quick and dirty' fuel installations started leaking badly at the starting grid, and that the YP-47M (then registered as NX4477N, with race number 42) was scratched," Brabham concluded.[7]

The "XP-47M" and three YP-47Ms were involved in intensive flight testing in 1943 and 1944. This all became a bit complicated because there was congruent overlap with emergency development in progress on the long-range XP-47N. There was no official XP-47M, probably because of some "off-the-cuff" discussions with the AAF Plant Representative or with the project officer about the unofficial "X" airplane. By change order to the contract, P-47D-27-REs (AC42-27385 and -27386) were taken from production and converted to become YP-47M-REs Nos. 1 and 2. By that time there was evidently a strong need to accelerate testing of the high-speed M version, so the last airplanes in the original block of 615 airplanes became the two YP-47M-REs. The test program was set up with two main objectives on the agenda:

1. Provide a "hot-rodded" interceptor version of the P-47D as an economical substitute for the P-47J. Trade off some performance gain for quick transition on the assembly lines without slowing base production rates.

2. Fully develop the potential of the War Emergency Power (WEP) rated engine in the R-2800C series for ultimate introduction in the regular Thunderbolt production scheme. (Wartime requirements had a tendency to change overnight because of enemy actions.)

Engineering development and flight test programs related to the P-47M progressed satisfactorily throughout the balance of 1943 and the early part of 1944, well before D-Day actions triggered the missile responses. Unfortunately, the timing for satisfactory completion of YP-47M flight testing was bad.

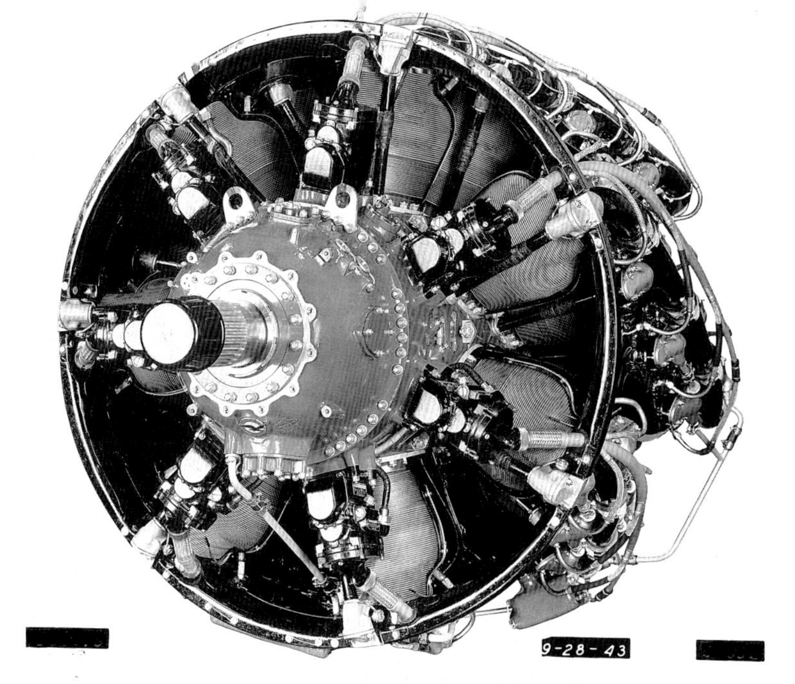

ABOVE AND BELOW : The massive 28-cylinder P&W R-4360-13 non-turbo engines installed in two XP-72 prototypes had ejector exhaust stacks and no turbo installation. That early series engine was rated at 3000 hp for takeoff, and the planned -19 engine was to employ a shaft-driven centrifugal-flow compressor to provide a high-altitude stage blower. Like the experimental XP-47J, it was intended that contrarotating propellers be installed to absorb the projected 3500 horsepower at 25,000 feet altitude and to provide ground clearance. (United Aircraft Corp.)

North American P-51D production was accelerating rapidly, providing the AAF with a single-engine fighter boasting a 1000-mile radius of action and an acceptable reliability factor. Pilot transition was relatively easy and it honestly gave many pilots the sense of affection so many had felt for the Supermarine Spitfire. To cap the entire matter, General Jimmy Doolittle with his great sense of pragmatism had made a decision (concurred with by his fighter commander, M/Gen. William Kepner) to reduce logistics problems by concentrating on one fighter type (almost, but not quite) exclusively. Coupled with all that in the immediate time frame was the need for hell-bent-for-leather expansion of the reborn Ninth Air Force with its IX FC component, destined to become the flying artillery for Operation OVERLORD. Doolittle knew far more about the field-level needs of fighter group commanders than a multitude of lower-echelon generals who should have been in a state of great compatibility with those directly in confrontation with the enemy.

The Allison-powered P-51 Mustang came to a period in its life when it was destined to become a terminated program – even as production of Bell P-39s continued apace? – because there was no place for it at all in the AAF fighter program. In direct parallel, the P-47D production effort was headed toward oblivion. This was unknown to any but the highest level executives in Republic Aviation. In military circles the decision to terminate all Thunderbolt production some time in 1944 was close at hand. Any need for large quantities of limited-range fighters was rapidly

[7] This was a portion of one of several man-to-man interviews conducted with Lowery Brabham in 1973, prior to his untimely death. He also responded in writing to many pages of questions.

evaporating, and *range* was suddenly the biggest thing. At the single production source for Lockheed Lightnings out at Burbank, California, full-scale production of that old timer was, if anything, accelerating. In fact, they were hammering away at building 15½ of the newest P-38L-1-LO model per day to meet demand for fighters and reconnaissance airplanes everywhere on earth. (It merely proved General Kenney had been correct from day one.) The latest models had boosted controls (for excellent roll rates), more than 3000 horsepower on tap, and a combat radius-of-action somewhat in excess of 1500 miles. Combat speed had risen to the 440 mph realm. The major limitation on range had become one of pilot endurance. Production rates would reach their peak in August 1944. After all those years, the War Production Board finally saw the light and authorized a second production source for Lightnings, retooling the Consolidated Nashville (Tenn.) plant for exclusive production of P-38Ls.

Somewhat beyond comprehension, the General Motors-Fisher plant in Cleveland, Ohio, had been awarded a massive contract for thousands of P-75As, a totally unproven (in combat) fighter, and the orders were about to increase to approximately 4000 airplanes. Why? Mostly on the promise it would have a 1500-mile combat radius.[8]

More or less working in secrecy, starting in November 1943, Republic had designed a set of wings with various stub inserts for installation on that enigmatic old bird, the P-47C-5-RE, serial number AC41-6601. Fitted up with the new wings (sans fuel cells) featuring 27-inch stubs (left and right), it then became an unofficial "XP-47N" which never ever appeared on official records. It was mushroom time on Long Island, if you get the meaning. Internal company documents referred to the aircraft as P-47C-707. Actual flight testing began by the first week in June and test results were sent to Fil Gilmer. Copies were distributed to Kartveli, Hart Miller and other selected persons down the line. The date: June 6, 1944. "D" (in this case) was looked upon as Decisive Day at Farmingdale. The first flight date for P-47C-707 will probably never be known. But, more anon. (See Chapter 17 for the Long Range story.)

SUPERBOLT

Ray Higginbotham. That name should have as much impact as the name Alex Kartveli. But it more than likely means as little as the name Mike Gregor did before the true story of Seversky Aircraft was committed to print. Higginbotham was Chief Powerplant Engineer at Republic Aviation during the war years, and probably until he retired in ill health. It was under his guidance that the XP-47J, XP-47M, XP-72 and XF-12 propulsion systems were developed. Whether or not he was the instigator or prime force behind the programs is hardly relevant. His guid-

[8] Combat radius, for the USAAF, was generally defined as takeoff and climb, cruise to destination, combat for 5 minutes at WEP, combat for 15 minutes at military power, cruise to home base, descend and land with 30 minutes reserve fuel remaining.

BELOW AND OPPOSITE: No Thunderbolt ever looked better than the XP-72 prototypes. Powerplant Project Engineer R.R. Higginbotham's team outdid itself in producing the beautifully cowled engine that eventually led to the cowlings employed on the large XR-12. If the P-72 had gone into production in the P-47N format, it would have been hoped that the cowl and spinner would have been retained. As shown, but with a -19 engine and blower, it was expected to see a true airspeed of 504 mph at 25,000 feet. **(Republic via George Colchagoff)**

ance shows clearly. The XP-47J program had the good sense to borrow back from the German engine industry what they had borrowed from American engine manufacturers. His use of the fan in place of cuffs to help cool the radial engine was long overdue and owed nothing to the NACA or to Air Materiel Division's Powerplant Lab. America, the inventor of the NACA cowling and the Townsend ring in the late 1920s and early '30s, could never seem to make large streamlined spinners work with our radial engines, even after nearly every prototype (especially the Brewster XSB2A-1) was first exhibited with such promising features. One wonders at the problem's pervasiveness in America, because use of spinners in Germany, Japan and Russia was widespread and evidently succesful. Messier in France combined them with tight cowlings, but to no real avail because the engines were unreliable at best.

America's key propeller makers made great strides in hubs with efficient pitch control, but they forgot about blade effectiveness. The Germans must have looked at Holland; then at U.S. propellers and said, the Dutch are right. You can't run a windmill with toothpicks. By 1941 they had adopted that philosophy and it showed on Focke-Wulf, Junkers and other aircraft. Kurt Tank's Ta 152C and Ta 152H late model fighters were domineering examples of combined spinner and paddle blade propeller advancement. (Equipped with windmill-wide propeller blades and nitrous oxide [GM-1] injection systems and sailplane-like wings, the Ta 152H was among the highest flyers in the war with a service ceiling of just 48,550 feet.)[9]

But, back to Republic. Higginbotham's team outdid itself with the propulsion system and cowlings applied to the *Superbolt* XP-72 and perhaps even more so with the four engines on the XF-12. Nobody produced better looking, more efficient radial engine cowlings in all of WWII. That XP-72 designation puts it in company with such aircraft as the Fisher XP-75 (it deserved better), and perhaps like the North American XP-78 (redesignated XP-51B) it would have been returned to the proper fold. Begun as a project in 1943 to take advantage of a potential successor to the R-2800 engine – Pratt & Whitney's Army/Navy R-4360 – development could not progress faster than the engine. The engine intended for use in the XP-72 was only military rated at 3000 hp for takeoff and 2400 at 25,000 feet, but that compares more than favorably (at an early stage) with the 2100 hp produced at 30,000 feet by the R-2800-61 used in the very fast XP-47J. If a military power rating of 3500 was expected in due time, a WEP rating for a good R-4360 could easily be viewed as at least 3750 horsepower. The XP-72 was to be unique in having a variable-speed remote centrifugal supercharger, engine driven through a fluid coupling and an extension shaft. Exhaust gases from the 4360 cubic-inch displacement Wasp Major engine were ejected via the NACA-developed system employed on several Curtiss experimental fighters. Engine cooling air flow was expelled through a tight ejector slot, with flow assisted by the exhaust gas flowing from a series of exhaust stacks tightly grouped as well at the same slot position.

The Wasp Major almost failed to make it to the starting line, and there is reason to believe it did not play a significant role in World War II simply because of divided engineering effort at East Hartford, even before America entered the conflict. Divided effort is mainly attributable to the master planners in the Army Air Corps, but the United Aircraft hierarchy must share the responsibility because of far-reaching management decisions. In the

[9] Republic P-47B pilots of the 56FG in 1942, please take note. With wing spans of 47½ feet and gross weight of less than 11,000 pounds at altitude, the service ceiling was 48,550 feet, in 1945.

In the same timeframe of Hawker Tempest Mk II/V/VI planning in the U.K., at least two American fighter manufacturers were racing on parallel tracks. In 1942, Pratt & Whitney hung a 3000-hp, 18-cylinder R-4360-10 (probably) powerhouse on the front of a Vought F4U-1, BuAer serial 02460, and that led to Goodyear's F2G-1 Corsair after adequate FG-1 production rates had been achieved. As Boeing was building the similarly powered TBM-size XF8B-1, out on Long Island Republic was fabricating two P-47D clones with Wasp Majors up front. **(United Aircraft C-961)**

early 1930s, Frederick B. Rentschler, chief executive officer for Pratt & Whitney, had resigned to assume other responsibilities in the parent U.A.C. As his successor at East Hartford, he chose Don Brown to become president. Rentschler's longtime friend and associate, George Mead, remained in the top engineering job. He had been in the running for the presidency, but did not honestly feel he could fulfill the responsibilities of the position. However, he was opposed to the selection of Don Brown. All the technical and sales problems of the mid 1930s, aided and abetted by those dreaded Depression and Recession years took their toll on Mead. The engine manufacturer, not unlike Wright Aeronautical, had suffered serious master rod bearing failure problems, nearly putting the military version of the R-1830C Twin Wasp into the records as a failure.

Wright was hardly in better shape, the performance of engines in the 1935-36 pursuit competition doing little to enhance their reputation for reliability or ability to perform as advertised. Add to those situations the seduction of the AAC by General Motors-Allison with the V-1710 and prospects for survival were hardly likely to generate enthusiasm. Mead's health was seriously impaired, but he carried prime responsibility for the company's engineering programs as vice-president of engineering. With the Air Corps sponsoring a Hyper engine program centered on development of high-performance liquid-cooled engines, pressure was brought to bear on P&W to develop a high-power inline engine compatible with official planning doctrine. Not wishing to endanger its long-term, productive relationship with a vital military customer, and with its internal planning controlled by an advocate of inline engines, P&W acquiesced. By coincidence the company was working on a low-profile program to develop a 2300-horsepower liquid-cooled engine for some Navy project, despite basic BuAer opposition to such developments. Actually it had started life as an inline aircooled engine to be used in weight-carrying applications (torpedo bombers and flying boats). At time of project initiation, Navy disenchantment with liquid-cooled engines of any kind was well known.

In making the move to the Prestone-cooled configuration, Mead decided to utilize sleeve valves in place of the almost standardized poppet valves. Only in England was there real interest in sleeve valves, and they had rarely found any application in American engines. Mead had traveled to England in the 1937-38 period to study those British developments and overall progress in British aviation, especially Royal Air Force programs. Visits to the facilities of major engine producers such as Bristol, Napier and Rolls-Royce intensified his interest. At the time, Napier was deeply into development of their Sabre 24-cylinder, flat H configured engine with liquid cooling. Rolls-Royce was heavily committed to its big 24-cylinder, X-type Vulture engine. (That engine was essentially composed of two of its soon-to-be ill-fated Peregine V-12s, one mated over the other on a common crankcase. The RAF was already deep into its Shadow Factory scheme for production of military aircraft, having foreseen the probability of war with Germany far more clearly than the Prime Minister did. Of all the liquid-cooled engines in the high-power category which Mead was privileged to see, he had no inkling only the Sabre was going to rise to the challenge and have a significant place in history. The only really successful large radial aircooled engines produced in the United Kingdom came from Bristol's factory, and even their success factor was spotty.

Upon his eventual return to East Hartford, Mead launched the design effort aimed at serving as successor to that minor Navy project. His idea was to develop a powerful inline engine for the Air Corps and the Bureau of Aeronautics which would fulfill each service's specific requirements. It was laid out as a 24-cylinder upright H-type sleeve valve, liquid-cooled configuration, obviously influenced greatly by British progress. (The political climate in the United States, as pointed out previously, was anything but conducive to licensing arrangements for production of foreign products. World trade was almost a dirty word in the '30s decade.)

Mead's new Pratt & Whitney X-1800 project – just by some rare coincidence matched to Wright's Tornado designation at the time – was intended to produce 2400 hp for takeoff from 2600 cubic inches displacement. The military designation for it was XH-2600. Obviously it was intended to be far more powerful than Wright's Cyclone 14, initially (and unintentionally always) rated at 1600 hp with not much hope of growth. Its navalized version, generally identical but with larger cylinder bores, was designated H-3130 and was expected to produce 2650 horsepower initially. Advanced features of the X-1800 design included an ability to be installed in either the vertical or flat position (horizontally opposed pistons). It also featured individual cylinders in place of the more common block construction, and built-in aftercooling on the induction system.

Although Frederick Rentschler was dubious of the liquid-cooled engine program, primarily because it would dilute resources available for concentrated engineering effort, he refused to interfere in P&W's autonomous operation, a seriously flawed mistake in judgment. However, some sort of divine guidance intervened and Don Brown became seriously ill at the same time George Mead was so sickly he had to conduct operations pertaining to the X-1800 project from his home. Concurrently, Chief Engineer Leonard "Luke" Hobbs decided to launch a major effort to develop a multi-row, large displacement aircooled engine as a successor to the very rapidly maturing R-2800 Double Wasp. It was his contention and belief it could be accomplished without generating excessive cooling or parasitic drag. The company was quite at home with the aircooled radial engine concept. His research indicated most of the disadvantages could be overshadowed, if not overcome, by the attributes of a four-row radial arrangement, using a convoluted and heavily baffled cylinder arrangement, sort of a glorified corkscrew concept. Mead was forced by poor health to resign in 1939, while Don Brown passed away in the first half of 1940. Managerial pressures were taking a toll. Rather logically, Luke Hobbs took over complete direction of all engine company projects, but he made a decision not to interfere with the progress being shown on Mead's pet liquid-cooled engine program. As of that time, Air Corps and Navy contracts were funding development.

Corporate decisions put Rentschler back in the saddle as Chairman of the Board (a very active one) for the corporation. Since Hobbs and Rentschler concurred in development ideas – rather open opposition to liquid-cooled engine development – they must have rejoiced when the Vought-Sikorsky Division was able to reveal the great potential of their new Vought XF4U-1 Corsair and its P&W R-2800 engine early in 1940. Powered by the new XR-2800-2 or -4 Double Wasp, the airplane had flown over a measured and marked course at a speed of 405 mph. Of course Rear Admiral John Towers, a great pioneer in naval aviation and then Chief of the Bureau of Aeronautics, believed he was correct when he told event witnesses "they had just seen a demonstration of the fastest and most powerful fighter in America at that time." (Perhaps he meant on that particular day. He was obviously unaware of the embarrassment to which his staff officers had exposed him when they failed to do their homework. The Lockheed XP-38 had exceeded the magic 400 mph mark more than a year earlier on approximately 2200 horsepower, or perhaps it was closer to a realistic 2000 hp, depending on which set of figures were released by Allison.)

A rather minor-rated event took place in the meantime, which probably had a much greater impact on American air policy in World War II than anyone could possibly have believed until well after VJ-Day. A more or less informal visit by M/Gen. Hap Arnold to East Hartford triggered the germination of ideas leading directly to development of the Republic XP-47B Thunderbolt and ultimately to the XP-72. In company with Rentschler, General Arnold studied the X-1800 design, viewed the R-2800 Double Wasp in operation on a production test stand and capped the visit off with an open-minded discussion with the Board Chairman. Arnold was known for being a believer in candor, and Fred Rentschler was in a candid mood. He laid it on the line. No matter what attributes the X-1800 liquid-cooled engine might have, it was never going to become an active participant in the war. (Isn't it interesting that so many people could be certain of our eventual participation in WWII, while others honestly believed we could avoid foreign entanglements.) Quite frankly, Rentschler told Arnold, it would take astronomical expenditures of engineering hours and machine tool commitment to produce the X-1800 without any valid promise of success.

He pointed out the availability, at that moment, of what would shortly be a well-qualified 2000 hp engine, while at the same time there were no liquid-cooled American engines producing even 60 percent of such power. In fact, no liquid-cooled engine in the 1500 hp range was even ready to move from drawing board to the machine shop. Finally, Rentschler blurted out his feelings. "If we're going to straighten out Pratt & Whitney's tangled engineering programs, we should be permitted to cancel out on your liquid-cooled engine contracts right here and now. We'll even do it at our own expense," he volunteered, feeling victory within his grasp.

Whatever was motivating Arnold then, he agreed. "Your absolutely right," Arnold barked out. "Why the hell didn't you come to me about this a long time ago?" That clinched it. When told about the Hobbs sponsored 3000 hp aircooled radial and how its very existence as a development project hung on Navy acceptance, Arnold grinned and said, "Now we're getting somewhere. You initiate your papers for cancellation. I'll see that they're approved in a hurry. Then (you) get started on the big 3000 horsepower radial job."

Did BuAer object? Not at all; in fact they were happy to see the big radial engine follow-on to the R-2800 effort with the abundance of large aircraft projects in the design stages certain to be underpowered unless something like the "spiral" engine came along soon. The Navy had a penchant for big aircraft, and flying boats never seemed to be large enough. For the first time in five years, P&W got its engineering talent aimed in the right direction, concentrating on a relatively few major projects. Even George Mead managed to shake off his illness, showing an outstanding ability to bounce back and contribute greatly to the war effort during the next four years.

The X-Wasp, as the new R-4360 Wasp Major was originally known, benefitted from the concept of using components from other projects at the earliest stages in order to reduce time required to get it off the drawing boards and running on the test stand. Double Wasp cylinders were utilized, accounting for the same overall diameter, and the team adopted everything possible to expedite the program. In due time a new cylinder design evolved to enhance maximum cooling capability. The departure which made the engine unique for some applications was the idea of using a remote second-stage blower, not just at two speeds but multi speeds. It became a high-altitude engine without a gas turbine supercharger.

Although the R-4360 design was soon overshadowed by rapid development of turbojet and, eventually, turboprop engines, the company went on to build more than 18,000 Wasp Major engines between 1943 and 1955. It was very large and complicated, but a real marvel of design. In approximately the same timeframe, P&W's factory units delivered 129,505 engines during the war period. In addition, licensees constructed another 226,183 engines, while the new P&W Missouri plant built

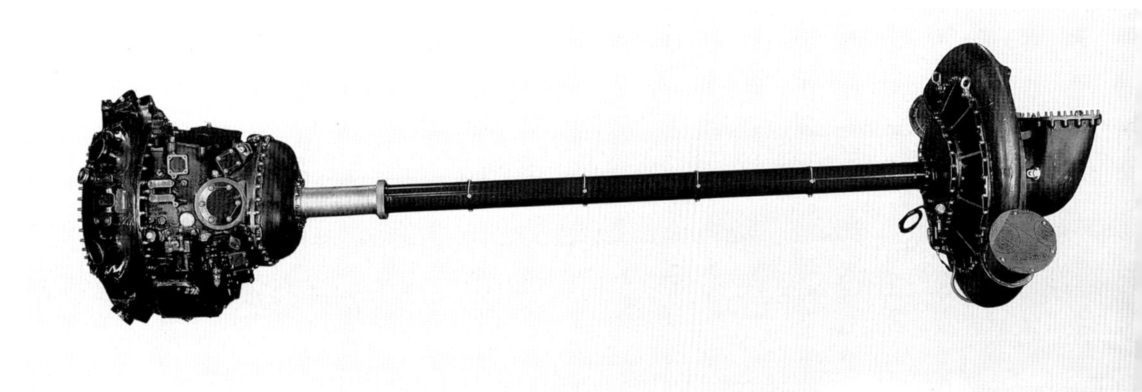

Although the photograph belies its size, that unusual centrifugal blower/supercharger was almost 5 feet in diameter. Connection to the R-4360 accessory case was to be by means of a fluid coupling and that long extension shaft. The ultimate production engine was intended to produce no less than 4000 horsepower, giving the P-72 fighter a top speed of approximately 540 miles per hour. (UAC)

7981 of the R-2800C engines, mostly for use in Thunderbolts. All told, some 363,619 engines from the Pratt & Whitney family were produced in the wartime environment.

That "off the cuff" meeting in 1940 launched the XP-72. About three years later, on 31 July 1943, Republic Aviation sent a letter to Air Materiel Command in which performance calculations and delivery schedules were quoted for P-47D-10, P-47D-15, P-47J (production version with bubble canopy) and P-72 airplanes. The germ of an idea leading to the XP-47J project in July 1942 was almost a year behind the initial sketches created for the Wright Tornado-powered XP-69. Model designations established at Wright Field in the early days of the war ranged from XP-65 to XP-77. Of that group, the XP-72 was the only fighter to be powered by the new X-Wasp. By basing the new design on most components of an existing airplane, Republic could reap the benefits of minimal interruption of production schedules in the event of advanced, high-performance concept acceptance. While the XP-69 and XP-72 projects were proceeding almost in parallel, the time came when the engine situation closed out the XP-69 entirely. The XP-72 project languished until engine development reached a respectable level of progress. Just two and a half months before the July status letter was mailed, the decision to drop the XP-69 was formalized by contract cancellation on 11 May 1943. The action was precipitated by a Wright Aeronautical decision which also seriously impacted Lockheed's big XP-58 Chain Lightning project. Wright made a stop-work decision to halt all activity on the XR-2160 Tornado engine. (Evidently they had already ended work on the X-1800/XR-1800 version in favor of the more powerful unit.) Lockheed's XP-58 had been forced to switch to the Tornado engine when P&W acted to cancel Mead's pet X-1800 project. Ultimately, the Chain Lightning was redesigned to accept Allison's V-3420,[10] the same engine being used in Fisher's single-engine XP-75.

Republic simply explained why it had become more practical to abandon the XP-47J project, which required as much as 70 percent tooling revision, and move to the P-72A which was projected to be as much as 35 miles per hour faster. They calculated the P-72A could be placed in production on about the same terms as quoted for the slower P-47J, an airplane with far less growth potential. Projections from Republic also indicated performance of the P-72A at altitudes above 30,000 feet would far outstrip P-47J performance expectations. The company stated its concern about continuing both programs simultaneously, pointing out an interval between programs could not be less than nine months. Republic wanted to complete the XP-47J flight test program in accordance with the contract, at the same time devoting all engineering effort to the more promising XP-72s (two airplanes) and the P-72A in definitized format.

An AAF contract for the two XP-72 flight prototypes had been put into effect on 18 June 1943, and the first of these (AC43-6598) made its first flight on 2 February 1944, using a single-rotation 4-blade propeller of unspecified origin. It had an exceptional diameter according to Lowery Brabham, requiring three-point takeoffs and landings without exception. (In photographs, the propeller blades would appear to be Hamilton-Standard design, but there is no emblem on the blades and cuffs were not usually seen on H-S blades used on P-47s.) The second XP-72 prototype, which was intended to feature either an Aeroproducts or Curtiss electric contrarotation propeller, was completed on 26 June 1944, but it was introduced to trouble early in its career. During one of the first test flights aboard the No.2 airplane (AC43-6599), Brabham encountered a destabilizing problem which was promptly associated with the contraprops. (The problem was virtually identical to the troubles Hart Miller encountered with flying the little known "Double Twister" a couple of years earlier. The one Brab insisted did not exist.) He also reported that application of full rudder generally resulted in the control surface locking in the full deflection position. (Wasn't anybody alerted to the need for setting the fin in a neutral position?) Other than the problems relating to rudder lock and destabilization tendencies, Brabham said the XP-72s were "beautiful flying airplanes with amazing perfor-

[10] It sometimes appeared as if chaos reigned with regard to decision making about which engines to develop and what airplanes to put them in. Perhaps the Lockheed XP-58 suffered most. First it was to have two Continental IV-1430 engines while the smaller XP-49 was to get two P&W X-1800s. Next, the XP-58 was designed with the two P&Ws while the Continentals went to the XP-49 (more logical, for sure). With cancellation of the X-1800s, the XP-58 was redesigned to get Wright Tornado XR-2160s. With the end of that project, two Allison V-3420s were finally installed. The same airplane had gone from 1430 cu.in. engines to 3420 cu.in. engines before it ever left the ground. Why in the world didn't they install P&W R-2800s, already in production and have a pretty fair airplane? Everybody was afraid to jeopardize his career.

***LEFT AND OPPOSITE: When the second XP-72 appeared, it had contrarotating propellers, eventually to be Curtiss or Aeroproducts units. One P-47B, the "Double Twister" flown just once (by executive C. Hart Miller) had been equipped rather surreptitiously with similar dual props (confirmed by witnesses in Flight Test) in 1942 or early 1943. It is safe to say that no satisfactory mechanisms were developed prior to VJ-Day. Test pilot Ken Jernstedt made an emergency belly landing in this XP-72, and it was never repaired.* (Carl Bellinger Coll.)**

mance."

The No.2 airplane had to be written off in a crash landing as a result of an in-flight fire. Test pilot Kenneth Jernstedt, a former AVG fighter pilot and one of Brab's top testers, was flying the XP-72 at about 32,000 feet when he was faced with fire, one thing no pilot ever likes to see. While sliding the bubble canopy back, he nosed down, planning to bail out quickly. But at that high altitude the fire lacked enough oxygen to be sustained; the flames disappeared. Jernstedt then remembered he was out over open water, so he turned toward the New England coastline. He spotted an Army field he believes was in Connecticut, something that dovetails well with separate facts. It was probably one of the fields at which the 56FG had been based in 1942, either Bridgeport or Windsor Locks. Coming in deadstick with six monster blades still churning up front, he bellied in rather spectacularly. In one of those, "Why didn't I stay in bed days," a Vought-Sikorsky test pilot crash landed an F4U on the same field within 5 or 10 minutes of the time Jernstedt skidded to a halt. Some young AAF major, perhaps the Duty Officer, began thumbing madly through the regulations to find out how to deal with an AAF and a Navy airplane crashing on his field, but both with civilian pilots at the controls.[11] The identity of the exact field in Connecticut is really immaterial, but the incident is a gem. Unfortunately, that landing put the XP-72 into Class 26. In later years it was contributed to the Boy Scouts for training.

Neither of those prototypes ever had a supercharger of the type associated with the R-4360-19 engine. Both airplanes flew with -13 "semi-production" engines, and they were equipped with two-stage, variable-speed mechanical supercharging equipment. The big performance booster really was to come into its own at very high altitudes. Dry weight of the -13 engine was 3685 pounds; the -19 engine slated for production with the remote blower and extension shaft was rated at 3889 pounds, but be reminded the remote unit was about 5 feet in diameter. The early engines were rated at (Military) 3000 horsepower at S.L. and 2400 horsepower at 25,000 feet. Ultimately the ratings would have been 3650 hp and 3000 hp respectively, using full output from the variable-speed, non-turbo supercharger.

Test pilot Carl Bellinger attained a top speed of 480 mph at sea level in the No.1 prototype. However, he was quick to point out there was no mention at all of using a WEP rating where manifold pressures are far higher than those seen at military power ratings. It is only fair to point out this 480 mph speed at sea level compares very favorably with the best speed attained with the sleek prototype Hawker Fury I which was getting the maximum possible power from a very late model Napier Sabre VI engine. The Fury I was not, in that form, a viable combat airplane. Bellinger flatly stated there were no flights with the XP-72 in excess of 500 mph. Since he, Ken Jernstedt and Joe Parker did most of the testing, it can be presumed no piston engine airplane of the WWII era ever flew faster than the speed attained by Mike Ritchie in the XP-47J.

One final comment: The manufacturing finish on both prototype XP-72s was superior. And in spite of the size of that 4360 cubic inch engine, the beautiful cowling actually made it seem no larger than the engines used in Focke-Wulf's Fw 190s. It was very nicely scaled to the P-47 dimensions.

[11] In his letter and conversation with me, Jernstedt said he had first met Charles Lindbergh in the men's room at Grumman. His reaction, I stated, was exactly the same as mine had been when I met Lindbergh in the men's room at the Willow Run Bomber Plant in 1942. At that time, Lindbergh had just flown in with the first P-47 I had ever seen. (It obviously had to be a P-47B.) Yes, Lindbergh was a "very tall guy."

CHAPTER 12

BUBBLES, RANGE AND WEAPONS

Among the unfathomable, unanswered serious questions of World War II, some relate directly to the fate of the Republic P-47. It is necessary to address at least some of the most significant of these to expunge any truly unfair defamatory criticism about the range of this workhorse fighter and fighter-bomber:

1) When Col. Franklin Carroll's Air Materiel Division team faced the Republic conference representatives summoned to the Wright Field offices in June 1940, they had a mandate from General Emmons' Combat Command to provide a new, viable, contemporary fighter aircraft. The specifications did not say Interceptor in so many words, but the requirements inferred it. Republic's P-44, in mockup form at that time, was to have a design gross weight of 8700 pounds. The calculated time to climb to 15,000 feet was 4 minutes, and takeoff overload fuel capacity of 260.5 gallons was specified. It was duty bound to be a 1940-class interceptor, unaffected by results of air war in Europe. Powered by the same basic P&W Double Wasp engine, the projected XP-47B was to weigh in at 12,000 pounds, climb to 15,000 feet in 5 minutes, and have fuel capacity slightly expanded to 315 gallons. Of course it had to be physically larger with accompanying drag rise and only about an 8 percent increase in power. If you increase gross weight by 40 percent, wing area by 30 percent and power by only 8 percent (not to mention drag rise increase due to wetted skin area), are you likely to have an interceptor? More power pulling more weight plus more drag reduces climb speed, and it did not turn out to be merely one additional minute to reach 15,000 feet. (On a good day, the XP-47B was ultimately to take at least 2.7 minutes of additional time when compared to the P-44, or a 50 percent increase.) Certainly some 8 percent more horsepower would eat up much of the skimpy fuel supply?

Is that the reason nobody cared to mention the airplane's mission? Why build a new, powerful fighter if nobody defines, even gives thought to, the "job description?" Evidently the Emmons Board, for all its great work, did not think to tell what they intended to do with the new marvel. And apparently nobody, including Kartveli, asked.

2) According to legend (or more), Alex "Sasha" Kartveli deplored the idea of hanging external stores on his birds, and we do know he was more than a bit vociferous about it. Did he actually wield so much

Exactly three months after the first P-47D-25-RE was completed, the first of 611 updated P-47D-27-RE fighters came off the Farmingdale assembly lines. Just two days later, on 6 May 1944, the last two airplanes in that block (Nos. 612 and 613) were completed as high speed YP-47Ms (Republic c/ns 4962 and 4963). Externally the P-47D-27s and YP-47Ms were indiscernible from the D-25 airplanes but the M models were several notches faster, even if they were a bit heavier. **(Author's Coll.)**

Just weeks after the 1944 New Year's celebrations came to an end, unpainted P-47D-22-REs were moving down the Farmingdale production lines at a rate of nearly one every 45 minutes. Conditions in Building 17 were uncluttered and efficient looking. Factory c/n 3714 is the first airplane on the right. (C.A.M./Ed Boss)

BELOW LEFT: Revealing some rearrangement of flight instruments and switches, the cockpit of a P-47D-30-RA is reasonably representative of all D-25 through D-28 blocks and nearly 3000 Thunderbolts in the P-47D-30 production run. The latter models flowed mainly out of Evansville, looking pretty much like this to the pilots, aircraft mechanics and technicians occupying the metal seat. (NASM)

power he could dictate tactics to the Combat Command?

3) When both Bremen bomber raids resulted in unacceptable losses in mid 1943, did everybody sit on their hands and fail to clamor for escort fighters? Did they think things were going to improve on long-range bomber penetrations? We all know VIII BC commanders had significant power and, at the moment, prestige. Why didn't they descend on General Hunter like chickens on a June bug?

It was inevitable somebody would ask why nobody in the 56PG(F), working up on the first operational group of P-47Bs, rang the alarm bell about range deficiency. The answer was only too obvious. Their specific mission was to operate as an interceptor force against any enemy attempting to attack the New York City and New England industrial centers where a large part of the aircraft, engine, propeller and machine tool manufacturers were bunched together. The pilots were far more interested in climb rate (poor), speed and maneuverability. Some of the earliest immediate prewar commanders of such entities as the Eastern Air Defense Command, the First Air Force and the First Interceptor Command were not even flying officers. Their claims to the jobs came from lengthy experience in command functions and administration, making them totally unqualified for technical and combat tactics decisions. Reorga-

Former AVG fighter pilot Ken Jernstedt became one of Republic's experimental test pilots as that organization expanded rapidly under the impetus of war. He is shown flying one of the first P-47D-25-REs (AC42-26428), an airplane with a completely new profile and numerous technical changes. Jernstedt is shown flying over the south coast of Long Island during the first or second week of February 1944. In the background is the barrier reef that is known as Fire Island, and beyond that is the Atlantic Ocean on a grey day. (Carl Bellinger)

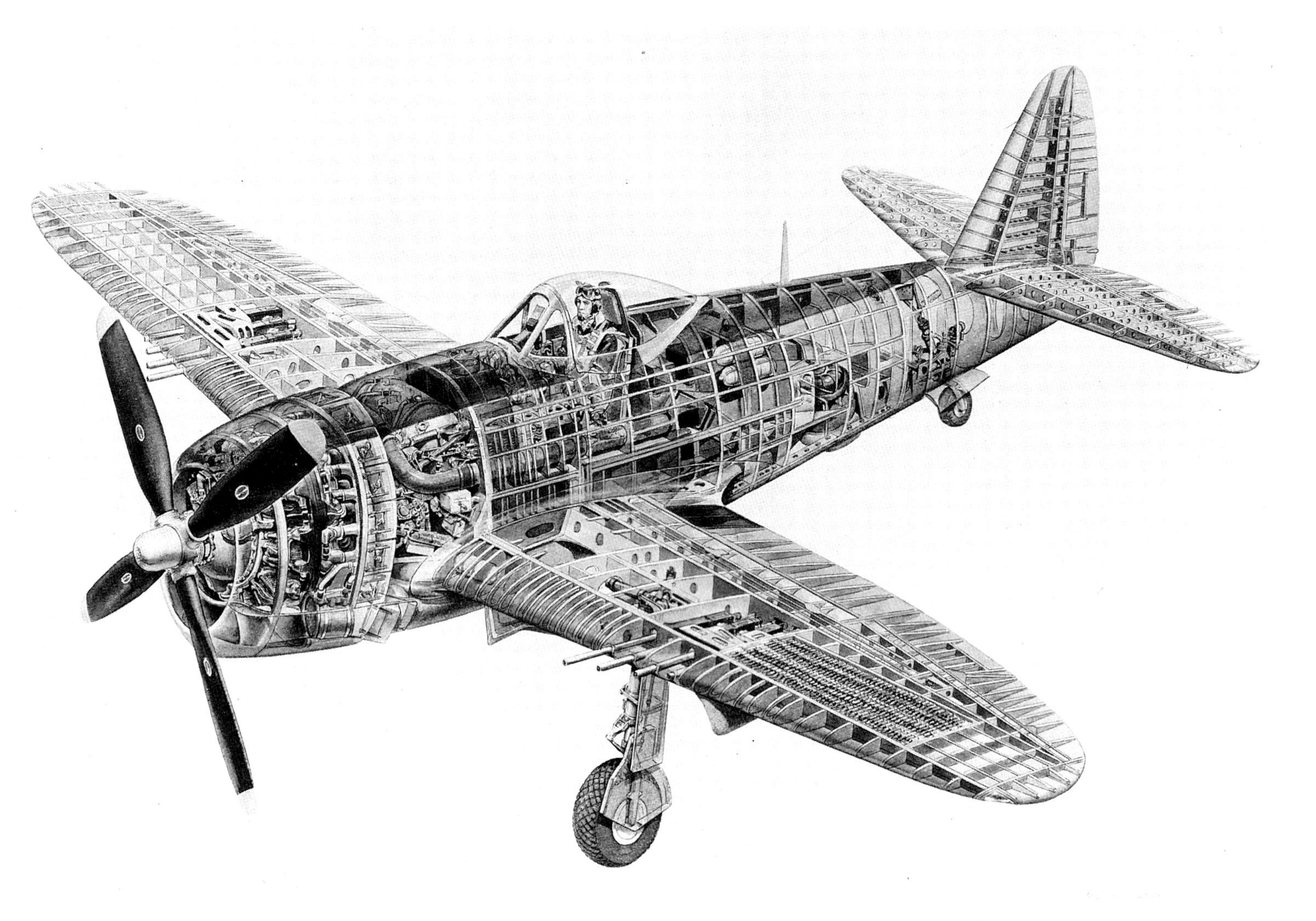

Excellent constructional detail is revealed in this "skinned" bubble canopy P-47D-25-RE. Rather interestingly, this type of production illustration was developed in immediate pre-war days in Great Britain. It was rapidly exploited by companies like Douglas and Lockheed when work forces had to be expanded tremendously. **(Mfr. H-6486)**

nizations were heaped upon reorganizations, and hardly a month went by without some upheaval.

Then consider the responsibilities piled high on new, youthful commanders and their own confidence in decision making – especially when one small confrontation or miscalculation could mean stagnation or even demotion. Hubert "Hub" (NMI) Zemke was in the midst of all this, but had risen rapidly from captain to lieutenant colonel when still in his twenties. Not one thing in Air Corps training had prepared such youthful men for anything like group command. Less than five years earlier, he had not yet officially received his second lieutenant bars and was loosely categorized as a cadet, uniform and all. But he was flying around in Curtiss P-6E biplane fighters as an Air Corps "regular." By the early months of wartime 1942, Zemke was commanding the first Thunderbolt combat group, the 56PG(F), responsible for defending the entire Eastern Defense Command area. Realistically, it meant First and Second Air Force defense because the latter was concentrated primarily on training. He was responsible for building his squadrons, coping with the machinations of service testing a brand new fighter, and developing interception tactics for a 400-mph fighter, something far afield from anything the command center at Mitchel Field, New York, had ever contemplated.

Was Lt. Col. Zemke likely to even think of combat radius of action over Europe? Any such thought had to be tied to the still unwelcome requirement for hanging a selection of ornamentation on a high-speed aircraft. In those truly hectic days where a false alarm air raid over Los Angeles County threw Southern California into panic disarray, what young fighter pilot was likely to even think about range extension unless he was responding to a direct question?

In far away California, the situation was considerably different. Col. Benjamin Kelsey was not only riding herd on the Lockheed P-38 project as trail boss, he was teamed with Lt. Cass Hough in the intensive planning effort for Operation BOLERO. He personally flew a P-38 over much of the raw pioneering stretches of territory to be traversed by the first squadrons of fighters in history ever to be ferried across the North Atlantic. Kelsey had also been central to launching the initial authorization for Lockheed's Kelly Johnson to develop high-speed external drop tanks, planning to get General Arnold's permission *after the fact*. Certainly he had conspired with young Johnson to develop long-range tanks before the Pearl Harbor attack since design work on the P-38F model was completed as of 1 January 1942 and the first production airplanes with a 2000-pound stores capability pylon under each wing centersection were being delivered to the AAF in March 1942. A dozen different tank designs had been tested, and two basic versions were already moving down assembly lines: the primary 165-gallon laminar-shape tank and the 310-gallon version. (By 1943, many of those same primary tanks were seen underwing on dozens of Thunderbolts.)[1] Those immediately recognizable drop tanks looked right, were right from the start. They were an absolute necessity if American airpower was going to defend Alaska and the Aleu-

ABOVE: Dropping the landing gear, Jernstedt pitched the P-47D-25-RE back and forth to show the airplane with landing gear extended. He was soon joined by a few more test pilots in identical fighters for a few formation pictures. That distinctive Seversky-Republic wing planform, essentially unchanged since about 1931, was still serving its masters well. (Carl Bellinger)

In one of the rare periods where no employees were in sight, the scene inside Bldg. 17 at Farmingdale early in 1944 reveals an ultimately neat, efficient subassembly and final assembly area. USAAF P-47D-23-RE and RAF Thunderbolt I (razorback) production ended in April 1944. That is not the initial P-47D-25-RE in the foreground. The first D-25, c/n 3966, rolled out the door on February 3rd. Some major redesign in the firewall and cockpit areas is revealed. We can't explain the production mix. (RAC)

There can be little doubt that the AAF was less than aggressive in improving propeller designs in the face of known German Air Force developments early on. Messerschmitt, Focke-Wulf and Junkers combat aircraft were appearing on the scene with radically revamped propellers even as America went to war. But under intense pressures from VIII FC, both Curtiss-Wright and Hamilton-Standard forged ahead in designing improved blades. Wider, more efficient hollow steel blades such as these were being shipped to the ETO just 8 months after design work was initiated. H-S Hydromatic propellers were also equipped with the wider paddle blades by the time P-47D-25-REs were ready for production. Test pilot Gerard Demunck shows off the Curtiss Electric type propeller. (NASM)

tian Islands; they were useful and they were there when needed. Without them, Operation BOLERO could not have succeeded, at least with respect to fighter aircraft. Later on, in 1943, a clutch of P-47 Thunderbolts made it across the North Atlantic route to the United Kingdom. Lo and behold, they were able to perform that feat on the strength of a pair of P-38 tanks on each fighter, despite the use of multi-strut, high-drag mounts.[2] (Cass Hough's Technical Operations group at Bovingdon would have solved that problem in a matter of minutes.)

Republic's P-47B was a strong aircraft, and the P-47C series was perhaps even tougher. With hundreds of Bell P-39s and Curtiss P-40s flying around in the eastern half of the United States, more often than not with external drop tanks, it is surprising nobody in the AAF seemed concerned about the lack of such appendages on Thunderbolts. They all had become accustomed, it seems, to viewing P-43 Lancers sans external loads of any kind.

Ah! It had to be something more plausible...like too many trees in the way. Nobody could see the forest.

As discussed in Chapter 10, Republic eventually supplied one so-called drop tank design in the form of the 205-gallon "cow's udder" affair. Two airplanes had been involved in the flight testing of the unit at an unknown date, long before Republic Airport began to look like a parking lot for fighters. It had to have been in late summer 1942, perhaps not until autumn. Apparently no attempt was made to do any wind tunnel testing of airplanes, but some tests were conducted in November on a 1/6th scale model of a P-47C in a Langley Field NACA wind tunnel. Contrary to expectations, the model was configured for wing-mounted external tanks. Republic had not bothered with any tunnel testing of the 205-gallon, 4-point attachment affair. It was a rather unfavorable contrast to the Lockheed effort about a year earlier. Service use of that unpressurized, ungainly, sometimes undroppable monstrosity was (thankfully) brief. Nobody attempted to send the Republic tanks to the Fifth Air Force, and that non-action had its good reactions. General Kenney's people designed and fabricated the metal 200-gallon "Brisbane" tank, tested it successfully and Ford Motor Car Co. produced them in quantity.

They say pictures don't lie, and that is borne out by some pictures in this book depicting airplanes with virtually every change normally seen on a production P-47D-23; however, the very visible tail number on one overturned Jug reveals it to be a relatively old P-47D-2-RE. The airplane was equipped with three 108-gallon drop tanks, it had a bulged keel, and the cowling was up to D-23 standard. Quite obviously it had benefitted from a major depot rebuild. The moral of the story is, don't judge a book by its cover. What sometimes seems to be one thing may be something else.

Thunderbolts eventually carried many different drop tanks in numerous combinations, ranging in capacity from 75 gallons to 310 gallons, with the XP-47N carrying two of the larger-size tanks on range determination test missions. Eglin Field test pilots

There was great excitement among all RACers at Farmingdale on 20 September 1944 when "Ten Grand Thunderbolt" was rolled out of Bldg. 17. It marked completion of a P-47D-30-RE (AC44-20441), the 10,000th T-bolt produced at the New York and Indiana plants. The day may have been dull and overcast, but the spotlight shone brightly on Republic Aviation. The same company, under the name of Seversky Aircraft, had been a tiny manufacturer on the brink of financial collapse just five years earlier. In shameful contrast, old-line Curtiss Airplane Division at Buffalo (starting in a dead heat with Evansville) produced only 354 Thunderbolts in an 18-month period. When Curtiss production was cancelled, Republic was turning out over 600 airplanes per month! This airplane soon found its way to the MTO and assignment to the 79FG. (C.A.M.)

[1] If anyone is inclined to believe this author is in awe of the late B/Gen. Benjamin Kelsey, they will not get an argument. If any living person can claim to have studied this man's lifetime achievements more thoroughly, he is a rarity and would share my opinions completely. Kelsey was a super-achiever. Few men have ever contributed more (unselfishly) to their country than this man. Another in the same category, Gen. James Doolittle, did not hesitate to confirm his admiration for Ben Kelsey during a 1970s interview. He was one of the truly great men of aviation.

[2] That is what happens when somebody (unknown) attempts to reinvent the wheel. Wouldn't it have been much more logical for a supply officer to requisition two dozen sets of P-38 pylon mounts? Adaptation would have been simple, especially if they did not have to be expendable except in an emergency. Simple wooden or metal fillers would have taken care of any significant difference in lower surface airfoil shape.

Republic Aviation certainly had something to be proud of in building 10,000 Thunderbolts by 20 September 1944 in two plants that did not even exist as the Battle of Britain ended. This 79FG (12AF) P-47D-30-RE, logically named "10 Grand Thunderbolt" by Republic employees, revealed just how dedicated most Americans were to winning the war. The Farmingdale-built Thunderbolt served in the MTO with the 79FG. Most Curtiss-Wright employees were probably just as dedicated to that same goal, but either Curtiss management at all upper levels was incompetent or motivated by greed. The Curtiss Airplane Division performance in producing exact copies of P-47Ds beginning in 1942 was dismal, even tragic. When compared to Germany's Messerschmitt and Focke-Wulf production achievements, the C-W and Brewster performances deserved war crimes trials. (Mfr. H-4891)

recommended against combat missions with two full tanks of that size. Inadvertent release of one tank during takeoff could cause the aircraft to roll over abruptly when one 2000-pound tank separated. One of the finest drop tank designs of the war was the 108-gallon Bowater-Lloyd paper tank developed by Cass Hough and Bob Shafer at Bovingdon. It created relatively acceptable drag, could be pressurized with ease, and duplicates constructed of metal were totally interchangeable. Wider versions of the same tank could carry 150 gallons, and flat-section tanks could carry from 150 to more than 200 gallons of gasoline.

Officially – contrary to much published information – P-47C-2-RE and subsequent airplanes could be equipped with shackle equipment permitting installation of a bomb (or tank, obviously) on a centerline belly mount and one bomb under each wing. The fuel system was not, at that time, set up for wing tanks. However, that could hardly have been considered a major detriment. From the time P-47D-5-REs were in production, the Type B-7 bomb shackle installation was suitable for use on the crash skid (keel). Some of the P-47C-2-REs could accept installation of the Type B-10 shackle and its adapter (pylon mount), and later models/blocks were either delivered with the shackles or could be retrofitted in the field. The B-10 unit was suitable for carrying loads up to 1600 pounds, and was considered safe up to 2000 pounds. Beginning with P-47D-35 and P-47N-10 airplanes (modification center blocks), a new, improved Type S-1 shackle could be used on fuselage or wing adapter positions.

Photographs of Republic P-47s dispatched to the USSR on the Lend-Lease program seem to be rarities. One wonders if any such photos ever originated in the Soviet Union. This drab Thunderbolt is a P-47D-27-RE, and we can determine from the AAF tail number that it was test flown toward the end of May 1944. (Mfr. H-2898)

In reviewing the progressive translation from "razorback" to "bubble" configuration in the T-bolt line, all investigations are blocked by path omissions, not deliberate walls of silence. Why did it take from May 1941 until November 1942 for Republic Aviation or Materiel Division to do any wind tunnel testing on wing-mount drop tanks (and none at all on centerline drop tanks or stores pylons on the wing)? Considering the rapid-fire reaction time to obvious needs out at Lockheed, com-

With its slotted flaps lowered, a new P-47D-25-RE passes over the Fairchild test hangar and parking lot on final approach to landing at Republic Airport. This particular airplane was probably assigned for aerodynamic testing, judging from the fact that no guns were installed. (Carl Bellinger)

LEFT AND MIDDLE: Air Materiel Division requested that the Armament Test Center at Eglin Field conduct exhaustive tests on the new P-47D-25-RE to determine performance with various combat loads. Initial tests were conducted with 1000-lb. bombs on the wing pylons and the metal 110-gallon belly drop tank alternated with an additional 1000-lb. centerline bomb. **(George Colchagoff)**

parable situations at Farmingdale (except for the 1944 thousand-mile radius-of-action needed by Hart Miller) withered on the vine or were sidetracked for interminable periods. There had to be some primary reasons. In 1973-74, I approached the key sources with these and other appropriate questions.

Over a period of eighteen months, the following persons were interviewed, in person whenever possible, by mail and telephone where some serious difficulties intervened:

Stuart Moak - Then current president of the P-47 Thunderbolt Pilots Association.

Carl Bellinger - Former chief test pilot for Republic Aviation

L. L. Brabham - Former chief test pilot and director of flight operations for Republic.

Ray Higginbotham - Former chief powerplant engineer for Seversky and Republic Aviation.

C. Hart Miller - Former assistant to Seversky and Republic presidents and a vice-president of contracts and service engineering (also involved in top-level engineering and flight testing).

B/Gen. Benjamin Kelsey - Retired commanding general of Logistics Command at Wright Field with a lengthy career in the USAF.

L/Gen. Laurence Craigie - Then a consultant at Lockheed-California Co. and retired USAF with many commands over a ifetime.

M/Gen. Marshall Roth - Retired USAF, former project engineer on the P-47 Thunderbolt.

L/Gen. William E. Kepner - Former commanding general of VIII Fighter Command, and a man with decades of key assignments in the Air Service, Air Corps, Army Air Forces and the USAF.

Dr. Walter C. Beckham - WWII fighter ace; postwar scientist.

Some of these men – in fact most of them – were in good health, cooperative and the majority with very good memories. In brief discussions with retired Col. Howard Bogart, however, the poor man was suffering terribly from a disease called shingles. General Roth was not a man to retain records and some of his recollections were soon determined to be faulty. After letting me know in strong terms he had been "ripped

This Republic P-47D-16-RA was apparently equipped with a pair of Lockheed 165-gallon drop tanks and a metal 75-gallon centerline drop tank at the Evansville Modification Center as part of yet another range-extension effort. The entire P-47 Thunderbolt effort at the Indiana facility was remarkable since it was located deep in middle America's farm country. There was, essentially, no cadre from which to build a large manufacturing team such as was enjoyed by Curtiss Airplane Division in Buffalo. Yet the Evansville product was far superior, and the production rate...compared to Curtiss's...was near phenomenal. **(Republic)**

RIGHT AND BELOW: Flown without a centerline store, the P-47D-25-RE test airplane was initially flown with two 110-gallon drop tanks mounted on the wing pylons. They then switched to two 165-gallon Lockheed drop tanks for comparative testing. All such tests were flown with the tanks fully fueled and with full machine gun ammo loads. Evidently a test was conducted with two 310-gallon Lockheed P-38 drop tanks installed, and although it was accomplished it was not considered to a safe procedure. Inadvertent jettisoning of one tank on takeoff would most likely prove fatal. (George Colchagoff)

off" by several "writers" or was considered by some to be "senile," General Kepner – then 82 years young – proceeded to give me an "earful." To say the least, it proved to be enlightening. Coming from other commands, he did not have the answers to questions about prior omissions, but he was a one-man Blue Ribbon committee in the rectification processes. By that time, in 1974, Alexander Kartveli and his wife, as I recall, were in poor health at advanced ages. Ray Higginbotham was very definitely in poor health.

Hart Miller provided his "one man's personal opinion" which may uncover a key clue to the problems, perhaps not. Some time after our 1974-period lengthy interview, he wrote, "I really have only one basic difference with the story (of the Thunderbolt) as you have written it. In my opinion, Mark Bradley had ten times as much influence on both the development and success of the Thunderbolt as Mish Roth." He stated "Mark was involved from the word Go." (In all candor, it must be reported that General Bradley, in retirement in the 1970s, invited writer Gary Pape and me to his home for an interview. He proved to be far less than cooperative, not permitting us to record anything worthwhile. Even the very ill Col. Bogert made every attempt to be cooperative, but was unable to continue. Gen. Bradley was the logical man to provide answers to the external tank questions, and we certainly would have asked.)

Outside of writing a complete investigative book on the internal machinations affecting the progress of P-47 engineering changes, that would seem to sum up my feelings, but only with respect to the engineering and manufacturing aspects. General Bradley had nothing at all to do with the operational, tactical success of the Thunderbolt. Whatever shortcomings there were – and they were numerous – in the P-38, P-47 and P-51, General Kepner and many of his group commanders in the ETO worked well with General Doolittle to overcome them for the Eighth Air Force. With reestablishment of the Ninth Air Force on 16 October 1943 under the command of M/Gen. Lewis H. Brereton to take a major part in Operation POINTBLANK, there was a definite strengthening of the battle-oriented product and tactics improvements. That it all came together is reflected in the successes of 1944-45.

There lies the answer to the severity of problems facing all combat pilots throughout 1943. Like it or not, the men at the helm are either right or wrong for the job. With the wrong "captains" in charge, little or nothing gets done and progress is minimal. Without regard to individual political likes and dislikes, the early 1930s proved President Hoover's policies to be all wrong for conditions; President Roosevelt could certainly ruffle feathers, but his "medicine" was sorely needed to drag America out of the terrible ravages of the Great Depression, probably the worst ever to engulf our nation.

With the departure of B/Gen. Hunter, the external fuel drop tank situation seemed to take shape at flank speed. If the T-bolts could not come from the factory with wing and centerline mount provisions for external stores, there were numerous retrofit and modification programs in place to rework the aircraft. It

Early in 1944, with knowledge that numerous landings of invasion forces in the ETO and Pacific theater would become routine, tests of the M-33 chemical tank installation on P-47s became a priority operation. These large canisters could be used to lay smoke screens, of course, but dispensing of other chemicals was also possible. (AFM)

soon became commonplace to encounter P-47D "early block" aircraft with nearly all the attributes of a P-47D-22-RE or a P-47D-23-RA before those production aircraft could reach the ETO. Also by the autumn of 1943, the 4FG, 56FG and 78FG were advancing their new combat tactics to a point where aerial victories over a tough enemy were climbing rapidly. We were out of the doldrums. Suddenly there were multiple choices for drop tanks. They included, but were not limited to, the 75-gal. teardrop tank, the Bovingdon 108- and 150-gal. paper and metal drop tanks, 150/165- and 215-gal. "flat" belly tanks, a 110-gallon teardrop tank in metal, and 165-gallon Lockheed tanks (widely used in the MTO and CBI theaters).

The wing pylon mounts (adapters if you prefer), fitted with B-10 shackles, really brought the Thunderbolt to life. While the potential for extended range had been there all along, because of the Jug's ability to lift extraordinary loads, the potential was not recognized until the pylons were installed. A modest rework of earlier P-47C versions, clear back to the -2 block, revealed they could carry up to three drop tanks, or various combinations of bombs and external tanks.

Curtiss Propeller finally got around to changing their blade design to provide a better bite into the air, especially at high altitudes. Moving away from the "toothpick" tip, they created a "high-activity" propeller with broader blades. Almost from the start they were referred to as "paddle blades." Virtually all P-47s so fitted experienced a speed increase of as much as 10 mph, but the main attribute was an improvement in climb rate of approximately 600 feet per minute. Whereas the more powerful D models took about 20 minutes to reach 30,000 feet, there was a much better expectation for getting there in about 13 minutes with the new unit. A later version employed blades with extended trailing edges, giving an even better bite into the air. Beginning with the P-47D-22-RE, Farmingdale-built T-bolts began to appear with Hamilton-Standard Hydromatic propellers using paddle blades very similar to those used on the first paddle blade Curtiss propellers. Although the H-S Hydromatic hub unit was a shorter, fatter design than the electrical pitch change mechanism used on the Curtiss Electric propeller, there was little if any effect on airplane performance.

Typically, a P-47C or D without a drop tank had a flight duration of about 1 hr. 40 min. to 2 hr. 5 min.[3], but using one centerline 108-gal. tank, flight duration was extended to 3 hr., even 3 hr. 15 min., not allowing for combat time. That equated to about a 200-mile radius of action. The large 205-gallon belly tank, in use for a brief period, typically only added 30 miles additional penetration depth. Even the little 75-gallon P-39 belly tank could extend the duration time from the minimums to 2 hr. 30 min. up to a maximum of 2 hr. 50 minutes. One of the best combinations was the use of two 108-gallon cylinderical tanks on the B-10 shackles, providing a duration of up to 5 hr. 30 minutes.

The Emmons Board specified installation of six (alternate, eight) .50-cal. Colt-Browning machine guns. All were to be located in the wing, and it was easier to provide for eight guns as standard than to provide unused space for two guns and ammunition. Specifications may have called for 267 rounds per gun, up to 500, but 400 rounds of API ammunition came to be favored by Ninth Air Force pilots.

Triple tube 4.5-in. folding fin rocket launchers proved to be powerful destructive weapons when they hit a target. However,

[3] Based almost entirely on 4th, 56th and 78th Fighter Group records for 1943.

A pair of P-47D-23-RA Thunderbolts is shown undergoing final preparations for smoke dispensing tests. The nearest fighter is shown fitted with the older, pre-war M-10 tanks. Mechanics are shown making final adjustments on one of two M-33 large-capacity chemical tanks installed on the second airplane. (AFM)

***At least one bubble canopy P-47D was tested at Farmingdale with new zero-length rocket launching installations for 5-inch HVAR (high-velocity aircraft rockets) weapons. The airplane was also equipped with the newest Curtiss Electric paddle-blade propeller. While only two of the new rockets could be carried inboard of the tank pylons on the D models, they proved to be more effective and had few, if any, negative effects on overall performance of the fighters.* (NASM)**

they were frequently not very accurate except against a large target such as a freight train. The clustered tubes of the 80-lb. M10 Bazooka launcher were made of a plastic composition, while the alternate M15 assembly was made of magnesium and weighed 86 lbs. A steel M14 model weighed in at 190 lbs. and was likely to find a better home on a Jeep or a truck. Emptied tubes, especially the steel-tube launchers, could prove to be a serious handicap if not jettisoned. When the 5-inch HVAR (high-velocity air rocket) was developed late in the war, some P-47D-30 (and later) models were equipped with so-called zero-length launcher pads, two for each rocket and four pads inboard of the stores pylon. Eventually many P-47Ds were configured like P-47Ns, accommodating five rockets on each wing.

For short range operations, two 1000-lb. bombs could be carried on the wing pylons along with a 500-lb. bomb on the fuselage shackles. It was no problem to develop weapons loads to suit the attack mode on a particular target. Large numbers of RAF Thunderbolt Is and IIs operated from Burmese airfields in hot, high-humidity conditions from short, unpaved airfields. A routine weapons load would be 500-lb. bombs on the two pylons and a 75-gal. (U.S.) teardrop tank on the fuselage centerline mount. One type of external store, not seen frequently, was the chemical tank, used primarily for laying smoke screens in support of friendly troops, or a beach landing party. The M-10 tank was smaller and created less drag, while the large M-33 unit was about the size of a 1000-lb. bomb.

With delivery of the first P-47D-25-REs, the Air Proving Ground Command (APGC) at Eglin Field, Florida, conducted detailed tests, using various combinations of bombs and assorted drop tanks. The Jug had made a complete transformation from interceptor to flying artillery and anti-tank weapon. The only disturbing part is the loss of nearly a year as the bomber escort it was intended to be, and that fault could be traced to leadership/decision making failures.

As for the adaptation of the clearview bubble canopy to the Thunderbolt, the legend or facts about Kartveli's serious objection to the change can hardly be accepted unless somebody working closely with him can provide absolute verification...and that is becoming closer to impossibility with each passing week. If it did not occur to Sasha Kartveli, any responsible engineer could have strongly suggested use of a precision wind tunnel model. It could have been created with a removable upper fuselage section, allowing insertion of a razorback spine and canopy or a bubble canopy unit. Any eventual loss of speed was minimal and may have been attributable to the flat plate windshield in place of the knife-edge vee windshield.

The "Down to Earth" attack policy adopted by Gen. Kepner had an ideal support weapon in the guise of a P-47D. The shouted "Achtung, Indianer" (a colloquialism for AAF pilots) must have been reverberating throughout Germany in 1945. In loose translation, it meant "Lookout, Americans." There was no time for more words.

CHAPTER 13

STRIPES

Great Britain, in the first half of 1944, was suddenly awash with stripes. "Stars and Stripes Forever" was a tune heard more than infrequently. Master Sergeants, First Sergeants and Technical Sergeants wore a plethora of stripes. Three up and three down in the two former cases; three up and two down in the latter. Actually the "down" stripes were not stripes at all, but rockers. But not many soldiers ever said that...if he had six stripes he was a Master Sergeant or a First Sergeant. If either of them – especially the grizzled old timer who must have been God – said "Jump," you did jump. He was the one with the diamond below the top three stripes...the "Top Kick" or First Sergeant. The five-striper was the guy who got the real work done, usually. Techs and Staff Sergeants (3 up, 1 down) were shop heads, line foremen, crew chiefs who carried a lot of responsibility for "keeping 'em flying." One stripers picked up butts. Not really; they just believed that. Guys who knew they were good, kept

Cocooned, sprayed, and doctored with the best techniques and materials available in late 1943, some of the first North American P-51B Mustangs and late-series razorback Republic Thunderbolts (P-47Ds) arrived almost daily at ports such as Belfast, North Ireland. With tankers like this one heavy with oil or high-octane gasoline riding low in the water, Atlantic storms often caused serious damage, increased corrosion potential and even washed airplanes overboard. Work crews, probably from Langford Lodge (AAF Sta. 233), are shown removing tiedowns in preparation for transferring these Project UGLY fighters to road transportation. (USAAF)

BELOW RIGHT: Capt. William Tanner played football in school and looked the part of an All American. He did a couple of tours in the ETO and became a 5½ airplane ace. He flew with the 350FS of the 353FG on that first tour. Bill may not have been renowned, but like so many others, he performed above the call of duty many times. (William F. Tanner)

out of trouble and had "been in grade" for some time stenciled their stripes on fatigues. All others usually sewed them on. Nobody was an "Airman" in those days. "Pick up that butt, *Soldier*!" was what you heard, not *Airman*. If a 6-striper wore a campaign hat, you knew he was God. MPs with one or two stripes just looked like they had six of them.

But the real STRIPES were yet to come. They appeared on June 5, 1944, as if by magic on nearly everything that could fly. Even some DeHavilland Dominies and Beech C-45s had them. They either had to be black and white or white and black. No exceptions. On camouflaged or natural metal P-47 Thunderbolts, the first white stripe was to start 8 feet 9 inches inboard from the tip. Orders specified five stripes of equal 20-inch width totally encircling the left and right wings...white, black, white, black and white. A series of five nearly identical stripes encircled the fuselage, but they were to be 18 inches in width, and the first white stripe was to begin at the tailwheel door opening.

Who was the brilliant person coming up with the adopted concept of nearly error-proof recognition, and how did he ever sell it to the top generals? How could they order all that white and black paint and print up tons of instructions for painting the aircraft without some spy tipping the Nazis about the invasion? Just think of what might have occurred had the Germans known all details and painted many of their fighters in a similar manner. But British or American, all Allied aircraft had the markings and none of the enemy did. As a backup measure, P-47s which were at depots at the right time had 55-inch or even 60-inch diameter star and bar national insignias painted on the wing lower surfaces as a recognition aid.

On D-Day, Operation OVERLORD invasion day, it looked like the British Isles themselves should have been sheathed in black and white stripes. They were to be seen everywhere you looked. And like a good-looking tie on a man, they added some pizzazz to the airplanes. If any trigger-happy GI fired at those airplanes, he must surely have been given a Limited-Service assignment.

One often overlooked facet of the recognition marking program was the basic scarcity of photographs which survived the war to depict participating aircraft with the full set of specified markings. Once the Allies had established a firm beachhead on French soil, the Ninth Air Force's close-support fighter-bombers began to operate from airfields located a few miles from the moving front lines. Sometimes it was from battered, captured *Luftwaffe* air bases; at other times, Engineering Battalions following closely behind the advancing tanks and infantry, used their equipment to carve out landing strips, mostly very crude but useful and almost habitable. Not wishing to make highly visible targets of their airplanes to aid the Germans in any way, the Allied aircraft group commanders were directed to delete the stripes from all visible upper surfaces. That meant deletion from upper wing surfaces and the upper half of the fuselage rather quickly. In fact, it was initiated in less than two weeks after D-Day and was generally *fait accompli* throughout Europe by 1 August.

During the first hectic days of the Continental invasion,

OPPOSITE: Even as the P-47 Thunderbolt improved greatly as a dogfighter, its mission capabilities as an attack-bomber – the British preferred strike-fighter – increased far beyond anything ever contemplated at the 1940 Wright Field meeting. Bearing letter perfect theater markings and a brand new set of fuselage and wing "invasion stripes," this P-47D was fitted with two wing-mounted 1000-lb. bombs plus a centerline 500-pounder. It is well to observe that this Jug was not even powered by a "C" series R-2800 engine, nor did it have the high-activity paddle blade propeller. (USAAF 54973AC)

***Although the Grumman Hellcat benefitted from having a P&W R-2800, it most certainly would have fared poorly in ETO combat without a better critical altitude rating. Grumman, with BuAer acquiescence (at least), eventually tried the unproven Birmann turbosupercharger unit in the XF6F-2 (3rd airframe) with the Double Wasp early in 1944, although a version of that project had begun in 1942 or earlier. It was wasted effort. They would have been far ahead in just copying their neighbor's General Electric turbo installation. Few people ever knew of the turbo project.* (Grumman)**

when one day might see as many as 14,000 sorties flown by the Eighth and Ninth Air Forces and the Royal Air Force, there were far more camera subjects than photographers available to click shutters. Combined military cameramen and private media photo journalists accompanying the huge contingent of invading troops into *Festung Europa* constituted but a tiny fraction of cameramen who would cover far less important media events in the final decade of the 20th century. Pilots, aircrew and maintenance personnel were so busy, few had the time to even think about recording the scenes on film, if they could find any. By the time anyone had real opportunities to exercise shutters with any regularity, upper fuselage and wing striping had become a thing of the past. There is great likelihood few, if any, photographers even conceived of the stripes disappearing within a matter of weeks. All of these factors combined to limit the number of pictures of aircraft displaying full Invasion markings which would be available in postwar years.

Objectivity requires an unbiased view of changes which took place upon assumption of major commands by new commanding generals subsequent to the August 1943 Regensburg and Schweinfurt bombing raids. With arrivals of B/Gen. William Kepner to take charge of VIII Fighter Command and M/Gen. James Doolittle moving smoothly into the Eighth Air Force command slot, radical changes occurred in the ETO air war. It was as if the "corporation" was moving out of Chapter 11 bankruptcy into its most profitable year in history.

POINTBLANK, the Combined Bomber Operation, was in trouble in October 1943, and the Royal Air Force essentially held the top slot overall, clashing with the rising stature of Eighth Air Force importance in the ETO. Reorganization was obviously needed to meet the requirements of a new Supreme Allied Command to carry out the forthcoming OVERLORD effort. Cutting through all of the administration and operations lines, we find that command of United States Strategic Air Forces in Europe (originally USSAFE, but changed to USSTAF on 4 February 1944) was delegated to Gen. Carl Spaatz. Doolittle was in command of the Eighth Air Force and, although VIII Bomber Command ceased to exist, VIII Fighter Command continued. Authority for creation of USSAFE took effect on 1 January 1944, with Spaatz formally assuming command on 20 January. He was, in effect, head of strategic operations of the Eighth and Fifteenth Air Forces. Operational command of the Ninth Air Force, reactivated on 16 October 1943, was under the Allied Expeditionary Air Force.

With a couple of Lockheed P-38 groups and one North American P-51 group having been acquired through the medium of exchange, the 8AF entered 1944 with its first long-range escort capability since November 1942 when the existing P-38 groups moved south to participate in Operation TORCH. After more than a year of delay, the ability of the AAF to carry out General Henry Arnold's directive about gaining control of the air, issued in the strongest possible language, required the Army Air Forces to: *Destroy the enemy air force wherever you find them, in the air, on the ground and in the factories.* General Doolittle's determination to provide long-range escort for the heavy

***Although the official U.S. Army caption would have us believe that this picture was dated 13 April 1944, it would have to be presumed that a great many P-51B/Cs and razorback P-47Ds had been forgotten for weeks. By April the field should have been spotted with at least a few dozen bubble canopy P-51Ds and P-47D-25s at what appears to be a major Fourth Echelon depot, probably Burtonwood or Speke (BRD). The first of 385 a/c in that Thunderbolt block was completed on February 3rd, and the company rarely missed a beat on rolling out sixteen airplanes a day at that time at Farmingdale. Total checkout and shipping time from the U.S. should been around 45 to 60 days. Many Douglas A-20s are to be seen in the background.* (U.S. Army SC189698)**

As England came to look more and more like a huge parking lot for aircraft, weapons and vehicles, it soon became evident that deliveries were outstripping processing facilities at LOC, BRD and AAF depots such as Burtonwood. For unknown reasons, the cocooning material on several P-47D canopies had either been ripped off for viewing some cockpits (likely) or had not been properly applied. The pattern of airplanes affected would seem to point to the first reason. (U.S. Army)

and medium bombers was an absolute "Must" since he was in complete agreement with Arnold's wishes on that score. After he and M/Gen. Lewis H. Brereton, CG of the Ninth Air Force ironed out the details of long-range escort requirements vs. tactical support of the Allied ground forces, the machine began to operate in the desired manner.

Buildup of IX Fighter Command was rapid, with up to nine fighter groups coming aboard by year end, 1943. The never-anticipated ground support role into which the P-47 Thunderbolt moved was a revelation as to the airplane's true capabilities. In the final quarter of 1943, VIII FC Thunderbolt groups were beginning to receive much improved versions, the P-47D-11 through -16 models. Earlier versions were being rebuilt or modified in AAF depots in the UK, and the most potent razorback Jugs, -20 through -23 blocks, were in the pipelines. The three-droptank Thunderbolt had become a reality. As Ninth Air Force fighter units were committed to the ground support effort, they came under control of IX and XIX Air Support Commands. The XXIX ASC was soon to join them.

With many 8AF groups still equipped with P-47 and P-38 fighters pending arrival of sufficient supplies of P-51B, C and D models, the VIII FC, under M/Gen. Kepner, launched into a "domination of the air" assault against the Luftwaffe with an intensity not seen or envisioned previously. In fact it was the embodiment of the principles which Major Alexander Seversky had been advocating since the beginning of the 1940s decade, and had broadcast to the world in his "Victory Through Air Power" preaching.[1] By February and March 1944, VIII FC fighters were destroying German aircraft at an unprecedented rate. Although enemy production output continued to set new records, the Nazis could hardly have any confidence about prevailing in the war if the Allied Air Forces fighters continued to make such inroads on their operational aircraft supplies and production rates.

It all really began with "Big Week" on 20 February 1944, the first mission on which the Eighth AF dispatched 1000 heavy bombers to attack German aircraft production centers. Something under 900 bombers actually found their targets or targets of opportunity. On 1 February 1944, VIII FC had a total of twelve fighter groups assigned, including the 20th and 55th Fighter Groups equipped with a mixture of P-38H and P-38J fighters. One P-51B group, the 357FG had just transferred (the previous day) to the 8AF in exchange for the Thunderbolt-equipped 358FG which moved to the Ninth Air Force. Doolittle and Kepner were well aware of the perils of the forthcoming "Big Week" venture in which a thousand long-range bombers could not count on more than 125 Lockheed P-38s being somewhere along the routes trying to protect them. (Col. Jack Jenkins, commanding the 55FG, spoke of the failure of bomb groups to maintain tight, protective formations. On one particular day, he said the bombers were strung out along a 60-mile path with about sixty P-38s – certainly no more than 100 of them – trying to fend off *Luftwaffe* attackers, numbering into the hundreds at any given moment.) The generals also knew the attacks to repel the bombers would be every bit as determined during withdrawal as they were during bomber penetration. The IX FC was not a factor in the situation since most of its groups had P-47s and P-38s based in England. Thunderbolts of the VIII FC were, by then, operating with as many as three 108-gallon drop tanks,

[1] Anybody can hazard a guess about the leanings which led to the adoption of the victory through airpower concept by the 8AF and VIII FC. Doolittle was a veritable "mustang" and WWI pilot. Kepner was a Regular ground army man who had moved into flying as an Air Corps pilot considerably later in life. You might have suspected him of opposing Gen. Billy Mitchell's concepts, but that idea could be 100 percent wrong. Although Doolittle never voiced his feeling about Mitchell with any ferocity, anyone who knew the man would bet hard cash the Eighth Air Force CG was a staunch supporter of the defrocked general's theories. It is equally unlikely Spaatz advocated the Mitchell and/or Seversky ideas. Spaatz, like Eisenhower, was evidently a great trouble-shooter.

With something in the neighborhood of 8000 Thunderbolts having rolled off the assembly lines by D-Day, attrition from accidents was expected to be high. Even as aerial battles raged mightily over the Continent on June 14, a P-47D-22-RE assigned to the 2nd Service Group pranged in Iceland when the landing gear failed to extend. Up to that time, "Big Bastard" was evidently in pristine condition. That leaden sky extended over 90 percent of Germany on that day, so France and Belgium caught the brunt of attacks. (USAAF)

Capt. Robert Johnson of Lawton, Okla., and his ETO crew chief were depicted with what the ace described as his final 56FG combat T-bolt mount, a P-47D-21-RA (AC43-25512, coded LM★Q) named "Penrod and Sam." Johnson had moved over to the 62FS and this fighter was assigned to him on 22 April 1944. S/Sgt. J.C. Penrod served the captain well, most certainly giving him great confidence in the reliability of his Thunderbolt; the "Sam" came from the captain's middle name, Samuel. The captain, unlike some others in the 56FG, retained the eight .50-cal. gun armament on every mission. His officially recorded victories in aerial combat totalled 28 after his return to the Z.I. on 6/6/44, subsequently revised to total 27. (C.A.M./Ed Boss)

reaching to their limits to provide penetration and withdrawal support for the bombers.

During the final four months of 1943, VIII Bomber Command had expanded tremendously in their determination to carry out the mandate of Operation POINTBLANK, the CBO. Starting in the same general period, General Kepner did everything possible to make up for lost time and a wrenching lack of trained and equipped long-range fighter squadrons. His men were committed to fighting with what they had and making the most of it. It was to be another "Fight at Odds." One can't help but wonder if General Arnold's words about destroying enemy air power wherever they could be found was his own wording or he was driving home the ideas of Doolittle and Kepner. The latter's "Down to Earth" assault scheme proved to be a good one and at the right time. In carrying out "Big Week," the heavy bombers (B-17s and B-24s) acted as a magnet for the German fighters; the AAF VIII FC was anticipating that, and they gave it all they had. Destruction of *Luftwaffe* fighters reached a pinnacle in the process.

Contrary to many beliefs about the "massive" size of VIII FC on 1 February 1944, the configuration of the command is provided here in detail, with groups shown in the order in which they had been assigned to the Eighth Air Force:

ACQUISITION SEQUENCE	FIGHTER GROUP	ASSIGNMENT DATE	FIGHTER TYPE
1	4FG	12 Sept. 1942	P-47C/D
2	78FG	29 Nov. 1942	P-47C/D
3	56FG	12 Jan. 1943	P-47C/D
4	353FG	7 June 1943	P-47D
5	352FG	6 July 1943	P-47D
6	355FG	6 July 1943	P-47D
7	356FG	25 Aug. 1943	P-47D
8	20FG	25 Aug. 1943	P-38H/J
9	55FG	16 Sep. 1943	P-38H/J
10	359FG	19 Oct. 1943	P-47D
11 out	358FG	20 Oct. 1943>	P-47D>
12	361FG	30 Nov. 1943	P-47D
11 in	357FG	31 Jan. 1944	P-51B<

> = to 9AF; < = from 9AF

Note: The 364FG, flying P-38Js, joined VIII FC on 10 Feb. 1944, followed by the 479FG, flying the same type, on 14 May 1944.

In the ETO College of Strategic and Tactical Knowledge, "Bold Stripe" Day was scheduled as a blind date. And come it would. And did. During the strenuous, combative final 90 days of 1943, VIII FC was certainly fighting with the odds against it. But two things were there to inspire the fighting men on: M/Gen. Kepner in command and the trickle-to-deluge of external drop tanks. After the lethargic reign of General Hunter in which there was no spirited attempt to rebuild the force of long-range fighters to defend the heavy bombers already in training at bases like Pyote, Texas, and scheduled to carry out the mandates of POINTBLANK, Bill Kepner was a real shot in the arm. With

OPPOSITE: Now and then some airplane mystery seems to defy resolution. This critically sharp crash scene involving a 353FG Thunderbolt illustrates such an enigma. The AAF caption incorrectly interpreted the airplane call number, showing it as a P-47D-26-RA which does not fit a razorback Jug. The well-defined tail number (28380) has been researched in depth as AC42-8380. This 28 April 1944 picture, then, reveals the enigma: We have a P-47D-2-RE with wing pylons and a late (D-23) cowling and engine installation, neither of which was on any production P-47D-2 airplane. Only two explanations seem valid: the fighter was rebuilt with a powerplant and major components from one or more late-model P-47D writeoffs, or it was totally rebuilt at a depot. Both tires appear to be flat/blown and the paper drop tanks seem to be undamaged. Evidence shows that the pilot survived, possibly uninjured. (AAF 69562AC)

the arrival of Doolittle as CG to encourage and support Kepner and the bomber wings, it was a whole new ballgame. The kickoff on 20 February established the pace.

Republic's P-47 could not win the ETO endurance race, but in the buildup of the Ninth Air Force a new crew of pilots were about to spur the Thunderbolts to new and vigorous heights of success. By a wide margin, the most prominent and dominant Allied fighter in the ETO was the Jug. By a typical count of noses, they outpointed combined P-38 and P-51 airplanes by a massive number. In addition to the fighter groups in VIII FC, IX FC had eight fighter groups on line. (See Chapter 14.)

Distortion of history is a criminal act. If it is written and has not been researched deeply, it can do irreparable harm because correction of damage is more difficult than original construction. As an example, the argument about WWII airplanes exceeding the speed of sound should never have become an argument at all. Proper research would have revealed it never came close to happening. (This book has taken the greatest of pains to put that old wives' tale into the ground forevermore.) Another argument stems from any writings in which the only *real* fighter in the AAF for World War II comes through as the Merlin-powered North American P-51. Considering what happened with the Hispano-Suiza 20-mm cannon, and the rapidity with which it came into production in America, we can only presume the British Purchasing Commission did not really understand the problem they were attempting to correct. Had they specified a Merlin engine for the as-yet-undesigned fighter they ordered, Packard could have manufactured them from an earlier starting date. Proper Mustangs would have existed for at least another year. But P-51B/C/D airplanes did not fight in the ETO or the MTO or any other place unti December 1943. Griffon-powered Mustangs should have existed by 1944. Merlin and Griffon-powered Lockheed P-38s should also have fought but for some major oversights. (That is why aircraft companies had Preliminary Design departments. It doesn't seem that Air Materiel was the roadblock to progress. Old company loyalties and ties seemed to have had more to do with decision making at the top levels than any desire to improve quality of fighting machines.)[2]

Probably one of the busiest places in the Allied world in April 1944 was B.A.D. (Burtonwood Air Depot) located in Lancaster, not far from Liverpool with its great port facilities. A newer depot was established at Warton in the same industrialized area. Late model razorback P-47s like this D-22-RE version (foreground) were assembled as required, given correct basic markings and sent off to appropriate fighter group stations. The camouflaged LM★V is an older P-47D-6-RE that was just processed through overhaul. In the background, there is an assortment of Mosquitos, Havocs, B-17s and others. Even one P-38J is in the pack, although more of those could be found at BRD/Speke and Langford Lodge. (Roger Freeman Coll.)

The distortion mentioned earlier merely added fuel to the kindling. The following quotation from a major 1960s aviation history book was planted in many minds and was probably used as research data later on: "*On January 11, 1944, eight hundred*

[2] Nobody need look further than the battle tank and artillery situations to understand. German *Panzer* divisions received improved armor far faster than American units did. Most new M-3 tanks still mounted spinoffs of the WWI French 75-mm cannon on D-Day. It would seem logical, if nothing else, to have copied the German 88-mm gun and ammunition in 1941 and put it in production with improvements. The U.S. Army had squandered a terrific opportunity in the 1930s when the Christy tank was rejected. The Japanese proved copying (with improvements) was remarkably successful. Best example: aircraft carriers and their airplanes. And had it not been for Gen. "Black Jack" Pershing pushing to get the .50-caliber machine gun in WWI, we would not have had it in WWII. However, it was rather old fashioned when compared to newer German weapons; Aberdeen Proving Ground and the arsenal did little to improve the products for two decades. The rate-of-fire and muzzle velocity should have increased by 50 percent in two decades.??

Rhubarbs often revealed that the enemy was shooting back with a vengeance. This P-47D-23-RA assigned to the 358FG was on such a mission when flak took out a couple of cylinders, rearranging the cowl lines. A hot oil bath could not do much to aid the pilot in attempting to return to base. But this one managed to do a great job in bringing his fighter home without additional damage. (Ray Bowers)

American bombers, escorted by Mustangs, struck three major German aircraft factories. On the same day the Mustangs claimed over 155 German airplanes in the air."

Pure, unadulterated FICTION. What did the official AAF record say about that day? It records that 663 heavy bombers (B-17s and B-24s) attacked targets in three major cities, with "fierce opposition...estimated at 500 fighters...encountered, with 60 heavy bombers lost." Jimmy Doolittle could not have been very happy at the end of that day. What else is wrong, wrong, wrong in that so-called *history*? Inasmuch as the Ninth Air Force only had one Mustang fighter group operational through January 1944 – the 354FG of IX FC, flying P-51Bs in support of the Eighth Air Force heavy bombers – how could Mustangs have escorted heavies to at least three different major target centers? The new 357FG was not yet operational for combat, and it most certainly had not yet been transferred to VIII FC maximum group strength. Would anyone, aware of combat conditions, believe something short of a maximum group strength of 75 Mustangs – the actual total was 44 – with little more than 30 days in operational combat status could escort nearly 700 heavy bombers to assorted targets while shooting down *155 Luftwaffe intercepting aircraft...nearly one out of every three e/a involved in the day's combat!?* If one group of fighters had to escort bombers to more than one target, it meant splitting the group in half and operating with

***One splendid P-47D-6-RE (LH★J) named "El Shafto" was evidently shafted by gremlins. Every pilot who has suffered this indignity can commiserate with the 350FS pilot, especially considering the area adjacent to the taxiway on which it happened. "Everybody, but just everybody, in the 353FG will know" was probably the plaintive cry. Rather interestingly, this -6 model was equipped with external stores pylons.* (AAF)**

At a meeting with Supreme Allied Commander, General Dwight Eisenhower, not many hours before Operation OVERLORD was launched, the top 8AF commanders were photographed for the historic occasion with their chief. Pictured (l.-r.) are Lt.Gen. Carl Spaatz, Lt.Gen. James Doolittle and Maj.Gen. William Kepner. (AFM)

about seven fighters in each partial squadron.

Here is just one bit of *corrected* history. The Thunderbolt-equipped 56FG was assigned to that 11 January *Ramrod*, with half the group going to Halberstadt/Oschersleben, the other half furnishing penetration escort to Brunswick. In actual fact, the 56FG "A" aircraft escorted the "Big Friends" to bomb an Fw 190 factory at Oschersleben. Near Minden, the leading bomber elements were bounced by at least fifty Me 109Gs and Fw 190As in a fierce attack. Seven bombers in the first box were knocked down on that initial pass. Thunderbolts (not Mustangs) of the 56th responded by dispatching ten e/a. Two probables and seven damaged aircraft were also claimed. Lieutenant Gus Schiltz of the 63FS scored his second triple kill that day. Escorting other bombers to Brunswick, the 56FG "B" aircraft did not encounter any German aircraft, so their mission proved to be uneventful. On that same day, the 354FG of IX FC, flying P-51B Mustangs, claimed 11 e/a shot down and no fewer than 8 probables. What P-5ls could have accounted for the other 144 *Luftwaffe* fighters claimed as aerial victories?

In actual fact, the 4th, 78th, 353rd, 352nd, 355th, 356th, 358th, 359th and 361st Fighter Groups – all still flying those Jugs – dispatched 499 Thunderbolts that day to escort the bomber groups. And a check of 20FG and 55FG mission reports reveals they were providing escort over the target areas with 49 of their twin-boom Lightnings. Is it any wonder many young people grew up with distorted viewpoints of what actually occurred in the air war over Continental Europe? Just for emphasis, more P-38s were on escort duty that day than P-51 Mustangs, and there were eleven times as many Thunderbolts protecting the bombers. Official records were there all the time. Failure to use them creates a fraud.

O, what a tangled web we weave/When first we practice to deceive. – Scott.

And just to add one parting touch, the first two 700-plane heavy bomber raids were launched on 29 and 30 January. On the 29th, no fewer than 763 bombers, led by Pathfinders, attacked Frankfurt/Main while 46 others hit a target of opportunity (T/O)...Ludwigshafen. German intercepting fighters and flak downed 29 of the bombers, but the total VIII FC escort tally for the day for penetration and withdrawal (main target) was an impressive 42 kills. Of those, the 56FG claimed 6 e/a destroyed. On the very next day, 701 bombers struck at Brunswick, bombing through the overcast with radar. A block of 51 bombers went to T/Os, including the city of Hannover. While 20 bombers were lost, VIII FC aircraft – mostly Thunderbolts, for the record – claimed destruction of at least 45 Messerschmitts and Focke-Wulfs. Rather remarkably, total 56FG losses for the two days amounted to one a/c missing in action. Was the Thunderbolt a first-rate fighter? Assuming most of those claims were verified, one wonders if any RAF Spitfire group ever had a 45:1 kill/loss ratio against fighters...not bombers or transports.

Serious heavy bomber losses at the beginning of January 1944 were quickly followed by nearly a month of very bad weather over the Continent. As a result, VIIIth Air Force bomber missions were seriously curtailed. Toward the end of the month, that began to change. In the meantime, General Eisenhower moved into the slot as Supreme Commander, Allied Expeditionary Forces. That certainly telegraphed a message to Hitler, if nothing else. Other organizational changes

Everybody subject to orders issued by Hq. at WIDEWING would certainly have removed the Invasion stripes atop wings and fuselages within 7 to 10 days after D-Day. If anyone still doubts that bubble-canopy P-47Ds were then operational, here is Lt. Col. Gabreski heading out for what probably was a D-Day mission (freshly painted stripes, over and under) on June 6 in his new P-47D-25-RE. The airplane had been painted by the 56FG or the 33rd Service Group maintenance people and the pilot had certainly checked out in the newest model, all of which took time. (USAAF)

occurring during the month included assignment of General Nathan Twining to the post of CG, Fifteenth Air Force. General Carl Spaatz took command of United States Strategic Air Forces in Europe with HQ at Bushy Park (WIDEWING), and Doolittle moved into his new post commanding the Eighth Air Force. The actual command in the MTO assumed by General Ira Eaker was CG, MAAF, with Spaatz and Air Chief Marshal Tedder moving to England. In this same general timeframe, the Eighth AF finally reached its authorized strength of 40 heavy bomb groups (officially as of 6 June). Rapid fire changes were occurring as OVERLORD became a reality, the greatest invasion in history.

At about the time Doolittle took command of 8AF in England, a new report indicated growth in Luftwaffe defensive forces threatened the CBO. Concentration on fighter production, plus factory dispersal, resulted in increased manufacturing rates, with forces being bolstered by fighters being transferred from the Eastern to the Western front. In February, disruption of enemy communication lines and overall destruction of German air power – amounting to gaining command of the air – became high priorities on a revised Combined Chiefs of Staff directive. Bombing intensity, starting about the end of January, increased tremendously in the following three months. Typically, more than 1000 heavy bombers were dispatched on 20 February to bomb German fighter aircraft production centers. Ninth Air Force reorganizations were occurring rapidly to cope with the major expansion proceeding.

According to accounts by B/Gen. Francis "Butch" Griswold, the primary role of VIII FC had been, up to early March 1944, protection of the Eighth AF heavy (and medium) bomber forces involved in carrying out POINTBLANK. Long-range escort may have been a priority, but until substantial numbers of Mustangs came into the picture, the extremely limited numbers of P-38s available as of 1 January 1944 made that priority a farce. General Arnold's New Year statement about destruction of enemy aircraft in every venue, including on the ground, was turned into action by Generals Kepner and Griswold early in March. After each group had successfully accomplished its protective mission, it was given full authority to go "Down to Earth" in an all-out assault against German air power. But let us not forget, as will be seen in the next chapter, such action was already the primary role of the Ninth Air Force as provided for in its Tactical Air Support makeup. Operating from the United Kingdom, IX Fighter Command was in the process of developing strafing, dive-bombing and skip-bombing techniques for their fighters, already being established in the fighter-bomber role.

On the morning of D-Day, Capt. Dewey E. Newhart's P-47D-15-RE (LH★Q; properly Q bar) was fitted out with a local production 108+/- gallon drop tank and a pair of 250-lb. bombs. On the port side, Newhart's marking was "Mud 'n Mules" but the round blank on the starboard side indicated that marking work was incomplete. "Arkansas Traveler" had been applied earlier. Capt. Newhart was lost on 12 June while flying Lt. Col. Wayne Blickenstaff's airplane and leading the 350FS on a Rhubarb. (USAAF 69085AC)

(Even in official reports near the end of March 1944, VIII FC Thunderbolts were being referred to as fighter-bombers.)

As of 13 April, Eisenhower's SHAEF assumed command over all AEAF, RAF, Bomber Command, and USSTAF, and even Allied Naval Forces. The bombing campaign was being stepped up radically, weather cooperating. On 12 May, Eighth AF launched 800 heavy bombers against German targets as the Fifteenth AF put a nearly equal number into the air, their greatest effort to date. With about 430 Luftwaffe interceptors attacking the 8AF formations, 46 of the heavy bomber force airplanes were lost. (Losses not reported for the 15AF units.) Even with very bad weather blotting out primary targets through a lengthy period in June, Eighth Air Force managed to launch 1314 heavy bombers on the 14th, incredibly followed by 1225 heavy bombers the very next day. The German oil industry had, by command of Gen. Spaatz, been given top priority a few weeks earlier, but bad weather had been a feature of the period. With a break in the weather, Ninth Air Force put 1000 fighter-bombers into the air on 20 June.

It wasn't just "June breaking out all over," according to the Broadway tune, there was a world of stripes to be seen...everywhere. The only thing which could have been better was a red, white and blue motif instead of black and white. Nevertheless, striping proved to be one of the more successful, if infrequently recognized, projects of WWII.

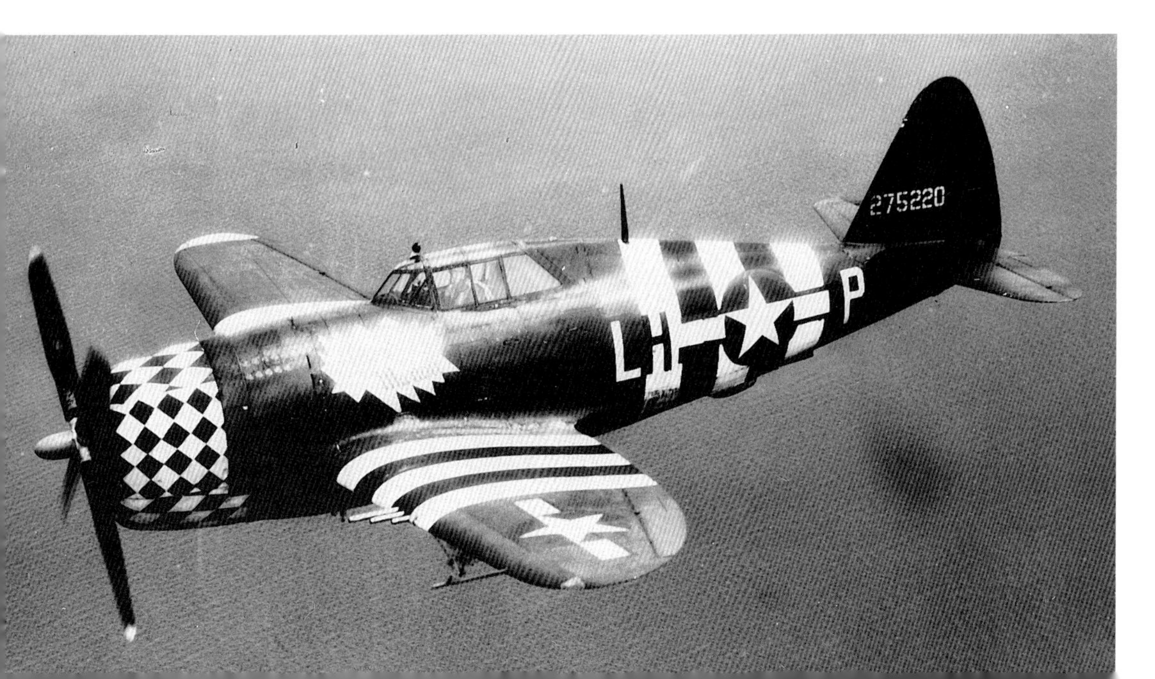

What appeared to be a simple noseover was certainly anything but that. A missing right flap; a 108-gal. belly tank full of holes; a badly distorted left wing trailing edge; a large slash in the canopy glass; medics and firefighters very active, and the canopy is obviously jammed, all of which screams EXPLOSION! This 358FG Republic P-47D-20-RE, carrying the CP★D code and full OVERLORD markings most certainly had a bomb fall off and explode as the aircraft was taking off or landing. With firefighters spreading foam on gasoline, a crewman on the wing is talking to a trapped (and injured?) pilot as medics stand by. Transferred from 8AF to 9AF at end of Jan. 1944, they were operating from this Marston-mat strip on the south coast on D-Day. (R. L. Cavanagh Coll.)

OPPOSITE BELOW: A paucity of comments would lead anyone to believe that nobody ever noticed how few WWII in-flight photographs have surfaced which show fighters of VIII FC and IX FC with full invasion striping; i.e., stripes on wing upper surfaces and top half of the fuselage. There were probably two key reasons: a.) Action was so intense that few thought of taking pictures of wingmen and, b.) an order came down in the first week after D-Day to remove stripes from all upper surface areas. One of the rarities is this P-47D-11-RE coded LH★P and named "Maid Marion" (although reflections blotted that out). This 350FS Jug had a badly mismatched cowling arrangement and a paddle blade propeller. (Wm. A. Tanner)

BELOW: Having already become a double ace plus in the 62FS, Lt. Fred Christensen attained special celebrity status on 7 July 1944 with six confirmed aerial victories on one mission, a Ramrod to Leipzig. Having joined the 56FG as a replacement pilot in August 1943, Christensen is shown with his P-47D-25-RE "Miss Fire" (LM★C) soon after the especially successful July shootdowns. (AAF 12413AC)

BELOW: Although it could hardly be classified in the weapons category, the field conversion of a War Weary P-47D-11-RE was extremely interesting. With no intention of creating a "crew trainer" in the true sense, 56FG mechanics managed to convert one of their older fighters to carry VIPs and other personnel for a variety of reasons. Named "Category E," this Doublebolt provided full accommodation for a second crew member. It appears to have been equipped with a late series engine and a paddle-blade propeller. Col. Schilling was the back seat occupant when this picture was taken. That pre-historic looking device in the foreground is the battery cart for starting. (56th Fighter Group)

Shortly after the 56FG moved to Boxted (AAF Sta. 150) in April 1944, three of the key officers in the group were photographed walking on the airfield perimeter track with a new record breaker, Lt. Fred Christensen (r.). The lieutenant had just become the first fighter pilot in the ETO to shoot down six e.a. on just one mission (7 July 1944), had a final total of 21½ aerial victories to his credit. Pivotal men in the famed 56th were (l.-r.) Col. Hub Zemke, Lt. Col. Dave Schilling and Lt. Col. Francis Gabreski. The latter officer went MIA/POW just days after this picture was taken. (56FG)

RIGHT AND BELOW: A couple of 78FG T-bolts assigned to the 84FS displayed full group and temporary invasion markings, but on entirely different backgrounds. The late-series P-47D-22-RE "Zombie" showing code WZ★V had been delivered sans camouflage; the P-47D-11-RE lacked a personal name but was coded WZ★K, all applied over camouflage paint that had received many coats of wax in its career. (Ray Bowers)

There were times when Thunderbolts did not seem overly large, but for some reason Major Lucian "Pete" Dade, Jr., looked more like he was driving a Grumman TBF torpedo bomber in this early June 1944 photo. At that time he commanded the 62FS. It is possible that this picture was taken on D-Day because the invasion stripes look fresh on Pete's P-47D-25-RE. A few months later, Dade became group commander and frequently was seen at the controls of a P-47M. His official victory record was five enemy aircraft destroyed in combat. (Robt. Cavanaugh Coll.)

The look on Capt. B.I. Mayo's face seems to reflect that nicely lettered "No Guts-No Glory" sign on his natural-finish P-47D flown in a 78FG squadron. At the time, Mayo had symbols of two confirmed victories over e.a. exhibited on his fighter. (78FG)

BELOW: A 353FG mission on 13 August 1944 was not conducted without some expected difficulties. "Smoocher," a P-47D-21-RA of the 351FS, had taken a hit, just barely made it to the airfield. An AAF wrecker was moved into place to salvage the fighter. The pilot was uninjured in the crash landing. (USAAF 69593AC)

This P-47D-23-RA was among the last of the production P-47 "razorbacks," produced at Evansville and operated by the 82FS of the high-scoring 78th Fighter Group based at Duxford in August 1944. The pilot flew up close to a B-17 on the return to the U.K. and was photographed by a crewmember of the bomber. (USAAF)

BELOW: Woodsman, spare my trees. Lt. Roy T. Fling, flying with the 56FG, evidently tried to "clearcut" a forest with his P-47D. With all oil cooler and supercharger intercooling air cut off, not to mention a great percentage of cooling air for the P&W Double Wasp engine, Fling managed to bring his mount home. Unfortunately, this intrepid aviator was subsequently KIA. (AFM)

ABOVE: Sometimes pilots got overly aggressive in their strafing runs and tried to bulldoze their way out. This P-47D plowed through a stand of trees rather than climb into a hail of withering flak sizzling overhead. One or two trees fought back. The 56FG pilot from the 62FS had to belly in at Manston minus hydraulics and with only minimal fuel remaining. One of Disney's "Snow White" dwarfs is featured on the starboard engine cowling. (Ray Bowers)

Production of bubble canopy K-type P-47D (-25 and subs) Thunderbolts was rolling at high pitch by mid June 1944. Quite unexpectedly, the censor awakened and obliterated constructor's numbers for the first time in months. However, a few partials were to be seen and led to easy identifications. Beginning with the D-28 block, technically identical models began to flow from Farmingdale and Evansville concurrently. Royal Air Force allocations of Thunderbolt II (P-47D-28s here) fuselages are to be seen in the foreground. (Mfr. H-9012)

BELOW: Oh, but to possess an original Kodachrome of this magnificent scene of visiting 78FG Thunderbolts on temporary relocation at Bassingbourn (AAF Sta. 121). The P-47D-27-RE, MX★S, usually flown by (then) Maj. Joseph E. Myers, in the foreground is depicted with two 108-gal. metal wing tanks and a 150-gal. belly tank, obviously set up for a maximum-range Ramrod in midsummer 1944. Second in line is a P-47D-22-RE, "Miss Behave," the second-last -22 built. There were very few razorbacks operating with the group at the time. Bassingbourn was the home of the 91BG. (AAF 55432AC)

Some days it did not pay to get up. That paper l08-gallon drop tank may be the clue that a short Ramrod mission was scheduled but ended abruptly with inadvertent brake application during taxi out. It usually subjected the crew to extra work changing the propeller, engine and part of the cowling. That was cause for scowling. (Ray Bowers)

BELOW: In the valued opinion of Col. Hub Zemke, Col. Frederick C. Gray of the 78FG was one of the finest group commanders in the ETO. The group got off to a bumpy start when all of its P-38 Lightnings and many pilots were stripped from VIII FC for Operation TORCH replacement duty in February 1943. Switching to some of the first P-47Cs to arrive in the ETO, the group lost two commanders in one month – July 1943. Freddy Gray took command on 22 May 1944 for the critical Invasion months of that year, and his squadrons earned a DUC for action in support of airborne troops in Holland (September). The 78th was the only 8AF outfit to fly all three main fighter types (P-38, P-47 and P-51) in combat operations. (AAF 72335AC)

BELOW: During Col. Freddy Gray's watch as commander of the 78FG, it became routine to apply camouflage paint in a standardized pattern to all P-47Ds received in natural finish from the factory, be it Farmingdale or Evansville. MX★E was a P-47D-25-RE that had been stripped of its wing pylons. (G.L. Letzter)

One of several T-bolts bearing the name "Dove of Peace" came to a temporary operational halt in this field some time before D-Day when the 353FG commander, Col. Glenn E. Duncan, lost power. His current mount at that time was a P-47D-21-RE coded LH★X. The first two commanders of the 353rd had gone MIA and POW, in that order, and on 7 July it was Col. Duncan who became an MIA pilot. Since it was this group that pioneered P-47 dive-bombing, glide-bombing and skip-bombing techniques standardized by 8AF and 9AF units, it was common for their aircraft to sustain flak damage. Col. Duncan was able to evade capture, returning to command the group again in April 1945 after some R&R. Notice that in a black and white photo, early black and yellow checkerboard cowling markings looked just like 78FG markings. (Ray Bowers)

Alexander Kartveli, unhappy when belly drop tanks were appended to his pride and joy, must have been absolutely distraught when additional mounting points were added inboard of the wing pylons that he had opposed. With the arrival of 4.5-inch Bazooka rocket launchers for Army ground troops, a new weapon came into being very quickly; it was the triple-tube launcher installed on simple supports immediately beneath the quad machine gun installation on each wing panel. This particular add-on weapon seen on a [illegible] Jug provided the P-47s with yet another combat dimension. (Fred LeFebre)

Following a successful strike mission in the morning hours, the 56FG headed off on another Rhubarb/Rodeo combination at 1609 hours on 5 September 1944. Arriving in an area northeast of Frankfurt, they attacked Gelnhausen's airfield where more than fifty aircraft were spotted. The 62 and 63FS claimed 47 aircraft destroyed on the ground plus 16 damaged. Two hangars destroyed may have contained other aircraft. The two missions that day resulted in claims of 78 e.a. destroyed on the ground and in the air plus 19 damaged. Total 8AF claims for the day were 172 destroyed and 87 damaged. The excellent strike photo of Gelnhausen Airdrome was taken during the action. (AFM)

ABOVE AND RIGHT: As the assault on Festung Europa was carried on with great intensity by the 8th and 9th Air Forces and the RAF after D-Day, this 361FS Republic P-47D-22-RE was headed out on a Ramrod or a Rodeo deep in France or even into Germany. "Angel Eyes" carries two 108-gallon metal drop tanks, the design virtually duplicating the original paper drop tanks developed in May 1943 by the Air Technical Section at Bovingdon. (Campbell Arch.)

When the XP-47B was designed, nobody ever conceived the idea that production versions of the high-altitude interceptor would ever stoop to intercepting locomotives and boxcars. First Lieutenant Kenneth Chetwood (seen here with his P-47D-22-RE "Maggie IV") was one of the original members of "Bill's Buzz Boys." The 353FG's Lt. Col. Glenn E. Duncan (Chetwood's commander) had proposed development of dive-bombing techniques to Maj.Gen. "Bill" Kepner. The general approved formation of a unit to test the procedures. The famed "Down to Earth" policy was born, dooming the Luftwaffe. At the time the picture was taken, Chetwood had destroyed fourteen Nazi locomotives. (William Tanner)

While one armorer swabs out the gun barrel of one of the eight .50-cal. Brownings on a 78FG Jug at AAF Sta. 357 (Duxford), two others string the expendable-link cartridges in their channels in the outboard wing structure. The large size of the piano-hinged access door is evident in this photo. Col. Hub Zemke had one unpleasant experience of returning from a combat mission with most of the starboard door (remains) in this position after a rather vicious onslaught by a Luftwaffe fighter. In this position, the door serves as a massive asymmetrical dive brake. (S. Haddock Coll.)

Most people associated with WWII aviation history believe that great numbers of P-47D-25-REs were produced, but that would be incorrect. Of all the production blocks of bubble canopy D models produced at Farmingdale, the -25 block proved to be the smallest with only 385 built. One unexplained mystery: with numerous technical changes and a major appearance change, why were these airplanes still "D" models instead of being "K" models? North American's P-51B/C models were virtually indistinguishable, and P-5lDs only differed from "K" versions in propeller installation. Both companies had dual fighter production factilities, so the answer did not lie there. Whatever dictated the situation has been lost in the sands of time. (Art Krieger Coll.)

Not many days after Gabreski flew his D-25 airplane on an OVERLORD mission, the Jug (HV★A) was leading a pack of 61FS Thunderbolts that included everything from a D-22-RE to a D-28-RA. Expressing the status of the 56th Fighter Group, it is notable that all but one of the fighters had been "camouflage" painted, evidently to individual pilot specifications. Whatever any publication states, there can be no doubt that bubble canopy Jugs were flying over the Continent in June 1944. Gabe's "A" may be seen above and leading HV★H, with a lone razorback D nearest the camera a/c. (USAAF)

RIGHT AND BELOW: A few pilots of the 56FG tried out the 4.5-inch triple rocket tube installation with mixed results. Capt. Donovan Smith of the 61FS and his group commander, Col. Dave Schilling, removed their wing tank/bomb pylons when carrying the tube clusters. Since the key ro of the 56th was bomber escort and air-to-air combat, they quickly learned that the rocket armament in that form could be a penalty, not an asset. Smith's P-47D-26-RA, coded HV★S and named "Ole' Cock III," is shown at Boxted.

Although Alexander Kartveli may never have welcomed the "bubble" member of his Thunderbolt family with enthusiasm, pilots must have looked upon the improvements as they would on a new car model. Nostalgia aside, the clearview canopy version looked right. Better vision through a flat-plate windshield and a 360-degree field, a saving in fabrication manhours and no real loss in top speed – some claim it was faster – made the new models more popular than anticipated. Lt.Col. Philip E. Tukey, Jr., new C.O. of the 356FG based at Marltesham Heath, flew this camouflaged (but waxed) P-47D-25-RE. Tukey left the group before they switched to P-51 Mustangs in November 1944. (R.L. Cavanagh Coll.)

One experimental installation that seemed to offer merit for the ground attack mission was a pair of 20-mm Oldsmobile-built Hispano cannons hung on the bomb/tank pylon shackles to provide explosive shell firepower. Col. Fred Gray's aggressive 78FG mounted the cannons on a single P-47D-25-RE at AAF Station F-357 (Duxford) about 24 October 1944. This complete installation had been created by Technical Operations at Bovingdon, including streamlined fairings that were never installed. Unfortunately, drag and weight reduced top speed by nearly 50 mph; without adding significant punch, they became a detriment in air-to-air combat. The aircraft group code was HL★Z. (USAAF)

OPPOSITE BELOW: Certain to create confusion among the "experts" in aviation history is this new P-47D-25-RE that was assigned to Capt. William Tanner and promptly named "Prudence V" by that ace. The reason that this Jug, LH★O (AC42-26472), is a conundrum centers on the fact that all D-25s supposedly were equipped with Hamilton-Standard Hydromatic propellers. This airplane obviously has a Curtiss Electric prop, complete with cuffs. This photo evidence came from...the 353FG pilot. (William Tanner)

Quite unlike his predecessor in VIII Fighter Command, M/Gen. William Kepner participated in numerous missions (24 recorded). Two-thirds of those missions were flown in this Thunderbolt named "Kokomo," a P-47D-25-RE. At the time it was photographed it was a Division Control aircraft. The general, not wishing to incur the wrath of Eaker or (subsequently) Doolittle, was probably assiduous in avoiding actual combat missions. (U.S. Air Force)

ABOVE: A time came when it was almost unusual to see a 56FG bubble-canopied P-47D in natural metal finish. "Suzie," a P-47D-28-RE operating with the 63FS, was one of those rarities. This fighter was pictured flying at relatively low level without any external load. The pilot and location were not identified. (AAF)

ABOVE RIGHT: With dive-bombing techniques developed by the 9AF and pretty much adopted as standard for all ETO-based Thunderbolts in 1944, death and destruction was heaped on Luftwaffe airfields, all forms of German ground, rail and water transport and, of course, Wehrmacht and Panzer forces. A German troop and support equipment train was interdicted at a depressed curve at Gisers, France, by five P-47s with these results. (AAF)

RIGHT: For unknown reasons this P-47M-I-RE was at Newark Airport after VE Day was history, never having been shipped to the ETO. It had been completed by mid October 1944 and was a low-time airplane as of 30 April 1945. These "sprint" Jugs were capable of attaining a speed of 473 mph under optimum conditions at a critical altitude of 32,000 feet. With improved propeller blades like those installed on most P-47D-40-RA airplanes, P-47Ms might have exceeded even those figures. (Author's Coll.)

Flying in precision formation under the leadership of Maj. Harold "Bunny" Comstock, 63FS flights maintained precision formation for this excellent portrait over a landscape dotted with cottonball clouds. "Happy Warrior," coded UN★Z, is a P-47D-28-RA flown by Comstock (?) in Colonel Zemke's absence in mid 1944. Although it appears that two aircraft have the code letter R, one is actually R̲ (R bar) to denote the second P-47 in the squadron with the suffix R. (This original print is marked "UN-Z leader, P.J. Conger," and it may be that Comstock was actually in UN★V) (USAAF)

A pair of P-47M-1-REs, with Lt. Col. Lucian M. Dade, Jr. (more widely known as Pete Dade) in the lead aircraft, take off into hazy skies and most likely headed for the Continent. P-47Ms rarely had wing pylons installed to carry tanks as long as they were assigned the task of chasing V-1 missiles over England. An earlier report that the M models were P-47Ns fitted with D series wings was incorrect. More accurately, they were P-47D-30-REs fitted with new R-2800C-series engines (initially -57s, but some received different dash number engines) and the G.E. CH-5 turbosupercharger. Each of these rather gaudy P-47Ms carries a pair of 108-gallon metal droptanks. (AAF via NASM)

A team of the "Good Samaritan" air/sea rescue Thunderbolts, a WW (War Weary) P-47D-15-RE in the lead, head for their respective English Channel stations (below that cloud cover, of course) where they will await appropriate "May Day" calls. Equipped with a 108-gallon drop tank and smoke markers under the fuselage, plus dinghy packs on the wing shackles, they would remain on station as long as weather and fuel permitted. Marked as 5X★A, etc., the 25 or so Jugs were from Detachment B of the 65FW (later becoming the 5th Emergency Rescue Squadron.) (Roger Freeman Coll.)

Marked with red white and blue cowling rings and yellow tail bands, this war weary P-47D-5-RE and two dozen brothers served in the 65FW, Detachment B, as emergency rescue service aircraft under Gen. Doolittle's 8AF. With extra fuel tanks, dinghy packs and smoke markers, they patrolled over the English Channel from May 1944, speeding to the assistance of any downed airmen. The national insignia used by detachment P-47s were undoubtedly the largest used on any fighter. The circle was not even continued onto the aileron. All of the rescue P-47Ds were re-engined with new R-2800-59 powerplants. (Ray Bowers)

CHAPTER 14

DOWN TO EARTH ACTION

There is widely held belief that the "Down to Earth" concept adopted by VIII Fighter Command in March 1944 was an original idea put forth by M/Gen. William Kepner or possibly by B/Gen. Francis "Butch" Griswold, by that time commanding VIII FC. Actually, the words about carrying the air war to Germany in a manner which would gain Allied control of the air over *Festung Europa* came from a speech delivered by Lt.Gen. H.H. "Hap" Arnold. As 1944 dawned, he spoke of carrying out the battle against the *Luftwaffe* in every venue, including on the ground. Not only that, the Ninth Air Force had been restored to life in a massive way to serve as the tactical air weapon in carrying out Arnold's expressed wishes. At the same time, the top priority job of VIII FC, according to the published words of General Griswold, was to "escort the Fortress and Liberator bombers in their mission to succeed with Operation POINTBLANK." From the outset of its planned rebirth as directed by Arnold on 25 August 1943, the

Major Wayne Stout, leader of the 509FS, shields his eyes against a late afternoon sun as an Army Intelligence officer takes photographs. Considering the conditions of field operation in a combat zone, the P-47D-28-RE is remarkably well maintained. The location at that time was described as A-64 at St. Dizier, France. (William Swisher)

BELOW: Although it was the VIII FC of the 8AF that brought fame to Republic Thunderbolts, after D-Day 1944 it would largely fall to the 9AF and 12AF groups to keep the P-47 mystique alive and well. The Doolittle-Kepner policy aimed at making the North American P-51 the prime long-range escort fighter aircraft of VIII FC was the result of a dispassionate decision to simplify logistics and center on a new state-of-the-art fighter. This immaculate P-47D-28-RA (6V★R) with blue cowl ring belonged to the 36FG's 53FS at Brucheville or Le Mans, France. The Engineers who prepared this forward airstrip with its Marston mats did an exceptionally fine job. (U.S. Army Signal Corps)

Ninth Air Force/IX FC was to function as a tactical air force under command of M/Gen. Lewis H. Brereton. Reactivation of 9AF in the United Kingdom after lengthy service in the MTO occurred on 16 October 1943. Headquarters was established at Sunninghill Park close to Ascot, the famous race track. Just as AJAX was the code name for VIII FC headquarters, GANGWAY became the code name for Ninth Air Force HQ. B/Gen. Victor Strahm was still Brereton's Chief of Staff at that time, as he had been for a lengthy period in the MTO.

B/Gen. Elwood R. "Pete" Quesada commanded IX Fighter Command from the outset and Col. Benjamin Kelsey was committed to move into the slot as his deputy in the immediate future. (That never happened because General Doolittle specifically asked for Kelsey's return to Bovingdon to command the 8AF Technical Operations unit.) Quesada's HQ was established at Middle Wallop, but only one tactical reconnaissance group comprised the command at that time. Activation of fighter wings began in earnest in November. Although the initial approach was to build the force with North American Mustangs, General Doolittle's wishes impacted the idea, with the resulting heavy assignment of Lockheed P-38s and Republic P-47s to the command. The first P-47 T-bolt group to join the new IX FC was the 362FG equipped with P-47Ds.

OPPOSITE: It may have been referred to as a "full vision" canopy, but any pilot who has sat behind that 2800-cubic-inch radial will be glad to tell you there is no way you can see around it. Taxi strips on Continental airfields were frequently narrow (and crowded), and getting a wheel off into the mud could hold up a mission. The answer was to use a guide riding the wing. Multiple ace Major Glenn T. Eagleston of the 354FG only watches the taxi strip peripherally, concentrating on hand signals from his enlisted "wingman." It would appear that Eagleston was returning from a mission as nothing is carried externally. He was the highest-scoring ace in the Ninth Air Force with 18½ confirmed aerial victories. His nicely decorated P-47D-30-RE (FT★E) was pictured at St. Dizier (A-64) or Rosieres-en-Haye (A-98), France, in December 1944. (AAF)

So it was, IX Fighter Command was moving into the "Down to Earth" role, albeit not very rapidly. Through January 1944, the only operational group was the 354FG, equipped with P-51Bs and assigned to the heavy bomber escort role in support of the 8AF and Operation POINTBLANK. As if for emphasis, the new (not yet operational) 357FG, equipped with P-51 Mustangs, was traded to the 8AF in exchange for a fully operational Thunderbolt group, the 358FG. This new group proceeded to fly its first mission as a component of the IX FC on 3 February 1944, a little over a month before Kepner and Griswold took up the cudgel to pummel every Nazi object caught on the ground, moving or static. A clear pattern had been established; Mustangs were not going to populate many groups in either the Ninth or Twelfth Air Forces. Originally designed to serve as high-altitude interceptors, P-47 Thunderbolts were about to become, literally, "ground pounders," but not in the usual sense as applied to the infantry. Personnel in those two air forces were about to live in much the same manner as infantrymen, artillerymen and tankers. Spit and polish were going to be something men might yearn for, not to mention civilized meals.

As the new groups assigned to the Ninth Air Force were

Focke-Wulf Fw 190 fighters like this Fw 190A-5 were produced in great numbers by the Germans in widely dispersed component factory units. This series of the Fw 190 had a top speed of about 382 mph at 20,000 feet, but versions could attain 418 mph on override boost (1 min.). Typical speeds of 400+ (w/injection boost) were attainable, but even with two 66-Imp. gallon drop tanks installed, range was less than 1000 miles (second photo). A fan-cooled radial engine and paddle blades on the prop were in use in 1941.

being brought up to operational status that winter, anyone who had failed to apply himself to navigation studies, formation flying and fighter proficiency during the previous months in the Z.I. was in for a rude awakening. The "Jerries" were anything but happy with the tides of war at the time, and if the largest invasion in history became a successful reality those Germans were not going to favor looking the prospects in the face. Perhaps worse, the weather in Western Europe in winter could provide many unpleasant flying conditions at the most inopportune moments. It was the sort of adventure few were likely to look upon with fond memories as time passed.

Meanwhile, the RAF was pounding Germany every night, and 1000-plane raids were no longer the rarest of rare events. Eighth Air Force heavy bomber group strength was finally up to planned levels, and production of North American P-51s needed for escort duties was already at peak levels. The command decision had been made to not only allow VIII FC fighter pilots to go down on the deck and strafe targets of opportunity but to encourage such action whenever the escort duty assignment ended for any given mission. That major policy change went into effect early in March 1944. In fact it became rather commonplace for command decisions to come down wherein the mission assignment called for a low-level strafing mission to be conducted against specific targets. One such assignment, among the earliest, was levied on the 352FG – flying out of Bodney – on 11 March. The mission of the group that day was a low-level sweep over the Pas de Calais area of France. What started so unobtrusively was destined to become what many 352nd pilots would eternally consider to be their most memorable and dangerous mission of the entire war.

The group, commanded by Col. Joseph L. Mason, had originally gone into combat in 1943 flying P-47D-5-RE razorback Thunderbolts, while the most modern versions in operation with the group were P-47D-16-REs as March came in like a sabre-toothed lamb. Rumor had it that replacement P-51B and D Mustangs were scheduled to replace the entire complement of Thunderbolts within days. Indeed, the first seven P-51B Mustangs had gone to the 486FS on 1 March, providing the mechanics and pilots with a mix and match aircraft situation.

During the morning briefing they learned their objective was to gauge the location and strength of German defensive flak batteries. Not only did they have a great opportunity to evaluate the defensive power in the area, they nearly proved it was impenetrable. The group put up thirty-eight P-47Ds that day, and they began to encounter ground fire while still some distance from the coastline. By the time they made landfall, they were swathed in a virtual curtain of accurate flak. Any thought of completing (or even beginning) their strafing assignment disappeared in seconds in the face of such accurate shooting. Nearly every aircraft suffered battle damage immediately, with the 487FS losing Lieutenant Harold S. Riley while Lt. Ralph Hamilton suffered wounds. For pilots normally involved in escorting heavy bombers and engaging enemy aircraft at altitude, the reception was unexpected and akin to finding yourself walking in a minefield.

At a later date, Major William T. Whisner – 15.5 victories plus 5.5 in Korea – flying P-47D-5-RE (AC42-8404) HO★W recalled the episode with some distaste:

"That mission must rank as one of my most memorable because I even remember the date, March 11th. We flew the mission in minimal weather because the 'powers that be' wanted to check out the flak defenses in that area. Each squadron was assigned its own target. We went in low, never getting over 500 feet after takeoff. We flew down across England and the Channel, and even before we pulled up over the cliffs of France we were taking fire from their 88-mm shore batteries. We hadn't fooled anybody. I don't know what kind of intelligence the Germans had in England at that time, but...I'm telling you, brother, we kept getting fire until the whole thing had been

It might be a good bet to assume that the condition of those perforated steel planks – known everywhere as Marston mats or PSP – on this advance airfield on the Continent indicates they were just installed. The P-47D-22-RE getting ready to roll is watched closely by the Caterpillar tractor driver from the Engineer Corps. (9th Air Force)

Lieutenant Donn Madden's "Hazel Honey-2" heads out from St. Marie Dement, France, in August 1944 for a strike mission in support of ground troops. This 36FG Republic P-47D-26-RA is equipped with a 150-gal. belly tank and a couple of the useful GP (general purpose) bombs. The tail legend is "Easy's Angels" on a yellow background that matches the cowling behind the blue nose ring. Strangely, the P-47D razorback following this 23rd Fighter Sqdn. appears to have a white tail and a cowl to match, something more likely to be seen in the 5AF areas. (Donn Madden)

aborted. I had never seen such ground fire in my life. We never even got to our targets. Ours (the 487FS) was an airfield, and as we approached the area, 'Wheels' Riley on my left went down. He went right straight in, making a furrow a half-mile long.

"Tracers were coming from all directions and all hell broke loose. Soon afterwards, we aborted and got out of there. I guess we accomplished our mission because we sure found out there were heavy, very heavy anti-aircraft defenses in the area. We also learned something else (not expected) – they sure knew we were comin' and had 'baked us a cake.' In addition to (losing) Riley, Bill Schwenke of the 328th squadron was killed in a Channel crash while trying to get back to Bodney." (Whisner was the recipient of the Distinguished Service Cross, w/OLC, and the Silver Star during WWII.)

Lt. Schwenke's fighter was hit as he and Jack Thornell were strafing Targets of Opportunity. When interviewed considerably later, Major John F. Thornell, Jr., – 17.25 aerial victories – a member of the 328FS and flying P-47D (PE★I), named "Patti Ann", was able to provide an eye-witness account of the entire episode:

"We crossed the enemy coastline at 08:58. I was flying Red Two and we continued on course at almost zero altitude. Along the way I fired at gun positions, a radio station, at enemy troops and a B-17 laying in a plowed field with its wheels up after a crash landing. It appeared to be in fair condition. I gave it a long burst...noticed it had First Bomb Division markings – triangles on it wing and tail.

"Then we approached a gun position on the hill, and that's where Schwenke got hit. I saw strikes on his engine and when we crossed out over the coastline together, his ship was smoking. He called to me to say 'I'm bailing out...get a fix[1] on me.' We climbed to about 1000 feet and he went out. His chute opened immediately and he hit the water OK. I made a 360 and came back slow and low as his ship sank. I saw his chute go down, but no dinghy. We stayed over the oil slick for about 45 minutes, then came home. An ASR Walrus[2] was over the spot when we left."

---

[1]Get a position fix (map location coordinates) for any possible rescue action.

[2]Air-Sea Rescue Supermarine Walrus single-engine biplane amphibians did yeoman service in rescuing pilots and others from the Channel throughout the war. A 1930s design, it looked more like a Model T antique than a warplane. But it was nearly ideal for the job it had to do.

What could have happened? For reasons which will never be known, Lt. Schwenke's Mae West lifejacket failed to inflate and the boots, clothing and equipment must have pulled the young pilot down to his death. If nothing else, his demise revealed just how different was the "Down to Earth" war from aerial battles between airmen attempting to gain the upper hand over a challanger mounted on an aircraft perhaps the equal of, or better than, your own. And surely the source of ground fire which put the Thunderbolt out of action was never seen. It was nearly the equal of the difference between naval warfare and infantry battles. In the ETO and the MTO, as no other place in the world, enemy anti-aircraft weaponry was probably the best in the world.

On that same morning venture, Lts. Ralph Hamilton and Robert Berkshire (487FS) were most fortunate not to have suffered Schwenke's fate. Both of the P-47 pilots flew straight into murderous flak barrages. Their fighters were mauled by light arms fire and small-caliber flak.

Many years afterward, Ralph Hamilton recalled the events of that morning:

"I believe the roughest mission for the 352nd Fighter Group was the March 11th low-level sweep near Amiens. J.C. Meyer (the squadron commander) said, before we took off, that we could possibly lose half the squadron and to be prepared for a rough day. It was, and we did have losses. I picked up several 35- (sic) or 40-mm hits before I could get out – and I wasn't in there more than 15 minutes. Some pilots were lost, including one in the drink...never recovered. After returning to Bodney we began checking our lifejackets and found someone had removed some the CO_2 capsules (cartridges) which inflated the jackets.

"Headquarters later told us the purpose of the mission was to locate the position and number of flak guns in the area. If that

In stark contrast to the Icelandic accident that befell "Big Bastard," another P-47D flown by Lt. Karl T. Hallberg (9AF) returned from a strike mission with a hung 500-lb. demolition bomb. Unfortunately, it was fuzed or it may have been one of those Vertoten early series RDX bombs that could be triggered internally. Unaware, the pilot touched down and the bomb fell off the shackles. Exploding below and and slightly aft of the cockpit, it demolished the aircraft. Now, the good news. Lt. Hallberg, saved by the protective armor plate and P-47 structure, soon returned to active duty. (AFM)

was the purpose, we certainly carried out our mission."

Lt. Berkshire barely made it back across the Channel in his badly scarred P-47D-2-RA ("Brutal Lulu") (AC42-22515) coded HO★U. "After crossing the coast I became separated from my flight just as the order to withdraw was given. At one time I was forced to shoot up a gun emplacement directly on my course to make it through. There were many soldiers around them (the guns), some of whom certainly became casualties.[3] I also crossed the south end of an airdrome, but did not see any targets...too busy trying to dodge the heavy crossfire from their ground guns. I took four direct 20-mm hits, three in the tail and one in my left wheel well. After that it was difficult to take evasive action because the controls were hard to move. I flew the remaining fifteen miles or so over the tidelands, at one time following a tidal channel with flak on both sides firing at me all the time. I managed to get back across the Channel and landed at Manston."

Some group pilots found opportunities to hit some targets of opportunity as they worked their way westwawrd to the French coastline. A few gun positions, flak towers and other assorted ground targets were damaged, but it was small consolation for the pilots who survived the mission. Bill Whisner, speaking for some of the others, summed up feelings about the mission:

"We suffered tremendous battle damage that day. If my memory serves me correctly, only my plane and one other made it home unscathed. And there wasn't a really high purpose to be served in finding out what the air defenses in the Pas de Calais were like. It was a damned fool thing to do since it accomplished absolutely nothing from the standpoint of offense against enemy targets."

For the pilots who flew the mission on March 11th – and for the man who took responsibility for ordering it, General Jimmy Doolittle – it had been a bitter pill to swallow. Some years later, Lt. Robert "Punchy" Powell got into a conversation with Doolittle at an Eighth Air Force reunion, and the general wanted to know which group Powell had served in.

"I'm proud to say I flew with the 352nd Fighter Group," Powell responded.

"Well," the general said, "I owe you and your fellow pilots a big apology."

"Why, sir?" Powell inquired.

"Do you remember a low-level mission over the Pas de Calais area a few months before D-Day?" the late General Doolittle asked.

"I certainly do, sir. We got the hell shot out of us," Powell replied.

"That's why," Doolittle responded. "I ordered that mission, and it was a terrible mistake."

Bob Powell compressed his lips and could only nod concurrence.[4]

***A typical P-47D-25-RE cockpit looked pretty much like this development Jug in 1944, although an automatic fuel cock drive was being tested on this airplane. Most of the instrument panel and portions of the consoles are included in this picture.* (NASM)**

When "Big" Miller climbed onto the wing of his Thunderbolt, which answered to the name "Red Raider," on 15 April 1944 he knew it was going to be a down-to-earth mission, but he did not expect a water treatment and over-exposure soaking in the process. The big 200-pounder from the wilds of 1930s Wyoming dropped into the cockpit of his equally large P-47D-5-RE (AC42-8511) with the bucking bronco emblazoned on the forward fuselage. His aircraft bore the code PE★W on the aft fuselage. Lt. Fremont Miller was leader of Yellow Flight on what was to be a strafing mission against a *Luftwaffe* airfield at Diepholz, Germany. There is no way in which the memory of that mission could ever be driven out of Miller's mind. Well, let him tell part of it in his own words:

"I was flight leader of Yellow Flight and had the "spare" with me, making it a five-ship flight. Normally the spare was sent home before we crossed onto the Continent if no one had to abort, but since this looked like a fun mission, I let him tag along.

"The four flights separated several miles from Diepholz, and my flight was to cross the field from south to north. There were no e/a parked at our end of the field, so I concentrated my firing

---

[3] Wasn't that a target of opportunity? And wasn't that what it was all about? Had the lieutenant missed the "Kill or be Killed" lectures in the Z.I.? To pass up any gun emplacement would logically allow it to "take out" the next aircraft unlucky enough to pass over.

[4] Thanks, Bob, for allowing me to report on this very important discussion.

***Set up for boresighting the guns under difficult and primitive conditions – most likely in Belgium or France – a Republic P-47D-22-RE reveals its Invasion stripes and 8AF tail markings, but it appears that most of its camouflage paint has been removed in the field. Vestiges of fuselage paint and a dark cowling remain, but there are no squadron identification markings.* (AFM)**

A small German town astride a stream and a country road presented a peaceful backdrop for a low-flying 1st Tactical Air Command P-47D-28-RE Thunderbolt bearing the logo "That's Urass" on the cowling and a multitude of kicking asses (most appropriate) on the cowl flaps. Consider the fact that it was but one of hundreds of fighters that punished Nazi forces everywhere in their crumbling Festung Europa for 10 months. General Hoyt Vandenberg commanded the Ninth Air Force in those critical months. (This Jug was from the 367FS, 358FG.) (AAF 57285AC)

on several buildings. Just as I pulled up, one of them exploded, catching me in the blast. Some of the debris went through my airplane.

"My left wing had a big hole in it, with the metal skin flapping. I soon realized that something had broken my oil line (in the forward fuselage - Auth.) between the supply tank and the engine. Hot oil was pouring out and covering my canopy. Knowing I would be in trouble when the 30 gallons had run through the engine, I headed for home. Looking back I saw a big cloud of smoke. I don't know what was in those buildings, but it must have been bombs, ammo or fuel storage.

"I never was too great on flying instruments, but I had to rely on them now. Opening the canopy brought hot oil into the cockpit, so I had to close it. I finally got up to 10,000 feet and crossed the coast, heading out over the North Sea. My plane wasn't flying too great either. The engine was running rough and the left wing kept pulling down. By crossing controls (right stick and left rudder) I managed to keep it headed home.

"It wasn't long after I tuned into the Air/Sea Rescue channel for a fix on my position that oil quit running over my canopy, so I was able to open it and fly visually. All I could see was water. Five minutes later the engine quit completely and I had another from ASR asking if I was going to ditch or bail out. I told them I would bail out at 1500 feet. The voice said, 'Keep talking so we will have a fix on you to the last minute.' I started reading off the altimeter numbers as I loosened my safety harness.

"This was my first experience at bailing out, and I tried to remember all the advice I had been given...I climbed out on the wing, sat down and slid off the trailing edge. I counted ten, made sure I had the ripcord handle and gave it a good tug. When the chute opened, the chest buckle came up and hit me in the mouth. I thought I had broken a tooth. While checking this, I remembered to unfasten the leg straps before I hit the water, but I was in before I remembered to inflate one half of the Mae West, get out of the harness and then inflate the other half. By then the parachute had settled over me and I had to work to get it off. Then I inflated the dinghy, but had some trouble getting in it. First I tried it over the small end, but water poured into the dinghy. So I tried the large end; it flipped over on top of me. The second time I made it over the small end." (Miller had been warned about getting into the dinghy quickly in cold water before the muscles became immobilized. The North Sea in April was extremely cold.)

Clouds and fog moved in fast that day, and although Miller heard a motor launch within a few hundred yards, they neither saw nor heard him. It turned dark, he settled down for the night, knowing planes and ASR would be back looking for him the next day. It was very rough during the night. Next day was dark and cloudy, so nobody was flying. Even a bird was sitting on a piece of driftwood nearby...not flying either. Miller had no water to drink, but he did have three 1-inch squares of butterscotch candy and five milk tablets. He rationed his meals: one-half of one candy square per day.

The 17th was just like the day before. The dinghy was getting soft, and the big bird – he reckoned it was a buzzard – came back "to see if I was alive. I was determined to outlast him." The

One of our favorite paint schemes on a Jug was the "Chief Ski-O-...?" imagery on this P-47D-27-RE (G9★J) of the 509FS flying with the XXIX Tactical Air Command of the 9AF. It is also frustrating not to know the full name legend (last word not discernible) and the color scheme. Pictured at airfield Y-29, Asche, Belgium, by a P-38J crew chief, it awaited its turn to deliver an assortment of general purpose bombs to Germany's Wehrmacht fighting fanatically against the 4th Armored Division not many miles from this base. (Albert Meryman)

third night was even worse. Stormy winds howled all night; he was cold and soaked through. In the morning it was clear and the bird was gone. (It was probably a sea gull.) Planes flew over, but nobody saw him. Late that afternoon, a P-47 came over, very low, and Miller waved frantically. That big, beautiful T-bolt circled, climbed to an altitude where he could call ASR, and it was not long before one of the powerful British Power Boat ASR boats, or possibly one of the Vospers, hauled Miller aboard, ending his 76-hour ordeal in the water.

When he was covered with blankets and given some hot broth, they obtained his squadron information and radioed ashore. His squadron was contacted, but in typical style, the answer came back as "We haven't lost anyone for three days." However, Major Earl Abbott took off in a DeHavilland Tiger Moth biplane and flew to the docking location. Miller's legs and feet were so bloated and black, some of the AAF doctors recommended amputation. But the English doctor looking after the fighter pilot stuck to his guns and refused. By May 2nd he was back at Bodney, unable to walk. S/Sgt. George Wilcox fashioned a mobile wheelchair out of bicycle parts so that Miller could get around a bit. By June 5th he was returned to the Z.I. for further treatment. By the end of summer, he was back and prepared to fly his 60th mission. Nobody could figure out how he survived on that half-inflated dinghy for more than three days in the North Sea.

It was but one of many amazing and heroic episodes of men at war. Most go largely unrecognized, but that does nothing to diminish the importance of such events.

During the first days of Operation OVERLORD, Maj. A.W. Cortner took his razorback P-47D into battle and scored the first "kill" for the 23FS. Unfortunately, the major was KIA during the early weeks of that 1944 Battle of France. (Donn Madden)

OPERATION BASEPLATE (BODENPLATTE)

Almost exactly a half century ago, *Unternehmen BODENPLATTE* ushered in the New Year of 1945 with a *Luftwaffe* air assault aimed at totally disrupting air operations of the Royal Air Force's 2nd Tactical Air Force and USAAF's Ninth Air Force operating from bases in Holland, Belgium and northeastern France. Nobody seems to have definitive facts on the raid, originally planned for mid December. Perhaps some organizations believe we should let sleeping dogs lie. Or, how about the possibility of letting lying dogs sleep. As late as a quarter century after the assault took place, the German operation was being reported as *HERMANN*. Far worse, Allied losses were reported as "minimal" and "minor." Just to adjust the focus a bit, let us list far more accurate totals derived from detailed studies of Allied and Axis records. The 2nd TAF (RAF) was operating from four seriously damaged German airfields, crammed with more fighters and fighter-bombers than could be utilized safely or efficiently. At least five of perhaps sixteen (or nineteen) Ninth Air Force fighter groups were in-

ABOVE AND BELOW: Leaden skies that often prevailed in the terribly cold winter of 1944-45 in Western Europe made close-support missions against German armor and assault troops a trifle hard on the nerves. Low-lying clouds often hid hill and small mountain tops, making penetrations in force extremely dangerous. "The Dutchess" was a P-47D-28-RA, generally flown by Lt. John Reynolds of the 411FS. This scene is at Le Culot, Belgium, where the 373FG took up residence late in October 1944. (William Mullins)

volved in the action of 1 January, and the 352FG (Bluenose Bastards of Bodney) Mustangs were at Asch on detached service to aid the Ninth in beating back the Germans after the breakthroughs that created the Battle of the Bulge.

KNOWN LUFTWAFFE, RAF, AND NINTH AIR FORCE LOSSES AS OF 1990[5]

Luftwaffe: At least 200 aircraft (but at least 30 more aircraft were lost on withdrawal because of pilot inexperience, Allied anti-aircraft, and "friendly fire."

2nd TAF: A minimum of 524 Typhoons, Spitfires and Tempests were written off completely.

Ninth Air Force: With the 9AF operating from at least six airfields in the area, only four supplied fairly reliable figures. Known losses of 410 aircraft, including C-47s, are probably at least ten percent or more on the low side. No P-38 losses have even been reported.

Eighth Air Force: Losses were negligible.

The Luftwaffe launched the attack just after 09:00 with Messerschmitt Me 109G, Me 109K, and Me 262A fighters, Focke-Wulf Fw 190As and Fw 190Ds, and Junkers Ju 88Gs (acting as lead ships for the inexperienced teenagers flying most of the fighters). Few of the young pilots had more than 60 hours total flight time logged. It has been estimated the attacking German aircraft comprised a force of at least 800 *Jagd* (fighters), *Jabo* (fighter-bombers), *Kampf* (battle, or bombers) and *Schnellbombers* (fast bombers, as Me 262s).

Allied Intelligence did not have a clue about the forthcoming *BODENPLATTE* attack, just as they had failed to even anticipate the German retaliatory raid against shuttle-bombing B-17s, part of the 8AF Operation FRANTIC mission to Poltava (Russia) on 21 June. The *Luftwaffe* assault at that time destroyed at least 47 of 73 Fortresses at the airfield. Most of those remaining sustained serious damage. It can only be viewed as another dismal intelligence-gathering performance.

Probably the most decisive air battle of the New Year's Day assault was a marvelous example of 8AF and 9AF, plus RAF integrated operations – especially since it was unplanned – of the entire Battle of the Bulge era. It took place at Asch (Airfield Y-29), Belgium, and involved the 487FS (Code HO★) of the 382FG "Bluenosed Bastards of Bodney," (perhaps subsequently named, by the Germans, as the "Bluenosed Bastards of Bastogne") of VIII FC and the 366FG of the IX FC.

The legendary commander of the 487FS was Lt. Col. John C. Meyer. His squadron was under no orders to fly on the holiday, so the P-51Ds were to stand down. Meyer, well versed in history, psychology and enemy tactics, anticipated a New Year's Day raid by the Germans. Just like Patton, he understood the enemy's thought processes. He requested 9AF permission (on Dec. 31) to fly a CAP (combat air patrol) on the morning of January 1st. Certain of approval, he ordered his pilots to delay celebrations and ground crews to have a dozen Mustangs ready to go by 09:00 on the holiday.

Approval to fly came in the early morning. Across the airdrome, the 366FG of IX TAC were preparing their P-47D Thunderbolts for an early day attack on *Wehrmacht* positions at the "Bulge" perimeter. Two flights of 390FS Jugs were to be first out. They were preceding the Bluenosers by about 15 minutes, slamming into the air with their full loads of weaponry when anti-aircraft bursts caused Lt. Jack Kennedy (not in PT-109) to shout an alert. As the 487FS taxied out for takeoff by flights, the 390th – led by Capt. Lowell Smith – threw a resounding block against the oncoming Focke-Wulfs and Messerschmitts. It disrupted the Germans just enough to allow Colonel Meyer's squadron to make a mass takeoff run. Of course some of the huge attacking force was only momentarily delayed by a mere eight P-47s, so the Mustangs had to literally fight their way off the ground. Both Col. Meyer and Capt. Whisner of the 487th scored by shooting down Fw 190s during climbout. Meyer scored twice that morn-

[5] A thorough, multi-faceted research effort must be pursued by a talented researcher to uncover all existing factual data about Operation *HERMANN* and Operation *BODENPLATTE* if we are ever to know the real facts. Every published data compilation to 1990 is in major conflict with all others. The best British authors have missed major USAAF participants in the battle, and losses on the Ninth Air Force ledger are totally guesswork. Most American authors seem to have forgotten the RAF was involved, massively. The most important factor: How many *Luftwaffe* pilots, with the list of senior veteran pilots exclusively oriented, were permanently lost in the operation. Although the Allies lost at least 1000 aircraft, pilot losses were minimal. Massive production capacity allowed the AAF and RAF to replace all aircraft with new ones in a few days, actually a beneficial accomplishment.

ing, Whisner shot down four before he had to land his battered P-51 with one aileron flapping and no oil left in the tank.

The total victory tab for the 487th that morning was 23 aircraft. All ground personnel had a ringside seat for the battle, with fifteen of the enemy aircraft crashing within a mile of the Y-29 airfield. Pilots of the 366FG obviously scored too, but I have not been able to determine how many they knocked down.

One other highlight of the day was a visit by Generals Carl Spaatz, Hoyt Vandenberg and Elwood "Pete" Quesada to congratulate all for their efforts. With well over 1000 Allied and German aircraft turned into junk in northeastern Europe that morning, the bulldozers of the Engineers must have been very busy for days.

There is really no way, within the confines of these book covers, to properly describe the work done by Eighth and Ninth Air Force fighter and bomber pilots, ground crews and all supporting cast members. They flew in weather conditions that grounded birds. A very dangerous enemy did not know when he was whipped. So he fought on with determination and blind faith.

In many ways it was the best of times and, concurrently, the worst of times. No doubt about it, War Is Hell.

When snow laid a white mantle over most of Northern Europe in one of the coldest winters then on record, the fighter-bombers of IX Fighter Command under the leadership of Brig. Gen. Elwood Quesada continued operations in all but the most drastic weather conditions. Here, a pair of Thunderbolts – caught by a fast shutter that "stopped" the propellers – head out on a Rhubarb from (probably) Chievres, Belgium, passing behind a line of 388FS Jugs awaiting their call to action. (AAF)

Brilliant sun brightens the snow as a relatively new P-47D-30-RE assigned to the 386FS warms up prior to an attack mission against nearby Wehrmacht and Panzer divisions. They were attempting breakthroughs in what came to be known as the Battle of the Bulge. Certainly not long on experience, the squadrons of the 365FG proved their mettle in combat by earning two DUCs in very short order. As airfield defenders were soon to learn, light anti-aircraft weapons such as these multiple .50-caliber machine guns were not unnecessary decorations. On 1 January 1945, the Luftwaffe launched a widespread attack known as Operation BODENPLATTE (Baseplate) that destroyed far more RAF/RCAF and 9AF aircraft than were destroyed at Pearl Harbor. However, losses of something more than 200 attacking aircraft were devastating to the Luftwaffe. (AAF 16215AC)

The importance of a "wing rider" to guide pilots as they taxied on narrow strips on Continental airfields, especially after a winter snowfall, cannot be overestimated. The pilot of this 1st TAF Jug, a P-47D-27-RE, could not possibly judge the proximity of his wheel to the edge of that taxiway. Here, the guide rides far out on the wing so that both wheels can be observed as the 10FS aircraft moves toward a dispersal area at Toul, France. (AAF)

BELOW: It would be safe to state that few aerial photographs of 405FG, 510FS airplanes have appeared in print; the outfit was probably too busy attacking German targets to record events and equipment. This P-47D-16-RE, 2Z★M, flown by Capt. Charles Mohrle, had been pretty well beat up when the rare in-flight photo was taken. After just a few more of its 50+ missions, it had taken so many hits that it was grounded. Most missions flown by the group were "down and dirty" strafing ventures. "Touch of Texas" was a reminder that West Texas has an abundance of rattlers. (Charles Mohrle)

This remarkable (unretouched) gun camera film frame came from Capt. Charles Mohrle's airplane, showing a successful but dangerous attack on a moving train near Argentan, Normandy, France. The 510FS pilot pressed home the attack with determination, scoring direct hits on the locomotive cab and boiler with numerous API (armor-piercing incendiary) bullets. At least seven APIs can be seen hitting the cab in just this one frame of the film. Many moving trains were "Q ships" with false boxcars containing a deadly array of flak weapons. (Charles Mohrle)

ABOVE: As troops assigned to the 5th Armored Division were moving across Germany in pursuit of the faltering Wehrmacht, a P-47D-28-RE carrying the code A8★E that identifies it as a 391FS aircraft of the 9th Air Force was struck by defensive fire while on a strafing mission. The Rhubarb was being flown in support of the 9th Army in April 1945. Damage to the engine and oil system (see evidence on tailplane) caused the pilot to make a quick emergency landing near Peine, Germany. The curious American troops naturally had to take a look and also verify that the pilot was not injured. (U.S. Army)

LEFT: When the sounds of battle had died away in Europe, a P-47D-28-RE bearing the name, "Little Bill," was pictured in a setting thought to be Frankfurt-am-Main, Germany. Fuselage stripes indicate that this was probably Maj. Clary's 397FS airplane, then being flown by Lt. R. M. Martin. Code letters were D3★J, and it is evident that the pale blue cowling colors of the 368FG have faded badly. In the aftermath of war, even the gunsight appears to have been removed. (G. L. Letzter)

A major bridge spanning the Seine River at an unidentified city was dropped by American tactical bombers as invasion forces surged through France in 1944. Whether the bombers were P-47s or medium bombers escorted by Thunderbolts was not divulged, but it is a fact that the fighter-bombers did take out many bridges by using dive-bombing tactics developed for Jug operations by a 9AF group. Since the war ended, many stories have stated that P-38s and P-47s accelerated too fast to be successful in dive bombing. In fact, both types were extremely successful with such tactics in the MTO and later in the ETO. Target position and flak defenses actually determined which type of attack would be the most successful. (AAF)

Stripped of its guns in expectation of some mock aerial combat with a Lockheed P-38 (to alleviate post-war boredom, no doubt), a yellow-tail and cowl P-47D-30-RA from the 23FS, 36FG, cruises out of Kassel over the German countryside that shows no sign of the lengthy war just ended. The photo was taken from a Northrop P-61A Black Widow night fighter. The large national insignia on the wing of 7U★L is unusual (probably an ex-8AF airplane), and the black cowl ring has been deleted. (Robert Durkee)

Displaying the famous Seversky wing planform as it moves forward past the Northrop P-61A carrying the photographer, a squadron or group commander's P-47D-28-RE shows its 3T★P code letters and the leader's fuselage stripes generally omitted in wartime conditions that had prevailed. The 36FG Jug was assigned to the 22FS, a 9AF fighter based at Kassel, Germany, in the autumn of 1945. It carried a small cartoon and the name "Dutch's Devil" on the cowling. "Snapping" was painted on the fuselage below the bubble canopy. (Robert Durkee)

Enjoying the pleasures of cloud chasing without the attendant bit of trepidation that accompanied such cumulus and cirrus formations when combat was a threat, a 404FS pilot cruises serenely between cloud levels while flying out of the base at Furth, Germany, in 1945. This 371FG Jug carries the code 9Q★U, a full red cowling with alternately painted cowl flaps that prevailed after hostilities ended and the name "Always Marge!" However, there was no red fin and blue rudder. (Robert Durkee)

BELOW: In a picture probably taken in 1946 at a former Luftwaffe airdrome in Germany, a late-model P-47D rests with its 108-gallon droptank virtually touching a portion of Marston mat acreage. This squadron or group commander's fighter carries a non-combat code TD16 of the type that came into general use in Occupation days. (Ray Bowers)

Abandoned in front of the camouflaged hangar belonging to a former tenant involved in flight test work and evaluation of captured enemy aircraft, a P-47D looks rather forlorn. This yellow-tailed fighter bore complete Luftwaffe markings, but lacked a propeller and it had a broken tailwheel strut. The photograph was taken at Gottingen, Germany, on 12 May 1945. (AFM)

When an American fighter group moved into the abandoned Luftwaffe airfield at Göttingen, Germany, the captured and German-marked P-47D pictured in the aerial view was found to be non-operational. The tailwheel strut was broken or collapsed, the propeller was missing and other components were in need of replacement or repair. All machine guns were operational and ammo boxes were full. (AAF 57465AC)

While nobody can legitimately or logically claim that AAF and RAF airpower won the war in Europe, the Allies could never have prevailed over the Axis powers without gaining air superiority. In the end it all came to this. Some of the Luftwaffe's best – Focke-Wulf Fw 190A, Junkers Ju 88, Messerschmitt Me 262A and the ever-present Me 109G – ended their combat days in the silence of a junkyard, and not very neatly at that. It matters not that the fighter in the foreground might have been the "best" of all time. A terribly complex series of decisions and events brought it and its kin to this end. Germany, 1946. (James Kunkle)

Designer Kurt Tank eventually concluded that necessary high-altitude performance needed for the Fw 190 series was not to be achieved with the BMW 801 radial engine. He turned to the powerful supercharged Junkers Jumo engine. It may come as a surprise to many that the Fw 190A was lacking in good performance above approximately 25,000 feet. Its ultimate replacement was to have been the high-performance, high-altitude Focke-Wulf Ta 152H and medium-altitude Ta 152B and C versions. The Fw 190A series fighters weighed about half the gross weight of a typical P-47D and had at least 70 percent of the rated power. This interim type in the series was the Junkers Jumo 213A-powered Fw 190D-9 with improved altitude performance over the radial engine models. Top speed of the best Fw 190As: 418 mph. (Author's Coll.)

Messerschmitt's Me 210A model was to replace the slow Me 110, but an intensive development program failed to correct major defects in flight characteristics. Far newer than Lockheed's P-38 design, the 210 was not even close in performance. Its Me 410A and B successors were much improved but could not handle P-47Ds or the newer P-51Ds at all. By late 1944, P-38H and J Lightnings were doing more jobs for the Allies in the ETO/MTO than the Me 210/410 was designed to do for the Luftwaffe, yet failed to accomplish. The Messerschmitt was closer to de Havilland's Mosquito in concept, execution. A captured Me 410B was photographed in Germany. (James Kunkle)

Ultimately symbolic of what Nazi Germany came to: Aachen, 1945. It was essentially an exploded, burnt-out shell. Most sections of Germany, and much of the Continent that had existed under Nazi domination, fared no better than Aachen under Hitler's protective dominance. Although the P-47 Thunderbolt only created a share of this destruction, the existence of that tough fighter helped to produce those empty streets and hollow shells. (Author's Coll.)

CHAPTER 15

LA DOLCE MAAF VITA

Operation TORCH certainly was not the largest of major theater operations during the war, but it surely was a major effort. And one of the most confusing relative to organization. Even M/Gen. J.H. "Jimmy" Doolittle could not avoid a snide remark about it in a February 1943 conference. Speaking of Twelfth Air Force status, with its units, personnel and equipment having been transferred to the Northwest African Air Forces (NAAF), itself under Mediterranean Air Command (MAC), he wondered why on paper and actuality the Twelfth "seemed to have disappeared." He then opined that once such matters as courts-martial had been wound up, the "skeleton" of that air force – "the name only" – would have to be returned to the Z.I. for a reincarnation or be decently interred by War Department order. Even Spaatz had to ask Eisenhower about the status, being informed HQ, Twelfth Air Force would be continued as administrative headquarters for U.S. Army elements of NAAF. Without going into more detail, the Twelfth served mainly to mystify all but a few experts at HQ. Near the end of 1943, it came to life again.

The 325th Fighter Group, soon to be under the command of Lt. Col. Robert Baseler, arrived in North Africa with Curtiss P-40Ls aboard the USS *Lexington* as part of the Ninth Air Force, then was assigned to the Twelfth Air Force, then to Doolittle's Northwest African Strategic Air Force (NASAF) eventually becoming the single fighter unit with P-47s in the Fifteenth Air Force. You get the picture.

During the early months of 1944 the 15AF was operating primarily with P-38 fighters plus the one group of P-47Ds, while the Twelfth Air Force was being supplied primarily with Thunderbolts. Soon the 325FG made a transitional move to new P-51Ds, ensuring the 15AF had the longest-ranging types for escort, namely P-38s and Merlin-powered P-51s. After divesting themselves of older P-51 and P-39 fighters, all but specialty units concentrated on Thunderbolts for their tactical operations.

On 11 October 1943, Colonel Baseler buzzed the 325FG base at Mateur with a new Republic P-47D-10-RE, beginning the group transition to Thunderbolts, but only a short time later the group moved to Foggia,

The 325FG, under command of the very tall Lt. Col. Robert Baseler (ace), gave up its Curtiss P-40 Warhawks on 18 September 1943 and awaited arrival of new P-47D-6 and D-10-RE Thunderbolts in North Africa. They would become the first operational P-47s in the theater, scheduled to operate out of Mateur. However, shortly after Baseler flew the first of the contingent of Jugs into that base on 11 October, the arrival of the rainy season forced movement of the group to Soliman. Ace Lt. Frank "Spot" Collins is seen piloting No.79 and in the distance is group C.O. Baseler leading a flight of similar P-47Ds in November when the group was transferred to the Fifteenth Air Force. **(Robert Baseler)**

A fantastic Rodeo was set up with the aid of Colonel Baseler for January 30, a mission on which the men of the 325FG scored 38 confirmed victories against German and Italian fighters, bombers and transports. On that single mission, "Herky" Green had six e.a. added to his score. Baseler's group – he led the mission – caught the defenders in their normal bomber interception routine (according to plan), and when six AAF heavy bomb groups moved in after the Rodeo they were able to destroy 70 additional e.a. that never left the ground. Baseler is shown in his P-47D-16-RE Thunderbolt at Foggia No.1 in February 1944 following that beautifully timed and executed mission of January 30. (Robert Baseler)

Italy. Originally received with no provisions for drop tanks, the airplanes were soon modified with standard Republic pylon mounts and bomb shackles. The standardized drop tanks used on all 325FG airplanes were Lockheed 165-gal. units, most likely because the 325FG moved into the new Fifteenth Air Force where the other three fighter groups had Lockheed P-38s. The tanks were readily available in the MTO and adaptation was straightforward.

The 325FG "Checkertail Clan" moved to Lesina, Italy, when the base at Foggia 1 became extremely overcrowded with heavy bomber groups. Lesina, at the time of the move in the last days of March, was the most advanced base in use by the 15AF. Right in the midst of the move operation, Lt.Col. Chester Sluder and members of his 318FS got into a battle with numerous Messerschmitt Me 109s taking out seven of the Germans with the loss of three T-bolts. Lucky man for the day, for several reasons, was Lt. John R. Booth. Tangling with several of the Messerschmitts, he shot down two. One of the two remaining

The 6'2" Lt. Col. Robert Baseler is seen at Foggia, Italy, in the last days of 1943 (he said) before the group made another move to Celone, better known as Foggia No.1. On his right is ace Frank "Spot" Collins; Lt. Don F. Kerns, Lt. James A. Jones and an unidentified pilot are on his left. Even though the irrepressible Baseler was moved up to staff at 306th Fighter Wing HQ in February, he managed to retain his P-47D, leading missions when he could and finally giving command of the 325FG over to Sluder on April Fool's Day. (Robert Baseler)

Seated in his P-47D bearing the name "Shimmy," Lt. Col. Chester A. Sluder was Bob Baseler's right-hand man from the early days in North Africa. Operating out of Soliman, Tunisia, in the autumn of 1943, the 325FG airplanes were not equipped with wing pylons and evidently lacked provisions for belly droptanks. Some records state that all D-6 and D-11 Jugs were replaced with D-15 and D-16 models in January 1944 after the move to Foggia, Italy. The facts refute the records and create recognition problems. Baseler's D-10 and others were actually retrofitted to carry belly and wing stores, including 165-gallon P-38 wing tanks, and they also had the famed checkertail markings applied. It then became virtually impossible to know if they were D-6/D-10 or D-15/D-16 fighters unless you knew the serial number. **(Chester A. Sluder)**

Me 109s made a head-on pass at Booth. The German either panicked or froze, crashing into the tough Jug. In the collision, Booth's P-47D lost 4½ feet off the right wing; the Me 109 went out of control completely while the Jug hardly wavered. In its gyrations, the German fighter collided with his own wingman, triggering a massive explosion. That's a touchdown with a 2-point conversion anywhere.

Col. Baseler was moved up to 306th Fighter Wing Headquarters as Assistant Chief of Staff, A-3, to B/Gen. D.C. Strother. In his own inimitable style, Baseler took his P-47D, "Big Stud" with him. (Earlier, as a squadron commander, Bob had managed over a period of time in Tunisia to "acquire" two different Messerschmitt Me 109F/G fighters. In one of them named "Herman", he had "attacked" Col. Ralph Garman's 1FG base right at chow time. With his first (slow) pass, he made sure everyone knew it was a Messerschmitt. Accelerating into a 360 arc, he proceeded to "buzz" the chow line at very low altitude, creating something of a wild panic. Later on, Garman had a few unprintable words to unleash at Baseler for that intrusion.) The Thunderbolt which Baseler took with him was not for joyriding. He either made individual sorties or flew with the 325FG led by Col. Chet Sluder, his former Deputy CO, on aggressive missions. In fact, he was directly involved in planning a very special *Rodeo* designed to totally upset the routine bomber interception strategy used by the Germans at a five-base complex in the Udine and Villaorba areas. On 30 January 1944, the 15AF put up six heavy bomber groups, to be protected by sixty 325FG Jugs. Crossing up the German routine, the 325th, led by Baseler himself, arrived via a different route, ahead of the bombers and very early, catching dozens of the enemy aircraft on the ground or in the process of taking off. All the P-47s had approached at an altitude of less than 50 feet above the Adriatic before climbing to stack up between 15,000 and 19,000 feet.

The Checkertail Clan scored 38 confirmed victories and got six probables for the loss of only two Thunderbolts that day. When the heavy bombers hit the base, they destroyed an additional 70 enemy aircraft on the ground. Capt. H.H. "Herky" Green, at

Captain (at the time) Herschel H. "Herky" Green was the highest scoring ace in the 325FG and reigned as tops in the 15th Air Force by scoring 18 confirmed aerial victories. His P-47D-10-RE had received retrofit treatment in the field to enable it to carry up to three external fuel tanks. This airplane, named "The Star of Altoona" on the right side, may have had a different name on the opposite side, but records indicate the unit number in the group was 11. Maj. Lewis W. Chick, Jr., the 317FS commander when this picture was taken, was succeeded by Maj. Green in that post on 25 March 1944. **(Robert Baseler)**

the front of "A" Flight which included F/Os Cecil O. Dean and Edsel Paulk plus Lt. George Novotny, led them to a field day battle in which the four shot down no fewer than fifteen e/a. Green was credited with four Junkers Ju 52s, a Macchi MC. 202 and a Dornier Do 217. True, it was an atypical day for victories, but it gives a brief insight into the type of battle which could take place in Italy.

Thunderbolts would replace nearly all North American P-51 (Allison-powered) and A-36 attack-fighters in the Twelfth Air Force when it was resurrected as the tactical fighter-bomber force under MAAF. General Ira Eaker, displaced by M/Gen. James Doolittle as CG, 8AF, came down to the MTO at the time to take command of the new MAAF and 15AF. Like their compatriots in the Ninth Air Force, ETO, the men and airplanes of the 12AF were scheduled to serve as flying artillery in the MTO.

"Herky" Green was photographed at Foggia after that spectacular Rodeo of January 30. On his initial attack, he shot down four Junkers Ju 52/3m troop transports in less than 15 seconds. All four exploded. He then chased and dispatched a Macchi MC.202 fighter. After leaving the scene of destruction at Villaorba, he caught a lone Dornier Do 217 light bomber and shot it down. (Robert Baseler) (The author's father commanded an Air Service Group at one of the many Foggia bases during this same period.)

With transfer of Gen. Ira Eaker to the MTO at the beginning of 1944 as CG, Mediterranean Allied Air Forces (MAAF), the Twelfth Air Force was reactivated and remanned. (The following groups and squadrons comprising the P-47 Thunderbolt organizations may not have been fully equipped with that type aircraft at the start, but they were all primarily flying Jugs in the thick of battle during 1944. Two groups, the 33FG and 81FG, didn't have any P-47 equipment until they were transferred to the Fourteenth Air Force in February and March 1944.):

TWELFTH AIR FORCE (As of 31 January 1944)

XII Air Support Command (ASC)
- 27FBG
- 57FG
- 79FG
- 86FBG
- 324FG

XII Fighter Command (FC)
- 350FG

At Lesina, Italy, not long before the P-51Ds superseded the P-47D razorbacks in service with the 325FG, Lt. Carswell describes a deflection shot victory over an enemy fighter. The P-47D-16-RE "The Jenny "A"' was the newest type Thunderbolt flown by 325FG pilots during the seven months or so that they had been the equipment of choice. Fifteenth Air Force Thunderbolts were replaced by North American P-5ls with the tactical mission Twelfth Air Force absorbing all MTO P-47 fighters.(Chester Sluder)

Remaining under the Fifteenth Air Force, the 325FG flew their Thunderbolts until conversion to North American P-51s began. Back in October 1943, General Arnold had issued orders limiting deliveries of long-range Mustangs and Lockheed P-38s to the ETO, specifically for bomber escort duties. (At that time, at least as winter approached, Gen. Eaker had become very upset when he learned P-51B and C models were being delivered only to the Ninth Air Force. Command HQ made certain those fighters would not be assigned to tactical operations, but would function as an arm of VIII FC on bomber-escort operations.) Protests from General Kenney in the SWPA, and from other Pacific Ocean Area commanders must have

***Tail markings on 86FG Thunderbolts in the 12th Air Force may not have been the most extravagant – Robert Baseler's 325FG, the Checkertail Clan, had achieved that goal – but the tail striping certainly gained plenty of attention. Seven narrow horizontal red bands were painted on the horizontal and vertical tail surfaces, generally matching the cowling nose ring, cowl flaps and propeller cap except in the areas of top cowling anti-glare paint. The normal tail numbers were transferred to the fuselage side panels immediately forward of the assigned squadron numbers. This P-47D-28-RA belonged to the 526FS.** (Fred C. Dickey, Jr.)*

been loud and long.[1] One other 15AF fighter group, the 332nd, had been flying P-39s and P-40s. Although they were to get P-51Bs or Cs, that Arnold directive may have led to assignment of P-47s for a time in 1944. The group seems to have jumped back and forth between Mustangs and Thunderbolts well into 1946.

The performance of the Twelfth Air Force in Italy and into Central Europe in 1944-45 in the tactical role was one of the most important factors in driving the German *Wehrmacht* and *Luftwaffe* northward and, eventually, out of Italy – primarily by means of unconditional surrender. Thunderbolt actions were a key to immobilizing German ground transport and destruction of German airpower. In the long run, it was Allied airpower which destroyed Nazi ability to wage war in Italy. In attempting to flee, they found their ability to cross the Po River and escape had collapsed under Allied domination of the air. They were trapped.

The Allies had bombed all Po River permanent crossings into oblivion by July 1944. Immediately following a lengthy reconnaissance on 21 April 1945, the Germans had attempted to escape northward by using ferries and pontoon bridges to cross the Po. Virtually every fighter-bomber, light bomber and medium bomber in the MAAF was thrown into the fray, day and night, to wipe out crossing efforts. The attacks went on relentlessly until April 24. Destruction was so widespread, nobody could possibly estimate how extensive it was. The commanding general of XIV Corps, 14th German Army Group, Gen. von Senger, said of the raids, "That is what finished us. We could have withdrawn successfully with normal rear guard action despite the heavy pressure, but due to the destruction of the ferries and river crossings we lost all our equipment. North of the river, we were no longer an army." The only reason the war in Northern Italy did not end then was grounding of all heavy and medium bomber action in the face of terrible weather. American fighter-bomber operations were limited, almost ceasing by 2 May 1945 because even the number of targets was severely reduced. On that same day, the Germans signed terms of unconditional surrender and the war in Italy was at an end.

The part played by Republic Thunderbolts in the MTO is far too complex to describe in any detail in this single chapter, even ten chapters. Air battles involving both the Twelfth and Fifteenth Air Forces raged on in spite of terribly difficult operational conditions, terrain and weather. The Germans put up a tremendous defense until they could no longer obtain supplies and replacements. Airpower was literally pulverizing supply routes.

---

[1] Although we have not discovered documentation which would have initiated the action, there exists an interesting coincidence pertaining to Lockheed P-38 output. Burbank production efforts reached maximum possible levels throughout 1944 and into 1945. Finally, after what seemed an eternity, the Consolidated-Vultee factory at Nashville, Tenn., delivered their last Vultee A-35B Vengeance in June 1944. Gen. Knudsen's WPB finally woke up to the folly of building essentially non-contributing, ridiculously obsolete Attack-category airplanes in what was supposed to be all-out war. Nashville conversion to Lockheed P-38L production took months, with the first four Lightnings being accepted in January 1945. Final total delivered: 113. Total Vultee Vengeances delivered: 1529. There can be no valid excuse for not having made the conversion by January 1943 when the first Republic P-47Cs were being shipped to the ETO. When B/Gen. Hunter lost his P-38s to TORCH in November 1942, he should have pointedly alerted Gen. Arnold to the essentials associated with the Operation POINTBLANK demands which had not evaporated. The WPB was certainly culpable. Threats would not have induced General Kenney to accept Vultee A-35s for combat in the SWPA. Any P-38, beginning with the F model, could carry twice the bomb load a far greater distance than any Vengeance ever built, twice as fast with much greater firepower. And no 2-man crew was needed.

***This P-47D-25-RE assigned to the 66FS at Grosseto is marked with an emblem indicating that the war bond signup of Republic Aviation employees had paid for the 45th such P-47 financed in that manner. All of the cowling emblems applied to 57FG aircraft were so professional that either one talented artist was kept busy full time painting them, or expensive transfers were obtained somehow.** (Art Krieger Coll.)*

An Evansville-built P-47D-28-RA Thunderbolt flown by a 66FS pilot in the MTO shows off unit recognition markings used by that 57FG outfit. Black-banded, wide yellow stripes on the wings and vertical tail were for quick identification, but the squadron emblem carried on both sides of the cowling – featuring a pugnacious mongrel bird – became rather famous in time. A red cowling band, like that seen on many 56FG airplanes, was up front and the propeller hub usually had the same treatment. A simple numbering system virtually duplicated the one adopted early on by the 325FG. (Wayne Dodds)

A tale of heroism on the part of one Twelfth Air Force pilot represents the performance of a multitude of pilots who flew combat missions in the MTO, and rather unexpectedly it is tied directly to our Brazilian ally.

The Medal of Honor was awarded posthanously to 1st Lt. Raymond Larry Knight, a Twelfth Air Force P-47 Thunderbolt pilot. Ostensibly the last award of a Medal of Honor to any aviator in World War II, it was also the only such award to any P-47 pilot in the MTO. Lt. Knight was in the Mediterranean Allied Air Force in the single fighter-bomber group – the 350FG – operating with XII FC in Northern Italy. As the war in Italy was nearing a final crescendo in the Po Valley region in late April, 1945, the 350th was operating with a fourth squadron, 1° Grupo de Aviação de Caça, *Força Aérea Brasileira* [FAB or Brazilian Air Force]. (That is an important detail in explaining a long-standing mystery relating to the markings carried on Lt. Knight's Medal of Honor fighter.)

The Battle of the Po Valley, as described briefly intensified greatly on 21 April. By the 24th, the Germans were fighting for survival. Lt. Knight was a participant in three aggressive raids on enemy airfields and other targets of opportunity that day. He was already the recipient of six Air Medals, two Purple Hearts and had been recommended for the Distinguished Flying Cross, the last for a successful attack on a railroad bridge at Castelnuovo on 9 January 1945. Having dropped his bombs with good effect, he turned to a large ammunition dump. In repeated strafing attacks, Knight's APIs blew up two buildings packed with explosives. Much of the debris struck him and his aircraft, probably a P-47D-26-RA. He managed to fly the badly mauled aircraft back to Pisa. Apparently the airplane was no longer serviceable and through negotiations with the Brazilians, a camouflaged P-47D-27-RE (AC42-26785) which had been allocated to the FAB under Lend-Lease on 15 January was returned to the USAAF's 350FG. This fighter, painted in a Brazilian camouflage scheme, was given the markings 6D5 (in white) on both sides of the forward fuselage and cowling, a 346FS checkerboard scheme on the rudder, and some personal markings including the name "Oh Johnnie" painted (white) on the left fuselage panels near the cockpit. (The same name had been applied in black letters to Knight's earlier P-47D, coded 6D2, but that aircraft was not camouflaged.)[2] A correct profile of the camouflaged Medal of Honor airplane – executed by Bob Boyd – appears in this book.

Lieutenant Bill Hosey was flying the morning mission on 24 April with Knight who had volunteered to lead a 3-plane mission

[2] A well executed painting of "Oh Johnnie" 6D5, plus a profile painting of the same airplane, published some years ago, are technically in error. Evidently an earlier "Oh Johnnie" was a P-47D-26-RA or a D-28-RE because it was equipped with a Curtiss Electric propeller. The normal Curtiss-patent hub cuffs show up clearly on the badly damaged aircraft. The airplane bore the code 6D2. Serial number is unknown. Apparently the artist did not realize that the replacement aircraft was a different type because it was manufactured with an H-S propeller and had a different block number. The camouflaged 6D5 is undeniably a P-47D-27-RE as proven by the serial number combined with a closeup of Raymond Knight standing by the nose of the latter airplane. It shows a Hamilton-Standard Hydromatic propeller (sans cuffs).

The quality of this picture and location/date of origin leads us to believe that the person behind the camera when "Ponnie" was committed to film had to be Capt. Fred Bamberger, intelligence and photo officer with the Twelfth Air Force at the time. Flying with the 64FS out of Grosseto, Italy, this P-47D-28-RA had just been fitted with a new rudder. Like other P-47s assigned to the 57FG, it carried its (scorpion) squadron emblem unofficially on the cowling. (Author's Coll.)

***ABOVE AND RIGHT:* When the pilots of the 350FG received replacement fighters in the form of bubble canopy P-47Ds in autumn 1944, they generally performed their duties as flying artillerymen, although there were some medium bomber escort missions. Beginning in December, they were operating out of Pisa. Capt. Arthur Scramm was not only adept at handling his own P-47D, he managed to record many of the formation flights with great talent in spite of bumpy conditions that were generally encountered over the Apennines with peaks reaching as high as 9560 feet. He flew with the 345FS which marked their fighters with a single black-banded yellow lightning flash. Some of these Jugs were so new that they did not even have time to add unit markings of any kind. On the other hand, there is graphic evidence that P-47D-28-RA 5C6 had already completed 100 bombing sorties. (Arthur Scramm/AAF)**

against an airfield at Ghedi, the first of three missions launched against *Luftwaffe* targets on the 24th. It was a wild venture in which Hosey's P-47 was damaged and he was given a shower of hot oil in the cockpit. Some got into his eyes, so he was grounded by the flight surgeon. Knight went off on a second mission that day, and upon returning he described events to Hosey. Even single attacks against airfields are known to be extremely dangerous, and to make repeated attacks against them on any day is essentially suicidal. Hosey stated he made a serious attempt to dissuade Knight, but Ray had seen something on his final pass and he was determined to destroy it. Pilot and airplane attrition in the 350FG was extremely high at the time in the face of intensive fighting.

On the first mission of the 24th, Lt. Knight had ordered the other pilots to remain at altitude while he descended to evaluate the target situation. Flying through a hail of flak at high speed, he spotted eight enemy aircraft hidden under heavy camouflage. Returning to his flight, he briefed them on the situation and they returned to the attack at low level. Knight managed to destroy five of the e/a on his pass, and two additional e/a were destroyed by the others in his flight. It was during this attack that Lt. Hosey's T-bolt was hit, subjecting the pilot to an oil bath.

Upon his return to the Pisa airfield, Knight volunteered to lead another flight of P-47s to reconnoiter Bergamo airfield, another heavily defended enemy base near Ghedi. In view of the 100-percent effort by Twelfth Air Force, already in its fourth day, he received permission to proceed. Again leaving the flight in order to evaluate the situation, he made a high-speed dash on the deck through a veritable curtain of flak. Although his Jug sustained very heavy damage, he had discovered an entire squadron of hidden twin-engine bombers and single-engine fighters. After leading the flight of 346FS attackers in a strafing attack, Knight – with his characteristic determination and intensity – thundered across the Bergamo airfield on no less than ten successive passes. Although his aircraft sustained additional damage on two passes, he succeeded in completing the destruction of six 2-engine aircraft and two fighters. Members of his flight had destroyed four other multi-engine types and one fighter. Having survived this second mission, the flight returned to base. His crew serviced and patched his P-47D as rapidly as possible, knowing the lieutenant was determined to continue his assaults.

Early in the dark morning of the 25th, he returned to the Bergamo attack scene with three other 350FG pilots to finish the job. The four-plane flight made a blistering attack, catching one e/a on the runway. Although they managed to destroy three of the twin-engine aircraft, intense gunfire from multiple sites virtually shattered Lt. Knight's aircraft. He managed to gain altitude, determined to return what had been a nearly new fighter so that it could fly again after repairs. With the controls seriously damaged, the fighter failed to respond when caught in violent air currents over the rugged Apennine Mountains of Northern Italy. It was reported he had finally attempted to bail out at minimum altitude, but his effort was too late.

***Looking like it had attempted a Kamikaze attack on Ploesti oil tanks, this 347FS Jug staggered back to base after attacking a target near Brescia, Italy. As Lt. Edwin L. King was completing a strafing run, his P-47D took a direct hit in the lower section of the R-2800. With the entire canopy and windshield bathed in hot oil, the fact that Lt. King was able to safely find the ground is truly remarkable.* (AAF 62824AC)**

With the Allied forces poised to cross the Po River to establish a bridgehead, it was known those *Luftwaffe* aircraft were intended to carry out desperation attacks which could have inflicted great casualties on Allied troops and equipment.

For his gallantry in action, Lt. Raymond L. Knight was awarded (posthumously) the Medal of Honor. His heroic attacks on remaining German aircraft damaged the enemy's ability to fight to such an extent, they surrendered unconditionally within just a few days.

The *Força Aérea Brasileira* had established the 1° Grupo on 18 December 1943, placed under command of Air Major Nero Moura on 3 January 1944.

After an intensive training program on Republic P-47D Thunderbolts, concluding at Suffolk AAB in New York, the FAB squadron was transported to Livorno, Italy. Moving quickly to Tarquinia airfield, pilots and crews received their new P-47D-25-RE fighters, soon supplemented with P-47D-27-REs, -28s and -30s. The unit was assigned as a fourth squadron of the 350FG, the single day fighter group assigned to XII Fighter Command. Five other Thunderbolt groups operated by the Twelfth Air Force were assigned to XII Air Support Command (ASC) in Italy.

Brazil's FAB squadron went into action for the first time on 14 October 1944, becoming the first (and only) South American force to enter combat operations in Europe against Axis forces. In the intense winter of 1944-45, the battle of the Po Valley was notable for the desperate fight put up by the German military after the Italian surrender. In that Northern Italian Campaign, the FAB unit flew 445 missions, mostly ground attack involving some 2546 flights.

No doubt the part played by Twelfth Air Force fighter and fighter-bomber groups operating in Italy was a key factor in driving the Nazi war machine from Italy. Republic Thunderbolts proved to be an optimum assault weapon because of their tremendous firepower, ability to withstand damage from some of the most powerful anti-aircraft weaponry in the world and a bomb-lugging capability unmatched by any other single-engine high performance aircraft.

***A Twelfth Air Force P-47D is pictured landing, probably at Grosseto, Italy, late in 1944 or early 1945. The only markings that give any clue as to the group or squadron are painted wingtips and the name "Eddie & Irene" on the starboard cowling. The pilot's landing approach appears to be near perfect.* (Fred Bamberger)**

The 10,000th Republic P-47, AC44-20441, wound up in the war assigned to a tactical role. It carried on in yeoman style, primarily aiming its weaponry at German ground targets. This particular P-47D-30-RE was the recipient of one of the latest design Curtiss propellers featuring the widest paddle blades fitted to Thunderbolts on operational status. No pilot identification was provided, but he was a member of the 79FG operating in Italy. (Kevin Brown)

BELOW: One WWII publicity photograph that has created many false impressions showed Brazilian (FAB) and Royal Air Force (RAF) P-47D-27-REs flying together over Long Island in January 1945. Both aircraft had camouflage paint schemes and carried official FAB and RAF markings, respectively. However, Triple Seven never was delivered to the FAB, being assigned to the AAF in July 1945. Brazilian markings were removed and replaced with standard USAAF insignia, but it is not known if the camouflage paint was removed in the process. (Mfr. G-147)

BELOW: Carrying an unapproved squadron emblem on its cowling, a late-series P-47D operating with the 350FG in the XXII TAC out of Pisa, Italy, temporarily reposes on the ubiquitous Marston mat that was one saving grace of field operations. The number 7 and the large letter A identify the aircraft as belonging to the 347FS. (NASM)

OPPOSITE: A blizzard of publicity accompanying the wartime release of this picture stated that the P-47D-28-RE was one of those provided under Lend-Lease to the Brazilian Air Force. It was taken on charge by the Força Aérea Brasileira (FAB) on 28 October 1944, eventually carrying the unit identity A6 on the cowling. This T-bolt (AC44-19663) was sent to the MTO to serve with the FAB unit. The AAF 350FG stationed in Italy had the Brazilian 1° Grupo de Aviação de Caça attached, serving as a fourth squadron, and that organization flew its Lend-Lease aircraft in combat operations out of Pisa. (Mfr. H-3280)

Armed to the teeth with 500-pound bombs, two rocket tube clusters, and full ammo bays, this P-47D features the fighting red gamecock emblem identifying it as a team player with the 65FS that had been called to active duty on 15 January 1941. The 57FG, like Hub Zemke's 56FG, had been on station at Mitchel Field and in several New England bases before heading overseas. Unlike the famed 56th, this group did not receive any P-47s until 1944. (NASM)

Purged of its weaponry, including machine guns, this P-47D-27-RE that bears a well worn legend of "The Ox Cart" on its starboard fuselage panels must have suffered unseen damage or it was just plain worn out. Apparently based at Grosseto, Italy, with the 57FG and assigned to the 65FS, it displayed the large, professional and approved squadron emblem on both sides of the cowling. Of interest is the fact that only emblems of the 65th and 66th Fighter Squadrons had official AAF approval, but all 64FS fighters displayed theirs in the same fashion. However, the approval for that emblem was not forthcoming until 1956. Perhaps Maj. Gen. John K. Cannon, C.G. of MATAF, sanctioned it. (NASM)

ABOVE: A closer view of the P-47D-30-RA engaged in hedge-hopping over German farm country as it escorted a B-25J reveals that it was named "Norma," and bore the 65FS fighting red rooster insignia on the cowling. The fighter also carried two 110-gallon external tanks on the wing pylons plus a 75-gallon belly tank. Nazi capitulation had ended the war in Europe and these airplanes were evidently in process of being transferred to new owners stationed in Germany with the Occupation Forces. Literally thousands of dangerous sorties had been flown at low level, frequently in terrifying weather conditions. (via James Harp)

BELOW: A pair of Força Aérea Brasileira P-47D Thunderbolts, each carrying a 75-gallon droptank but without bombs on the racks, were photographed taking off from a Northern Italy airfield, probably on a training flight. The FAB operated exclusively with the XXII TAC in the Twelfth Air Force and as a fourth squadron with the very active 350FG, primarily based at Pisa. (AAF 56926AC)

LEFT AND ABOVE LEFT AND RIGHT: Three Força Aérea Brasileira Thunderbolts, each carrying a different assortment of external stores, were photographed at Pisa during a period of intensive air operations. The fighter with three external drop tanks was the personal aircraft assigned to the commanding officer, Lt. Col. Nero Moura (No.1). Bearing the code D5 and showing 93 missions flown, a P-47D-28-RE was assigned to 2nd Lt. Meira. Rockets, bombs and a 75-gallon belly tank adorn the P-47D-25-RE (C1) flown by Blue Flight leader, Capt. Fortunato. (T. Savage via Hagedorn)

OPPOSITE TOP: Displaying full FAB markings, this 1°Grupo de Aviação de Caça Republic P-47D-25-RE (AC42-26758) is shown operating from a former Axis airport which had been heavily damaged in earlier Allied attacks. Although all FAB Thunderbolts received via Lend-Lease were camouflaged, the Brazilians flew some unpainted "exchange" Jugs. In turn, the AAF 350FG airplanes were not known to have been painted. However, Lt. Raymond Knight's Medal of Honor Thunderbolt (6D5), named "Oh Johnnie," was a 350FG, 346FS airplane, but factory-applied FAB paint (and the Air Corps serial number) guarantee that it had been delivered to the Brazilians. It is probable that Lt. Knight's own P-47D, having sustained major battle damage during a prior strike, was replaced by a D-27-RE airplane (AC42-26785) that had been transferred from the 1° GAvCa at Pisa. (Dan Hagedorn Coll.)

ABOVE: After having been subjected to some arid and frequently mountainous terrain of Italy for a year, the North American B-25J crewmen and the P-47D-30-RE pilot from the 57FG probably wondered where the war had been fought in this area as they flat-hatted across German landscapes. That No. 54 airplane was interesting because of its camouflage paint scheme. But here again is another airplane that had been allocated on Lend-Lease to Brazil, only to be reallocated to a group in desperate need of replacement fighters as losses to enemy flak mounted. Nobody is likely to ever know how many Allied ground force lives were spared solely by the ground-attack sorties carried out by Ninth and Twelfth Air Force fighter-bombers. (via James Harp)

Of all the Thunderbolts operating in the MTO, some of the most flamboyantly marked were those of the 79FG. All had blue vertical tailplanes marked with an assortment of yellow lightning flashes that varied in shape and size. (See color sections. "The Trojan Warhorse" was literally a flying signboard.) Evidently the blue paint used was of poor quality, possibly color-mixed in the field so that color was inconsistent and tended to fade. (Fred Bamberger)

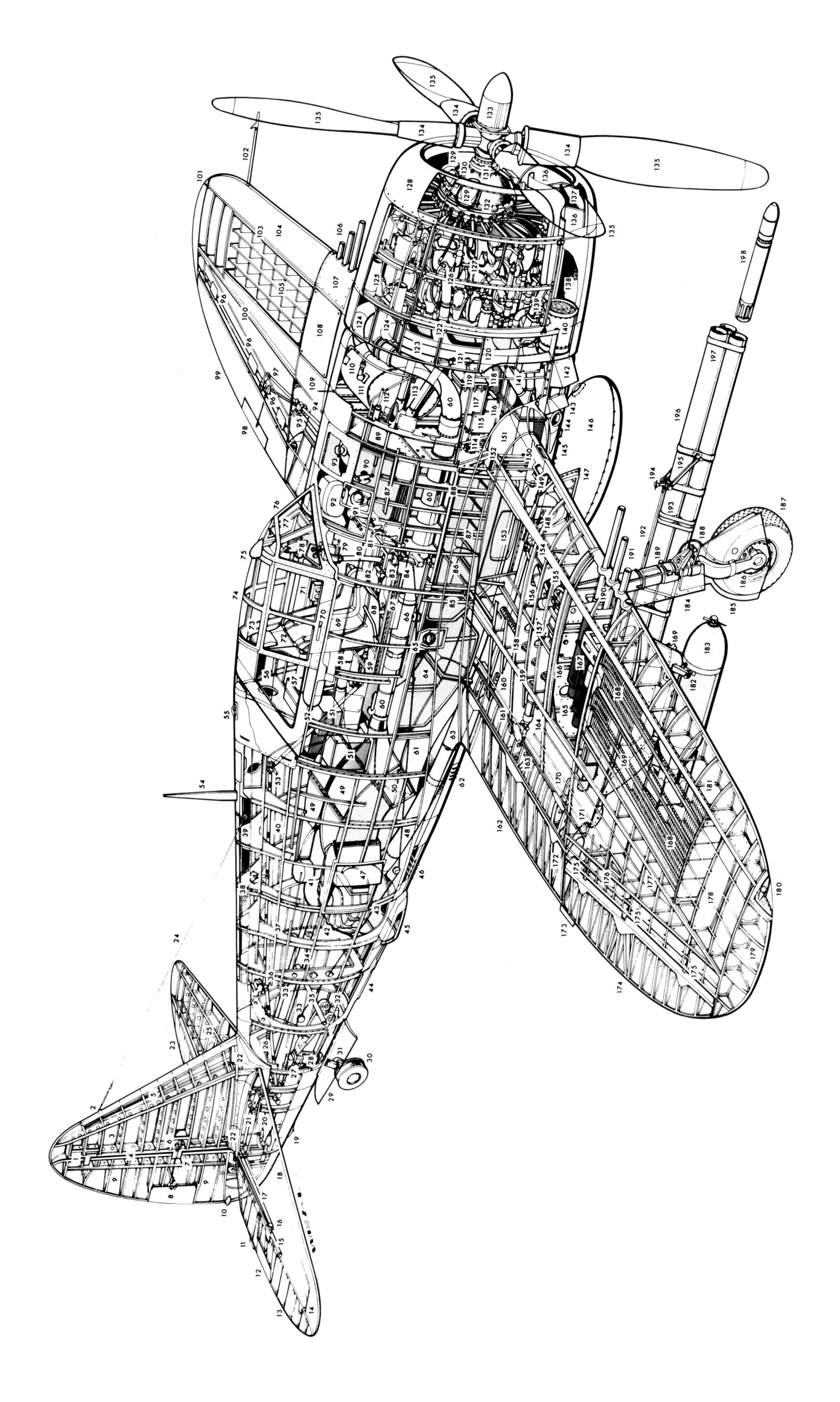

Republic P-47D-10-RE Cutaway Key

1. Rudder upper hinge
2. Antenna wire attach point
3. Vertical stabilizer (fin) ribs
4. Rudder post and fin aft spar
5. Fin front spar
6. Rudder trim tab actuating mechanism
7. Rudder center hinge
8. Rudder trim tab
9. All-metal rudder structure
10. Tail navigation light
11. Elevator fixed tab
12. Elevator trim tab
13. Right elevator structure (metal covered)
14. Elevator outboard hinge
15. Elevator torque tube
16. Elevator trim tab actuating mechanism
17. Trim tab actuating drive chain
18. Right horizontal stabilizer
19. Aft fuselage jack point fitting
20. Rudder control cables
21. Elevator control rod and linkage
22. Fin aft spar attachment points
23. Left elevator (metal covered)
24. Antenna wire
25. Left horizontal stabilizer structure
26. Tailwheel retraction worm gear
27. Tailwheel shimmy damper
28. Tailwheel oleo strut
29. Tailwheel doors
30. Retractable, steerable tailwheel
31. Tailwheel fork
32. Tailwheel mount and pivot assembly
33. Rudder cables
34. Rudder and elevator trim control cables
35. Aft fuselage lifting tube
36. Elevator control rod linkage
37. Semi-monocoque all-metal fuselage structure
38. Fuselage dorsal spine ("razorback")
39. Antenna wire lead-in insulator
40. Fuselage stringers
41. Supercharger air filter
42. G.E. turbosupercharger
43. Supercharger turbine case
44. Supercharger compartment air exhaust vent
45. Turbosupercharger exhaust hood (shroud)
46. Aft fuselage air exhaust louvers
47. Intercooler exhaust air doors (l. & r.)
48. Engine exhaust pipes
49. Intercooler exhaust air ducts
50. Supercharger intercooler unit
51. Radio transmitter-receiver units
52. Sliding canopy track
53. Elevator push-pull rod linkage
54. UHF radio antenna mast
55. Formation light
56. Rear vision Plexiglas panel
57. Oxygen bottles
58. Supercharged air (to carburetor) pipe (LH side)
59. Elevator rod linkage joint
60. Supercharged air (to carburetor) pipe (RH side)
61. Intercooler inlet air duct
62. Central fuselage cooling louvers
63. Wing root fillet
64. Internal auxiliary fuel tank
65. Auxiliary fuel tank filler cap
66. Rudder cable pulley
67. Cockpit floor and support
68. Seat adjustment lever
69. Pilot's seat assembly
70. Canopy external emergency release handle (RH shown, LH opposite)
71. Trim tab controls
72. Pilot's seat back armor plate
73. Headrest
74. Cockpit sliding canopy
75. Rearview mirror and fairing
76. V-shaped windshield assembly
77. Bulletproof glass
78. Reflector gunsight
79. Engine control quadrant, LH console
80. Control stick
81. Rudder pedals
82. Oxygen regulator
83. Elevator control quadrant assembly
84. Rudder cable linkage
85. Wing aft spar-to-fuselage attachment fittings
86. Wing-fuselage lower bulkhead section
87. Main fuel tank
88. Forward fuselage structure
89. Stainless steel firewall assembly
90. Cowl flap valve
91. Main fuel tank filler cap
92. Anti-freeze fluid tank
93. Hydraulic reservoir
94. Aileron control rod
95. Aileron trim tab control cables
96. Aileron hinge access panels
97. Aileron, trim tab control linkage
98. Aileron trim tab (left wing only)
99. Frise-type aileron
100. Wing aft spar
101. Left wing navigation light
102. Pitot-static tube/head
103. Wing front spar
104. Wing L.E. stressed-skin panel
105. Machine gun ammunition troughs (4)
106. Machine gun barrel blast tubes
107. Gun barrel access panel
108. Machine gun servicing access panel
109. Forward gunsight bead
110. Oil feed pipes
111. Oil tank
112. Hydraulic pressure line
113. Engine mounting upper support struts
114. Engine control correlating cam
115. Anti-icing fluid pump
116. Fuel level transmitter
117. Generator
118. Battery junction box
119. Storage battery
120. Exhaust collector ring
121. Cowl flap actuating cylinder
122. Collector ring engine exhaust stacks
123. Cowl flap area
124. L. & R. supercharged air duct elbows to carburetor
125. Exhaust air tubes
126. Cowling support ring (Typ.)
127. Pratt & Whitney R-2800-63/-59 Double Wasp 18-cylinder radial engine
128. NACA cowling nose ring
129. Magnetos (Scintilla or G.E.)
130. Propeller governor
131. Propeller shaft seal housing
132. Engine reduction gear case
133. Propeller electric pitch change mechanism cover (hub)
134. Curtiss propeller blade cuffs
135. Curtiss constant-speed electric propeller (4-blade, 12 ft. 2 in. diameter)
136. Oil cooler duct inlet
137. Turbosupercharger/intercooler air intake duct (two)
138. Cooling air inlet ducting
139. Oil cooler plumbing
140. Oil cooler (two)
141. Engine mounting lower support struts
142. Oil cooler temperature control door
143. Deflector
144. Exhaust waste gate
145. Centerline drop tank/weapons shackles
146. Auxiliary drop tank (75-gal. metal)
147. MLG inboard wheel well closure door
148. MLG inboard door actuating cylinder
149. Gun camera (right side only)
150. Cockpit air-conditioning air inlet (right side only)
151. Wing root fairing
152. Wing front spar-to-fuselage attachment fitting
153. Wing inbd. rib MLG wheel well recess
154. Wing front sparcap
155. MLG strut pivot forging
156. Landing gear hydraulic retraction cylinder
157. Aux. (MLG mounting) wing spar
158. Machine gun bay warm air duct
159. Wing aft sparcap
160. Wing flap inboard hinge assembly
161. Aux. (flap mount) wing spar inbd. section
162. NACA slotted landing flap
163. Wing flap center hinge assembly
164. Landing flap actuator hydraulic cylinder
165. Four .50-caliber Browning machine guns
166. Gun bay inboard stub rib
167. Ammunition feed chutes
168. Ammunition troughs
169. Underwing fuel tank/weapons pylon
170. Wing flap outboard hinge assembly
171. Flap access door
172. Wing flap rib (typical)
173. Aileron fixed (adjustable) tab (RH wing only)
174. Frise-type aileron structure
175. Aileron hinge (steel forging) assemblies
176. Aux. (aileron mount) wing spar outboard section
177. Multi-cellular wing construction
178. Wing outboard ribs
179. Wingtip assembly structure
180. Right navigation light
181. Leading edge rib sections
182. Bomb stabilizing adjustable supports
183. M-43 500-lb. demolition bomb
184. Landing gear strut fairing
185. MLG wheel fairing
186. MLG axle support arm
187. MLG wheel/tire assembly
188. Hydraulic brake line
189. MLG hydraulic shock strut
190. Machine gun barrel blast tubes
191. Machine gun barrels (staggered)
192. Triple rocket launcher slide bar
193. Launcher tube center strap
194. Triple rocket launcher front mount
195. Mount support struts
196. Type M-10 triple-tube 4.5-inch rocket launcher
197. Launcher tube front strap
198. M-8 folding fin 4.5-inch rocket

CHAPTER 16

FAR, FAR AND AWAY

Springtime, 1940. Personnel at Selfridge Field, Michigan, began to see the first of thirteen Republic YP-43 interceptors which would be undergoing the rigors of service testing there. It is probable one or two of the YPs were retained at Materiel Division, Wright Field, Ohio, for testing because there was no 'X' (Experimental Model) airplane ordered. Perhaps two of the new aircraft were operated out of nearby Patterson Field by the Production Engineering Section. That newly organized section of Materiel also retained jurisdiction over the YP-43s which would be flown as part of the accelerated flight testing program by the First Pursuit Group. War or no war in Europe, Selfridge Field was hardly less peaceful than New York City's Central Park on a 1940 weekday.

Suddenly, great numbers of Americans had their ears glued to their Philco or Majestic radios every evening and well into the wee hours of the night. They could not believe what Edward R. Murrow and H.V. Kaltenborn were saying about the Nazi invasion of Holland, Belgium and France. The "Phoney War" was over and soon forgotten. *Blitzkrieg* was the word on everybody's lips. Britons were rousted from lethargy as they never had been before and most likely never would be again.

France, especially its Army, was in chaos; they were trained to fight a plodding WWI, not a Lightning War. Belgium and Holland never did really anticipate being involved, again, in a major invasion, preferring to look the other way as Central European countries simply vanished from the scene. Stiff-collared British Prime Minister Neville Chamberlain, the very epitome of Etonian stuffed-shirt pragmatism – a non sequitur, I presume – was so hidebound he could not believe Adolf Hitler would look him in the eye and lie about anything. But he was the man at the helm. (One chokes on saying "man in charge.") The 120,000,000 Americans on the North American Continent were fascinated, as with a cobra, but they really only wanted to look at that far away continent of Europe obliquely.

Far Away! Americans, in those days, hardly knew what far

A much better breed of Thunderbolt than the earlier P-47D-2-REs appeared in the north of Australia at Eagle Farm about the middle of the Australian summer season, a bit after Christmas Day 1943. That new breed included P-47D-16-REs with underwing attach points for drop tanks and bombs in the 1000-pound realm, plus other performance bonuses. The presiding 58th and 348th Fighter Groups were in combat in New Guinea when this model was taken on charge. Lt. Col. Neel Kearby became one of the AAF's top scoring aces at the controls of this type of fighter, having found the combination to its attributes quickly. (Author's Coll.)

away meant. If you lived in Santa Monica or Long Beach, Los Angeles was 'far away.' But it would not be long before they would learn first hand what it meant. Who knew much of anything about New Zealand and Australia. Anyone venturesome enough to vacation in Australia had to be wealthy and most likely bored with your basic travel objectives. Even American businessmen rarely, if ever, traveled to Japan. The only people who really knew a lot about the Philippine Islands were U.S. Navy personnel and a handful of Army and Air Corps types. The closest thing to a Republic YP-43 at Clark Field was one of those Curtiss P-36s that had finally been purchased in some quantity in 1937. (When the Curtiss-Wright Corporation recovered from the 1936 shock of being trampled by that upstart Major de Seversky and his pitifully small factory out there in a place called East Farmingdale, their lobbyists pulled Curtiss' chestnuts out of the fire.) Basically, a bunch of obsolete Boeing P-26s comprised the defensive force out there in the western Pacific in support of the group of Curtiss P-36As in 1940. The tiny Philippine Air Force was still operating some Keystone B-3 and B-6 biplane bombers. And, like Alaska and the Aleutians (what/where were they?), those Western Pacific islands (P.I.) were out of sight and out of mind. Venturesome Pan American Airways was struggling to establish pioneer flight paths across the Pacific, but it was hardly thinking of tourist travel to Japan.

The really far away places were India – probably about 12,000 miles from San Francisco – Burma, Siam, Australia and Antarctica. To people in North Dakota, Georgia, Alabama, those earth spots might just as well have been on Mars. In fact, thanks to writer-actor Orson Wells, Mars was far more familiar. Closer, but still far away, France had collapsed. The weaponless British Expeditionary Army was back in England, and the first skirmishes of the Battle of Britain had been fought in late spring 1940.

But by springtime twenty-four short months later, it would be difficult to find a literate American who did not have some familiarity with those places. How many Americans knew, back in 1940, that Australia was the largest island in the world? That it was ten times as large as the second biggest island, New Guinea? Or that the latter island was three and a half times bigger than Great Britain? All of that changed in 1942. Americans would soon be living on all three of those islands, not comfortably or necessarily by choice.

The geographical scope of those places is dwarfed by the Pacific Ocean-Far East area in which Americans would be forced to do battle with its enemy. From the Aleutian Islands to the southern shores of Australia, and from the West Coast of the United States to the western borders of China and India, Americans were going to have to fight under water, on the surface of many oceans, on previously unheard of ground and in that vast airspace. Only in the relative quiet of the Aleutians (after 1942) were Republic P-47 Thunderbolts unheard, unseen.

Control of the air in that unbelievably vast region was never gained by one predominantly successful type of airplane. Early on, Curtiss P-40s and Lockheed P-38s lead the AAF league in the air battles. A few fistfuls of Thunderbolts, Spitfires and Hurricanes

Caught by the camera shutter a split second before touchdown, a Republic P-47D-2-RE (no wing pylons) was one of the first T-bolts allocated to the Fifth Air Force in the SWPA. Gen. George Kenney, commanding the 5AF, only wanted P-38s because they were the sole fighters with range capabilities to fulfill his requirements. The early P-47Ds had no external fuel tank provisions in the summer of 1943, so they enjoyed no better combat radius of action than the Curtiss P-40s. Just as bad at the time was the sluggish rate of climb. The Jug shown in this photograph was landing at Eagle Farm Aerodrome, Australia, to complete its first post assembly flight. (AFM)

were involved and gave yeoman service, but range was a major consideration. Early P-47s arriving in the SWPA for the Fifth Air Force had voracious fuel appetites and no external tank provisions at all. Failing to gain support from Air Materiel Division because of higher priorities for other theaters, the Fifth Air Force people developed their own tank design (just as the 8AF had in the ETO and as the 15AF people in the MTO adapted their Jugs to carry plentiful Lockheed P-38 tanks. The 200-gallon drop tanks[1] produced in Brisbane, Australia, by Ford Motor Co. were not pretty, but they came close to doubling the combat radius of action. And it was Lt.Col. Neel Kearby's tactical operations style and aggressiveness which made the Thunderbolt such an efficient killing machine in his hands. Of course the performance of Kearby and his 348FG improved General Kenney's disposition greatly, although he was – logically – one of the greatest advocates of the Lockheed P-38. It was as if the Lightnings had been designed from the outset for the war in the Aleutians and in the Southwest Pacific Areas.[2] (What was needed was more 3-man teams like George Kenney, Ben Kelsey and Neel Kearby.)

[1] The so-called Brisbane tanks would never have received any standing ovation or approval from Air Materiel Division. It mattered not that they were efficient, enabling the P-47s to do the job General Kenney demanded. (In the ETO, you could bet Kenney – had he been in command there – would have flatly rejected Republic's 205-gallon "cow's udder" tanks and would have immediately authorized Lt. Col. Cass Hough's Bovingdon team to develop efficient tanks as early as April 1943.)

[2] The biggest failures affecting the long-range escort success of P-38 Lightnings in the ETO were not aircraft design faults. Planning and command decisions were the biggest factors. The type was not "second-sourced" from the outset; the volume-production Packard V-1650 Merlin proposals for use in P-38s as early as 1941-42 were ignored by the WPB; and, of course, when the P-38 groups in England were diverted to Operation TORCH late in 1942, B/Gen. Hunter did nothing to demand transfer of existing P-38 groups from the Z.I. to the ETO. As a result, nine to ten precious months were lost. (With early second sourcing, Kenney might have received all the P-38s he needed.)

Beware of rashness, but with energy and sleepless vigilance go forward and give us victories – A. Lincoln.

Banking away from the camera plane, a 348FG Thunderbolt reveals its classic wing shape that would never attain the notoriety and fame P-47s found in the ETO. With a far smaller stable of fighters in hand, and at least four different types to service, it is likely that logistics problems half a world away from New York proved to be a serious handicap. With a far greater combat radius of action, the Lockheed P-38 proved to be the main choice of General Kenney. Incidentally, it is not generally known that the 348th group used the all-white tail markings on P-47Cs flown in New England, prior to departure for the SWPA. This has led some to write that P-47Cs were used "down under," a terribly false premise. Those early series Thunderbolts, lacking external drop tank provisions, were totally unsuitable for coping with the expanses of the SWPA. **(AFM)**

Major operational users of Republic P-47s in the Fifth Air Force during the war were the 8FG, 35FG, 49FG, 58FG and the 348FG. It was hardly an easy life, fighting a war with a new fighter type some half a globe away from the manufacturing facilities, maintenance training and service and supply commands. But just think of all the great things these young men were learning about geography!

Far, far and away was the nemesis in fact for fighters of the Seventh Air Force. This Central Pacific Area force, initially based in the Hawaiian Islands, was another great potential home for Lockheed P-38s pending ultimate development of the P-47N type and North American P-51Ds. And, with the arrival of P-38Ls in considerable numbers as the longest ranging fighter, the AAF was in the cat-bird seat.

For a long time in the central Pacific, it had to be largely a Navy, Marine and Army island-hopping war. Long-range escort of Seventh Air Force bombers was essentially out of the question. Bell P-39s and P-400s, along with any of the Curtiss Warhawk P-40 models assigned to VII FC were absolutely committed to shore patrol and not a lot more. None of those types was ideally suited for conducting long-range operations over hundreds of leagues of boring but dangerous water. If they and the early 1944 arrivals of Thunderbolts wanted to reach those "far away places with strange sounding names," they had to hitch a ride on one of many aircraft carriers, courtesy the U.S. Navy. Ultimately, in mid 1944, some fairly decent numbers of razorback P-47Ds joined the ranks of "flying artillery" pioneered by their distant cousins in the VIII, IX and XII Fighter Commands a bit earlier. Even later, the P-47Ns with long legs were getting off small ground at gross weights in the twelve-ton neighborhood, albeit not without some painful losses. Unexpectedly, the P-38 tanks they carried were rarely carrying fuel for the R-2800-C Double Wasps; more often than not they were filled with hot-burning Napalm, an effective weapon of war against the entrenched and stubborn Japanese enemy forces.

Seventh Air Force combat groups which pledged allegiance to the Republic P-47 for most of the Pacific War were the 15th, 318th, 413th, 414th, 507th and 508th Fighter Groups. Few pilots flying in those groups ever saw the high altitudes for which the Thunderbolt was intended, at least on combat missions. And, in fact, long range was not the usual key to their success. The Jugs had become perhaps the ultimate dive bombers of World War II. That 100-degree per second rate of roll was rarely necessary, the speed of

LEFT AND OPPOSITE: A newly assembled P-47D-21-RA took off from Eagle Farm to complete a post-assembly checkout flight, but something went very wrong. The landing gear failed to extend, or did not lock down, and the surprised pilot found himself alighting on full 165-gallon drop tanks. They emptied almost immediately, but luckily there was no fire. It appears that all fuel moved outboard. It is obvious that the 81st Repair Group was going to have some extra work. Taking a lead from the early-arrival 348FG, all Fifth Air Force, V FC Thunderbolts had white tails and, usually, white wing leading edges. Those large P-38 drop tanks were available and somebody must have been well aware that Bob Baseler's 325FG in the MTO had used them with great success. **(AAF 4989, 4990)**

more than 450 mph at 25,000 to 35,000 feet was only seen by a few in the Seventh Air Force, and merely a handful of pilots "remained seated" to test themselves with the combat radius of action of up to 1100 miles (or more). For that small favor, they were probably thankful.

Only one Thunderbolt group, the 80th, could technically be considered as a part of the Tenth Air Force in the China-Burma-India theater of operations. Trained in the Z.I. to fly Thunderbolts, and expecting to see combat in the ETO, they went far away to the other side of the world and much too close to the equator to be comfortable. How they must have envied their near cousins assigned to the 78FG operating in the British Isles. Not only that, they did not become acquainted with new P-47s, as they had expected, for some prolonged period. What they got instead was some Curtiss P-40 Warhawks. (They might well have turned to Uncle Sam and said, "Are you mad at me or something?") They eventually did get into P-47D Jugs and perhaps had a better feeling about their war, if that was at all possible.

Unusual situations resulted in the creation of the 1st Air Commando Group in India in late summer, 1944, comprising but two squadrons. They were the 5th and 6th Fighter Squadrons (Commando).

Changing conditions ensured that they only flew combat operations for about seven months before being inactivated at the beginning of November 1945. They flew almost all missions from Indian bases.

The 80FG campaigned for a considerably longer period, principally in India-Burma and Central Burmese campaigns.

With the opening of a major Japanese Army offensive in Burma in March 1944, No. 221 Group of the Royal Air Force was reinforced with eight fighter squadrons transferred from Bengal. However, those squadrons were equipped with the typical short-range fighters supplied to the RAF in that period, so they had no real fighter capability with which to strike at enemy air bases. The USAAF's 459FS of the 80FG, flying Lockheed P-38s, had to be called upon when airfield strikes were necessary. Lightning air superiority over enemy fighters and fighter-bombers had forced the Japanese to move back, well beyond range capabilities of the RAF fighters, primarily Hawker Hurricane IIcs. Production of P-47Ds had risen to such high rates by spring 1944 that the RAF was enabled to obtain a substantial number of late-model razorback P-47Ds, redesignated as Thunderbolt Mk.Is... in all, the Royal Air Force received 239. Following close on their heels...rudders if you prefer...a further quantity of 587 bubble-canopy P-47Ds was allocated to the RAF, with most of these Thunderbolt Mk.IIs serving in Burma, generally supplanting the Mk.Is in time. Large numbers of these two versions were camouflaged aircraft out of Farmingdale, N.Y., at about the same time the Brazilian squadron was being equipped with Jugs for duty in Italy.

Most Thunderbolt Mk.Is and some Mk.IIs were fitted with Hamilton-Standard Hydromatic propellers, but seal problems eventually resulted in all new deliveries being equipped with Curtiss Electric propellers. Most of the earlier aircraft were retrofitted with the Curtiss units.

Although the Thunderbolt type was not perfect for operations in Burma – what airplane could be in WWII?– it was well suited for the job at hand. The RAF pilots, for the most part, had little trouble making the transition from the docile Hurricane to the big ol' Jug.

(Refer to Appendix C for a comprehensive listing of RAF squadrons operating in Burma and elsewhere. Refer to Appendix B for listing of Republic P-47D Aircraft Allocated to the Royal Air Force.)

As everyone must be aware, the Fourteenth Air Force was faced with daunting problems to say the least, and range was not the least of them. Their P-47 Thunderbolts served them pretty well, and it is really too bad they could not have flown squadrons of P-47N versions, enjoying 100-degree per second roll rates and at least an 1100-mile combat radius of action. Nobody should ever forget that China was about as remote as it was possible to be a half century ago. Logistics was a subject most people would not even want to think of encountering half a world away from Evansville or Farmingdale. Once again, the P-38 Lightning was a valuable tool, and that seemed to be especially true with the F-5 recon version. On the face of things, one might have said, join the 14AF and really see the world...a world totally (or nearly so) like so much fiction to the multitude of Americans circa 1941. The task of just learning to spell most of the names of bases in which they were ensconced must have been tiring. And it would not have been surprising to find they had flown Seversky P-35s in combat, considering all other types they flew in the war.

Nobody need ever feel that he did not do his part in the war while flying combat with the 33rd Fighter Group, and especially the 58FS. And the 81st Fighter Group and its three squadrons seems to have been everywhere and into every possible battle, be it Morocco, Algeria, Sicily, Italy, India or Fungwanshan, China. (Now, class, how many have ever heard of that one?)

Can you imagine what occurred when anyone serving in those units was asked, "What did you do in the war, daddy?"

"Well, son, if I did nothing else, I flew P-47 Thunderbolts, and what more could anyone ask."

With some twenty confirmed victories in hand, Lt. Col. Neel Kearby was moved from command of the 348FG to the V FC of the Fifth Air Force. It may have been the Christmas holiday season when actor John Wayne was on a USO tour to Australia and was photographed with fighter ace Kearby. (Republic)

Subtropical conditions guaranteed that few good wartime P-47 photos would come from Fifth Air Force units, but this is one of the better views. The P-47D-21-RA carries proper 310FS markings for a 58FG airplane, except that the AAF serial number has been overpainted for some (security?) reason. However, it shows through. The 310FS commander's aircraft – with unit marking H34 – was lost with a 311FS pilot at the controls. (Eugene Sommerich)

Taking off from an unidentified Australian airfield that most likely was Amberly Field or Eagle Farm, four P-47Ds with white tails from either the 58th or 348th Fighter Group head for a formation assembly. All carry the standardized 165-gallon P-38 drop tanks on the wing pylons, giving each as much as 330 additional gallons of fuel and nearly doubling the non-encounter range. (AFM)

ABOVE: Sasha Kartveli never, ever had a dream that his heavyweight P-47s would operate from aircraft carriers, especially those tiny escort carriers that were converted from Maritime Commission tanker hulls. However, when effective strike aircraft were needed in the battle of the Marianas following the securing of a large portion of Saipan – the island was not free of organized resistance until 9 July – Thunderbolts of the 318FG were launched from the decks of the USS MANILA BAY and the USS NATOMA BAY by 23 June 1944. They landed on a newly constructed airfield. The former carrier is shown under attack by four Japanese aircraft as it approached Saipan. (AAF 52905AC)

ABOVE: In a view from the island of an unidentified escort or light carrier, a few of some of the dozens of razorback P-47Ds that were transported to Hawaii in 1943 illustrates the manner in which they were prepared and secured. Headed for Hickam Field, they were initially offloaded at Ford Island. These "hogs" were probably assigned to the 318FG and may have eventually flown from other carriers to a base on Saipan. (AFM)

RIGHT: A 73FS Thunderbolt, bearing markings similar to those applied to the ETO P-47s early in 1943 except for the vertical fuselage band, flashes over the bow of the carrier. Only the most minor incidents marred the group launchings in a well-orchestrated operation. The pilot of No.29 is performing in the best Navy style, head forward from the headrest and flying the airplane off the deck. (BuAer 238686)

One of the Republic P-47D-11-REs is shown running up to 52 inches Hg. (manifold pressure) immediately prior to being catapulted from the USS MANILA BAY. The Jugs had been fitted with novel catapult sling attachments and holdback fittings at Bellows Field, Hawaii. Each aircraft carried a 75-gallon drop tank for the short flight to the island. In mid July, the 333FS launched from the USS SARGENT BAY to join the rest of the 318FG on Saipan at Aslito airfield, renamed Isley Field within a short time. (AAF B-52837AC)

LEFT: When the 318FG Thunderbolts were catapulted from escort carriers to make the short flight to Saipan, the pilots probably had little idea that they would remain in the "pleasant" environment of Saipan for about 10 months. This P-47D-15-RE (No.78), showing yet one more depiction (small) of the "Dallas Doll" cowgirl-in-blue (barely discernible), awaits only the fusing of one bomb before setting off to attack the Japanese on Tinian. (AAF)

RIGHT: With her tail held high, "Lady Ruth," a P-47D-15-RE assigned to Lt. Howard Kendall (nicknamed "Big Stoop") from the 19FS awaits completion of bore-sighting and gun harmonization procedures at Aslito airfield (Isley Field). This old-line squadron dates back to 1917 in France, had been based on Hawaii from 1923. The 318FG operated a handful of P-38Js from Kagman Field, Saipan, for long-range strikes. (Dave Menard Coll.)

BELOW: As the crew chief is seen busy filling the 75-gallon belly tank carried by P-47D "Lady Ruth," armorers work on the .50-cal. machine guns and the pilot stands impatiently on the port wing. All personnel appear to be clean and neat in contrast to the reports of dirty clothing and poor personal hygiene caused by a lack of adequate water and weather conditions. Could this have been a media event? (NASM)

ABOVE AND LEFT: Equipped with a pair of triple tube Bazooka-type rocket launchers mounted inboard of the wing pylons, P-47D-20-RA "Miss Mary Lou" was readied for a close-support assault on enemy positions. Combinations of bombs, rockets and napalm tanks carried by the 318FG Jugs proved to have deadly effect on the enemy troops. (NASM)

Two 19FS Jugs taxi out at Isley Field, decked out with pairs of 500-lb. bombs and the 75-gallon belly drop tank, ready to take that short hop to Tinian Island. These high-altitude fighters rarely had to see 5000 feet on the altimeter. Their real nemesis was small-caliber anti-aircraft fire. Control of the air in the Marianas Islands was held by AAF and Navy pilots. (AFM)

Lockheed P-38s dominated the scene in the SWPA; then, after Republic P-47s helped Army and Navy forces take the Marianas Islands, other Lightnings flew long-range strafing and escort missions to soften up Iwo Jima. With hundreds of P-51s now available for escorting Boeing B-29 bombers from that island, Japan's fate was fast being sealed. Here a trio of hot pilots try to overshadow each other in flying a compact formation. Col. James Beckwith of the 15FG was identified as flying the P-51D. Phew, that is too close for comfort. (56685 USAF)

The pilot and crew chief assigned to "Big Squaw" proceed through a cockpit check during a temporary lull in the action. Designed as an interceptor fighter with high-altitude capabilities, the big P-47 Thunderbolts were several notches better at the dive-bombing job than the Luftwaffe's Junkers Ju 87s and even larger Ju 88s. The P-47Ds enjoyed many more important features including far heavier gun firepower, no requirement for additional crew members and they had up to double the combat speed. (R.L. Cavanagh Coll.)

RIGHT AND OPPOSITE: Accidentally or purposefully – we really don't know – the photographer of this P-47D-23-RE, No.66 (located at Barrackpore, India, in March 1945) pictured a D-21-RE with a negative-like image on the same day. As a result, we have a fine comparison of reversed markings on nearly identical fighters, one camouflaged and one that was delivered sans paint. The only fly in the ointment seems to be that the dark airplane has the AC serial numbers in black, making them all but impossible to see. While that was early official policy, addition of high-visibility stripes negated the logic of that policy. The group/squadron identities were not provided, but it is almost certain they were assigned to the 1st Commando Group. (P.M. Bowers)

Making their departure from an unidentified eastern base in India to support 81BS Mitchells in an attack on Japanese targets in Burma, two P-47D-23-RAs show that there was a certain amount of inconsistency in applying squadron markings. Virtually every Jug flying in the basic Tenth Air Force/Eastern Air Command operations carried a pair of 165-gallon Lockheed P-38 drop tanks. (Howard Levy)

BELOW: A team member of the 1st Air Commando Group of P-47D-23-RA fighters is shown departing from Barrackpore, India, probably for the regular base at Asansol in the last quarter of 1944. Five canted black bands on the fuselage identified all fighters and bombers instantly as being operational with the Commando group. (Howard Levy/75533AC)

Intentionally or otherwise, a fighter pilot was always gambling with fate in making a one-wheel, hot-landing touchdown, especially with external tanks still in place. This 1st Commando Group P-47D-23-RA (AC42-28153) could easily have blown a tire on impact. It is probable that other aircraft lined the runway (see Douglas C-47 in background) and a swerving, out-of-control fighter could take a catastrophic collision course. For those at this Eastern Indian base as of the 8 November 1944 date, the war in the Far East was still very active. (Howard Levy)

BELOW: If a 6 December 1945 date is correct, the appearance of this 1st Air Commando Group P-47D at Kiangwan Airdrome, Shanghai, China, suggests that it was to be turned over to Chiang Kai-shek's Chinese Air Force since hostilities were over. Appearance of the partial "slash" tail marking (see beneath tailplane) is unusual . Normally a 93FS, 81FG band extends upward from the horizontal tail to the fin tip; it rarely appeared on fuselage. That B-25J in the background had been totally disarmed. (P.M. Bowers)

RIGHT: In the process of being fully updated with the latest in approved markings for the FAM, this Thunderbolt displays the white petal cowl marking and the Mexican national insignia together with the distinctive tail stripes applied to 201 Escuadron P-47s. Ground support equipment seems to have been in short supply at Clark Field in August 1945. (USAAF 61243AC)

Another Thunderbolt, in the guise of a P-47D-30-RA that had been flown as part of the 1st Air Commando Group, appeared with several close relatives at Kiangwan Aerodrome in China shortly before Christmas 1946. Again it is presumed that the ultimate destiny for this fighter was assignment to the Chinese Air Force's 11FG. Equipped with the latest in propeller blades; the Jug appeared to be especially well maintained. Those are 110-gallon drop tanks under the wings. That low number (2) on the fuselage together with yet another narrow tail stripe would most likely define the P-47 as being assigned to an HQ officer. (P.M. Bowers)

LEFT: The P-47D-28-RA in the immediate foreground carries all of the markings of a 41FS, 35FG airplane in the AAF, but since all of the Thunderbolts in the background are Lend-Lease airplanes in the 201 Escuadron of the Fuerza Aerea Mexicana (FAM) its exact status is unknown. At the time it was photographed, it may have just been placed on loan with the FAM. The location was Clark Field, 13 July 1945. American propeller manufacturers had a strong lead in product development as the last year of the '30s decade dawned. However, they quickly fell about 5 years behind the Germans in blade design by the end of the year. (NASM)

BELOW: Bearing the tail stripes adopted by Fifth Air Force groups that had moved up to the Philippine Islands as part of the assault to drive the Japanese out, this P-47D-28-RE (No. 65) carries a "petal" nose marking attributed to the 40FS of the 35FG. Significance of the single black fuselage band, whereas the normal Philippine Invasion marking was two wide parallel bands, is unknown. Two 165-gallon P-38 drop tanks are mounted on this fighter which has the name "Violet" in script letters on the right side. (Author's Coll.)

ABOVE: Returning to base, a Republic P-47D-23-RA makes a smooth approach for a landing at an airfield in China. This Jug displays the markings of the 81FG which had done yeoman service in North Africa and Italy before being shifted to India and then on to China in the CBI Theater. Serving in the group were the 91FS, 92FS and 93FS, all equipped with P-47Ds beginning in 1944. (Author's Coll.)

Here, again, is another markings dilemma from the Western Pacific theater of operations. This P-47D is marked with just one wide black stripe on the fuselage, it has a "petal" cowling paint scheme indicative of a 35FG fighter, yet it exhibits a very freshly painted 58FG white tail. That probably was a carryover from the SWPA home of the 5AF along with the subsequently standardized prewar tail stripes. However, that normal blue vertical bar has been overpainted white. It appears to be a 58FG airplane transferred to the 35FG, even to the locale. (Eugene Sommerich)

LEFT AND BELOW: As the war in the Pacific reached a crescendo of violence, 81FG Thunderbolts like this P-47D-30-RA (#910) went through the daily routines of maintenance and then preparation for flight, time after time. The slash tail markings applied in India or China were certainly distinctive. It is a virtual certainty that this fighting machine found its way into General Chiang Kai-shek's Chinese Air Force at the end of hostilities. (McKay via L. Davis)

BELOW: One of a small number of P-47D-30-REs (Thunderbolt Mk IIs) used by 73 Operational Training Unit at Fayid, Egypt, is seen in flight over harsh landscape. At least one of the Mk IIs delivered to the OTU was painted black overall with silver striping and was flown by the chief instructor at Fayid. (Martin & Kelman)

BELOW: A somewhat dogeared P-47D-15-RA was photographed at Kiangwan Aerodrome, Schwangliu, China. That backslash tail marking indicates that the airplane was flown in the 60FS, 33FG assigned to the Tenth Air Force. After having fought well in the MTO, the group was moved to the CBI for transition to Thunderbolts. During the summer of 1944 the squadron was attached to the Fourteenth Air Force but returned to Burma-India (and the 10AF) after September. The tail number is all but obliterated.

Another razorback "hog" sits on a wet Chinese airfield in 1945, already having sprouted Chinese national markings. It was equipped with a pair of battered 75-gallon drop tanks. In the event, it looked rather forlorn and any USAAF markings that might have remained are unseen. (McKay via Larry Davis)

BELOW: A Thunderbolt Mk II (P-47D-28-RE) seen at the factory and awaiting acceptance trials was eventually assigned to 73 OTU in Egypt. It was the eighth D-28 delivered to the RAF in a group allocation of 170. It bore the RAF serial number KJ137. That pale blue aft fuselage band generally designated a Thunderbolt assigned to other than a combat unit. (Mfr. H-3181)

BELOW: Although the Royal Air Force deigned not to use any early series Thunderbolts in homefront squadrons, preferring to stay with the Hawker Typhoons and (late in the war) the Tempests, some 826 of the P-47Ds were used extensively in the unfriendly environment of Burma. Range, load-carrying capabilities and demonstrated toughness were factors in that assignment. This Thunderbolt Mk I (FL839) served with RAF squadrons in Southeast Asia, being a mainstay in the Burmese campaign. It was a P-47D-22-RE (originally AC42-25787) and is shown at Farmingdale prior to delivery. (Frank Strnad Coll.)

ABOVE: Maintaining aircraft in heat exceeding 100 degrees (F) with high humidity could not have been much more than bearable. Virtually all RAF Thunderbolt Is, normally equipped with Curtiss electric propellers, were used by the RAF for transitional training. Those equipped with H-S propellers were almost universally employed by the combat units. In service, the H-S props were affected by tropical conditions and tended to blow seals, but electrical corrosion problems plagued the Curtiss propellers. Three P-47D-22-REs sweep over FL804 (134 Sqdn) undergoing routine maintenance at a Burmese airfield, probably after they attacked targets in the south or for some media event. (Mfr. H-6803)

As pilots and their crews await orders to move out to attack targets as far south as Rangoon, three Mk Is do a flyover, most likely returning with empty drop tanks. HD173 in the foreground at Chittagong, Burma, has the propeller cap painted red, common to many 135 Sqdn. P-47Ds. Seen in November 1944, all fighters are seen with vertical and horizontal empennage identification bars extending onto rudder and elevators. The Air Ministry objected to the practice, ordering deletion from the moveable surfaces because they were convinced it affected control balance. (IWM)

Following a tradition established in 1943 by the MTO-based 325FG then commanded by Col. Robert Baseler, the drop tank of choice for P-47Ds operating in the MTO and the CBI was Lockheed's 165-gallon unit that carried 137 Imperial gallons of fuel. No location identity was supplied but the permanent building structures indicate that it was a major base in England or India. No unit markings of any kind can be observed on this aircraft.

BELOW: Apparently awaiting preparation for shipment overseas, this P-47D-30-RE, identified as KJ346, was photographed at Newark, N.J., on 16 February 1945. The pale blue aft fuselage band indicates that it was intended to be employed as a trials (evaluation) aircraft rather than being destined for combat assignment. A twin (KJ299) to this airplane was sent to the United Kingdom two months earlier, and it is presumed this airplane was headed for the same destination. Shortly thereafter, Thunderbolt IIs were being delivered without any camouflage paint. (H. G. Martin)

BELOW: One of the relatively rare (to the RAF) P-47D-27-RE Thunderbolt Mk IIs – only 30 of the mark delivered – appeared in one of the best T-bolt photos to come from the SEA theater of operations. This fighter, serial HD298, carries the markings of 30 Sqdn., RAF. Some sixteen operational squadrons flew P-47Ds in the SEA Command. Unseen squadron code letters forward of the roundel were RS. It was not uncommon to see RAF pilots taking off or flying their P-47s with the canopy open. This was in the Arakan area, November 1944. (Author's Coll.)

CHAPTER 17

THE LONG RANGERS

Quite beyond comprehension, one of the most important aspects of World War II aviation history in America was almost entirely overlooked or ignored until the July 1974 issue of *Airpower* magazine was published. The story grew from discussions with C. Hart Miller in the course of preparing a multi-part story about the development history of the Thunderbolt.

The far-reaching implications of General James Doolittle's decision to build Gen. William Kepner's VIII FC around the new long-ranging North American (Merlin-powered) P-51 were innumerable and threatening. Then there was Air Materiel Command's seeming infatuation with unattainable performance guarantees for the General Motors-Fisher P-75A, an airplane which began life as an assemblage of overproduced obsolete components. Some more pragmatic officials were able to prevail in the face of rather dismal performance. The P-75 concept was totally re-evaluated/redesigned around a new high-altitude, mechanically-supercharged, unperfected Allison V-3420 engine, producing somewhat less than the projected 3000 horsepower. With those credentials, it was bally-

This vertical view of the XP-47N in the summer of 1944 reveals the modifications made to the previously unchanged – since 1931 – wing planform created in cantilever monoplane form just about four short years after Lindbergh flew from New York to Paris. Compact 20-inch wing stubs had been inserted at the fuselage juncture and the normal wingtips had been changed to avoid reduction in the rate-of-roll. Generally overlooked is the fact that in initial form, the XP-47N lacked any sign of a dorsal fin. (AAF C-28768AC)

hooed as the forthcoming champion long-range escort fighter for *high-altitude*(!) bombers. (What it really was destined to be was a pitchman's pattern for Preston Tucker to follow in offering a new, advanced automobile – the Tucker – in postwar times.) Fisher engineers touted the P-75 as a fighter with a 1500-mile combat radius of action! That converts to about a 3600-mile range with reserves. The Procurement people awarded contracts for about 2500 of the P-75As almost before a prototype flew and were tottering on the brink of ordering an additional 1500 to 2000 in autumn 1944. But it is necessary to evaluate performance at Fisher (Cleveland) and G.M. on the project up to that time.

Designer Donovan Berlin, whose main claims to fame were the good Curtiss P-36A and the fair P-40, proposed a large, long-range fighter that could use components of aircraft already in production. As may be seen in this view of the Fisher – as in General Motors – XP-75 with its Douglas SBD aft fuselage and tail, Vought F4U landing gear, Curtiss P-40 wings and (naturally) a G.M.-Allison V-3420 engine, it looked like a swoose (half swan, half goose). Any way you cut it, the much earlier Republic XP-69 of similar proportions and far more logical design was a superior concept. It was abandoned before this abortion was conceived. General Motors built two, and a huge new factory at Cleveland, Ohio. **(Author's Coll.)**

Maximum demonstrated speed: 433 miles per hour in very light condition, about the speed capability of the Republic XP-47B four years earlier. However, XP-75 performance evaluation was not an APGC attainment, and by the time they were flying the more conventional P-75A version, speed had fallen to just over 400 mph. The 433 figure would hardly stand up to a serious challenge. Unfortunately, pressures – like subtle hints during telephone conversations – from the WPB and other areas obviously had more impact than dull, factual test reports signed by some captain.

THE BRINK OF EXTERMINATION...

A scheduled (but oh so confidential) occurrence during the latter half of 1944, with the war in Europe at a pivotal crescendo, was *total termination* of P-47 Thunderbolt production at the two factories completed only about 30 months earlier! (Of course it had been necessary for General Motors-Fisher to build an entirely new plant in Cleveland, Ohio, at a later date, supposedly to provide complete power "eggs" for B-29s. Somehow, G.M. deftly sidestepped that project over which they had no real control and would hardly gain more than a small fixed fee over costs.) The impact on Long Island, N.Y., and southwestern Indiana would have had far-reaching implica-

General Motors managed to get their former chairman of the board, General Knudsen, and his WPB to oversee procurement of six completely new prototypes – using no existing components – and no fewer than 2494 production examples. Not even featuring high-altitude supercharging and with an attained top speed of 404 mph (in 1944), this monstrosity was going to fight off Focke-Wulf's new Ta 152, Fw 190D, Messerschmitt's Me 262 and Dornier's Do 335A? That should have sent Sen. Harry Truman's Senate investigation committee into convulsions. The P-75A-1-GC was about 20 percent bigger than a P-47N on virtually identical horsepower (below 20,000 feet for the P-75A). If nobody else could make contra-rotating propellers work reliably, how was Fisher going to do it? **(Air Materiel Command)**

***Test pilot Carl Bellinger is seen at the controls of the first of 1667 Republic (Farmingdale) P-47N airplanes ordered on Sales Amendment SA-31 to contract AC29279. Committed to flight testing, the airplane had the company's typical chrome-yellow cowling and empennage. It took few flights in the XP-47N and the YP-47N to reveal a strong need for additional aft sideplate area. A neat XR-12 type ventral fin would have been more effective.* (Mfr. H-7817)**

tions.[1] One of the worst aspects would have been punishing Republic Aviation for having manufactured about 13,500 outstanding fighters (at the time) and, concurrently, rewarding Fisher and godmother (G.M.) for taking nearly three years to build less than a half dozen second-rate prototypes. At least some of those were assembled from parts built by others. It took only two prototypes to prove how fallacious the thinking leading to the project had been[2]. By then G.M. had its big foot in the door, and the only way to get it out was to win the war, fast.

This brings up a major point and some more viable options than the one chosen. Aren't military planners supposed to project into the future, not retreat? Why would they choose components from obsolete aircraft of pre-1935 concepts (i.e., Curtiss P-40,

[1] With the advantage of hindsight, of course, it is easy to conclude there had to be more than a powerful cadre of quislings in the WPB, or command control was out to lunch.

[2] Most interestingly, nobody in government ever stepped up to claim credit for the XP-75 conglomeration of parts idea. However, former Curtiss chief engineer Donovan Berlin stepped into a General Motors slot and fathered the Fisher XP-75, not forgetting to include a Curtiss P-40 set of wing panels designed about 1934 (for the Hawk 75 prototype).

***Some rather famous test/combat pilots were among this group of Republic personnel assigned to fly P-47Ns. Although chief tester Lowery Brabham is not in the picture, standing (l.-r.) are unk. mechanic, Ken Jernstedt, J. Croft, Frank Simpson, Parker Dupouy; kneeling are Carl Bellinger, Peter Collins, Fillmore Gilmer, and unknown. Jernstedt and Dupouy were both AVG combat pilots before joining Republic Aviation.* (Mfr. H-19667)**

ABOVE AND BELOW: Pictured beautifully in flight in the vicinity of Wright Field, a production P-47N-1-RE displays its fine features. As the great-grandson of the nearly unknown Seversky AP-4, this offspring was big and tough. Number 19 was inappropriately named "The Repulsive Thunderbox" by some irreverent person at the AAF's main test base. (Air Materiel Command)

Douglas SBD) and, in 1943, an undeveloped, unproven engine? They had a great long-range, combat-proven fighter with modern construction technology of flush-riveted skins and metal-covered control surfaces staring them right in the face, namely the Lockheed P-38. Can't you envision a smooth production changeover from the P-38H to the J configuration but with two Packard-built Rolls-Royce Griffon 2050 hp engines providing the power? Identical engines powered Spitfire fighters and high-altitude reconnaissance airplanes capable of operating efficiently at 44,000 feet *without turbo-superchargers*. With at least 4100 horsepower on tap in 1944, a P-38 Lightning with a proven ability to best German and Japanese fighters in every theater of war in which the AAF fought would have moved into the production lines with minimal disruption.

About ninety percent of the then-current airframe would have been able to accommodate the Griffons, and all 1944-45 Lightnings had a proven range capability greater than most pilots could withstand.

***LEFT AND OPPOSITE PAGE: A pilot's view of the left console and the instrument panel of a typical P-47N reveals that the airplane had, indeed, progressed amazingly from the day that the Army Signal Corps received its first aeroplane. The driver of this vehicle was a bit busier than the person driving your father's Oldsmobile.* (C.A.M. via Ed Boss, Jr.)**

If C.L. "Kelly" Johnson and Col. Ben Kelsey could not have produced a winner out of that combination with the greatest of ease, we had no right to win the war. The Fisher plant, if already constructed, could have manufactured parts for those Lockheeds. The millions wasted on the P-75 program would have gone far toward efficient production of useful combat aircraft.

One other even more viable solution was already on paper in the offices facing Conklin Street on Long Island. In fact it had been flown a year earlier as a half-hearted trial balloon, but for unknown reasons the germ of a good idea was not fertilized. That idea manifested itself in the form of an old P-47C-5-RE fitted with two 27-inch stubs between the normal wing panels and the fuselage. The idea was to find a way to carry 200 extra gallons of gasoline internally. After a few unimpressive flight tests, the project went dormant. For one thing, Republic was so busy getting up to speed in producing P-47Ds in 1943, developing the XP-47J and initiating work on the XP-72 design, they had little time to devote to range enhancement on any priority basis, and it wasn't funded. What a difference a year and Regensburg-Schweinfurt would make.

Prodded by several of the events mentioned, the need for large quantities of limited-range fighters was rapidly evaporated, and *range, range and more range* became the new rage.[3] Lockheed hammered away at manufacturing between 15 and 16 Lightnings (P-38L series) every day, six days a week, reaching a production peak in August 1944. North American happily churned out long-range, producible, Packard-Merlin-powered P-51Ds and Ks, the new darlings of the AAF. (Wasn't the AAF offered a great Merlin-powered P-38 two years earlier? Even Ben Kelsey wasn't able to penetrate to the core of the Corps with that great idea.) Unaccountably, the Bell organization up in Buffalo brought P-39Q Airacobra production to a halt in that same summer month, gaining some attention in the process. They merely stretched the machine into a Kingcobra (P-63A) format and began producing them at the rate of about 250 planes a month. (Did anyone happen to notice the XP-63, powered with a 1710 cubic inch, 1325-hp Allison, had almost exactly the same top speed as the Fisher XP-75-GC, which was powered by a 3420-cubic inch, 2600-hp Allison???) Three thousand Bell P-63As a year rolled from the lines during the Battle of the Bulge, and not one of them ever joined the Fighting Ninth Air Force. The AAF actually received something in excess of 3000 Bell P-63 *fighters*. But they weren't chasing Fw 190s or Me 109Gs or Me 262As around the Continental skies at all. Why produce thousands of fighters which would get whipped every day in the skies over *Festung Europa*? Did all Americans really want to win the war or did some prefer to build up a war chest for postwar competition?

Despite all the words relating to acceptable range capabilities of the P-47D, the fact stands it did not have a radius of action[4] of one thousand miles, the magic number in 1944 as the *Luftwaffe* began to sag. OVERLORD was proving to be the success hoped for, and short-range fighters would become a penalizing factor in the Western Pacific. (No P-63s there either.) Lockheed's P-38J-25s and all P-38Ls and North American's P-51Ds enjoyed combat radius capability of more than 1000 miles. In actual fact, the thousands of late series P-38s were the clear winners when it came to range. For unknown reasons, they were not supplied with the 310-gallon drop tanks in the ETO or MTO, although they were a standard operational item in the Far East/SWPA. Perhaps there was poor interchange of information, a real possibility. More likely, Hunter's negligence in demanding assignment of any new replacement P-38 groups to VIII FC eventually required almost panic skipping of the normal OTU phases of training. The 55FG and 20FG were committed to on-the-job training. Probably only Ben Kelsey would have been fully aware of the 310-gallon tank situation, but he had assignments elsewhere and was being prepared for the No.2 command job in the Ninth Air Force. Falsely attributing range honors to the P-51D in Europe, writers seem to conveniently forget all about those very large tanks and the extra fuel carried in the wing leading edge by P-38J-25-LO and subsequent airplanes in the L series. If any long-range testing of Fisher P-75As was conducted by APGC at Eglin Field or Orlando, those tests were surely never performed in competition with P-47s, P-38s or P-51s of the latest models. Never pictured with any external drop tanks as far as we can tell, the P-75A would had to have been a flying gas tank to attain the ranges claimed. In their desire to impress the procurement people, Fisher equipped the P-75A with ten .50-cal. machine guns, although the P-47s had proven eight were more than adequate. After all, at that stage the range situation was far more important than increased firepower.

Throughout the entire second half of 1943, as the fighter strength of the Eighth Air Force was building toward the target strength of 20 fighter groups for bomber escort, efforts aimed at expanding drop tank capabilities increased dramatically un-

[3] Trying to navigate one's way through the thicket of political expediency is more than difficult. If the short-range fighter was losing popularity, why didn't the Allison-powered Bell P-39 and P-63 programs dwindle or die? Why were some officials trying almost desperately to order thousands more Allison-powered P-75As when they couldn't even get one to meet guarantees? If the Allison V-3420 was worth a tuppence, why didn't Lockheed's XP-58 and Boeing's XB-39 programs prosper? Can anyone point to a V-3420-powered winner in any category? Hands! Please.

[4] As a memory refresher, combat radius of action, as defined for the USAAF, included takeoff and climb to assembly, cruise to destination, use WEP for 5 minutes of combat, fight for 15 minutes at military power, cruise to home base, descend and land with 30 minutes reserve fuel remaining. Of course that is arbitrary and idealistic, but it was a viable pattern.

der the watchful eyes of General Kepner. It was a remarkable contrast to what was going on in the first half of 1943 in VIII FC. However, few of the P-47Ds in the ETO when M/Gen. Doolittle took over command of the 8AF had three-tank (external) capability. The first P-47D-16s coming off the assembly lines were only being accepted in the Z.I. as Thanksgiving approached in 1943, and the initial P-47D-22 and -23 deliveries were not seen until early 1944. Those blocks of fighters were really the first to have good extended range capabilities, but even they were not capable of flying as far east as Berlin from the U.K. Only P-38s and P-51B-Ds demonstrated such capability in the spring of 1944. At the same time, depots and subdepots in Great Britain were staggering under the workload. Undermanned and undersupplied, there was little time for retrofit of large numbers of early model P-47Ds and perhaps some C models.

As 1943 came to a very combative end, there were only eleven operational fighter groups in VIII FC, and many of them had just barely become operational. Two of the groups were equipped with the long-range P-38Hs, unlikely to see many of the more powerful P-38Js in any real numbers until springtime 1944. In fact, the escort situation was so tenuous in 8AF, plans were in motion to bring the 1FG, 14FG and 82FG Lightning units back from the MTO, a plan soon found to be impractical under the demands of Operation POINTBLANK. The bulk of the other nine fighter groups in England were equipped entirely with "razorback" Thunderbolts.

Just as a reminder, Republic Aviation created an unofficial "XP-47N Mule" airplane toward the end of 1943. One 27-inch stub insert was added to each side of the fighter, increasing total span by 4½ feet. So, there was the old "707" razorback P-47C with the greatest wingspan seen on any Thunderbolt. It had, in the main, been created months earlier for aerodynamic testing and maneuverability, and was not equipped with any expanded internal fuel capacity. Flight testing did not begin until D-Day 1944. Standard wingtips and ailerons were employed.

Knowing very well what would occur at Farmingdale and Evansville if a set of longer legs could not be found for the best Thunderbolts just beginning to come into production, Republic quietly set about the task in November 1943. Of course AAF Project Officer Capt. George Colchagoff was a member of the closed society working on range extension, so on 19 May 1944 he issued an Expenditure Order (43C) in the sum of $101,000 for fabrication of one set of wing panels (with operating fuel tanks) to be used for extending the range of one YP-47M airplane.

Exactly one month later, Republic submitted a flight test report along with a proposal drawing PE-208 to AMC. Quite unexpectedly, the company dusted off the P-47C/"XP-47N" (AC41-6601) with the 27-inch wing stubs. Earlier (1943) flight testing had revealed maneuvering sluggishness with standard wing tips; therefore, a set of blunt wingtips replaced those. Drawing PE-208 depicted a D/M-style airplane with 18-inch wing stubs intended to carry 100 gallons of internal fuel on each side of the airplane. Two sets of wings would be required, because different aileron and planform shapes had to be tested. In his proposal letter, Mr. H. Lehne, assistant to Hart Miller in Military Contracts, proposed the use of an airplane certain to cause much confusion in the minds of everyone. He specified the No.3 YP-47M s/n AC42-27387 as the aircraft to be modified. (It must be noted here all company records and military records carry the aircraft as the XP-47N.) On the same day, Project Officer George Colchagoff put in a formal request for 90-day bailment of what then was actually the fourth YP-47M, AC42-27388. It was to be modified per verbal instructions from the Chief, Aircraft Projects Section. Flight tests were expected to take 20 hours following modification.

It seems obvious, although such is not stated specifically, there was great urgency on the part of B/Gen. F.O. Carroll for completing the necessary work. He wanted to use two aircraft simultaneously rather than switch wings twice on the No.3 airplane. Ultimately, after all testing was completed, the No.3 YP-47M retained a final set of long wings. M/Gen. Benny Meyers sent out a teletype on 3 July, notifying B/Gen. Carroll and Colonels Roth and Bogert of the action taken wherein the speedster became the one and only official XP-47N. After approximately 90 days, the No.4 YP-47M received its original set of wings and was retained on the M program. The unofficial "XP-47N" reverted to P-47C-5-RE status, retaining the more powerful R-2800 C-series engine. Its ultimate fate is undetermined. (In additional communications, the No.3 airplane was officially redesignated XP-47N, clearing up what for a few years had appeared to be some documentation mixup.) Total estimated/proposed cost of the modification and tooling program was quoted at $558,988.98. The company included a 4 percent fixed fee for all work involved.

For a mere fraction of money spent to gain the same end by creating the P-75, a years-long program, Republic had designed a far more viable fighter with nearly comparable range (never proven on the P-75A). The XP-47N had demonstrated far greater top speed, and it unexpectedly turned out to be more maneuverable than the P-47D-25 type fighter. In the process, usually overlooked, the Air Materiel Command had created the best fighter-bomber of the war, outshining everything (including Hawker's vaunted Typhoon) in range, vulnerability, speed and weight-carrying capability. The wing created by Mike Gregor and Alexander de Seversky when biplanes were still king of the roost was every bit the equal of a similar wing created for the Hawker Tempest a dozen years later.

Rarely was it possible to have a clear record of strictly classified insider conversations, but Hart Miller furnished this transcript because of its great importance to the Thunderbolt historical records. This telephone conversation between M/Gen. Bennett E. Meyers, CG of Air Technical Services Command,[5] and B/Gen. Franklin O.

[5] Vol. VI of the Craven and Cate book, *The Army Air Forces in World War II*, in an Appendix listing AAF Staff and Command assignments from June 1941 through August 1945 gives this as the official assignment. However, other references and the logic of conversations between Meyers and Carroll infers that Meyers was actually Commanding General, Air Materiel Command.

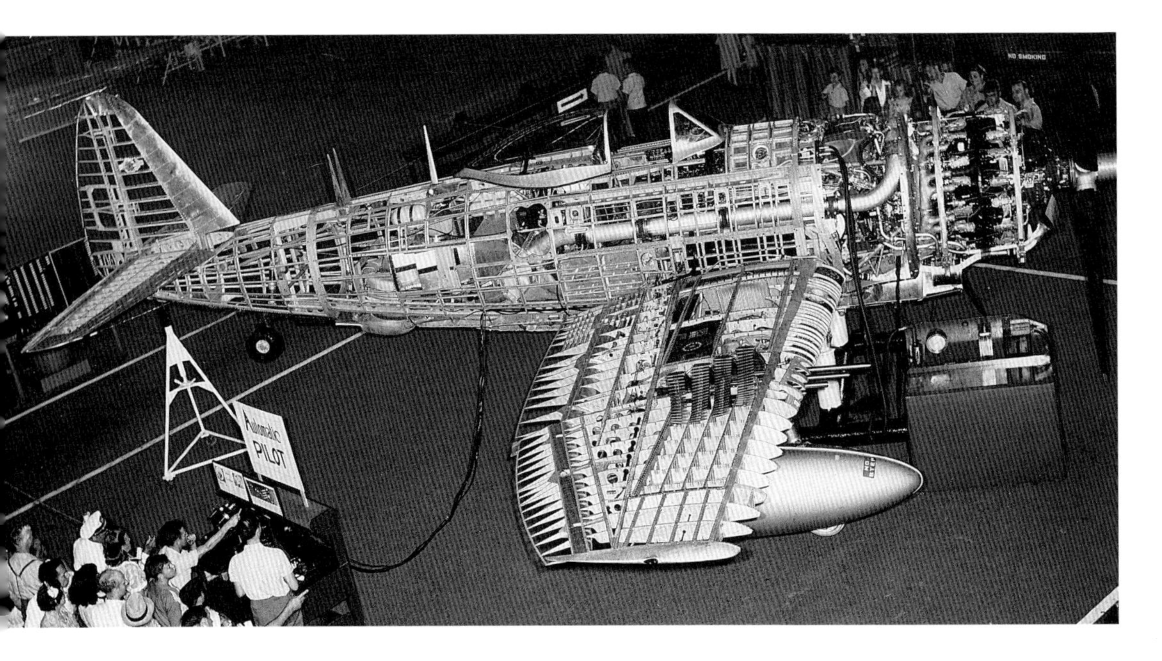

LEFT AND OPPOSITE PAGE: After building Thunderbolts for up to four years, most RACers had no idea of the complexity built into a P-47. Republic management pulled off a coup de théâtre with this "skinless hotdog" of a nearly complete P-47N-5-RE (minus the important dorsal fin). Only the starboard wing was fitted, but many controls were operable from the remote Automatic Pilot station. RACer Frank Strnad served as a T-bolt instructor for most of the war years. Like large numbers of P-47Ns flown in combat, this one carried a Lockheed-design P-38 drop tank. (It is a real crime that nobody with "clout" thought to preserve this wonderful training/ viewing device.) **(Frank Strnad Coll.)**

Carroll (AMC) took place just five days after the proposal letter was written and Colchagoff requested use of an additional test airplane:[6]

Meyers: I told Fred Marchev, today, that I wanted them to go ahead and build a wing if he could do it in three months time or less, that would give us 1000 miles radius on a P-47. You know about the cutbacks on the P-47?

Carroll: Yes.

M: I told them that as a matter of fact, in reality, it would put them out of business because we have no use for the 250-a-month we are going to build this year. The range is the only important thing now. He says he could build a wing in three months; try it anyhow. It would give us at least 1000 miles combat radius.

C: He was out here talking about what they might do. They were going in and make some sketches and bring them out (to Dayton).

M: He started to talk to me about a new airplane. I told him he didn't have time to do that; it was for something later on; if he wanted to build a wing right quick, there might be some sense to it.

C: Alright. Is he going to get up some drawings and come out right quick?

M: Yes.

C: I will get in touch with him.

M: Don't you think that is the right thing to do?

C: Yes.

M: I hate to see an airplane we have had this grand production on just walk out of the picture.

C: Personally, I think it is the best fighter we have.

M: In some cases, I think it is. It is too good to walk out on completely. Have someone call Republic on the wing.

C: Alright.

At the time, Benny Meyers was in New York; Franklin Carroll was at Wright Field. He initialed the transcript and routed it to Col. Howard Bogert and Capt. Colchagoff as well as civilian engineer, Paul B. Smith. The 1944 future of Republic Aviation most likely pivoted on that telephone conversation. The Long Island company immediately went into what seems to be a Lockheed "Skunk Works" mode, probably with some prodding (if needed) from Capt. Colchagoff.

In a 22 July telephone conversation between Col. Roth at Wright Field and Hart Miller at Farmingdale, the following excerpt was of special interest:

Miller: The long-range job has flown.

Roth: What kind of flying machine is it?

Miller: With the same amount of fuel, it can whip the present airplane. It has a better rate of roll and lighter ailerons.

Roth: When did you fly it?

Miller: This morning. It is just ready to go up again.

Earlier in the month, the company Design team had turned out Production Engineering Drawing PE-215 to show the new larger, stronger extended wing design with 18-inch stub sections "for the P-47M." (This reveals how important it is to have more than one piece of documented evidence in researching history. To be correct, the statement should have referred to the YP-47M, the actual aircraft involved. It is also true the designation was officially revised to XP-47N by Gen. Meyers' teletype of 3 July.) The drawings showed where and how the 18-inch stub sections were installed in each wing and how 9-inch blunt wingtips were installed in place of the standard 18-inch elliptical tips. A 100-gallon self-sealing fuel tank was inserted in each wing immediately outboard of the attach point at the fuselage. Aileron and flap changes were also required. Total internal fuel capacity had increased to about 570 gallons, up 200 gallons from production P-47Ds and the YP-47Ms. The major differences between this XP-47N and the P-47D-28-RE, other than the new wing, included the use of an R-2800-C engine producing an extra 100 hp at military rating and an additional 200 hp for the WEP rating. Total span had increased to 42 ft 6-1/16 inches, and the landing gear track was increased by 3 feet, resulting in a track of 18 ft 5 inches, a spectacular contrast to the track of a P-35.

The airplane left the ground for the first time on 22 July, three days short of the required completion date. Truly a "Skunk Works" like effort. It flew in light condition with only 250 gallons of fuel aboard, well distributed in the four tanks. Gross weight at takeoff was 14,500 pounds. Test pilot George Wheat, who submitted the complete test report on 28 July, flew in a P-47D-28-RE chase plane. Subsequently, two combat suitability flights were made with takeoff gross weight at 16,300 pounds, 900 of those pounds being additional structural weight involved in the modifications. Quite unexpectedly, the rate-of-roll (at minimum) was equal to that of the D-28 model (low speed) and far superior to the D-28 at every 50-mph increment including 350 mph. At 250 mph indicated, the rate-of-roll at 100 degrees per second was 15 degrees faster than the D, and at a speed of 300 IAS the XP-47N beat the newest D type by 21 degrees per second.

In mock combat with a D-25 loaded to a takeoff gross of

[6] The quotes are from an official transcription of the conversation.

14,200 pounds, the XP-47N at full combat weight was 2100 pounds heavier than its "foe" for the day. The XP pulled an additional 40 hp on takeoff and climb, pulling 1340 hp at a fuel flow of 120-140 gph. Dive and zoom capability of the XP proved to be better, and it could actually turn inside the D-25, constantly keeping the "foe" in the gunsight crosshairs. An actual formation climb and cruise flight was conducted with three D-25 fighters. The XP flew with two full 165-gallon Lockheed-type drop tanks, the all up weight being 18,455 pounds. After climbing to 25,000 feet, the aircraft flew in formation for one hour. The test was then repeated, but the XP was then equipped with an additional 110-gallon belly tank, raising the gross weight to 19,205 pounds.

A final long-range test involved a non-refueled flight from Republic Airport to Eglin Field, Florida, and return, a total distance of 2100 miles. For this test, two 315-gallon P-38 drop tanks were filled with 300 gallons each. Total on-board fuel was 1170 gallons, upping the gross weight to 20,166 pounds. In company with a P-47D, the pilot reached Eglin in 3 hours 44 minutes, and both external fuel tanks were dropped. Another P-47D which had been flown to Eglin the previous day then engaged the XP in mock combat for 20 minutes, using military power for 15 minutes and WEP for 5 minutes. Bad weather at Republic Airport required the pilot to land at Woodbine, N.J., after covering a distance of 1980 miles, using a total of only 1057.5 gallons of available fuel. The short flight of 120 miles to Farmingdale was made the next morning.

In high speed tests, the XP-47N prototype demonstrated a top speed of on 2800 WEP horsepower of 450 mph at 26,000 feet, while a P-47M-1-RE attained a comparable-condition speed of 465.5 mph. Best speed attained by a P-47D-25-RE on that same day was exactly the same as that demonstrated by the XP.

Subsequent to the testing, a decision was made to put the long-range model into full production, a contract being awarded for manufacture of 1800 examples at Farmingdale. In the final block of three hundred P-47N-25-REs, 166 were delivered when the contract was terminated after the Japanese surrendered. At Evansville, production of Thunderbolts continued through the P-47D-40-RA block on Project 95D, followed by a contract for 200 of the P-47N-20-RA version. The first airplane was delivered on 18 July 1945. That contract was canceled after 149 of 200 airplanes were delivered. The total P-47N contract at Evansville had been for 2200 airplanes.

Republic Aviation had met the challenge beautifully, actually exceeding every expectation of Kartveli and Miller. General Carroll had been pretty much on target in supporting the project, but everything he and Gen. Meyers had hoped for – and really had not expected at all – was exceeded by a substantial margin. The old P-47C "Mule" airplane just had not had its full day in the pool, or it would have shown the way by the end of 1944. Republic had gone the extra mile in making the attempt without even asking for funds. They had really made a good investment and were ready to go at the sound of the bell. One thing stands out strongly in all communications relating to the project: Alexander Kartveli's name never appeared once. If there was any spark plug in the entire P-47N effort, it had to be the tall man, C. Hart Miller. Neither the quintessential pilot, nor the world's premier business executive, he most likely had his detractors, but at Republic Aviation he was the real mover and shaker. If anyone should have generated a motion picture film based on his life, Miller would have been a top qualifier. Jimmy Stewart should have played the role, possibly even John Wayne. The efforts of brilliant Capt. George Colchagoff in boosting the progam alone cannot be discounted in any way.

By August 7th the XP-47N had managed to demonstrate a combat radius of nearly 1200 miles, and if anyone cared to look, the Lockheed P-38L-1-LO (with 1250 on order) was able to stretch out to more than 1500 miles radius. Suddenly, it wasn't the AAF airplanes failing to meet the challenge; indeed, the limiting factor had become the pilot. North American's P-51D/K series was, with that fuel tank behind the pilot, a "bear" to handle; it had little in the way of forgiveness. But once again, the Thunderbolt seemed to have set the pace for future "Rocky" legends. The promise of the Republic P-72 seems to have been overlooked by all. Undoubtedly, a production version with a full military power rating of about 3450 horsepower and a P-47N wing would have been seeing speeds of 500+ mph while carrying external loads never even proposed for any P-51. The August 7 date was important in other ways too. By then, General Carroll was thoroughly convinced the P-47N was a "go" airplane. Just thinking about it and his own perception of the Jugs must have given him a warm feeling. Suddenly the Fisher P-75A was an endangered species. Once again the power of General Motors – with a 100 percent interest in the airplane and its engine – fought off the dogs and kept the entire project from being summarily canceled. It was held in abeyance pending expected increases in power and promises of quick introduction to production rollouts. Even as they spoke, Pratt & Whitney's R-2800-TSC2G Double Wasp was cranking out a reliable 2600 hp (Military Power rating) while the high-altitude Allison V-3420, benefitting from more than 600 extra cubic inches displacement, was breathing hard at 2600 horsepower. Could anyone ignore the performance of the P-47N-1-RE, demonstrating a top speed of 438 mph at a critical altitude of 42,000 feet? In the meantime, a production contract for one hundred P-72As had been awarded, based partially on a guarantee of 3450 hp being available. Also recall the XP-47N prototype showing a rate-of-roll of 100 degrees per second on its third or fourth flight. The P-75A would have been lucky to roll at 50 degrees per second.

Major Frank Borsodi, acting chief of the Fighter Flight Test Branch (and a friend of Carl Bellinger, like so many others), and

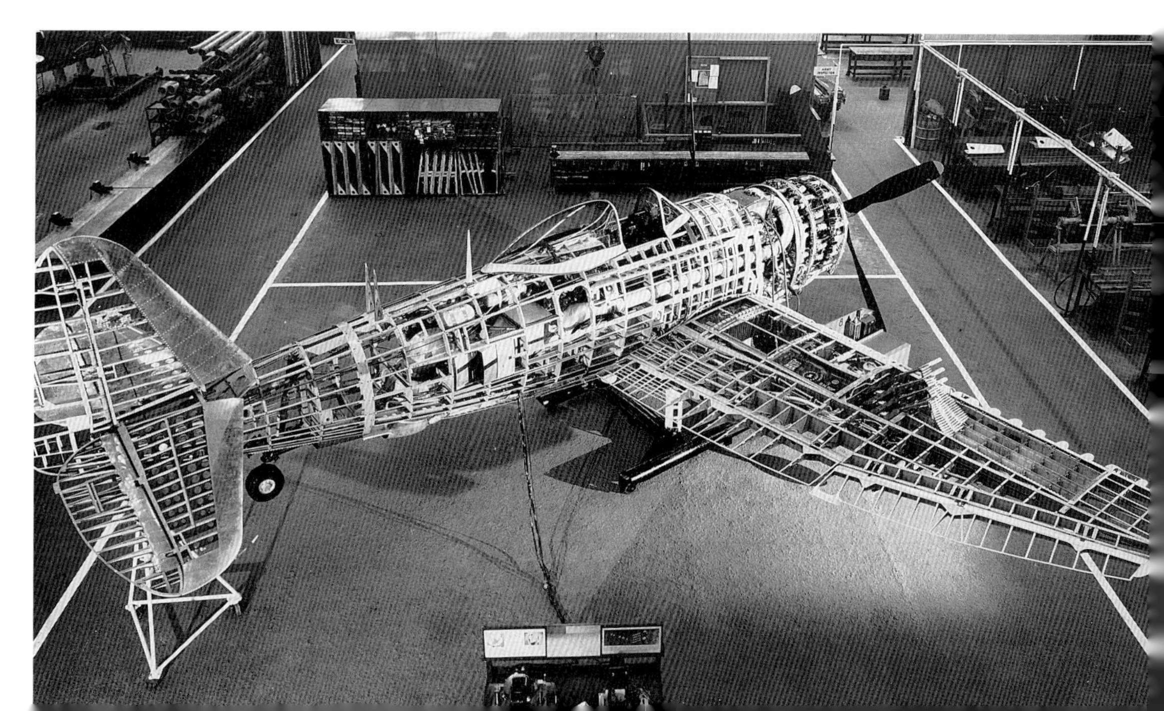

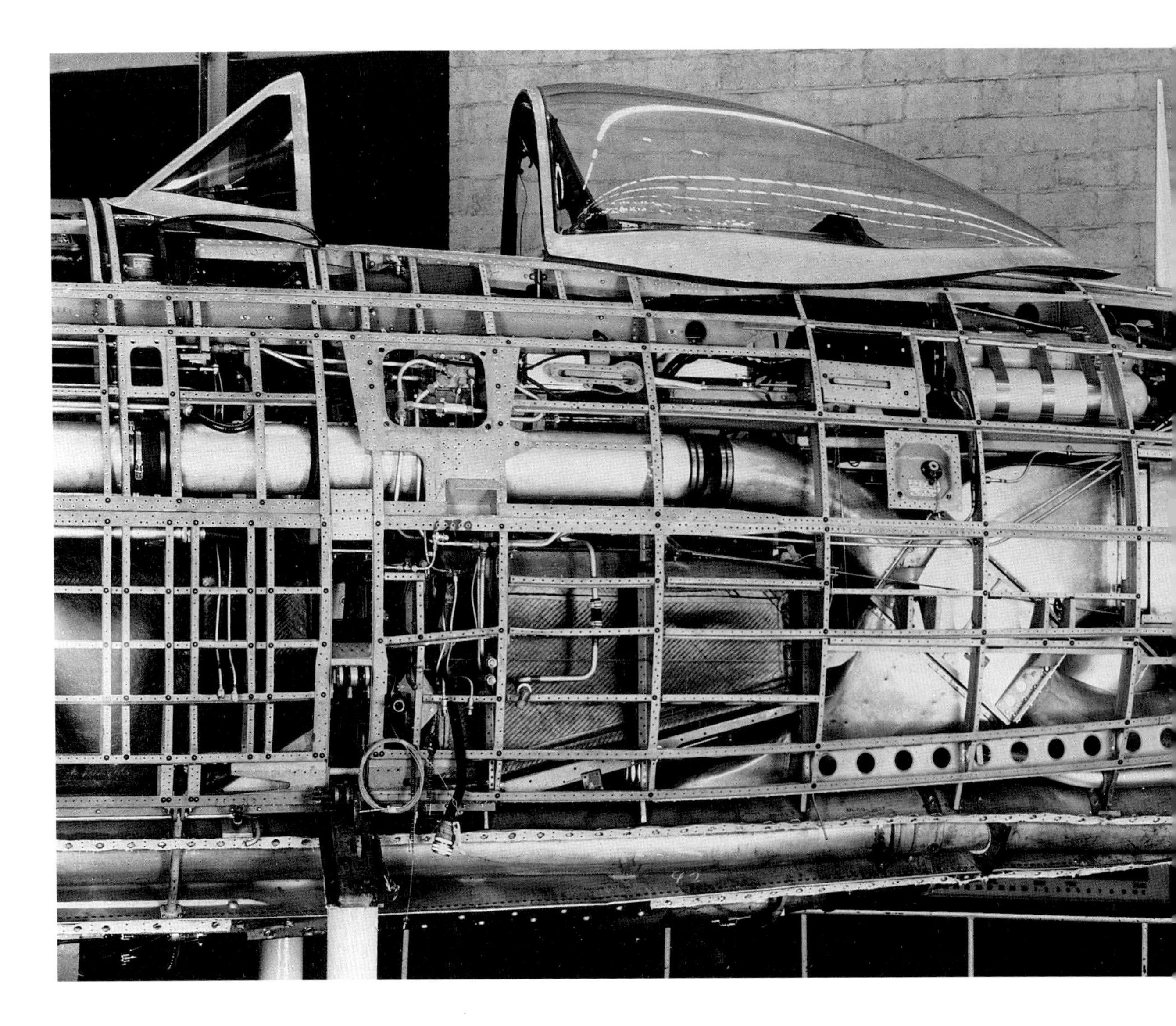

Capt. R.B. Johnston had conducted some very special flight tests at Wright Field. The series included taking off with full 330-gallon wooden drop tanks (official figures), one on each wing, and evaluating their performance. The tanks worked just fine, but Borsodi recommended against their use. Inadvertent release of one full tank during takeoff – unlikely, but considered possible – with an inexperienced pilot at the controls would probably prove fatal. The outboard positioning of the tanks was the critical factor. The major had made his evaluation flights at a takeoff weight of 20,080 pounds, but it should be recalled the Republic Aviation tester had taken the XP-47N off the ground at 20,166 pounds much earlier. (Carl Bellinger eventually coaxed one off the runway at a gross of 22,500 pounds.) Among other attributes, the P-47N in clean condition (no external stores) soon demonstrated a top speed of 453 mph at a rated power critical altitude of 38,750 feet. Rate of climb was not its greatest attribute, with it requiring 6.2 minutes to climb to 15,000 feet. However, using 2800 hp (WEP) conditions, a high speed of 467 mph was attained and the airplane climbed to the 15,000 foot level in just 4.6 minutes. Such performance was a serious challenge to the very fast P-47M-1-REs.

ABOVE AND OPPOSITE TOP: From the wingless side, a deskinned P-47N reveals what was packed inside this 11,000-lb. airframe (empty weight), at least in the aft two-thirds of the fuselage. The intercooler and ducting is far larger than the gas turbine supercharger. **(Frank Strnad Coll.)**

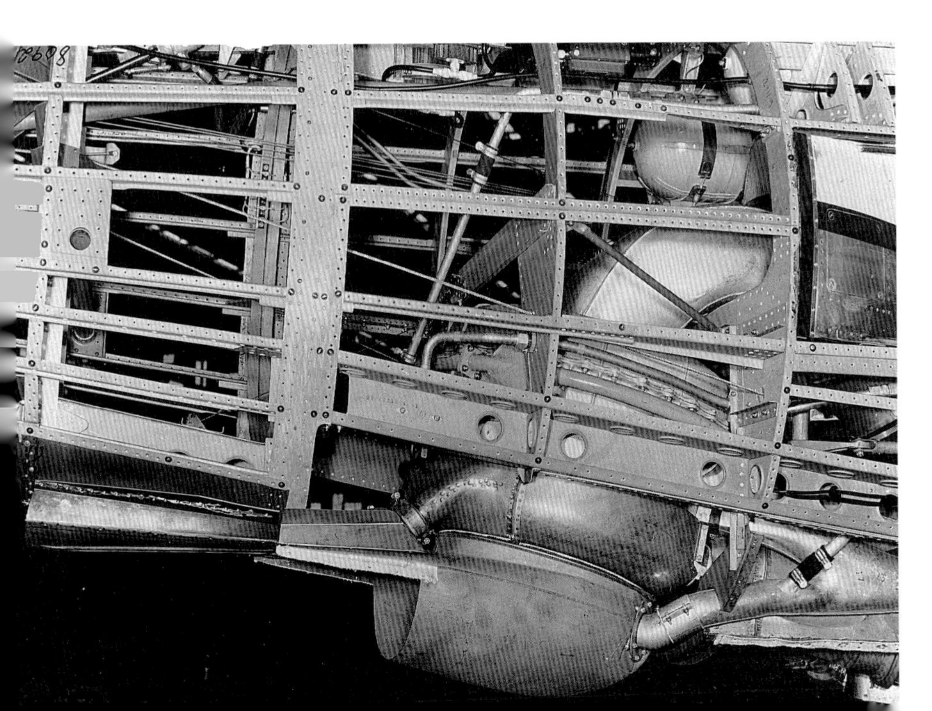

Viewed from the port side, the General Electric Model CH-5 turbosupercharger used on the XP-47J, P-47M and P-47Ns. (The CH-5 was also fitted to both XP-47Hs when the planned axial-flow superchargers failed to meet specifications.) The stainless steel shroud deflected hot turbine-driving exhaust gasses aft. The skeleton demonstration airframe represented a P-47N-5-RE. **(Frank Strnad Coll.)**

By September, the P-47N was in production, but events in Europe – namely the appearance of rocket planes, V-1 flying missiles, and what-have-you – brought about the crash program to modify airplanes on the line to be P-47M-1-REs in the month of October. Eventually, Republic delivered 130 of the fast hybrids to the ETO, with the largest batch going directly to the famed, high-scoring 56FG.

Although Michael Gregor played one of the most significant roles in the development history of the Thunderbolt, it seems unacceptable for him not to have shared in at least some portion of its ultimate success on a personal basis. Fate decrees certain people never attain the heights of fame and fortune to which they are justly entitled. As for Major Alexander P. de Seversky, he lived a long, productive and useful life, always remembered as a real patriot. The long-ranging P-47N had to be one of the performances he could savor, even if he could share that with few others.

With all the attention lavished upon the Allison V-3420, a very large and complex engine when equipped with the two-stage, two-speed mechancally driven supercharger, anyone would believe it was part of a large-scale production effort. The table to the right may surprise you.

An interesting comparison: With development of the Allison V-3420 and the P&W R-2800 starting at about the same time, notice the contrast in production quantities between the R-2800 and the V-3420. By 1945, both had almost exactly the same Military Power rating, 2600 hp, but the R-2800 had a WEP rating of 2800 hp, not to mention reliability.

AIRCRAFT ENGINE SHIPMENTS BY YEAR, MANUFACTURER AND MODEL*

	1940	1941	1942	1943	1944	1945
Allison						
V-1710	1141	6447	14905	21063	20191	5486
V-3420	2	1	-	30	112	10
Packard						
V-1650	-	49	7251	12295	22969	12571
Pratt & Whitney						
R-2800	15	1466	5429	7589	11565	8976
R-4360	-	-	-	8	25	110
Ford						
R-2800		264	6403	13337	24197	13436
Nash-Kelvinator						
R-2800	-	-	6	2692	9275	5135
Chevrolet						
R-2800	-	-	-	-	327	3955
Continental A&E						
V-1650	-	-	-	-	22	775

TOTALS

V-1650: 55910
V-1710: 69233
R-2800: 110887
V-3420: 155
R-4360: 143

*Source: U.S. Department of Commerce, Civil Aeronautics Admin., Office of Aviation Information - Division of Aviation Statistics; AIRCRAFT, ENGINE AND PROPELLER PRODUCTION - U.S MILITARY ACCEPTANCES 1940-1945.

Those were the engines powering what essentially had to be America's most powerful and important wartime fighters.

The war ended with things very much different than they had been in the final prewar days. Fighters could no longer operate safely and effectively without external drop tanks. Everything but the kitchen sink was hung underneath combat fighters. The role of the interceptor had become that of the long range fighter-bomber.

ABOVE: Wartime photos depicting combat area P-47Ns in formation flight are a rare sight. Compared with the ETO, FEAF media personnel were relatively scarce and it probably would have taken an act of Congress to get a 20th Air Force fighter squadron into the air just for photo journalism in the summer of 1945. At Eniwetok, Lt. Robert Forrest coaxed a borrowed Navy Curtiss SB2C into joining a formation of his 463FS comrades in their new P-47N-2-REs during a transition training flight. The Navy dive bomber had to struggle mightily to maintain position, but it soon lost. Unit markings had not been applied to the P-47s at the time. (Robert Forrest)

ABOVE: With his tenacious attitude never failing, Robert Forrest was ever willing to pursue a photo opportunity. When based at Ie Shima, within eyesight of Okinawa, he seized what must have been close to a final opportunity to record some 463FS friends tightly grouped in their P-47Ns over open water. This was Lt. Lee's flight, all aircraft carrying full combat markings and every one exhibiting nose art. It was September 15, 1945, and the war had ground to an end just two weeks earlier. Tail color was bright yellow with a pale blue triangle. (Robert Forrest)

Lt. Robert Forrest is shown in his P-47N-2-RE "Shell Pusher" loaded down with two 165-gallon napalm tanks, full ammo trays and a 75-gallon belly tank as he revs the R-2800 up to 52 inches manifold pressure prior to being dispatched on a mission out of Ie Shima in 1945. The markings indicate it was from the 463FS of the 507FG. They were awarded a DUC for operations over Korea on 13 August 1945 while flying Thunderbolts. The 463FS was allotted to the ANG, being redesignated as the 198FS in 1946. (Robert Forrest)

Initially assigned to operate with the 20th Air Force as escorts for Boeing B-29s, the 413FG and the 507FG actually joined the 318FG of the 7AF in flying strike missions. The 318th had been re-equipped with P-47Ns. Lt. Stout's 463FS T-bolt is shown in the runup area awaiting the ops controller's signal to release brakes and make the takeoff run at Ie Shima. Those 165-gallon P-38 tanks, filled with napalm, provided a substantial increase in local BTUs on impact. (Robert Forrest)

BELOW: It seems, upon seeing this picture, rather appropriate to state that the P-47N pilots in the 507FG were happy to have Streete signs. Major Alec Streete is shown energetically giving the "Take It" signal to the pilot of a P-47N-3-RE on the "vacation isle" of Ie Shima. Operating in Army Hot Day conditions at a gross weight that probably exceeded 10 tons was likely to induce some nervousness in the average fighter pilot. Full knowledge of how the N version evolved reveals that the fueled-wing version of the T-bolt could probably have been developed as much as a year earlier for ops in the ETO. All that was needed was a command decision in VIII FC HQ. (Robert Forrest)

BELOW: Republic Aviation test pilot Russ Hoagerwerf, pictured here with a P-47N named "Short Snorter" in the far reaches of the Western Pacific theater of operations (probably Ie Shima) posed with an admirable young female (painting) on 30 June 1945. Like his close associate, "Joe" Parker, Hoagerwerf was killed in a plane crash not long after he autographed this picture. (via Carl Bellinger)

ABOVE: It took a lot more than "shade tree mechanics" to perform maintenance on a P-47N in the field. Some 2800 cubic inches of engine was a pretty complex chunk of hardware to replace when working against time. One of the big G.E. magnetos had been removed, evidently leading to performance of a complete engine change. (Jake Templeton)

Hawker's Sabre-powered Typhoon, built as a powerful successor to the Hurricane type, began life with a very thick wing section having a chord/thickness ratio of 18 percent and an engine which was far from achieving the reliability of P&W's R-2800 in the same power class. A new "thin wing" with a root chord thickness ratio of 14.5 percent (about the same as Lockheed's P-38) and a wing planform that nearly duplicated the appearance of a P-47N or some Spitfire wings was mated to Typhoon-type fuselage. This was the new Tempest Mk V, initially flown on 2 Sept. 1942. No Tempest V became operational before January 1944 even though it was more of a model change than a new fighter. British press comments about P-47s notwithstanding, observe that the Tempest was BIG! (Mfr.)

RIGHT AND OPPOSITE TOP: Chief Designer Sidney Camm, obsessed with inline engines since the days of the Kestrel, detested "round" engines. But touchiness of the 24-cylinder inline Napier Sabre plus appearance of the Bristol Centaurus 18-cylinder radial and the need for a thin wing – if the RAF was to have any chance of combating high-altitude Luftwaffe fighters – forced the designer's hand. However, mating the thin wing to a Typhoon fuselage did not work out. No room for fuel in the wing meant adding a big fuel tank forward of the cockpit à la P-47. Result: an XP-47J sans a turbosupercharger. The big Tempest Mk II, destined for Far East combat, only became operational in August 1945. Hawker Tornado and Typhoon designs dated from 1937, but the "thin" wing Tempest activity was barely underway as 1942 dawned. The Thunderbolt wing dates back to 1931 Depression days, but was every bit as fast as the Tempest wing. (Hawker)

RIGHT AND BELOW: Two examples of the Luftwaffe enemy that Fisher P-75A pilots would have had to face in their 414-mph giant were the 472-mph Focke-Wulf Ta 152H and the equally rapid Dornier Do 335A depicted in these two photographs. Other enemies: the cost of each P-75A; hundreds of thousands of wasted manhours at Cleveland; a large new factory that could have produced at least a few thousand Merlin-powered (even Griffon-powered) P-38s, or Republic P-72s with extended wingspans. Another factor was exceedingly bad Air Materiel Division direction and management of engine development programs. With Packard producing excellent Merlins, why didn't Materiel drop Lycoming and Continental projects and have Packard produce Griffons? The other two could have subcontracted components. (Author's Coll.)

CHAPTER 18

ACES, HEROES AND OTHER GREATS

When all is said and done, it cannot possibly be the airplane which has done it all. Man created the machine, man serviced and cared for it, and at least in the 1930s and 1940s it was man who manipulated the controls. It took a great many of them – and a massive force of women aircrafters – to produce and ultimately deliver more than 15,000 of the big, complicated but extremely orthodox fighters. One of the most difficult things to believe and understand is the way in which the Republic Thunderbolt achieved its greatness while still carrying umbilical ties to 1930.

At last the long suppressed facts pertaining to the material parts played by Major Alexander P. de Seversky, more often thought of as simply, The Major, and the all-but-unknown Michael Gregor in creation of the P-47 have now been revealed. Even the reasons for the disinformation project have been brought to light. When powerful forces have created a myth through a propaganda campaign which would have brought a smile (and probably did) to the face of the Nazi, Joseph Goebbels, it is extremely difficult to turn the tide of belief. But it is well to remember the trial of Billy Mitchell did not do one tiny thing to change the truth and proof of his own convictions. In general, the public only believed he was convicted and punished. He sank battleships and cruisers; he made a significant number of hidebound U.S. Navy officials look bad. And he disobeyed a minor order for which he was convicted. Zealots are numerically a minority, but the world would be in a real, uncontrolled mess if some did not exist to be heard.

Russian immigrant de Seversky was quickly recognized by Mitchell as a genius with talents far exceeding those of many key persons in aviation. He was no "crackpot" inventor. The early support Mitchell showed the Major was as natural as rain. What it all amounted to, cutting through the obstacles, was two great minds on the same track. Let's make that at least three. Designer Mike Gregor was not political; he just had a very quick Russian temper. While the Major had some temper, it was no worse than Mitchell's, but he hated to be shackled by stubbornness or stupidity. The conviction and ultimate death of Billy Mitchell only served to convince de Seversky of the dangers of hidebound thinking and deception. Yes, he was zealous in his beliefs. But factual evidence of support for his beliefs existed, so how was

Leaning against his Seversky SEV-S2 racer after the 1938 Bendix Trophy Race was history, Frank Fuller (who placed second in the event after winning in 1937), talks with race winner Jacqueline Cochran. "Jackie" had flown a Seversky AP-7 to victory and then continued on to Bendix, New Jersey, to set a a new women's west-to-east record of 10 hours 7 minutes 10 seconds. Her performances certainly had a very positive impact on ultimate establishment of the Women's Airforce Service Pilots (WASP) organization in WWII. **(Mfr. 1503)**

Viewing one of the newest high-activity paddle-blade propellers from Hamilton-Standard, test pilot "Joe" Parker is seen with a chromate-yellow P-47D test airplane outside Hangar 1 at Farmingdale. Appropriately enough it was named "Yellow Peril." This is an almost totally unknown piece of nose art. (Frank Strnad Coll.)

he supposed to be "politically correct" and turn his back on unconstrained patriotism for his adopted country? His mouth and his pen could bring pain. The truth often hurts.

Aside from those elements, he was an exceptionally talented aviator, almost always pushing against the outermost limits. He never bailed out of any aircraft he was responsible for developing. In the decades of the 1920s and '30s, such performances were hardly the norm. His weakness...allowing enthusiasm for a product to overrule his business judgment and talents. He was not really comfortable when dealing with budgets, and his knowledge about the essentials of solid financial underpinnings and cost control was obviously minimal. Paul Moore had to be his accomplice, by not wielding his own financial clout or by avoiding confrontations at all costs. Out in California, Lockheed was dragged from bankruptcy by Robert Gross and a bevy of financial experts. Engineering talent and entrepreneurship came along afterward. They had it backward at Seversky Aircraft, and it ultimately was their undoing. The Major's wartime campaigns were quite successful in alerting the American people to the dangers inherent in allowing deception to hold a commanding position in military and political decisions. But he did far more for the war effort. Without the Seversky-Gregor wing, without the man's inventiveness and guts in forming an aircraft company in the face of the Great Depression; without the evolutionary elasticity of his airplane designs; and without his novel use of G.E.'s lagging supercharger developments there never would have been a P-47 Thunderbolt. Those are irrefutable contributions.[1] Major de Seversky, Seversky if you prefer, is an important milestone in illustrating the importance of mankind in our equation.

Certainly others played important roles in the history of Seversky's saga. Most prominent were Paul Moore, Frank Fuller,

[1] In an environment of tough financial times and a strong undercurrent of military opposition to de Seversky's policies and actions, Chief Engineer Kartveli trod the safe, guaranteed course by backing the XP-41 update of the final P-35 to the hilt. The flush-riveted, smooth skin construction techniques developed by Lockheed (and a few others) could not be avoided any longer. The same was true of flush-retractable landing gears – by then a mandate from from the military, if not from common sense. He would not only have failed to conceive of the turbosupercharger placement, he was on record as opposing the idea. At that time in history, the Lockheed XP-38 and Bell XP-39 supercharged fighters were deep secrets...very deep. Curtiss' failure to make the XP-37 and YP-37 work caused them to avoid turbosupercharging in their new XP-40 contestant. Consolidated, Boeing, North American and Douglas...the real "biggies" of aviation...stood MUTE! The Major was far out on a weak limb. So what occurred? *The handicapped Russian aviator emulated Kelly Johnson and pioneered without Air Corps guidelines.* Bell was mandated to use the supercharger by Ben Kelsey's specifications, but they did not persevere.

He is only a well-made man who has a good determination – Emerson.

In marked contrast to the youthfulness of most fighter pilots, the fatherly Lt. Col. Harold Rau was the initial commanding officer of the 356FG when they flew P-47s in the ETO, took over the 20FG at Kings Cliffe and flew Lockheed P-38s as C.O. when Col. Mark Hubbard took some flak hits and went MIA/POW on 18 March 1944. Still later on, when the group converted to P-51 Mustangs, Rau – who had been replaced by Lt. Col. Cy Wilson – returned one more time as C.O. when Wilson was shot down. It is probable that only Hub Zemke and Rau led fighter groups where P-38s, P-47s and P-51s were the assigned aircraft during their tours of command. Both men flew all three types in combat. (Harold Rau)

Although Capt. Howard Hively, known as "The Deacon" during his assignment with the 4FG, flew two different T-bolts (beginning with the P-47C AC41-6573 shown here), his major accomplishments in the 334FS came after he received a Mustang. Anyone who doubts the impact of psychology on performance under duress only has to review performances of the Thunderbolt-equipped 4th, 56th, 78th and 353rd Fighter Groups during the same months of 1943. Only the 4FG squadrons, nurtured on a diet of small Spitfires, failed to collectively embrace the big fighter. Most of the original 56FG in England had been weaned from P-40s to P-47Bs in the U.S.A., and Col. Zemke led the pilots to successful utilization. The 353rd was the last of those groups to get P-47s, but they performed well with that fighter type. (AFM)

Jr., Frank Sinclair, Evelyn Seversky, C. Hart Miller, Gen. William Mitchell, Jacqueline Cochran and, of course, Alexander Kartveli and Mike Gregor.

The importance of test pilots in developing and producing aircraft is yet another key factor. Other than people mentioned in previous paragraphs, Lowery L. Brabham, Joseph F.B. Parker, Carl Bellinger, Kenneth Jernstedt, Parker Dupouy, Mike Ritchie and Fillmore Gilmer all made significant contributions to the Thunderbolt development history. Somewhat incredibly only one man in that select group lost his life in a Thunderbolt, and it happened in the last days of a world conflict which had its beginnings almost exactly six years earlier. Joe Parker perished in a P-47N in a true "Murphy's Law" episode while attempting to solve a perplexing problem involving takeoffs at extremely high airplane gross weight. Factors included a minimum runway length, no overrun area, high humidity and the worst of Army Hot Day conditions. At the time, Parker was Director of Flight Operations at Republic Aviation in Farmingdale. Of course there were probably dozens of military test pilots who risked their lives pushing the flight envelopes to their limits in P-47s.

Speaking of risking lives it is vitally necessary to call attention to at least some of the pilots in the U.S. Army Air Forces who really made the Republic Thunderbolt the class airplane it proved to be. It has been, for a half century, almost a routine to concentrate attention on pilots of the mighty Eighth Air Force because of the high scoring record and the intensity and importance of the air war over *Festung Europa*. In due time, it was recognized they were there for but two purposes: to escort and protect the "heavies" or heavy bombers and their crews in carrying out the Operation POINTBLANK mission, and to gain control of the air over the Continent. The size, power and determination of the *Luftwaffe* to prevent both such occurrences led to the major air battles of the war. But with about 250 squadrons of P-47s operating throughout the world, heroic episodes were likely to appear anywhere. This book has gone far in its attempt to give credit where due in every venue.

VIII Fighter Command pioneered the use of Thunderbolts to command the air where such respected fighters as Curtiss P-40s, North American P-51As, Grumman F6F Hellcats and Vought F4U Corsairs were unacceptable or unlikely candidates for commitment. Obviously many of those types would be at a disadvantage in combating Messerschmitts and Focke-Wulfs over the German Fatherland. Such hopefuls as Bell P-39s and P-63s, Westland Whirlwinds and even Hawker Hurricanes and Typhoons really could not perform the bomber-escort role or dominate the skies against the formidable German Air Force. Even the redoubtable Supermarine Spitfire was – because of range limitations – *never* going to control the air over Germany, while the Hurricane and Typhoon lacked the altitude, range and maneuverability performance to begin with.

One important, if annoying, factor in this picture is the

One of the veteran pilots assigned to the 56PG(F) in the U.S. when Major Hub Zemke took over command was Capt. Philip E. Tukey, Jr., heading up the 63PS(F)*. Tukey is shown with his P-47C, probably at RAF Wittering because facilities were inadequate at Kings Cliffe. Fairness is important in evaluating fighters. Spitfires were short-range interceptors, and the P-47B/C was the AAF's version of that assignment. All P-47Cs and the early Ds had no provisions for either belly tanks or wing tanks. Pilots did not get the first 75-gal. and 108-gal. belly tanks on their fighters until summer was ending in 1943, or about seven months after they received their aircraft. Republic's 200-gal. tumor was unpressurized and could only be used half full. Lt. Col. Cass Hough and Capt. Shafer at Bovingdon were prime movers in getting 108-gal. and 150-gal. molded paper (and later, metal) tanks operational in 1943. (*=Mid-1942 designators during transition from Pursuit to Fighter.) (AFM)*

If Capt. David Schilling, commanding the 62FS in September 1943 and for many combative months, was unhappy with his P-47C, few knew it. Col. Hub Zemke rated Schilling as "very aggressive." For some ethereal reasons, aggressiveness was widespread in the 56FG in the ETO from an early date. Perhaps it was the hard-fisted boxer spirit in Hub Zemke that became contagious. In any event, in all of 1943 the Thunderbolt was there. Stripped of the P-38s that could have helped, VIII FC fought with new P-47s. With Gen. Kepner's arrival to replace Brig. Gen. Hunter came a couple of groups of Lightnings that had overtrained in the Z.I., but be aware that there were no Merlin P-51s until at least November-December 1943. None. **(Author's Coll.)**

reality of no more than sixteen fighter groups in VIII FC at peak level. Sixteen fighter groups were activated, but a maximum of ten were equipped with P-47 Thunderbolts as of 1 January 1944. By 6 June there were only four T-bolt groups in the Eighth AF, although the total of VIII FC groups remained at sixteen. In contrast, VIII Bomber Command was composed of no fewer than 40½ heavy bomb groups (plus 8 medium bomb groups in the Ninth AF). Whether or not VIII FC could and would have grown at a swifter pace as bomber losses escalated in the autumn of 1943 may have depended entirely on *suitable* fighter aircraft production rates established by Gen. Knudsen's War Production Board. By suitable, we mean P-38s, P-47s and P-51s, not Curtiss P-40s or Bell P-39/P-63 fighters...totally unacceptable for use in the ETO. Such was the shortage of escort fighters in the 8AF, there was every intention of transferring all three P-38 groups in Italy back to the ETO. Only in March 1944 was the final decision made to leave them in the MTO, very likely based on the amazing increase in aircraft deliveries to the ETO units in 1944. When the first P-47 groups were preparing to go operational in the first half of 1943, the supply system was barely able to maintain officially authorized delivery rates. In the second half of 1943, things improved tremendously with 2277 replacement airplanes delivered. However, 1257 of them were heavy bombers and only 723 were fighters, including some P-38s. By contrast, in July 1944 alone 2245 airplanes reached that theater, meaning the mid-1944 rate was six times what it had been in the last half of the previous year!

When the men of the 4th, 56th, 78th and 352nd Fighter Groups were struggling without range-extending drop tanks or tanks ill suited to the job, they could not look for any great influx of shock troops to aid them. But they managed to convince *Luftwaffe* pilots it was to be a fight they never anticipated. While the airplanes in hand were pretty good, it was guts, tactics and training which made the difference until help arrived in the form of useful and adequate drop tanks and aircraft changes which permitted the pilots to use them.

Pilots in the Ninth and Twelfth Air Forces fought a different kind of war, one unlikely to gain them many air-to-air victories over an enemy flying anything from a trainer to the latest piston engine or jet fighter. They were flying anti-tank guns, close support rocket launchers, airborne artillery, or Napalm-carrying flame throwers. Whether flying out of badly damaged enemy airfields, freezing pierced-metal airstrips where any bad landing could result in destruction of a complete line of bomb-laden fighters, or where a flight or a squadron might be forced to take off into a solid 300-foot ceiling in hopes of finding a hole to climb through, those pilots persevered. Shooting up trains which might be Q-ships on rails was yet another day's work, or in an attack on a scattering infantry column a pilot might be left with lifetime memories of half a hundred soldiers being cut in half or blasted into inhuman shapes by an avalanche of .50-caliber bullets.

Thousands of miles away in easterly or westerly directions, the same kind of pilots might take off from a tiny island airstrip and fly over hundreds of miles of water without seeing anything as large as a rock, wondering if he could find his way back to a safe haven if his fighter was badly shot up in a low-level attack or even if it was undamaged. In the Fifth Air Force there were pilots like Neel Kearby whose talent and aggressiveness might have made him the top ace in the AAF in time. But his allotment ran out during an attack on a bomber when, at very low level, his speed had dropped and he became a target for a single Japanese Nakajima Oscar fighter. Sometimes we just press our luck a click too far, and a very important story is ended in mid sentence.

Whether it was civilian Charles Lindbergh flying around in the far reaches of the SWPA in a P-47 or, even more likely, a

Duxford's 78FG pilots, according to the Duxford Diary published in 1945, destroyed approximately 667 Luftwaffe aircraft in the air and on the ground. Aces in the group included John Landers, A.M. Juchheim, Quince L. Brown, Eugene Roberts, Jack Oberhansly and Charles P. London (shown here). London flew P-47s in the 78FG, the only VIII FC group to fly all three main fighter types (P-38, P-47 and P-51) used by the AAF in the ETO. London flew with the 83FS on 14 May 1943, a date usually considered as a turning point in the aerial war's bomber-escort theory. (78th Fighter Group)

ABOVE: Fame managed to elude most in the armed forces during World War II, but it surely did not detract from the fine jobs the majority of them did. Lt. Donald J. Corrigan, flying in the 352FS of the 353FG in the ETO, was pictured with a newly assigned P-47D-22-RE "War Bond" fighter, funded by Republic employee bond purchases. It was the 21st such Thunderbolt paid for in this manner. Corrigan's group had only recently (June 1944) been commended by Gen. Doolittle for heroism in rescuing a bombardment wing from a savage attack by scores of Luftwaffe fighters. Corrigan had flown 60 combat missions and had been awarded the DFC. His crew chief, S/Sgt. C.E. Frye, Jr. is pictured with him and the unpainted Jug. (H-2705 via F. Strnad)

Lockheed P-38, Major General William Kepner dashing around England in his P-47D named "Kokomo" to maintain direct contact with his commanders, or Capt. Robert Forrest sharing his P-47N with another 507FG pilot on Ie Shima, all had a mission. It was to end the war with victory for the United States and its Allies.

The stars of the show were, among many others, the aces, heroes and great men and women sharing a place of honor in this memory-laden pictorial chapter. Hundreds, perhaps thousands, not included may have deserved far greater acclaim.

Perhaps the greatest of all honors live in the memory of the part we, as individuals, played in supporting roles in that largest of World Wars. The thousands of men and women who built, maintained or flew the Republic Thunderbolt know they were associated with something very special.

As the tension in the U.K. mounted with the approach of an unspecified D-Day, Lt.Gen. Carl Spaatz (head of United States Strategic Air Forces in Europe/USSTAF) and his top AAF generals in the ETO evidently toured several major bases. Col. Glenn Duncan (c.), commanding the 353FG at Raydon, hosted (from left) Maj.Gen. William Kepner (VIIIFC), Spaatz, Maj.Gen. James Doolittle (8AF) and an unidentified Brigadier. Duncan went MIA on 7 July, but evaded and reassumed command of the group on 22 April 1945. Spaatz assumed command of USSAFE/USSTAF at WIDEWING (Bushy Park), moving in on 20 January 1944. Doolittle set up his HQ at High Wycombe. (AAF via Jack Terzian)

BELOW: *Highest-scoring P-47 Thunderbolt ace in the MTO was Major H.H. Green, having destroyed 18 enemy aircraft during his tour of duty with the 15AF. All of "Herky" Green's victories were attained while he was flying the razorback version of the Jug. He is seen at Lesina, Italy, in the spring of 1944. One significant fact frequently overlooked in history is that the six highest-scoring American aces of WWII flew P-38 Lightnings, P-47 Thunderbolts, F6F Hellcats and F4U Corsairs, not P-5l Mustangs.* (via Chester Sluder)

ABOVE: *On the 22nd of February 1944, soon after the start of the 1000-bomber raids of "Big Week," Major Walter Beckham, at that time an 18-plane ace, joined other 351FS flights in strafing a heavily defended German airfield. During his high-speed pass with Lt. George Perpente flying as his wing man, flak found its mark and the P-47 was on fire. Beckham, then the leading ace in the 8AF, bailed out and became a POW.* (via Jack Terzian)

In the meantime, back in the United States, bubble-canopy P-47Ds were in full production. All deep animosities aside, it was certainly logical to see the progenitor of Republic Aviation, Major Alex P. de Seversky, ready to fly one of these newest models. His exact status with the War Department at the time of this event is obscure, but it seems that he was on the government payroll as an advisor to the Assist. Secretary of War for Air. Corporate policy at the time was to be vague about any connection. (C.A.M./E. Boss, Jr.)

Posing with his new P-47D named "Cookie II" and an unidentified lieutenant from the 2nd Armored Division, Major Shannon Christian was commander of the 351FS in the 353FG at Metfield. His earlier P-47D-2-RE bearing the name "Cookie" was lost when the group commander, Lt. Col. Loren McCollom, took a direct flak hit during a dive-bombing attack on a German airfield. Like Beckham, he was to become a prisoner of war. The very recognizable black and yellow checkerboard cowl markings of the group had not yet been adopted as standard. (AAF 68924AC)

BELOW: Perhaps Capt. Dewey Newhart was not destined to be an ace, but he gave all he had to defeat a determined enemy. Flying in place of the 353FG's Lt. Col. Blickenstaff instead of piloting his own P-47D "Mud N' Mules" on 13 June 1944, Newhart was killed in combat. In the 350FS he flew his P-47D LH★D with such great pilots as Col. Glenn Duncan, William Tanner and Ben Rimerman. (Author's Coll.)

BELOW: One bright autumn day in 1942 in Michigan, an unfamiliar aircraft turned into the pattern at Willow Run Airport adjacent to the new red brick bomber plant that was beginning to produce the first of thousands of B-24 Liberators. As this big fighter taxied onto the expansive concrete ramp, the powerful, muffled throb of its great radial stuttered, died and a tall, lanky pilot moved out onto the wing. I was impressed with this first view of a P-47 Thunderbolt, but was even more impressed when the pilot strode into the Ford Willow Run Bomber Plant hangar. It was this man, Charles Lindbergh (seen here in later months in the SWPA with Maj. Thomas McGuire, Jr., 475FG ace). Lindbergh, on assignment as a senior P&W technical representative, was at Willow Run to discuss Ford-built R-2800 engines with executives in Dearborn. It was certainly a banner day for any aviation-minded youngster in an aero engineering job.(475FG Assoc.)

Republic P-47N pilots like Capt. Robert Forrest, who flew "Shell Pusher" as a member of the 507FG, went to the Far Western Pacific region thinking they would be involved in flying escort for the B-29 Very Heavy Bombers (VHBs). Although technically assigned to the 20AF, the group and his 463FS were actually operating as part of the revamped Far East Air Force (FEAF). Reorganizations eliminated the area bomber commands, with the 20th Air Force essentially operating B-29s (and B-32s). Forrest's P-47N-2-RE took off on many napalm or bombing missions at a gross weight of more than 12 tons! He missed the DUC mission (not scheduled to fly) in which the 507th shot down 18 enemy aircraft for the loss of one P-47N. Lt. Oscar Perdomo (464FS) became an ace in a day. Bob Forrest's favorite fighter in WWII: Grumman F6F Hellcat, at least for flight characteristics. (via Robert Forrest)

BELOW: Seen at Ie Shima in late summer 1945, operating base of the 507FG, are (l.-r.) Lieuts. Wilson, Benway and Schmidt (back row), Edmunds and Gryska (front row) of the 463FS, FEAF. That gorgeous airplane behind (!) was (a P-47N) named "Chautaqua," probably carrying the number 109, but did anyone care? Seriously, though, pilots like these were really pounding everything in Japan and Korea in 1945, even if it didn't move. (via Robert Forrest)

BELOW: A featured model aircraft in Major Alexander P. de Seversky's office was not the P-35 but the prototype SEV-1XP seen in the background. According to memory, this picture was taken circa 1952. After a more than exciting career, the designer, pilot, entrepreneur, salesman and inventor passed away in 1974. Like all of us, he made some painful mistakes in life but that does not detract from his great contributions to aviation. (Author's Coll.)

Observe the turtle. He only makes progress when his neck is out.
Henry Ford

CHAPTER 19

AFTERWORLD

Most appropriately – or inappropriately and a few other fitting words, depending on your viewpoint – the Ides of August came on the thirteenth day of the month.[1] In 1945 it was as if the world had come to a halt, at least in the aircraft and engine industries. While tank and military truck builders could not have been particularly unhappy with events, there is no way aircraft and aircraft engine builders were going to switch to production of somewhere in the vicinity of 100,000 planes each year for civil

[1] According to the ancient Roman calendar.

Somewhat typical of wartime fighter trainers, "Flying Folly" – a P-47D-28-RA – was serving with the Training Command in New England when it was photographed in August 1944. While this example was gunless, it was not sightless. Flying in this clean condition, with only wing pylons to spoil the lines, a D-28 must have had pretty fair performance when well maintained. (AAF B40905-AC)

aviation use following cancellation of mammoth fighter, bomber and other aircraft contracts. The cuts necessarily came with swift vengeance. It was a return to the Pit and the Pendulum in many ways, at least with a modern connotation.

Fisher Body had struggled to turn out two P-75As in February and two more in May in their huge plant at Cleveland (Ohio) Airport, the end coming mercifully. As with other specialized aircraft, production of Martin B-26 Marauders had been halted in lock step with termination of hostilities in Europe. Those beautiful bombers had only one destiny awaiting them...being converted into pots and pans. The last four had been squeezed out in April, about 9 percent of the number built in the previous month. Brand new Lockheed P-38L-5-LOs rolled out of Burbank's Plant B-1, only to hear the whistle blow and a different blow fall. Later on we heard stories, among them tales of some sort of guillotine dropping on the tail booms as each shining fighter rolled from the assembly building. They did not even have a chance to be towed over to Bldg 304, Flight Test, for final touches. Since that made so little sense, we did not even bother to check for authenticity of the reports. They had really become property of the War Assets Administration since the AAF had no use for them.

Out in the far southwest corner of Indiana farm country, the last newly manufactured P-47N-20-RA was accepted in September. The assembly lines had slowed and stopped in August. The labor force at Republic's home plant in Farmingdale was decimated in August, but P-47N-25-REs were in production at reduced rates, finally seeing the last five sold off in December 1945. North American Aviation ended P-51D and K production with one of the latter out of Dallas in December ending the effort. The lightweight P-51H was soon to follow.

The world's tremendous Arsenal of Democracy disappeared with but a whimper. It had done a magnificent job, the likes of which will unlikely ever be seen again.

Military and naval aircraft began to collect in rows and bunches at rather remote locations. Kingman, Arizona, enjoyed wide expanses of flat, treeless topography and an arid climate, making it reasonably ideal for aircraft storage. For the immediate future, it became home for hundreds of B-17 and B-24 bombers, a couple of dozen B-32s (rendered useless by severing the upper engine mount on at least one engine, slamming the propeller into the dirt). The War Assets Administration (WAA) offered essentially brand new P-38Ls, with full fuel tanks, for a flyaway fee of $1200. (It didn't take test pilot Tony LeVier long to pick one out of the pack at Kingman and fly it directly back to home base at Burbank. During just a few events in a period of two years he made something like twenty times that cost in appearances, aerobatics and racing.)

OPPOSITE: Straying a bit from the formation, the commander of the 116 TFW based at Marietta herds "cloud sheep" over North Georgia with his F-47N-25-RE in this beautiful portrait taken in 1949 in the vicinity of Lake Chatuge. (The National Guard photographer, Jake Templeton, was the victim of one of those "hundred-year" snow storms in March 1993, very near the area where this photograph was taken in 1949. Forgotten for years, the original negatives were "gifted" to this chronicler just a few months before Jake perished. Without that act of kindness, the dozen and a half negatives almost surely would have been discarded by his remaining heirs.) (Jake Templeton)

Republic P-47 Thunderbolts were rarely to be found in WAA bases, although a handful of Curtiss-built P-47Gs – probably rather badly treated in their life cycle as trainers – were to be seen at places like Walnut Ridge, Arkansas. Those P-47Gs attracted a minimum of interested potential purchasers. A few of the brave but poorly trained pilot types with defiant self-complacency purchased things like thoroughly flogged YP-47Ms and Curtiss YP-60Es. They usually made matters far worse by amputating wing areas or adding dangerously located fuel tanks in hopes of winning air races. A few of the better pilots modified Bell P-39Qs and Curtiss XP-40Qs moderately,using much better engineering talent. Some people spent dozens of hours reworking Lockheed P-38s in every wrong way imaginable, creating ugly, usually slower monsters in the process.

Movie pilot and fixed-base operator Paul Mantz bought squadrons of Boeing B-17s to the wonderment of many. That sly fox drained thousands of gallons of aviation fuel from full tanks, then resold most of the Flying Fortresses for aluminum salvage. Somewhat later, with improvements in television and a resurgence of interest in the war, Mantz made more money renting those bombers out to film makers. Paul, canny type he was, bought some excellent P-51Cs and used the old "wet wing" concept to make long-range Bendix Trophy Race winners out of them. Old time stunt pilot Harold Johnson purchased an ex-AAF Lockheed C-69 Constellation and had a contract option to loop it at the Cleveland National Air Races. He formerly looped Ford Trimotors in many prewar air shows. Since he was evidently on Lockheed's payroll at the time (or had been), they convinced him it was bad policy to loop the Connie, at least in public. (Harold, more pilot and inventor than good businessman, had saved Lockheed's XP2V-1 Neptune when a large chunk of the vertical tail came off in a yaw test. The company had ordered him to bail out. Typically, being much in size and temperament like old prewar YP-38 test pilot Ralph Virden, killed by independence while flying the No.1 Yippee, he chose to stay aboard and managed a safe landing to save the aircraft.) Not only was Harold planning to loop the Connie, he planned to do 8-point rolls in it.

Why were there so few P-47 Thunderbolts passing into private hands? The AAF in early post-conflict days had no reason to operate Lockheed P-38 twin-engine fighters, especially with

It was Friday (TGIF), 11 January 1946. The first of those last six Farmingdale P-47N-25-REs was headed out on its Flight Test Inspection acceptance flight. This was to be a media event, and of course the Gremlins were present for their last gasps. A top corporation executive, C. Hart Miller, was evidently unfamiliar with one major characteristic of the P-47N. On his takeoff run down Runway 4-22 to the southwest, there was a normal characteristic momentary engine cutout as the throttle moved through the water injection gate. Miller, fully committed, tried to abort the takeoff. The brand new Jug went through the wire fence at the runway extremity, spun 180 degrees and came to a halt in the middle of Broad Hollow Rd., now a multi-lane highway. Miller was fortunate not to cross the road and hit a deep ditch. All company negatives were destroyed; only a few private photos eluded seizure, and this is one. **(Frank Strnad Photo)**

newer North American P/F-82 types becoming available as replacements for P-38s and Northrop P-61 night fighters. Therefore, both types rapidly disappeared from service. Many Thunderbolts were upgraded or refurbished at the Evansville modification center or major depots like Tinker Field, Oklahoma. Their most commonplace destinations then became one of several South American, Central American or Middle East air forces under mutual aid pacts. Some AAF/USAF units operated P-47s in a mix with P-5ID/H models. National Guard units east of the Mississippi usually were populated with Thunderbolts. Units in the western states were far more likely to receive P-51s. A major factor had to be the location of the original manufacturer, needed for supply of specialized services.

Just about five years after the last new P-47 Thunderbolts left the Farmingdale and Evansville manufacturing facilities – a period essentially equal to the entire design-development and production life of the fighter – aviation was in a tremendous transitional period. The piston engine fighter (F-82, F-51, F4U and F8F) was still a key item in the military services, but jets loomed large, creating an odd mixture. Added to the alloy, swept-wing jets were looming large over quite new straight-wing jets, but airframe design was rapidly outpacing U.S. jet engine development. And there were the unmanned aircraft, soon to be thought of only as missiles. Martin, with the aid of a German designer, had created a new tri-engine bomber, the XB-51, of most unusual design. The large and powerful assault airplane was intended to accomplish what an entire squadron of Thunderbolts would achieve on a ground-attack mission. All would be performed with a 2-man crew in an 80,000-lb. aircraft (at the prototype stage of development). In the same general time frame, the company came out with a near lookalike jet bomber, but it was pilotless. The designation YB-61 reveals some indecision about missiles, just as the military did with Northrop's Snark, designated XB-62 (eventually to become SM-62). Every one of these jet-powered bombers featured swept wings. And the prototypes of Thor, Atlas and Titan – the first of the wingless, tailless, rocket-propelled bombers – were already in testing programs. Missiles quickly became part of our everyday language. Progress in fighters was so rapid America was already fielding the Century-series supersonic fighters, prodded by the "Cold War."

In the idiom of General MacArthur, slightly paraphrased, old fighters never die, they just fade away. But like the Spad, Nieuport and Sopwith Camel of WWI, Republic's P-47 Thunderbolt will be revered and respected in 2044 when we celebrate the 100th anniversary of the D-Day Invasion. The airplane designed, essentially by an error of omission, as an interceptor made the outstanding transition into the role of escort fighter/ long-range escort fighter, and ultimately into the low-level ground-support interdictor role. It never was a good interceptor, but as a ground attack airplane it was virtually unbeatable. If any proof of the importance of that role is needed, just remember the Falaise Gap and the VIII FC "Down to Earth" policy which destroyed the German ability to control the air over their own homeland.

With the glory days a thing of the past, a half-dozen lonely P-47N-25-RE airplanes in varying stages of assembly bring up the rear on more than 9000 Thunderbolts built in Farmingdale's Building 17. The final RAC constructor's number was 1667 for the AAF's AC44-89450. In the background are Douglas C-54 airplanes being converted to civilian standards for American Airlines. Further west at the Evansville, Indiana, factory the customer's final serial number was AC45-50123. **(Frank Strnad Coll.)**

ABOVE: Republic's Chief Test Pilot Carl Bellinger posed proudly with the last P-47N-25-RE delivered by the company to the USAAF. It had served as a flight test and change control aircraft well into the post VJ-Day period as the military organization shrank rapidly. Carl went on to fly virtually all P/F-84 models, the XF-91s and various F-105s. A fine, very popular pilot, he was killed in a 1980s car/truck accident. (Carl Bellinger/ Mfr. H-9050)

BELOW: Numerous improvements incorporated in Thunderbolts during the war years were quite unknown to the majority of average citizens in the immediate postwar months. This TP-47D-40-RA, featuring the latest in propeller blade design, P-38 drop tanks and ten sets of zero-length rocket launchers, was a transient aircraft at Selfridge AAF, Michigan, in the summer of 1946. To many avid fans, it was splendor in the grass despite the dents in those drop tanks. (Warren Bodie Photo)

In glaring contrast to the beautiful F-47Ns in-flight views, perhaps it is appropriate that this sorry sight of Thunderbolts in a War Assets Administration scrap yard should prove to be a Curtiss (Buffalo)-built P-47G-19-CU (AC42-25198) and its sullen twin. It is generally conceded that no Curtiss-built P-47 ever went into combat anywhere. The Buffalo manufacturer's performance was certainly no better than Brewster Aircraft's and, like the latter outfit, Curtiss should have been shut down in preference to continuing the ongoing massive waste of manpower, materials and money. Their modern, massive facilities should have been turned over to Lockheed, Boeing, North American or Grumman for production of more valuable (proven in combat) aircraft, or at least components for them. The use of influence by T.P. Wright and Burdette Wright did not serve America well. Curtiss-Wright produced more useless prototypes during WWII than all other major aircraft manufacturers combined, while their Donovan Berlin moved to G.M./Fisher with his own Frankenstein monster, i.e. the XP-75. (A.U. Schmidt)

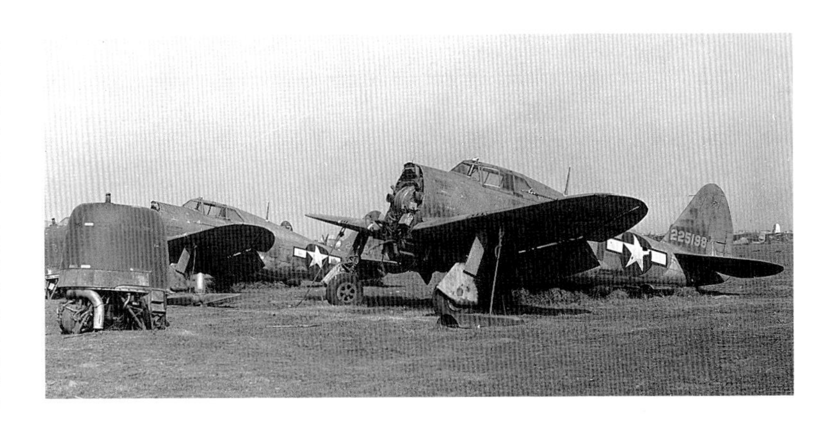

A 1946 encounter with Republic P-47Ns and North American P-51Hs at an old haunt named Selfridge Field brought on a real "double take." Col. David Schilling's famed 56FG was nicely dotted with fighters featuring red tails and nostalgia-generating prewar tail markings. Red, white and blue tail stripes were everywhere. Unfortunately, an order from some "higher authority" resulted in abandonment of this truly popular effort within a few weeks. (Warren Bodie Photo)

Tinker AFB had become a major recycling center/air depot for Republic P-47s after the war, so it was not unusual to have hundreds of them parked wingtip to wingtip in interlocking rows at the base. At least once the place took a direct hit from a tornado, wrecking many Jugs. But there was no bad weather report to warn them of another danger...falling aircraft. In this case it was an out-of-control Boeing B-29 that touched down with little advance warning, destroying and damaging numerous P-47s. Only a general date in the late 1940s is known. (Dave Menard Coll.)

An All Weather Flying Center had been established at Clinton County Airport in Ohio, not far removed from Wright Field, and it came under command of Col. Benjamin Kelsey early on. Aircraft ranged from AT-6s to the giant XB-19A, and functions ranged from automated flight to penetration of severe thunderstorms. Rather gaudy red and orange paint schemes were applied to the various aircraft assigned to the unit as on this P-47N-20-RE (AC44-89238). (Robert Volker Coll.)

LEFT: Colorfully decked out in All Weather Flying Center red and orange, this P-47N-20-RE (AC44-89238) was evidently the only Thunderbolt flown by the postwar organization commanded by Col. Ben Kelsey. Since Ben loved to fly, it is probable that he logged some hours on this airplane. Brilliant red and orange "plummage" served as anti-collision colors before arrival of Dayglo orange paint. (Dave Menard Coll.)

BELOW: In the summer of 1947, many radio-controlled drone aircraft were evidently being flown in the vicinity of Wright-Patterson AFB, so it was dangerous for civilian aircraft without radios to venture into that airspace. This F-47D-40-RA, photographed exclusively by the author, bore the legend CLEAR WRIGHT FIELD AREA in order to warn such strays away from the danger area. It was quite effective. (Warren Bodie Photo)

Some ingenious projects involved Thunderbolts in early postwar days, including serious efforts aimed at suppressing forest fires. The U.S. Forest Service borrowed some P-47N-25-REs, Boeing B-17Gs and even a B-29 to drop water-filled drop tanks on small forest fires to prevent them from becoming huge fires. A few USAF pilots on TDY to Missoula, Montana, primarily engaged in dive-bombing such blazes. Finned and fused water-filled 165-gallon drop tanks were launched into likely targets with some success. (U.S. Forest Service)

Thank God for the mechanics. Coming off farms, out of non-related factory work and out of high schools, boys and young men were very quickly trained in what it took to maintain complex machines. In the field, in sub-depots and in overhaul depots an F-47 powered by that potent P&W Double Wasp might well be stripped to this level, oftentimes with bent or torn structure to be eliminated first. It usually meant going without warm meals, a few hours of needed sleep and even passing up sick call. This is a Georgia Air National Guard F-47N in early postwar days, but it clearly illustrates part of the problem. (Air National Guard)

In lonely solitude, an Evansville-built F-47D-30-RA was still in fighting trim when pictured shortly after America finally got its long-awaited independent Air Force. A war-weary America also tended to quickly forget the importance of strength, allowing its mighty war machine to become decrepit. With only the rarest possible exception, the major and minor manufacturers of such wonderful victors failed to preserve so much as one example of their best works...even at no charge for the product. After V-J Day, any major aircraft builder could have easily arranged for permanent bailment or something equal. Visionaries were notable for their ability to remain invisible. (Warren Shipp)

Cruising over the beautiful North Georgia mountains near Hiawassee, Georgia and adjacent Lake Chatuge, four powerful F-47N-25-REs of the 128TFS – with the 116TFW commander leading – make a spectacular showing circa 1949. There has been a perennial ill-founded notion that no ANG units in the South ever flew Thunderbolts. Perhaps this view will change the unbelievers' minds. (J.C. Templeton via author)

Just a decade after Free French and Allied troops re-entered Paris, the French Armee de l'Air displayed this P-47D-30-RE, still in active service. In the difficult postwar period, the Armee flew an assortment of U.S., German and British types. (Serge Blandin)

With a salutary flick of its wing, a Republic F-47N was also saluting itself for a long, successful career in a world conflict of massive proportions. Imagine this: From its first tenuous Army fighter (pursuit) aircraft contract of 1936 until the final Evansville-built P-47N-20-RA was contracted for in 1945 (on Sales Amendment 54) Republic (nee Seversky) had built a handful more than 16,000 fighters. As noted, some even served Japanese masters, and few air forces worth mentioning did not fly Thunderbolts at some time or other. (Frank Strnad)

Upon his return from active duty in the Far East, Maj. Eugene Sommerich, a student of history and capable photographer, joined the Pennsylvania National Guard. Whatever fighter he might have been flying on a given day, he never failed to join up with some fellow pilot to burn up a roll of film. A New Jersey guardsman was the target on this flight. Later on Gene flew jets in Korea, even logging some time on a North Korean's MiG-15 that had been exchanged for a considerable sum of U.S. greenbacks. It is important to notice that Air National Guards had not yet been created. (Eugene Sommerich)

The brilliant white apron at what was probably Oscoda, Michigan, in 1946 served temporarily as a "typical" fighter base in the United Kingdom, circa 1944. All of those "purty" Thunderbolts of the P-47D-30 and -40 gender were gussied up with a lot of inappropriate combat markings (the scene was supposed to be England), looking more like the MTO. However, one of the brood seems to have proven that the brakes are very effective. With bent propeller for evidence, virtually all markings were quickly deleted. Huh! Who was embarrassed? The P-47 in the immediate foreground appears to be the designated player with duplicated markings. Anyway, the film cleverly titled Fighter Squadron, would have been a great hit for Robert Stack except for that ever-present movie master sergeant and his mouth! He wouldn't have survived for 10 minutes in any AAF unit I know of.

Carrying the authentic markings of the Nationalist Chinese Air Force in February 1955, four Republic F-47Ns were seen at Toa-Yuan Air Base in Formosa. A Douglas C-124A of the United States Air Force is seen taxiing by in the background. (Dave Menard Coll.)

In the vicinity of Teheran in May 1957, an Imperial Iranian AF Republic F-84G piloted by Lt. Raymond Kingston, USAF, flies wing on one of those elderly F-47D-30-RAs that had been delivered a few years earlier from Germany. Well, it was all in the interests of peace, probably, in a very different time frame. (F. Strnad Coll.)

Seen on what was the airfield at Neubiberg, Germany, circa 1950, a pair of incorrectly marked Republic F-47Ds (complete with frames on the bubble canopies) appeared in the film Berlin Airlift. Assigned to the 86FBG at the time, they most likely were components of the 525FBS. (R.L. Cavanagh Coll.)

Volunteer members of the Puerto Rico ANG's 156TFG restored this F-47N-25-RE (ex AC44-89320) to flying status in the 1960-70 era. Seen at the controls was Maj. Gabriel Peñagaricano (156TFG), joined by one of their newly allocated F-104C-5-LOs. The piston-engined fighter was decorated in the approximate colors of Col. Gilormini's wartime 350FG, 345FS Thunderbolt. (Although the tail marking is valid, the organization codes were applied incorrectly. Wartime arrangement was ★5A2 on both sides of the fuselage and about half size. No 12AF fighter is known to have used the cowling decoration applied to the PRANG fighter; only the 353FG airplanes of the 8AF carried that motif.) (Fernando Marrfro, PRANG)

An apparently pilotless F-47D on its way out of the 86FBG in 1949 powers up (without chocks) on one of the perforated steel mats so common to early post-war operations in Germany. Showing no signs of TLC, the Jug was equipped with three 75-gal. drop tanks that were intended to get it to then-friendly Iran. Hopefully there was a pilot with his head temporarily down to check on something that was amiss. (R.L. Cavanagh Coll.)

Mushing along to stay with Maj. Peñagaricano's beautiful F-47N and a slow camera plane, a Puerto Rico ANG Vought A-7D makes a fine comparison between WWII and "Cold War" strike aircraft. The latter could carry a weapons load exceeding that carried by a B-17 Flying Fortress. Colorful markings on the Jug can only be considered partially authentic for a wartime P-47D flown by Col. Gilormini, a member of the 345FS, 350FG during WWII. (Puerto Rico ANG)

GALLERY 1

The Big Landing Gear Mystery

One of the great mysteries in aircraft engineering involved nearly identical errors committed by three of the most talented aircraft designers in the world at a time when airplane development was in a spectacular period of flux and transition. Adding to the mystery is a seemingly total indifference to the situation on the part of aviation writers over a period of more than six decades. The three designers were Major Alexander de Seversky, John K. Northrop and England's Reginald Mitchell. In America, de Seversky and Northrop were the two outstandingly progressive engineers in development of all-metal stressed-skin monocoque fuselage structures, similarly fabricated cantilever monoplane wings and internally braced tailplanes. Major de Seversky's SEV-3 design of 1931-32 was constructed under duress of the Great Depression, so it was not flown until 1933. John Northrop's Alpha, Beta, Delta and Gamma monoplanes appeared sequentially during the 1929-1935 period following his departure from Lockheed Aircraft where he had developed wooden-structured airplanes along the same lines. Over in Great Britain, Mitchell hoped to meet a 1930 Royal Air Force requirement for a new interceptor by departing from the typical biplane concept and introducing an all-metal monoplane.

Now we have three designers, each separated by more than 2700 miles of land or ocean in a time of relatively slow and expensive communication when money was nearly impossible to accumulate beyond immediate living expenses. Although they each seemed to be moving far ahead of the pack in the field of advanced aircraft structures, each stumbled and fell over landing gear design in exactly the same manner.

Northrop produced his Alpha 1 monoplane first, but the landing gear looked like the most basic afterthought that could have been borrowed from any of hundreds of existing airplane designs. Based in a small factory facing on Empire Avenue in Burbank in the early 1930s, he seemed totally unaware that just a mile or so down that street his old employer (Lockheed) was successfully developing a fully retractable landing gear. When his Beta and Gamma designs following – well into the mid 1930s – he seemed to forget that landing gears contributed nearly half the drag generated by full-cantilever monoplanes. To reduce drag, he merely encased the wheels and struts in airfoil-shaped "pants."

Across the U.S.A. on New York's Long Island, Major de Seversky wanted to gain a foothold in military aviation procurements by winning a coveted "first" contract for a new-concept basic trainer. Seversky Aircraft converted the original SEV-3 amphibian to land gear by fitting structural pants that carried the wheel assemblies. For a basic trainer they were a bit "overstated" in lack of simplicity, but they paved the way toward a contract. The aircraft in this form was initially designated SEV-3L. When the Air Corps Project Office at Wright Field finally handed the Major a set of specifications, the demonstrator was reworked quickly to become the SEV-3XAR prototype. A contract for BT-8 trainers went to Seversky.

Perhaps de Seversky and Kartveli, the two Alexanders, were dumbfounded by the award, but what subsequently transpired was nothing short of a disaster. In a "hurry up" play to enter the 1935 Pursuit Airplane Competition, the Seversky team brought forth a spare fuselage and wing that had evidently been constructed when the company expected to pocket a contract extension from Colombia for three additional SEV-3M-WW reconnaissance amphibians after delivering the first three airplanes. They installed a brand new and essentially untried Wright Twin Whirlwind engine and unwisely decided to submit the pursuit (fighter) airplane in the existing 2-seater format, unaware that Project Officer Lt. Benjamin Kelsey was totally opposed to such configurations. Even more ludicrously, the two brilliant engineers ignored available evidence (even in magazines) and installed a slightly revamped version of the SEV-3L fixed landing gear! By that time it was universal knowledge that TWA, Varney Air Lines and the Secretary of the Navy were all flying Lockheed Altair and Orion airplanes (already outmoded and out of production) with flush retractable landing gears. Both Swissair and the Japanese had purchased some of those Lockheeds as early as 1932. In Europe, the Germans – greatly impressed by the Swissair Lockheeds – plunged into design of a competitive high-speed airliner that featured a reversed version of the Lockheed landing gear. The resulting Heinkel He 70a was a *tour de force* in design when it made its first flight in February 1933. With flush retracting landing gear, it attained a speed of 234 mph on only 637 horsepower, a speed unlikely to have been attained by the new SEV-2XP pursuit airplane when it first took to the air with (supposedly) more horsepower two years later.

It could not be considered misfortune when the SEV-2XP was "damaged" as reported. Attempting to play "catchup" or some such thing, Seversky converted the fuselage to a single-seat configuration. But with a "wet-wing" (fuel) centersection and no time on the clock remaining, de Seversky and Kartveli were left with but one retraction avenue open to them. Discarding the pants (officer, arrest them), they opted for merely swinging the wheels directly aft into the wing. With 75 percent of the wheel on each strut still exposed, they added a sparce set of fairings.

Surely there had to be a valid excuse for this terrible *faux pas*. No! And that is not stated with the safety of 20-20 hindsight. Glenn Curtiss adapted retractable wheels to his bamboo, cloth and wire E-75 biplane as early as 1911 to create an amphibious

airplane. In 1920, the Dayton-Wright RB-1 (also called the Baumann RB-1 for its primary designer) was a cantilever-wing racer with a retractable landing gear and a variable-camber wing. It raced in France in the Gordon Bennett Cup Race, falling victim to a broken rudder cable. In 1922, the Navy-sponsored Bee Line BR-I Special and no fewer than three Verville-Sperry (yes, the same Verville mentioned earlier) R-3 Army Air Service racers participated in the Pulitzer Trophy Race at Selfridge Field, Michigan. All four airplanes featured flush-retracting landing gears. One of those R-3 airplanes was refurbished in 1924, given a new engine with a 157 hp increase to 507 hp and it won the Pulitzer Trophy Race at Dayton, Ohio. That airplane with the Verville-design landing gear averaged nearly 215 mph over several laps of closed-course racing. Its best straight-line speed had to be in excess of 235 mph, some 70 years ago! And only 21 years after the first Wright Brothers flight of but a few yards.

The Detroit-Lockheed XP-900/XP-24 and its successor, the Consolidated Y1P-25, of the 1930-32 period featured landing gear arrangements that were virtual duplicates of the 1924 Verville design. And we should never forget about Grover Loening, developer of retraction systems that led directly to the Grumman biplane landing gear arrangements.

Over in England, the Supermarine Type 224 (F.7/30) interceptor designed by Mitchell was a very large single-seater featuring an inverted gullwing. But he too deigned to utilize a retractable landing gear of any kind, choosing to enclose the fixed landing gear within a set/pair of "trousers." Performance of the F.7/30 was, to say the very least, disappointing. That episode and a failed secondary design effort forced him to re-evaluate the entire fighter aircraft concept. As a result, he

There can be little doubt that Major Alexander de Seversky and his first chief designer, Michael Gregor, did an exquisite job of designing the wing and aircraft structure that were hallmarks of all Seversky airplanes. For the early 1930s, John Northrop and the Major were in the top five of the U.S. talent list. However, for reasons unknown, the Major and Alex Kartveli became enmeshed in retrogressive landing gear design. The first P-35 delivered to the Air Corps (and soon rejected) had these drag-producing wheel fairings. The tailwheel was normally fully enclosed, but evidently malfunctioned here in the 1937 picture. (Alex. de Seversky Coll.)

borrowed heavily from the Heinkel He 70a wing, landing gear and flush-riveted smooth skin to design the Type 300 Spitfire. With a simple, lightweight flush landing gear, the interceptor became an instant success and boasted a speed of about 360 miles per hour in 1936.

Sadly, as may be seen in the selection of photographs in this Gallery, Seversky/Republic was still building aircraft for combat with semi-retractable landing gears as late as 1940. Any excuse that might have been proffered could easily be shredded merely by illustrating the performances of Heinkel, Verville, Baumann, Mitchell, Sydney Camm (with the Hurricane) and at least a dozen others. It should be mentioned that Beechcraft's B-17 staggerwing biplanes built for private use featured fully retractable landing gear arrangements at about the same time that the SEV-1XP was declared winner of that almost farcical 1935-36 competition at Wright Field. It reveals that mankind frequently undermines his many brilliant moves and actions by committing errors that others could correct fairly easily.

For the avid history fan, the 1920 Gordon Bennett racer with retractable landing gear – the Dayton-Wright/Baumann RB-1 – is preserved at the Henry Ford Museum at Dearborn, Michigan.

Air racing in the U.S.A. reached a spectacular level in 1922 with Mitchell Trophy Race participation seen at Selfridge Field, Michigan, on 4 October. The U.S. Army Air Service entered, among others, three Verville-Sperry R-3 monoplanes with retractable landing gears and (unfortunately) overtaxed, underpowered Wright-Hispano engines. They lost out to small, powerful Curtiss Army and Navy biplanes. By 1924 this much improved and more powerful (by nearly 50 percent) V-S R-3 won the race at Dayton with a speed of 216.56 mph over a closed-course circuit. The flush retractable landing gear was a remarkable advance. (ATSC)

Constructed as a Lockheed Sirius with typical fixed, panted landing gear in the first Depression year of 1930, this airplane (the Altair 8A) was to become the first commercial and military type to have a flush landing gear when it was converted to this configuration in September 1930. The sliding canopy was also very innovative for the period. This landing gear design followed the concept put forth by Alfred Verville in 1922, if not earlier. Despite the pitiful financial condition of the Burbank company, the survivors proceeded with development. The new Altair was sold to the Air Corps and designated Y1C-25. (Harvey Christen)

As Detroit Aircraft tottered on the brink of collapse in late summer of 1931, the employees turned out two superb metal-fuselage Altairs, the Navy XRO-1 and the DL-2A demonstrator. There was the winner-quality retractable landing gear. This rare bird, completed in September, ultimately became Jimmy Doolittle's special Orion, the "Shellightning" that exists to this day in a Swiss museum. (Harvey Christen)

With a wooden Burbank-built Lockheed wing featuring a cleaner Altair-style landing gear that retracted absolutely flush, Detroit Aircraft turned out one prototype, YP-24, often referred to as the XP-900, and sold it to the newly established Air Corps. It demonstrated a pursuit plane speed of 235 mph at a time when the world's fastest fighters in development could barely exceed 200 mph. Purchased by the Air Corps, it was flown from Detroit to Wright Field, Ohio, on 29 September 1931. Unfortunately, Detroit Aircraft and Lockheed sank into bankruptcy a few days later. Had Lockheed's design team stuck closer to the V-S R-3 size – they had obviously copied the configuration – they would have had a 250-mph pursuit in 1931. But there was the pioneering landing gear. (Author's Coll.)

The Air Corps would not play ball with bankrupt companies. They gave all the marbles to Consolidated, and designer Bob Woods went the same route. By Thanksgiving Day 1932, the Buffalo-area company had built a slightly cleaner metal-winged version of the YP-24, now called the Y1P-25. And there is that updated Verville-cum-Lockheed landing gear! Just a bit of déjà vu. Production contracts were awarded for P-30 (supercharged) and A-11 (unsupercharged) versions, slightly simplified. (Mfr. 1836/11-22-32)

Air Corps cronyism, misguided management and a bit of French aviation leadership produced the P-30 of July 1934, only to see it become "Paradise Lost." Consolidated ownership decided to move to San Diego, delaying production for nearly two years. New AAC chiefs redesignated the pursuits as PB-2As, but all had fiddled as advantage slipped away. By 1936 there were the Bf 109, Hurricane and Spitfire showing the way. OK, score one for the Depression. (Mfr. 2220/7-5-34)

OPPOSITE PAGE, TOP: Design engineers have to be guided by certain axioms, and at least one of them has to be availability of necessary major components such as a powerplant for an airplane. It is also axiomatic that an original design by one can most certainly be improved upon by another. In Germany as of 1937 the State Ministry of Aviation began work on specifications for a new fighter. By 1939 a new Focke-Wulf Fw 190 prototype was ready to fly. Anybody who had ever seen a front view picture of the Hughes H-1 landplane speed record holder would likely have bet a week's pay that – viewed from the front – the prototype Fw 190 V1 was Richard Palmer's latest version of the Hughes racer. Kurt Tank's project leader, Sr. Engineer R. Blaser, most certainly saw the potential in the 1934 American design and followed the pathway established by Palmer. Even the BMW 801 twin-row radial engine was following a line of progression pioneered with success by Pratt & Whitney. We will never know why de Seversky and Kartveli went so far astray with the P-35 landing gear. (via Wm. Green)

BELOW: Richard Palmer, all is forgiven. You had the combination. But why couldn't you control the Hughes temper? On 13 September 1935(!!), this sorely misjudged aerodynamic beauty established a new World Landplane Speed Record (FAI approved) at Santa Ana, Calif., with a speed of 352.388 mph. And that was achieved with a P&W R-1535 Twin Wasp Jr., the same engine used in Northrop's 3A and the resulting Vought V-141/V-143. Rated horsepower: 725. As if that speed record was not enough, Howard flew it nonstop from Los Angeles to New York on 19 January 1937, averaging 327.5 mph for 7.5 hours in the process. See the flush landing gear; See Kurt Tank paying attention; See Air Corps generals and Hughes glaring at each other. Tank merely "adopted," improved the design. (Doubters, study automotive design! Or architecture. Tank just picked up the fumbled ball and rumbled on to the playoffs.) (Hughes Aircraft)

Here we have the Vought V-141 entry in the 1935 Pursuit Competition at Wright Field, but it did not arrive until the 1936 "rain date." Its fraternal twin, the Northrop 3A, had crashed in 1935. Whoever designed the 3A's retractable landing gear in El Segundo deserved accolades. If Kartveli and Major Alex de Seversky failed to read Aero Digest magazine and make the connection, the Japanese design teams at Mitsubishi and Nakajima did not make any comparable mistake. (ATSC)

An apparent infatuation with the bulbous landing gear fairings first installed on the initial production P-35 – and quickly rejected by Lt. Ben Kelsey as unacceptable – possibly caused Major de Seversky to overlook the shortcomings. Here he taxies the 1938 version of the AP-7 (NX-1384) demonstrator-racer at Floyd Bennett Field, NY, revealing excesses that the very youthful Clarence "Kelly" Johnson would have rejected out of hand. Movie films show severe porpoising of the AP-7 on climbout from take-off, most likely caused by those big "scoops" changing position as the gear retracted, slowly. Jackie Cochran flew this plane to victory in the 1938 Bendix Trophy Race shortly thereafter, but third place went to a much older Lockheed Orion with little more than half the power. Even that entry had a flush landing gear. Fourth place went to a Beechcraft staggerwing with one-third the power...and a flush landing gear! (Howard Levy)

GALLERY 2

TOO GOOD TO IGNORE

Seen in precision flight over the Eastern Michigan countryside, half a squadron of Seversky P-35s from the 27th Pursuit Squadron, 1PG based at Selfridge Field, present a pretty picture. This was in the still peaceful year of 1938 as Americans struggled to regain economic stability after years of grinding Depression and the setback of the 1937 Recession. In those days a military pilot was fairly well off at close to $150 a month in pay and allowances, but an Isolationist (at least anti-war) populace was hardly welcoming military personnel with open arms. An Army private's pay: $21 a month! (Army Air Corps)

RIGHT: Charles Brown in England was notable for his beautiful aerial photographs, especially the spectacular air-to-air views. But in 1938 it was most unusual for any American aviation photographer to talk a pilot into maneuvering as the pictures were taken. Not so when Major de Seversky was in the equation. Here we see the Seversky AP-7 climbing steeply in a zoom after passing over the 1939 World's Fair (including the Trylon and Perisphere) under construction in 1938. (J. Ethell Coll.)

BELOW: Although the Seversky SEV-S2 executive aircraft built for paint magnate Frank Fuller, Jr., looked like a virtual duplicate of any production Air Corps P-35, as factory c/n 43 it apparently preceded the first P-35 (c/n 44) on the assembly line and followed Shell Aviation Products' SEV-DS built for Jimmy Doolittle. Inasmuch as the latter airplane was damaged when Lowery Brabham collided with a wind-direction tee at Roosevelt Field, it was delivered about five weeks after the SEV-S2, seen here on the Farmingdale compass rose about 25 July 1937. Major differences from the AAC P-35 pursuits included a cut-down canopy and a more powerful R-1830SC-G engine with downdraft carburetion. Less than one month later, Fuller won the prestigious Bendix Trophy transcontinental race, beating Earl Ortman's special racer by nearly 2 hours. Racing number 23 was applied after this picture was taken. (F. Strnad Coll.)

Compared to the pursuit airplanes that appeared (or were to appear) at the 1935 Air Corps pursuit competition, the British Hawker F.36/34 Hurricane prototype (shown) and Willy Messerschmitt's Bayerische Flugzeugwerke A.G. Bf 109 V1 prototype were beauteous performers. Soon after the AAC competition was deferred to March 1936 for valid reasons, the Hurricane was demonstrating speeds unattainable by any of the Air Corps entries in any timeframe. Although that particlar 109 was no faster than Seversky's SEV-1XP, it was constrained by 30 percent less power at the time. (Hawker)

BELOW: Swinging that huge wooden Watts propeller, the prototype Vickers-Supermarine Spitfire is shown as it taxies out for a very early demonstration flight. First flight was on 5 March 1936, just as Seversky and Curtiss were about to do final battle for a 77-plane contract at Wright Field. In fact, no fewer than 310 of these Merlin-powered beauties were ordered into production on 3 June. Although wing shapes for the Seversky P-35 and Supermarine Spitfire were quite similar, the latter fighter was about 60 mph faster with about 5 percent more horsepower! (No doubt about it, the Spitfire is the Marilyn Monroe of aircraft.) (Flight)

The ill-fated Seversky AP-2, the first Farmingdale-built airplane to feature a flush-retracting landing gear, was evidently the fastest military airplane built in America when it took to the air for the first time early in 1937. In fact, only the Hughes H-1 special record breaker was faster, so the AP-2 (c/n 39) was also the fastest civil register aircraft in the USA capable of being used for everyday operations. After completing preparations for competing in the National Air Races, de Seversky crash landed at Floyd Bennett Field, N.Y., on 1 September 1937 when the landing gear buckled. It was a low point in that decade prior to the Major's removal as president of Seversky Aircraft in 1939. The fate of the AP-2 has never been determined, but the AP-2 Seversky – shown here at the Brooklyn airfield being towed to a hangar – apparently never flew again in America. (Howard Levy Photo)

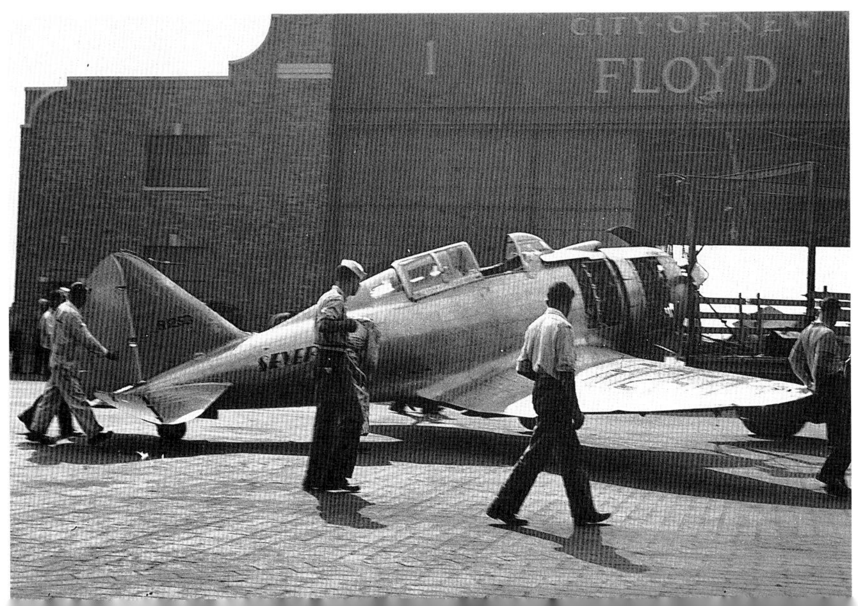

What ever became of ...? That is a phrase repeated over and over and everywhere. Sometimes we never really know because of errors in company and private records. That is true in the case of the Seversky (not Republic) EP1-68 "export demonstrator" and the AP-7 shown in this picture. The airplane undergoing runup at the Dade Bros. export facility at Mineola, L.I., has always been listed as the AP-9, but the Plexiglas turtledeck guarantees that it was the AP-7. Published reports have generally stated that the AP-9 went to the Dominican Republic, but all three of the above actually went to Quito, Ecuador in 1941. Very close examination of the wing centersections reveals the "sharpened" leading edges of the basic NACA airfoil applied to all new Seversky and (ultimately) Republic fighters, probably beginning with the AP-2. The revised airfoil was designated S-3 (for Seversky), was definitely applied to the Seversky EP1-68 and 2PA-BX, but Patent 2,257,260 was issued to Alexander Kartvelichvili (Kartveli) on 30 September 1941, assigned to "Seversky Aircraft Corp."!! (F. Strnad Coll.)

Three Reggiane Re.2000 Series III Falcons intended for service in the Italian Air Force were pictured outside the assembly building early in World War II. The Italian design team made little effort to conceal the fact that the design – structurally and aerodynamically – was based on the Seversky P-35 or EP1-68 airplanes. After all, weren't the Axis Powers going to become dominant in world affairs? Was Major Alexander P. de Seversky likely to sue? In a twist of fate, when the U.S. Government blocked delivery of half the Republic EP1-106 production destined for delivery to Sweden, Reggiane sold a like number of the Re.2000s to the Swedish government. One can envision the Italian physical gesture. (Author's Coll.)

Before war erupted in Europe, designers at the Italian Reggiane factory either plagiarized Seversky's P-35 airplane design, copied the Curtiss P-36 landing gear and installed a 14-cylinder radial engine to produce the Re.2000 fighter, or they surreptitiously purchased the Seversky AP-2 design. Whatever occurred, information was certainly suppressed. When the Flygvapnet failed to obtain all 120 Seversky EP1s on order, they bought 60 of the Reggiane types. The improved Re.2002 was not an entirely successful translation, especially in Luftwaffe markings. But it did emphasize how de Seversky and Kartveli struck out on the landing gear design at the SEV-1XP milestone. If the Major was not up to designing a flush-retracting land ing gear by 1935, a consulting contract with Alfred Verville or Richard Palmer could have solved the problem. (Author's Coll.)

BELOW: Squadrons of Republic P-43 Lancers from Selfridge Field's 1PG carried some unorthodox camouflage and large green crosses (not white, not red) when gathered on an unidentified new airport in 1941. Other Lancers temporarily based in Louisiana for war maneuvers carried white crosses, and many Navy/Marine Grumman Wildcats bore red crosses. The significance of the white underside paint has eluded us. (F. Strnad Coll.)

This never-before-published October 1940 photo shows the left console and a portion of the instrument panel of the P-44 mockup. It is of interest that there was no intention of creating an XP-44 prototype despite major changes in virtually every facet of the basic P-43 design. The engine change alone was equal to all of the differences between the Curtiss P-36C and the XP-40, so there had to be a calculated risk in ordering 80 production fighters. Compared to the cockpit of the XP-47B successor, the P-44 interior was rather barren. Generally overlooked previously in all publications, the entire cockpit structure from floor to canopy framing and from firewall area to the supercharger intercooler compartment was essentially the same for P-43, P-44 and P-47 airplanes. A gun-charging handle for the LH .50-cal. machine gun may be seen in the cutout above the switch panel. (F. Strnad Coll.)

Despite the RAF short-range interceptor concepts that prevailed before and during WWII until fighter-bomber versions of the Tempest Mk.II were developed (too late to be a factor), the Hawker Tornado and Typhoon fighters of 1940 were as large as the XP-47B of 1941. British derision of "massive" Thunderbolts was based on lack of knowledge about their home-grown products. The U.S. fighters were specifically designed to accommodate larger-framed Americans accustomed to flying long distances over a vastly larger country, so a spacious cockpit and long range were required. This first Bristol Centaurus-powered Tornado was a distant progenitor of the Tempest Mk.II. (Hawker)

After having employed sliding canopies since construction of the SEV-3 amphibian nearly a decade earlier, the Republic designers introduced a non-sliding canopy that imitated the Bell P-39 door arrangement to an extent. It was incorporated in the XP-47B and three P-47Bs. Even more likely, it may have been suggested by Col. Carl Spaatz after he encountered the Hawker Tornado/Typhoon prototypes during his Battle of Britain observations in 1940 while on TDY in England. It was not deemed an improvement on the canopy design used on all production P-43s. (F. Strnad Coll.)

A typical mass-production P-47D Thunderbolt was a remarkably attractive vehicle for an airplane that had to perform like a halfback or fullback (in sports). Considering the job that the Jug had to do on a daily basis in all kinds of weather, it was not a beast. A much more attractive brother, the XP-47J was a 500-mph-class fighter, but like the Tempest II and V airplanes it could not be made available in the required quantities when sorely needed. P-47Ds like this wartime trainer were the offensive linemen of WWII. Lookin' good out there. This particular airplane, No. 82, was active with an OTU in New England. (Mike Moffitt via D. Menard)

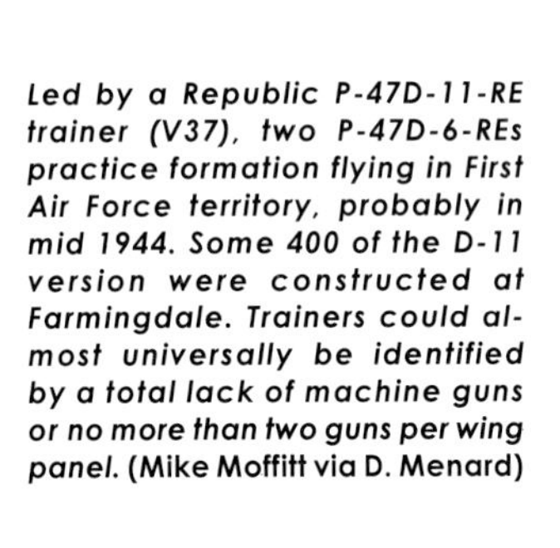
Led by a Republic P-47D-11-RE trainer (V37), two P-47D-6-REs practice formation flying in First Air Force territory, probably in mid 1944. Some 400 of the D-11 version were constructed at Farmingdale. Trainers could almost universally be identified by a total lack of machine guns or no more than two guns per wing panel. (Mike Moffitt via D. Menard)

As viewed from the tower at new Hangar 3, Runway 1-19 is bordered by a horde of bubble-canopy P-47Ds (and possibly some Ms) in this autumn 1944 panoramic survey of Republic Airport, looking southwest. An unanswered question: Was that runway inactivated? No Xs were painted at either end. Imagine the effects of a bad landing or a blown tire on a landing aircraft. Like the Thunderbolt II in the foreground, all camouflaged fighters on view were for foreign air forces. Two hangars at top center housed protective squadrons of P-38s (just joking, of course). AAF P-47 defense squadrons were based there. By the way, that was an active runway. (Mfr. H-4913)

Based on advanced planning by the AAF staff in Washington, two groups of P-47N Thunderbolts were to be assigned in 1945 for B-29 escort duty in the SWPA and Central Pacific areas of operations, backed by remaining P-47Ds operated by Eastern Air Command and the Fourteenth Air Force. Operating with the Twentieth Air Force, the 413FG and 507FG were dispatched to Ie Shima island near Okinawa to join forces with the 318FG of the Seventh Air Force. According to Capt. Robert Forrest of the 507th, that program never materialized, most of the P-47Ns being assigned to interdiction missions. Carrying a typical load of two 165-gallon P-38 drop tanks filled with Napalm plus a 110-gallon centerline drop tank, gross weight at takeoff on a normal Army Hot Day was in excess of 11 tons. This P-47N-2-RE named "Loretta Margie" displays the yellow-orange tail w/ blue triangle applied to 463FS airplanes of the 507FG. This pilot revs up to 52 inches manifold pressure prior to releasing brakes for takeoff. It was white-knuckle time for many of the pilots. (Robert T. Forrest)

Admired for its beauty of line and its very high speed, the Hawker Tempest I prototype must be appraised in realistic terms. First flown on 24 February 1943, it could attain 466 mph. But it was not a valid combat airplane in that form. First production Tempest Vs were considerably slower than massive numbers of in-service P-47s and P-51s as of late 1944, and P-38Ls could match Tempest speeds and at higher altitudes while flying hundreds of miles further. It is a fact that as December 1943 ended, only 36 Tempest Vs had been completed, with less than 50 percent delivered to the RAF. Designed specifically as a "battlefield air superiority fighter," it carried only four 20-mm cannon. No drop tank provisions for range extension existed. (A recent British book claims that these Tempests outgunned and could outspeed all contemporaries. It is a matter of record that eight .50-cal. machine guns or one 20-mm plus four .50s can throw a much greater weight of lead in a given period than four 20-mm cannons can. And contemporary P-47Ms were 475 mph airplanes, a speed unmatched by any Tempest V. Well, tunnel vision is difficult to correct.) Quantity, quality, attainment and timing are all components of a viable equation. (Hawker)

In postwar days, the Curtiss-Wright Corporation experimented with special supersonic blades fitted to a standard Curtiss propeller hub mechanism. These scimitar-shaped, extremely thin blades were tested on a Republic P-47D, although engineer-test pilot Herb Fisher marked the photo (incorrectly) as a P-47N. (Landing gear wheel doors and that inlet guarantee that it is a D model.) Maximum power operation had to be limited to seconds while strain gages were monitored. (Herbert Fisher/Curtiss)

OPPOSITE, BOTTOM LEFT: Probably the least publicized of all Seversky airplanes, except for the spring 1936 Cyclone-powered SEV-1XP, one of the 2PA-3B Convoy Fighters built for Japan was pictured at Seversky Airport prior to shipment to the Far East, supposedly to Siam. There was a long-standing general perception that Russian 2PA-L and 2PA-A airplanes were virtually identical. The perception was incorrect. All versions built for the Japanese had P-35 empennage components and P-35 landing gears, except for the unexpected fixed tailwheel. Ultimately contract revisions resulted in the 2PA-B3s retrofitted with long-range wings on all airplanes and retractable tail wheels. (Mfr/E. Boss)

That Seversky S-3 wing had to be one of the remarkable achievements of all time in aviation progress. The slightly earlier Northrop multicellular cantilever wing was a superior accomplishment in a period when biplanes dominated the scene. However, the airfoil selection made by de Seversky and Gregor combined beautifully to create a wing that a decade and a half later could fly smoothly at a speed of 500 mph (in the XP-47J and the XP-72). Designer Sydney Camm's 1940 wing shape for the Tempest fighter progression from the Typhoon was not thinner than the P-47 wing, but the more elliptical shape effectively reduced the British airplane's chord-thickness ratio. But remember, that was a full decade behind the Seversky design, which was also pioneering in stressed-skin all-metal cantilever, multi-spar construction. This Georgia National Guard P-47N was an astonishing evolutionary design, revealing that it could be beautiful as well as tough. (Jake Templeton - R.I.P.)

No aircraft company in history ever had the cheif executive pose with every aircraft type built by the manufacturer, except Seversky Aircraft Corporation. Alex de Seversky ran against the grain, as an entrepeneur, being photographed with airplanes he was proud of at every opportunity. When the competition at Wright Field was postponed until springtime in 1936, the Seversky SEV-1XP was returned to Farmingdale for additional rework, and the Major posed with it on 25 August 1935. (Mfr. #379)

AAF MANUFACTURING DATA

MODEL DESIGNATION	CONTRACT	QUANTITY	AIR CORPS (AC) SERIAL NUMBER
XP-47B (RE)	W535-AC-13817 C.O. 2	1	40-3051
RP-47B (RE)	W535-AC-15850	170	41-5895 thru 6064
P-47C (RE)	W535-AC-15850	57	41-6066 thru 6122
XP-47E (RE)	W535-AC-15850 C.O.	1	41-6065 (B model completed as E)
P-47C-1-RE	W535-AC-15850	55	41-6123 thru 6177
P-47C-2-RE	W535-AC-15850	128	41-6178 thru 6305
P-47C-5-RE	W535-AC-15850	362	41-6306 thru 6667
RP-47D	W535-AC-24579	4	42-22250 thru 22253
P-47D-RE	W535-AC-24579	110	42-22254 thru 22363
P-47D-1-RE	W535-AC-21080	105	42-7853 thru 7957
P-47D-2-RA	W535-AC-24579	200	42-22364 thru 22563
P-47D-2-RE	W535-AC-21080	445	42-7958 thru 8402
P-47D-3-RA	W535-AC-24579	100	42-22564 thru 22663
P-47D-4-RA	W535-AC-24579	200	42-22664 thru 22863
P-47D-5-RE	W535-AC-21080	299	42-8403 thru 8701
P-47D-6-RE	W535-AC-29279	350	42-74615 thru 74964
P-47D-10-RE	W535-AC-29279	250	42-74965 thru 75214
P-47D-10-RA	W535-AC-24579	250	42-22864 thru 23113
P-47D-11-RE	W535-AC-29279	400	42-75215 thru 75614
P-47D-15-RE	W535-AC-29279	250	42-75615 thru 75864
P-47D-15-RE	W535-AC-29279	246	42-76119 thru 76364
P-47D-16-RE	W535-AC-29279	254	42-75865 thru 76118
P-47D-20-RE	W535-AC-29279	249	42-76365 thru 76613
XP-47L-RE	W535-AC-29279 C.O.	1	42-76614 (250th P-47D-20-RE / taken from production)
P-47D-15-RA	W535-AC-24579	157	42-23143 thru 23299
P-47D-16-RA	W535-AC-24579	29	42-23114 thru 23142
P-47D-20-RA	W535-AC-24579	187	43-25254 thru 25440
P-47D-21-RA	W535-AC-24579	224	43-25441 thru 25664
P-47D-23-RA	W535-AC-24579	89	43-25665 thru 25753
XP-47K-RE	W535-AC-21080 C.O.	1	42-8702 (300th P-47D-5-RE / taken from production)
XP-47H	W535-AC-24579	(2)	(42-23279 and 23298) (conversion)
XP-47J	W535-AC-39160	1	43-46952
XP-47F	W535-AC-19378	(1)	(41-5938) (conversion)
P-47D-20-RE	W535-AC-29279-6	49	42-25274 thru 25322
P-47D-21-RE	W535-AC-29279-6	144	42-25323 thru 25466
P-47D-21-RE	W535-AC-29279-6	72	42-25467 thru 25538
P-47D-22-RE	W535-AC-29279-6	850	42-25539 thru 26388
P-47D-25-RE	W535-AC-29279-6	385	42-26389 thru 26773
P-47D-27-RE	W535-AC-29279-6	611	42-26774 thru 27384

MODEL DESIGNATION	CONTRACT	QUANTITY	AIR CORPS (AC) SERIAL NUMBER
YP-47M-RE	W535-AC-29279-6	2	42-27385 and 27386 (612th & 613th P-47D-27-RE / taken from prod.)
XP-47N-RE	W535-AC-29279-6	1	42-27387 (614th P-47D-27-RE / taken from production)
YP-47M-RE	W535-AC-29279-6	1	42-27388 (Final P-47D-27-RE / taken from production)
P-47D-23-RA	W535-AC-24579-7	800	42-27389 thru 28188
P-47D-26-RA	W535-AC-24579-7	250	42-28189 thru 28438
P-47D-28-RA	W535-AC-24579-7	1028	42-28439 thru 29466
P-47D-28-RE	W535-AC-29279-17	750	44-19558 thru 20307
P-47D-30-RE	W535-AC-29279-17	800	44-20308 thru 21107
P-47D-30-RA	W535-AC-24579-26	1200	44-32668 thru 33867
P-47D-30-RA	W535-AC-24579-33	600	44-89684 thru 90283
P-47D-40-RA	W535-AC-24579-33	200	44-90284 thru 90483
P-47D-40-RA	W535-AC-24579-34	465	45-49090 thru 49554
P-47G (CU)	W535-AC-24545	20	42-24920 thru 24939 (as P-47C-RE)
P-47G-1-CU	W535-AC-24545	40	42-24940 thru 24979
P-47G-5-CU	W535-AC-24545	60	42-24980 thru 25039
P-47G-10-CU	W535-AC-24545	80	42-25040 thru 25119
P-47G-15-CU	W535-AC-24545	154	42-25120 thru 42-25273 (4220 cancelled)
TP-47G-16-CU	W535-AC-24545	(2)	(42-25266 and 25267) (conversion)
P-47M-1-RE	W535-AC-29279-17	130	44-21108 thru 21237
P-47N-1-RE	W535-AC-29279-31	550	44-87784 thru 88333
P-47N-5-RE	W535-AC-29279-31	550	44-88334 thru 88883
P-47N-15-RE	W535-AC-29279-31	200	44-88884 thru 89083
P-47N-20-RE	W535-AC-29279-31	200	44-89084 thru 89283
P-47N-25-RE	W535-AC-29279-31	167	44-89284 thru 89450
P-47N-20-RA	W535-AC-24579-54	149	45-49975 thru 50123
XP-72	W535-AC-37879	2	** 43-6598 thru 6599

NOTE: ** The usually very accurate REPORT OF SERIAL NUMBERS ASSIGNED TO AIRCRAFT ON ACTIVE AIR FORCE CONTRACTS (dating to mid-1920's), generally known as AFPI Form 41, contains the following errors/omissions:

(1) There is no Republic XP-72 entry at all.

(2) The serial numbers assigned to the two XP-72s fall squarely within a block of 800 North American P-51B-5-NAs on contract AC-30479. It is a certainty that two P-51Bs in service had the same tail numbers the XP-72s carried.

P-47 PRODUCTION DATA	1941	1942	1943	1944	1945	TOTAL
DELIVERIES AND/OR ACCEPTANCES (By Plant)						
REPUBLIC (Farmingdale, New York)	1	516	3026	3901	1643	**9087**
REPUBLIC (Evansville, Indiana)	0	10	1131	3087	2014	**6242**
CURTISS-WRIGHT (Buffalo, New York)	0	6	271	77	0	**354**
TOTAL	**1**	**532**	**4428**	**7065**	**3657**	**15683**
ACCIDENT DATA (Continental USA Only)						
ACCIDENTS	N/A	106	958	1303	682	**3049**
FATALITIES	N/A	14	145	217	79	**455**
AIRCRAFT WRECKED	N/A	41	380	474	230	**1125**
ACCIDENT RATE (per 100,000 hours)	N/A	245	163	122	97	**127**
AVERAGE P-47 COST						
UNIT COST (In dollars)	113,196	98,033	96,475	85,448	84,897	

REPUBLIC P-47 THUNDERBOLTS ALLOCATED TO THE ROYAL AIR FORCE

THUNDERBOLT MK.I

MODEL	AC SERIAL NOS.	RAF SERIAL NOS.	QTY.
P-47D-15-RE	42-76324 & Unk.	FL731 & FL732	2
P-47D-21-RE	42-25421 thru 25438	FL733 thru FL750	18
	42-25449 thru 25488	FL751 thru FL790	40
P-47D-22-RE	42-25639 thru 25678	FL791 thru FL830	40
	42-25779 thru 25798	FL831 thru FL850	20
	42-25799 thru 25818	HB962 thru HB981	20
	42-25914 thru 25931	HB982 thru HB999	18
	42-25932 thru 25953	HD100 thru HD121	22
	42-26177 thru 26236	HD122 thru HD181	60

THUNDERBOLT MK.II

MODEL	AC SERIAL NOS.	RAF SERIAL NOS.	QTY.
P-47D-25-RE	42-26477 thru 26506	HD182 thru HD211	30
	42-26593 thru 26622	HD212 thru HD241	30
	42-26722 thru 26751	HD242 thru HD271	30
P-47D-27-RE	42-26885 thru 26914	HD272 thru HD301	30
P-47D-28-RE	44-19619 thru 19658	KJ128 thru KJ167	40
	44-19806 thru 19845	KJ168 thru KJ207	40
	44-19967 thru 20006	KJ208 thru KJ247	40
	44-20158 thru 20197	KJ248 thru KJ287	40
	44-20298 thru 20307	KJ288 thru KJ297	10
P-47D-30-RE	44-20308 thru 20337	KJ298 thru KJ327	30
	44-20488 thru 20527	KJ328 thru KJ367	40
	44-20628 thru 20657	KL168 thru KL197	30
	44-20738 thru 20797	KL198 thru KL257	60
	44-20817 thru 20846	KL258 thru KL287	30
	44-20877 thru 20906	KL288 thru KL317	30
	44-20947 thru 20976	KL318 thru KL347	30
P-47D-30-RA	44-90076 thru 90120	KL838 thru KL882	45
	Unknown	KL883 thru KL886	4
P-47D-40-RA	44-90335	KL887	1

P-47D aircraft undelivered: AC42-25426 (FL738), AC44-20750 (KL210), AC44-20835 (KL276) and AC44-20954 (KL325). Probable causes: written off in accidents or not shipped prior to termination of hostilities.

APPENDIX C SEVERSKY AND REPUBLIC FIGHTER

P-35/P-35A

1st Pursuit Group
- 17th Pursuit Squadron
- 27th Pursuit Squadron
- 94th Pursuit Squadron

4th Composite Group
- 17th Pursuit Squadron
- 20th Pursuit Squadron

8th Pursuit Group
- 33rd Pursuit Squadron
- 35th Pursuit Squadron
- 36th Pursuit Squadron

24th Pursuit Group
- 3rd Pursuit Squadron

31st Pursuit Group
- 39th Pursuit Squadron
- 40th Pursuit Squadron
- 41st Pursuit Squadron

35th Pursuit Group
- 18th Pursuit Squadron
- 20th Pursuit Squadron
- 21st Pursuit Squadron
- 34th Pursuit Squadron (attached to 24 PG)

49th Pursuit Group
- 7th Pursuit Squadron
- 8th Pursuit Squadron
- 9th Pursuit Squadron

50th Pursuit Group
- 10th Pursuit Squadron
- 11th Pursuit Squadron

58th Pursuit Group
- 69th Pursuit Squadron

53rd Pursuit Group
- 13th Pursuit Squadron
- 14th Pursuit Squadron
- 15th Pursuit Squadron

P-43 Lancer

1st Pursuit Group
- 27th Pursuit Squadron
- 71st Pursuit Squadron
- 94th Pursuit Squadron

14th Pursuit Group
- 48th Pursuit Squadron
- 49th Pursuit Squadron
- 50th Pursuit Squadron

20th Pursuit Group
- 55th Pursuit Squadron
- 77th Pursuit Squadron
- 79th Pursuit Squadron

55th Pursuit Group
- 37th Pursuit Squadron
- 38th Pursuit Squadron
- 54th Pursuit Squadron

58th Pursuit Group
- 68th Pursuit Squadron

66th Observation Group
- 97th Observation Squadron
- 100th Observation Squadron

73rd Observation Group
- 36th Observation Squadron

74th Observation Group
- 11th Observation Squadron
- 22nd Observation Squadron

76th Observation Group
- 20th Observation Squadron
- 33rd Observation Squadron

337th Fighter Group
- 98th Fighter Squadron
- 303rd Fighter Squadron
- 304th Fighter Squadron

<u>Royal Australian Air Force</u>
- No.1 Photo Recon. Unit

<u>Chinese Air Force</u>
- 1st Fighter Squadron

P-47 Thunderbolt

<u>Training Units - Zone of the Interior</u>

53rd Fighter Group
- 13th Fighter Squadron
- 14th Fighter Squadron
- 15th Fighter Squadron
- 438th Fighter Squadron

58th Fighter Group
- 67th Fighter Squadron
- 68th Fighter Squadron
- 69th Fighter Squadron

83rd Fighter Group
- 448th Fighter Squadron
- 532nd Fighter Squadron
- 533rd Fighter Squadron
- 534th Fighter Squadron

84th Fighter Group
- 491st Fighter Squadron
- 496th Fighter Squadron
- 497th Fighter Squadron
- 498th Fighter Squadron

85th Fighter Group
- 499th Fighter Squadron
- 500th Fighter Squadron
- 501st Fighter Squadron
- 502nd Fighter Squadron

87th Fighter Group
- 450th Fighter Squadron
- 535th Fighter Squadron
- 536th Fighter Squadron
- 537th Fighter Squadron

326th Fighter Group
- 320th Fighter Squadron
- 321st Fighter Squadron
- 322nd Fighter Squadron
- 442nd Fighter Squadron

327th Fighter Group
- 323rd Fighter Squadron
- 324th Fighter Squadron
- 325th Fighter Squadron
- 443rd Fighter Squadron

337th Fighter Group
- 303rd Fighter Squadron

338th Fighter Group
- 305th Fighter Squadron
- 306th Fighter Squadron
- 312th Fighter Squadron
- 441st Fighter Squadron

402nd Fighter Group
- 320th Fighter Squadron
- 442nd Fighter Squadron
- 452nd Fighter Squadron
- 538th Fighter Squadron
- 539th Fighter Squadron
- 540th Fighter Squadron

407th Fighter Group
- 495th Fighter Squadron
- 515th Fighter Squadron
- 516th Fighter Squadron
- 517th Fighter Squadron
- 635th Fighter Squadron

408th Fighter Group
- 455th Fighter Squadron
- 518th Fighter Squadron
- 519th Fighter Squadron
- 520th Fighter Squadron
- 639th Fighter Squadron

Combat Units

Fifth Air Force*
(Southwest Pacific)
8th Fighter Group
- 36th Fighter Squadron

35th Fighter Group
- 39th Fighter Squadron
- 40th Fighter Squadron
- 41st Fighter Squadron

49th Fighter Group
- 6th Fighter Squadron
- 9th Fighter Squadron
- 78th Fighter Squadron

58th Fighter Group
- 69th Fighter Squadron
- 310st Fighter Squadron
- 311th Fighter Squadron
- 201 Escuadron de Caza (Mexican Air Force)

348th Fighter Group
- 340th Fighter Squadron
- 341st Fighter Squadron
- 342nd Fighter Squadron
- 460th Fighter Squadron

Seventh Air Force*
(Pacific)
15th Fighter Group
- 47th Fighter Squadron

318th Fighter Group
- 19th Fighter Squadron
- 73rd Fighter Squadron
- 333rd Fighter Squadron

413th Fighter Group
- 1st Fighter Squadron
- 21st Fighter Squadron
- 34th Fighter Squadron

414th Fighter Group
- 413th Fighter Squadron
- 437th Fighter Squadron
- 456th Fighter Squadron

507th Fighter Group
- 463rd Fighter Squadron
- 464th Fighter Squadron
- 465th Fighter Squadron

508th Fighter Group**
- 466th Fighter Squadron
- 467th Fighter Squadron
- 468th Fighter Squadron

Eighth Air Force
(Europe)
4th Fighter Group
- 334th Fighter Squadron
- 335th Fighter Squadron
- 336th Fighter Squadron

5th Emergency Rescue Squadron
- (65FW Det.B before Jan 45)

56th Fighter Group
- 61st Fighter Squadron
- 62nd Fighter Squadron
- 63rd Fighter Squadron

78th Fighter Group
- 82nd Fighter Squadron
- 83rd Fighter Squadron
- 84th Fighter Squadron

352nd Fighter Group
- 328th Fighter Squadron
- 486th Fighter Squadron
- 487th Fighter Squadron

353rd Fighter Group
- 350th Fighter Squadron
- 351st Fighter Squadron
- 352nd Fighter Squqdron

355th Fighter Group
- 354th Fighter Squadron
- 357th Fighter Squadron
- 358th Fighter Squadron

356th Fighter Group
- 359th Fighter Squadron
- 360th Fighter Squadron
- 361st Fighter Squadron

359th Fighter Group
- 368th Fighter Squadron
- 369th Fighter Squadron
- 370th Fighter Squadron

361st Fighter Group
- 374th Fighter Squadron
- 375th Fighter Squadron
- 376th Fighter Squadron

495th Fighter Group***
- 551st Fighter Squadron
- 552nd Fighter Squadron

Ninth Air Force
(Europe)
36th Fighter Group
- 22nd Fighter Squadron
- 23rd Fighter Squadron
- 53rd Fighter Squadron

48th Fighter Group
- 492nd Fighter Squadron
- 493rd Fighter Squadron
- 494th Fighter Squadron

50th Fighter Group
- (to 1st TAF Nov 44)
- 10th Fighter Squadron
- 81st Fighter Squadron
- 313th Fighter Squadron

354th Fighter Group
- 353rd Fighter Squadron
- 355th Fighter Squadron
- 356th Fighter Squadron

358th Fighter Group
- (to 1st TAF Nov 44)
- 365th Fighter Squadron
- 366th Fighter Squadron
- 367th Fighter Squadron

362nd Fighter Group
- 377th Fighter Squadron
- 378th Fighter Squadron
- 379th Fighter Squadron

365th Fighter Group
- 386th Fighter Squadron
- 387th Fighter Squadron
- 388th Fighter Squqdron

366th Fighter Group
- 389th Fighter Squadron
- 390th Fighter Squadron
- 391st Fighter Squadron

367th Fighter Group
- 392nd Fighter Squadron
- 393rd Fighter Squadron
- 394th Fighter Squadron

368th Fighter Group
- 395th Fighter Squadron
- 396th Fighter Squadron
- 397th Fighter Squadron

370th Fighter Group****
- 401st Fighter Squadron
- 402nd Fighter Squadron
- 485th Fighter Squadron

371st Fighter Group
- (to 1st TAF Nov 44-Feb 45)
- 404th Fighter Squadron
- 405th Fighter Squadron
- 406th Fighter Squadron

373rd Fighter Group
- 410th Fighter Squadron
- 411th Fighter Squadron
- 412th Fighter Squadron

404th Fighter Group
- 506th Fighter Squadron
- 507th Fighter Squadron
- 508th Fighter Squadron

405th Fighter Group
509th Fighter Squadron
510th Fighter Squadron
511th Fighter Squadron

406th Fighter Group
512th Fighter Squadron
513th Fighter Squadron
514th Fighter Squadron

Tenth Air Force*
(China-Burma-India)
1st Air Commando Group
5th Fighter Squadron (C)
6th Fighter Squadron (C)

80th Fighter Group
88th Fighter Squadron
89th Fighter Squadron
90th Fighter Squadron

Twelfth Air Force
(Mediterranean)
27th Fighter Group
522nd Fighter Squadron
523rd Fighter Squadron
524th Fighter Squadron

57th Fighter Group
64th Fighter Squadron
65th Fighter Squadron
66th Fighter Squadron

79th Fighter Group
85th Fighter Squadron
86th Fighter Squadron
87th Fighter Squadron

86th Fighter Group
525th Fighter Squadron
526th Fighter Squadron
527th Fighter Squadron

324th Fighter Group
314th Fighter Squadron
315th Fighter Squadron
316th Fighter Squadron

350th Fighter Group
345th Fighter Squadron
346th Fighter Squadron
347th Fighter Squadron
1° Grupo de Aviação de Caça
(Brazilian Air Force)

Fourteenth Air Force*
(China-Burma-India)
33rd Fighter Group
58th Fighter Squadron
59th Fighter Squadron
60th Fighter Squadron

81st Fighter Group
91st Fighter Squadron
92nd Fighter Squadron
93rd Fighter Squadron

Fifteenth Air Force
(Mediterranean)
325th Fighter Group
317th Fighter Squadron
318th Fighter Squadron
319th Fighter Squadron

332nd Fighter Group
99th Fighter Squadron
100th Fighter Squadron
301st Fighter Squadron
302nd Fighter Squadron

Iceland Base Command
33rd Fighter Squadron

Royal Air Force
No.5 Squadron
No.30 Squadron
No.34 Squadron
No.42 Squadron
(formerly No.146)
No.60 Squadron
No.73 Squadron (OTU)
No.79 Squadron
No.81 Squadron
(formerly No.123)
No.113 Squadron
No.131 Squadron
(formerly No.134)
No.258 Squadron
No.261 Squadron
No.615 Squadron
(formerly No.135)

Armee de l'Air
Groupe de Chasse II/5
Escadre 4
Groupe de Chasse II/3
Escadre 4
Groupe de Chasse I/4
Escadre 3
Escadre 4
Groupe de Chasse I/5
Escadre 3
Groupe de Chasse III/3
Escadre 4
Groupe de Chasse III/6
Escadre 3

Chinese Air Force
11th Fighter Group

*5AF, 7AF, 10AF, 13AF, 14AF, 20AF became part of Far East Air Force on 7/14/45
**7AF OTU, Hawaii
***8AF OTU
****Trained in P-47s stateside, flew combat in P-38s.

IN MEMORIAM

HUBERT "HUB" ZEMKE
14 March 1914 - 30 August 1994

As *Republic's P-47 Thunderbolt: From Seversky to Victory* was going to press lengendary fighter leader Hub Zemke passed away at age 80. A wonderful man, full of life and enthusiasm, he seemed to be one of those great men who would live forever.

After getting his wings in June 1937, Hub moved through the ranks of command, going to England to observe air combat and Russia to teach pilots how to fly and fight in the P-40. In September 1942 he assumed command of the 56th Fighter Group, the first AAF unit to fly the P-47 Thunderbolt, and headed for England in January 1943. Flying its first combat missions in April 1943, the group quickly became a legend, scoring more aerial victories than any World War II fighter unit. Hub left the 56th in August 1944 to take command of the 479th Fighter Group, flying P-38s, then P-51s. He was the only man to command all three major American fighter types in combat. On 30 October 1944 Hub bailed out over enemy territory to become a POW. He was credited with 17.75 aerial and 8.5 ground victories.

Sadly, after writing the foreword to this book, he received only a few proof pages and never got to see the finished product. We will miss him greatly.

AIRCRAFT CHARACTERISTICS

EP1-106/P-35A

PHOTO: *Two initial production Republic EP1-106 fighters were rolled out onto the sod airfield which showed a light covering of snow in January 1940. Biggest variances from P-35 were in the greater overall length and the more powerful C-series engine.*

DESCRIPTION: Single-seat, all-metal monoplane pursuit-day fighter powered by one Pratt & Whitney R-1830SC3-G Twin Wasp engine. A single 3-blade Hamilton-Standard 10 ft. 6 in. diameter constant-speed Hydromatic propeller is installed. A retractable landing gear is fitted. Armament consists of 2x 7.9-mm KSP M/22 synchronized cowl-mounted machine guns; alternate armament consists of 2x 13.2-mm AKAN M/39 wing-mounted mgs. Ten 15 kg Swedish bombs carried on underwing racks.

DIMENSIONS: Wing Span/Area: 36 ft./225 sq.ft.
Overall Length: 26 ft. 10 in.

WEIGHTS: Gross Wt. (Normal): 5953 lbs.
Gross Wt (Overload): 6969 lbs.
Empty Wt. (Normal): 4608 lbs.

POWERPLANT: P&W R-1830SC3-G rated at:
Takeoff: 1050 hp @ 2700 rpm
Normal: 900 hp @ 2550 rpm @ 12,000 ft.
Military: 1050 hp @ 2700 rpm @ 7000 ft.

ARMAMENT: See Description

PERFORMANCE: High Speed (Actual): 306 mph
Cruising Speed: 290 mph
Rate-of-Climb (Actual): 2930 fpm
Service Ceiling: 31,000 ft.
Range (Max. fuel): 1122 miles

GENERAL: Normal Fuel: 130 gallons
Overload Fuel: 65 gallons

AT-12/2PA-204A

DESCRIPTION: Two-seat, all-metal monoplane Light Attack Bomber powered by one Pratt & Whitney R-1830SC3-G Twin Wasp engine. A single 3-blade Hamilton-Standard 10 ft. 6 in. diameter constant - speed Hydromatic propeller. Multi-spar cantilever all-metal wing. Provisions in rear cockpit for flexible machine gun or for instructor pilot. Fitted with dual controls. Prototype Seversky 2PA-BX (or 2PA-202) convoy fighter demonstrator was same size as the EP1-68 fighter demonstrator with comparable performance, depending on version of R-1830 engine installed. Swedish *Flgvapnet* requirements resulted in lengthened fuselage, increased wing span and mission change. Only two delivered; U.S. Government confiscated fifty undelivered 2PA-204A aircraft, converting them to trainers.

DIMENSIONS: Wing Span/Area: 41 ft./250 sq.ft.
Overall Length: 27 ft. 8 in.
Dihedral (Top Surface): 4 degrees

WEIGHTS: Gross Weight (Normal): 6606 lbs.
Gross Weight (Overload): 8580 lbs.
Empty Weight (Normal): 4813 lbs.

POWERPLANT: P&W R-1830SC3-G rated at:
Takeoff: 1050 hp @ 2700 rpm.
Normal: 900 hp @ 2550 rpm.
Military: 1050 hp @ 2700 rpm. @ 7000 ft.
Military: 1000 hp @ 2800 rpm. @ 11,500 ft.

ARMAMENT: 2x .30-cal. cowl-mounted synchronized Browning mgs.
1x .30-cal. flexible-mount mg in aft cockpit
2x Republic Mk. I internal bomb racks
6x Republic Mk. II extrnl bomb rack adapters

PERFORMANCE: High Speed (Actual): 293 mph @ 11,500 ft.
Climb to 13,124 ft.: 6.1 min.
Range: NA

GENERAL: Normal Fuel: 130 gallons
Overload Fuel: 183.2 gallons

PHOTO: *In the autumn of 1940, Materiel Division at Wright Field received its first Republic 2PA-204A, seized from the Swedish government purchase of 52 airplanes. Redesignated AT-12 to serve as an advanced trainer, the 2-seaters were as fast (at least) as the first-line P-35s still serving at Selfridge Field.*

YP-43

DESCRIPTION: The thirteen service test YP-43s were the direct outgrowth of the Seversky AP-4 which had not won the 1939 Pursuit Competition, but was the only aircraft entered able to perform at the high altitudes specified for competitors. It was a single-seat, all-metal monoplane powered by a turbosupercharged Pratt & Whitney R-1830-35 Twin Wasp 2-row aircooled radial engine rated at 1200 hp at an altitude of 20,000 ft. (Military rating). It featured flush-riveted, butt-jointed aluminum skins.

DIMENSIONS: Wing Span/Area: 36 ft./223.73 sq.ft.
Overall length: 27 ft. 10.5 in.
Wing Root Chord: 94 in.

WEIGHTS: Gross Weight (Normal): 7300 lbs. (design)
Empty Weight: 5477.7 lbs.

POWERPLANT: P&W R-2800-35 with G.E. Type B-2 s/charger rated at:
Takeoff: 1200 hp @ 2700 rpm.
Normal: 1000 hp @2550 rpm.
Military: 1200 hp @ 20,000 ft.
Curtiss Electric constant-speed 3-blade propeller with cuffs. Diameter: 11 ft.

ARMAMENT: 2x .50-cal. Browning cowl-mounted synchronized mgs.
2x .30-cal. Browning wing-mounted mgs.
6x 20-lb. Type M42 frag. bombs on 2x Type N5 racks

PERFORMANCE: High Speed (Mil. Rating): 351 mph (Guaranteed)
High Speed (Mil. Rating): 330 mph (Guaranteed)
Endurance at Oper. Speed (75% hp at 15K ft.): 2 hrs.
Time to Climb to 15,000 ft.: 6 min. (Guaranteed)

GENERAL: Wing loading: 31 lbs./sq.ft.. Normal fuel: 145 gals.
Overload fuel: 73 gals.

PHOTO: *The Republic YP-43 No.1, the first production fighter in the world to have a turbosupercharger installed remotely in the aft fuselage, was at Wright Field during service testing. Lengthened tailwheel strut was a partial corrective action to forestall ground looping tendencies.*

P-43A/P-43A-1

DESCRIPTION: (P-43A, top) Physically same as YP-43 but carried new markings and camouflage paint scheme. Equipped with P&W R-1830-49 engine. Performance variations from YP-43 were not much different than variations in speed between airplanes of same model. A quantity of 54 airplanes of the P-43 model was ordered to keep production at Farmingdale moving when production delays affected new P-47B production. The 80 Lancers ordered as P-43A were substitutes for the canceled P-44, an AP-4 series a/c outmoded by the XP-47B design.

DESCRIPTION: (P-43A-1, bottom) Physically unchanged from P-43/P-43A. Engine changed to Pratt & Whitney R-1830-57, and armament consisted of four .50-cal. Browning machine guns. Systems changes involved installation of self-sealing fuel cells. Armor plate protection provided. With utilization of Lend-Lease funds, 125 fighters were ordered for delivery to the Chinese Air Force. Ultimately, only about 51 were provided to China as fighters. Modification of 150 existing Lancers resulted in conversion to observation/tactical reconnaissance roles with cameras installed. Most airplanes were brought to P-43A standard with R-1830-49 engines. Numerous Lancers became mechanics' trainers as the airplane pictured.

DIMENSIONS: As YP-43

WEIGHTS: Approximates YP-43

POWERPLANT: Pratt & Whitney R-1830-49 Twin Wasp with G.E. s/c.
Propeller same as YP-43

ARMAMENT: As specified for YP-43

PERFORMANCE: Approximates YP-43 performance

GENERAL: All Lancers equipped with extension tailwheel in production accept for YP-43s. Retrofit installation on those non-camouflaged airplanes. Most early deliveries were to Selfridge Field.

PHOTO: *(Above) Seen at Randolph Field, Texas, while participating in 1941 southern maneuvers, this Republic P-43A appears to be halfway between natural metal finish and some camouflage. Flown down from Selfridge Field, it was marked with white crosses like the neighboring P-43 and Bell P-39 fighters.*

PHOTO: *(Below) With vestiges of snow in the background, a P-43A-1 is shown in 1942 colors, probably at Chanute Field, and possibly being used to train mechanics to taxi aircraft. Original tailwheel was to be fully retractable; revised unit only partly retractable.*

RP-43C

DESCRIPTION: Two only, as a/c pictured. Armed reconnaissance role. Four RP-43A-1s and four P-43Ds with appropriate camera installations were diverted to the Royal Australian Air Force in China. Engine models of R-1830 varied.

WEIGHTS: Gross Weight (Normal): 7476 lbs.
Empty Weight (Normal): 5975 lbs.
POWERPLANT: P&W R-1830-49
ARMAMENT: 4x .50-cal. machine guns
PERFORMANCE: High Speed: 356 mph at 20,000 ft.
Range (Normal): 540 miles
Range (Overload): 810 miles

PHOTO: *One of two very rare RP-43Cs, looking quite shopworn and bedraggled eventually came to rest in a War Assets Administration field. It was probably the personal attention of Lt. Col. Goddard that generated the photo-recon Lancers.*

XP-47B

DESCRIPTION: The XP-47B was an experimental prototype single-seat interceptor-pursuit airplane constructed for the U.S. Army Air Corps. It was a large all-metal, stressed-skin aircraft designed to meet requirements arising from results of the Battle of Britain. Military specifications required a speed of more than 400 mph at an altitude of 25,000 feet, use of a Pratt & Whitney R-2800 radial aircooled engine driving a Curtiss Electric four-blade constant-speed propeller. One General Electric Type C turbosupercharger was employed for high-altitude performance. Armament consisted of a minimum of 6x .50-cal. Browning machine guns mounted in the wings, with alternate armament to consist of at least 8x .50-cal. Browning guns. Normal fuel capacity (all internal) was 210 gallons carried in leakproof fuel tanks within the forward fuselage. The decision to have Republic Aviation construct the XP-47B prototype resulted in cancellation of contracts for one each XP-47 and XP-47A light-weight interceptors based on the Allison V-1710 engine. Design of the XP-47B was based on the new 18-cylinder 2800-cubic-inch displacement engine first installed in the Navy Vought XF4U-1 shipboard fighter and in the Air Corps Martin B-26 medium bomber. Design of the largest single-engine fighter (pursuit at the time) ever manufactured in America (and supposedly, the world) was based on requirements generated by the Emmons Board, named after M/Gen. Delos Emmons, CG of the GHQ Air Force in 1940. Essentially, the combat arm of the Air Corps was directing the Air Materiel Division to scrap the lightweight fighter concept in view of rapidly changing combat conditions elsewhere in the world, namely Europe. Specifications generated and improved at a Wright Field conference adhered closely to the critical report issued by the Emmons Board on 19 June 1940. By 15 August 1940, Change Order #2 against the XP-47/XP-47A contract was the purchase order for the XP-47B. In ending the light-weight fighter effort, they indirectly called finis to the Curtiss XP-46/XP-46A Allison-powered fighter programs. The semi-elliptical wing employed for the new interceptor was merely an enlarged version of the wing first used on Seversky's SEV-3 monoplane. The major differences were in the use of flush riveting, butt joint skin panels and the slightly modified airfoil shape designated as S-3. It was a joint de Seversky-Alex Kartveli shape, but the patent went to Kartveli after settlement of the former company president's lawsuit. The XP-47B prototype made its first flight on 6 May 1941. Contract coverage for 773 production P-47Bs was signed in September 1940.

DIMENSIONS: Wing Span/Area: 40 ft. 9 in./300 sq.ft.
Overall Length: 35 ft. 4 in.
WEIGHTS: Gross Weight (Normal): Design: 12,000 lbs.
Gross Weight (Normal): Actual: 12,500 lbs./Brabham
Empty Weight (Design): 8655 lbs.(+)
POWERPLANT: P&W R-2800-11 (Specified), -17 at 689 Board Review and -35/ P&W table
General Electric Type C-1 Turbosupercharger, Curtiss Electric Constant-Speed 12 ft. 2 in. dia. Propeller
ARMAMENT: 8x .50-cal. Browning machine guns
PERFORMANCE: Guaranteed per contract specifications: 400 mph*

GENERAL: *No formal speed runs ever conducted on XP-47B. Refer to P-47B Performance figures. Clamshell and door canopy. Specified fuel capacity was 315 gallons of 100 octane gasoline. Actual fuel capacity was 297 gallons. Specified endurance (normal power) was 1 hour at 25,000 feet. Overload fuel (the 105 gallons specified) provided 30 minutes additional endurance at normal power rating.

PHOTO: *Never previously published, this picture shows the prototype XP-47B rather crudely tethered outside Hangar 1 at the Company airfield. It was for high-power engine runs just a day or two before its first flight. Apparent 2-star emblem on fin is actually a small anti-spin parachute.*

P-47B

DESCRIPTION: Description essentially the same as XP-47B. Newer style antenna adopted for SCR-283 radio equipment. First two and final P-47Bs had XP-47B-type canopies and stub antennae. All others reverted to P-43 style sliding canopy. Externally the first two P-47B test airplanes were essentially duplicates of the XP-47B. Unofficially, at least one was a YP-47B at Farmingdale. Official rollout date of first airplane was 27 November 1941. Numerous production change items aboard. Fabric-covered control surfaces. Rudder and fin redesigned to delete exposed balance weight.

DIMENSIONS: Essentially same as XP-47B.

WEIGHTS: Gross Weight (Normal): 11,892 lbs. (Guaranteed)
Empty Weight (Normal): 8967 lbs. (Guaranteed)

POWERPLANT: P&W R-1830-21 rated at:
Takeoff: 2000 hp @ 2700 rpm.
Normal: 1625 hp @ 2550 rpm @ 6500 ft.
Normal: 1625 hp @ 2550 rpm @ 25,000 ft.
Military: 2000 hp @ 2700 rpm @1500 ft.
Military: 2000 hp @ 2700 rpm @ 25,000 ft.
Military: 1845 hp @ 34,000 ft. (Actual)

PERFORMANCE: High Speed (Actual TAS): 429 mph @ 27,800 ft.
High Speed (Actual TAS): 412 mph @ 34,000 ft.
High Speed (Actual TAS): 352 mph @ 5000 ft.
Climb to Altitude: 6.7 min. to 15,000 ft.
Climb to Altitude: 9.75 min. to 20,000 ft.
Climb Rate at 35,000 ft.: 450 fpm.
Service Ceiling: 36,000 ft.
Range (Overload): 450 miles @ 25,000 ft.

GENERAL: The first two production P-47Bs had exactly the same canopy design, the non-sliding, clamshell type, as the XP-47B. The final airplane in the B group also had the clamshell unit, but it became the pressurized-cockpit XP-47E. All standard B models were identifiable via the forward slanting antenna mast. Standard internal fuel capacity was 200 gallons in the main tank, 105 gallons in the auxiliary tank. One P-47B, s/n AC41-5942, was probably the most unusual (and best) looking fighter in the series. Equipped briefly with an Aeroproducts or Curtiss dual-rotation 6-bladed propeller and a rather large spinner, it only existed for one day and nearly cost C. Hart Miller his life. Number 5 and 6 airplanes lost within two months of each other, both with different types of empennage failures. Most P-47Bs redesignated RP-47B to indicate Restricted flight operating status, but those on flying status had retrofit action to install new metal-covered flight control surfaces. Empennage attach structure strengthened to proper stress requirements. Only one active duty AAF fighter group the – 56PG(F) – received P-47Bs and they only operated within the Z.I.. All P-47Bs were camouflaged.

***PHOTO**: It is doubtful if any larger number of P-47Bs was ever seen in one photograph than may be seen in this view. All of them appear to have the fabric-covered, early style rudders.*

P-47C-RE

DIMENSIONS: Wing Span/Area: 40 ft. 9 in. /300 sq. ft.
Overall Length: 35 ft. 5 in.

WEIGHTS: Gross Weight (Normal): 12,650 lbs.
Empty Weight (Normal): 9,575 lbs.

PERFORMANCE: Performance of this first production group of P-47Cs essentially duplicated characteristics seen in the P-47B. In effect they were improved B models but equipped in production with metal-covered control surfaces (not simply rudder and elevators) and the structurally improved empennage attachment points. No 8-inch fuselage insert was provided for the Q.E.C. assembly. (See P-47B)

GENERAL: Most, if not all, airplanes with this designation had the XP-47B/P-47B type turbosupercharger shroud fitted. Additional oxygen capacity was provided. Radio equipment changes accounted for a change in the antenna mast, no longer slanted forward.

***PHOTO**: One of the airplanes in the first group of 57 P-47Cs went to Wright Field for testing in late summer 1942. Most of the Operation BOLERO P-38Fs, B-17Es and C-47s had completed their transatlantic crossing. Meaning of #151 is unknown, but the airplane has a squadron commander's fuselage stripe.*

PHOTO: One generally unrecognized clue to P-47s with the 8-inch extended nose is the position of left and right waste gates being forward of the wing leading edge. Why Curtiss and Hamilton-Standard adhered to "toothpick" propeller blades and were so slow in producing wider blades as used in Germany is unknown.

P-47C-1-RE

DESCRIPTION: Pressures to get P-47 Thunderbolts into service in the ETO, to escort bombers, at least during penetration and withdrawal because range capabilities were no better than those of Spitfires, resulted in these fighters being compromised. Although they were the first P-47s to feature the 8-inch fuselage extension associated with the Quick Engine Change (QEC) package, they did not feature the bulged keel line. The latter change provided support/attachment features and plumbing connections for installation of a jettisonable external fuel tank or a 500-lb. bomb.

DIMENSIONS: Wing Span/Area: Unchanged
Overall Length: 36 ft. 1 in.

WEIGHTS: Gross Weight (Normal): 13,300 lbs.
Empty Weight (Normal): 9870 lbs.

GENERAL: Maneuvering performance of the long nose version was significantly improved by the 8-inch fuselage extension. Contrary to other information, it was merely a bonus and not the prime reason for the lengthening. No provisions on the wings for any external stores.

P-47C-2-RE

DESCRIPTION: Although the P-47C-2-RE series was the first to include provisions for installation of the 205-gallon unpressurized ferry tank, the unit was unsatisfactory as an interim combat drop tank. The bulbous tank could not be filled to capacity, would not jettison properly and contributed excessive drag. The P-47C-2 version, along with the C-5s, were the first Thunderbolts to enter squadron service in the ETO and to participate in combat missions. Otherwise similar to P-47C-1-RE. Many later retrofitted with the bulged-keel kit.

GENERAL: Numerous P-47C-2-REs were assigned to the 4FG, 56FG and the 78FG. Radio and antenna changes required for operations in British airspace.

***PHOTO**: One of the early arrival P-47C-2s is seen during runup, possibly at Speke Aerodrome near Liverpool in the early weeks of 1943. Revised shape of turbine shroud may be seen clearly.*

P-47C-5-RE

DESCRIPTION: Essentially the same as P-47C-2 aircraft, but block designation indicates incorporation of several equipment and technical changes from that block of airplanes. In operational service in ETO, many were retrofitted with bulged keel, Type B-7 bomb shackles for carrying teardrop belly drop tanks or bombs.

***PHOTO**: Numerous pictures of P-47C-5s show the airplanes without any visible radio antenna. Perhaps the whip antenna was used, but does not show up. Or it may have been in a period when the groups were awaiting new radio developments. These 62FS T-bolts display the definitive markings adopted as of May 1943 in the ETO.*

P-47D(RE)

DESCRIPTION: First four Evansville-built P-47Ds (Farmingdale P-47s were to be C models) actually had New York-manufactured fuselages. Although all P-47Ds in early blocks were to have additional cowl flaps, this and numerous T-bolts through at least the D-2 version had cowl flaps ending at cowl centerline. (Also, numerous D models rolled from the factory without bulged keel lines.)

***PHOTO**: Hardly anyone can differentiate between a P-47C and a P-47D without a serial number. This P-47D (RE) was one of first four from Evansville plant. It is evident that training numbers on side and on cowling had just been painted out.*

P-47D-1-RE

WEIGHTS: Gross Weight (Normal): 13,425 lbs.
Gross Weight (Maximum): 14,650 lbs.
Empty Weight (Normal): 9995 lbs.

DESCRIPTION: Contrary to nearly universal belief, few - if any - of the first 1000+ P-47D fighters through the D-4 block were delivered from the factories with the bulged keel and its basic set of bomb/ drop tank shackles. The P-47D-1-RE (AC42-7906) pictured was flown by Capt. Fred LeFebre as a member of the 353FG in the ETO. It had been modified with the bulged keel in a depot or in the field. (In approximately the same time frame, Capt. Eugene O'Neill's LM★O, a P-47C-5-RE, enjoyed the same configuration while no P-47Ds through the D-4-RA block in the Fifth Air Force (SWPA) were sharing those attributes. From the P-47D-1-RE (and subs.), all featured at least four additional cowl flaps below the propeller thrust line.

***PHOTO**: Capt. Fred LeFebre's P-47D-1-RE, "Chief Wahoo," was a nicely updated Jug when this picture was taken. A bulged keel had been retrofitted, giving the thirsty engine at least some additional flexibility. The captain was a member of the 351FS.*

P-47D-2-RA

DESCRIPTION: Sharing common ground for the first time, the P-47D-2-RA and D-2-RE models were essentially identical. The RA suffix letter indicated the fighters were built at Evansville, Indiana; the RE suffix letters were applied to all built in Bldg. 17 at Farmingdale, N.Y.. When Gen. Kenney could not squeeze more P-38s from Gen. Arnold, he accepted shipments of numerous P-47D-2 versions (from both plants). Eagle Farm in Australia provided successful field modification kits to install expendable drop tanks when Air Service Command and Air Materiel were unable to furnish either 75-gal. or 110-gal. teardrop external tanks. The rather flat teardrop tanks, produced in quantity by Ford Motor Co. and referred to as "Brisbane" tanks carried up to 200 gallons of useful fuel. These tanks used with good success by the 58FG and Col. Neel Kearby's 348FG in the SWPA.

***PHOTO**: A backlighted picture of a P-47D-2-RA emphasizes the unique shape of the T-bolt's wing. An unidentified airfield is seen rather clearly below, featuring a very complex perimiter track.*

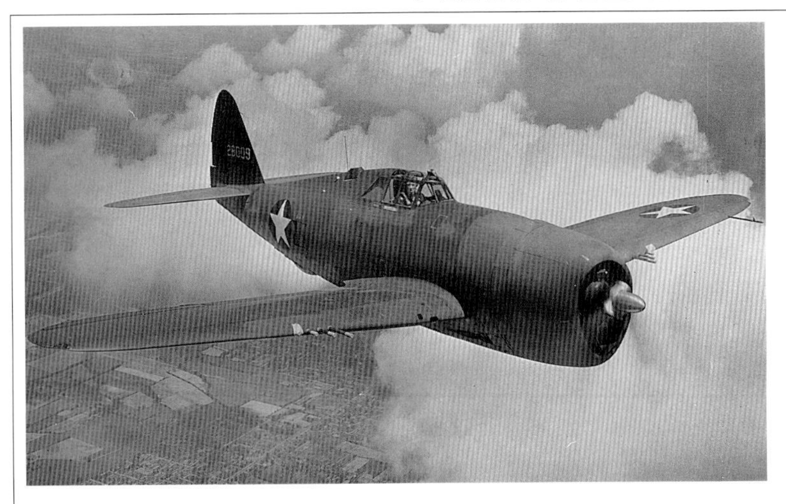

PHOTO: *Drifting through midwestern clouds, a P-47D-2-RE displays a few subtle differences. The radio antenna was a flexible whip unit and it was used in company with an RDF loop, the base of which is shown here.*

P-47D-2-RE

DESCRIPTION: It is important to recognize some salient points which have been almost totally ignored in all previous P-47 T-bolt coverage, i.e. the retrofit and depot modification effects upon external appearance and stores capabilities of P-47Cs and P-47Ds prior to D-5 or D-6 block. In the ETO (primarily), it was quite common prior to 1944 to encounter P-47Cs with later updates than some P-47D-4-RAs. Changes included the full bulged keel and installation of any number of different belly drop tanks, but primarily the 75-gallon standard metal teardrop tank and the 108-gallon Bovingdon-design paper (Bowater-Lloyd) or metal drop tanks. In the SWPA, it may have been more common to see a smooth-belly P-47D carrying a 200-gallon "Brisbane" flat drop tank, a rather quick-fix tank which was far superior to the 205-gallon "bloated udder" Republic ferry tank adaptation. If the military serial number (s/n) is not visible, it could be difficult for an expert to distinguish between a P-47D-11-RE and a modified P-47C-2-RE. The latter might even have water injection and a late-model cowl flap installation.

P-47D-4-RA

DESCRIPTION: Co-equal to P-47D-1-RE except for installation of G.E. C-21 turbosupercharger and water injection provisions. As produced, evidently none had the bulged keel line, but this one was modified. The D-4s were not to the P-47D-5-RE standard. No factory underwing stores provisions. Many sent to SWPA and to MTO. P-47D-3 and P-47D-4 produced only at at Evansville.

PHOTO: *Special fin insignia and absence of camouflage paint on this P-47D-4-RA in Italy indicate the pilot may have been a staff officer. Capt. Fred Bamberger captured the airplane as it was landing.*

P-47D-5-RE

DESCRIPTION: The Farmingdale plant built 299 of the P-47D-5-RE fighters plus one which was immediately converted to the bubble canopy configuration as the XP-47K. This version featured the G.E. Type C-21 turbosupercharger and factory provisions/installations for water injection. Factory installation of bulged keel with Type B-7 two-point bomb/tank shackle and drop tank fittings. Most of this production run was committed to the ETO. Although production installation of underwing pylons did not come from the factory until October 1943, approximately one dozen D-5 version were reworked to carry two 165-gallon Lockheed P-38 tanks for the trans-Atlantic ferrying mission. Future senator Barry Goldwater was one of the pilots for that successful crossing in August 1943. Each of the fighters was equipped with RDF (radio direction finding) loop antenna navigational aids. (No further ferry flights were attempted.)

PHOTO: *The term "jury rigged" seems very appropriate if applied to this P-47D-5-RE, one of a small group seen at Meeks Field, Iceland, on 9 August 1943. High-drag mounts could have offset half the contents of one tank at cruise. Engineering should have attempted to use low-drag P-38 pylons. Goldwater was flying either No. 6 or No. 9 in the crossing group.*

P-47D-6-RE

DESCRIPTION: Unconfirmed by available company records, it appears some factory strengthening of wings was introduced in the D-6 model. Details of longitudinal reinforcement/stores attachment bars are unknown, but they are evident outboard of the machine guns. If reinforcements had not been devised, the previous D-5 airplanes could not have carried 165-gallon fixed (not droppable) tanks for the 1943 Atlantic ferry flight. Full Lockheed tanks weighed approx. 1000 pounds each. First factory completion of a P-47D-6-RE was 1 July 1943, c/n D-851 (AC42-74615).

PHOTO: *Taxiing away from the compass rose at Farmingdale with a mechanic at the controls, a new P-47D-6-RE passes one more check point on the way to delivery.*

P-47D-10-RE

DESCRIPTION: Physical and functional description same as P-47D-5- and -6-RE airplanes in virtually all aspects. Progressive systems improvements in accordance with Unsatisfactory Reports (URs) from the field and production quality control improvements. Although information supplied by some sources indicates wings were not yet strengthened, retrofit action could provide bomb/ tank pylon instal.

POWERPLANT: P&W R-2800-63 rated at:

Takeoff: 2000 hp @ 2700 rpm.
Normal: 1625 hp @ 2550 rpm @ 25,000 ft.
Military: 2000 hp @ 2700 rpm @ all alt. thru 25K ft.*
*P&W Aircraft Data Sheets. WEP rating may be 2300 hp.
General Electric Type C-23 turbosupercharger; better turbine cooling.

GENERAL: High speed at altitude may have improved marginally by as much as 2 percent, but increased weight may have nullified most of that at lower levels.

PHOTO: *With his crew chief acting as a guide from a hi-viz wing position, the pilot of a 366FS, 358FG Republic P-47D-10-RE taxies toward his hardstand at a captured Luftwaffe airfield in Germany after a mission. Orange tail and red-with-white cowling.*

P-47D-11-RE

DESCRIPTION: As P-47D-10-RE/RA except for operation of water injection system.

PHOTO: *A new production P-47D-11-RE was pictured at the Farmingdale plant in Sept.-Oct. 1943. Antenna offset to left shows to advantage here.*

P-47D-15-RE

DESCRIPTION: Generally the same as P-47D-11-RE, but has wing strengthening for accommodtion of pylons capable of carrying bombs up to 1000 pounds or external drop tanks with capacity up to approx. 175 gallons.

PHOTO: *Red, white and blue cowled 5th ERS (Emergency Rescue Squadron) P-47D-15-REs with full D-Day invasion markings are shown in flight over East Anglia.*

P-47D-16-RE

DESCRIPTION: The P-47D-16-RE series was generally equal to the -15 airplanes in most respects.

PHOTO: *An 81st Repair Group pilot from Eagle Farm (Australia) has just released a 165-gallon drop tank from an airplane destined for the 58FG or 348FG of the V FC in the SWPA.*

P-47D-20-RA

DESCRIPTION: The major change in this series of airplanes was the substitution of the R-2800-59 engine with G.E. ignition system and tubular harness in place of -63 engines of the same horsepower. Changes made in cockpit heating and machine gun bays receive heat via ducting in lieu of electrical heat. Deletion of camouflage paint was scheduled to occur midway in this block. Other equipment changes were made.

PHOTO: *A formation of P-47D-20-RA trainers (4x .50 mgs) shows off special markings used for quick recognition. The petaloid cowl markings were most often seen on trainers, rarely on AAF combat aircraft overseas.*

P-47D-20-RE

DESCRIPTION: Data essentially the same as for Evansville D-20s.

PHOTO: *One P-47D-20-RE tested at Wright Field was evidently waxed heavily and was entirely without markings. It was at about this time all camouflage paint supposedly was to be deleted in production. It evidently failed to occur. Some airplanes in the P-47D-21-RA and P-47D-22-RE were still showing factory camouflage paint long after the established change point.*

P-47D-21-RA

DESCRIPTION: Only minor changes incorporated after -20-RA.

PHOTO: *Camouflaged and unpainted P-47D-21-RAs of the 318FG were to be seen in rather impressive numbers at Bellows Field, Oahu, Hawaii, on 15 May 1944. Narrow black fuselage band was evidently used on camouflaged airplanes as well, and there it was almost completely lost to view.*

P-47D-22-RE

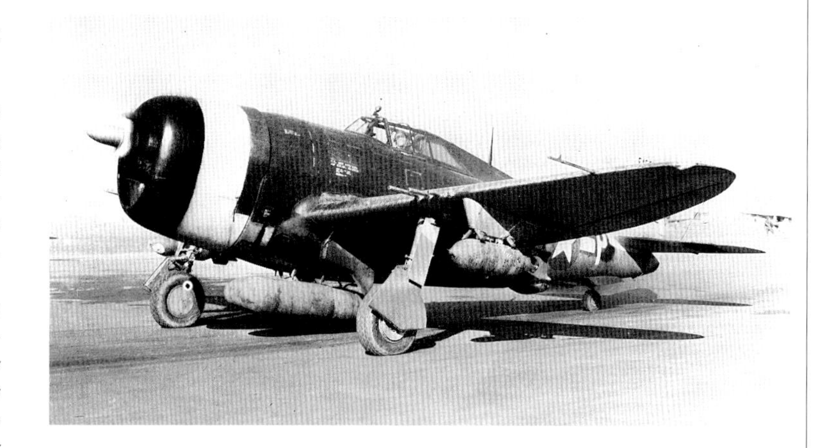

DESCRIPTION: A major change came into play with the D-22 block of fighters when Hamilton-Standard Hydromatic propellers with broader blades were adopted. Overall length with the H-S propeller was reduced to 35 feet 10 inches. The term "paddle-blade propeller" came into widespread use. Propeller diameter was 13 ft. 2 in., give or take a small fraction. Production with R-2800-59 engine continued.

PHOTO: *Some time after camouflage paint should have been deleted from production fighters, P-47D-22-REs like this 9AF razor-back Jug were in combat operations in the ETO (and probably elsewhere). Capt. Pat Pardridge, 23FS, 36FG was revving up, ready to hit the enemy where it would hurt most in October 1944. A crude looking 150-gallon belly drop tank would hardly win points for appearance. But it and the two 500-lb. bombs did their job and did not have to come back to the temporary base. Showing independence, this D-22 was in combat with a Curtiss propeller. Such changes would occur at depot overhaul because of shortages.*

P-47D-23-RA

DESCRIPTION: Equipped with a new Curtiss Electric propeller (using either the paddle blades similar to the H-S style or an even more advanced version with much wider, tapering blades), the P-47D-23-RA was the ultimate, final razoraback Thunderbolt design – the end of an era, the end of a style. In nearly every respect it duplicated the D-22, except for overall length. Other physical and functional data applied to Farmingdale and Evansville fighters. Application of camouflage paint was discontinued. All following data reflects differences from normal standards for P-47D-20s.

DIMENSIONS: Overall Length: 36 ft. 2 in.

WEIGHTS: Gross Wt. (Normal): 14,000 lbs. (approx.)
Gross Wt. (Max.): 16,475 lbs. (approx.)
Empty Wt. (Normal): 9950 lbs.

POWERPLANT: P&W R-2800-59 engine continued.
Propeller: Curtiss Electric 13 ft. 0 in. dia.

ARMAMENT: 2x 1000 lb. bombs on wing pylons.
1x 500 lb. bomb on centerline mount.

PERFORMANCE: Climb to Altitude: 5.9 min. to 15,000 ft.
Climb to Altitude: 8.1 min. to 20,000 ft.
Climb to Altitude: 10.6 min. to 25,000 ft.
Service Ceiling: 38,800 ft.

PHOTO: *A study of a new P-47D-23-RA (manufactured in March 1944) – equipped with a radio direction finder loop antenna and the latest in Curtiss Electric propeller blades – reveals how little Thunderbolts had changed since the XP-47B first flew. Yet, hundreds of improvements had been incorporated.*

P-47D-25-RE

DESCRIPTION: Installation of the same general type of plexiglass bubble canopy used on the latest versions of the Hawker Typhoon, Hawker Tempest and certain Marks of the Supermarine Spitfire changed the appearance of the P-47D-25-RE radically. Speed loss on average at maximum power did not generally exceed one percent. The canopy style was tested originally on the XP-47K, and the XP-47L was used to test the new canopy in conjunction with an increase in the internal fuel capacity. Production plans for the P-47L were revised, possibly because the planned changes did not seem to warrant a new letter designation. The P-47D designation was retained, new block numbers being assigned instead. The former very streamlined windshield was deleted, with a flat-plate armor glass windshield being substituted. Curving side panel plexiglass panels faired into the canopy.

DIMENSIONS: Overall Length: 35 ft. 10 in.
(Wing span unchanged)

WEIGHTS: Gross Wt. (Normal): 12,980 lbs.
Gross Wt. (Max.): 15,500 lbs.
Weight Empty: 9980 lbs.

POWERPLANT: Propeller: Hamilton-Standard Hydromatic, 13 ft. 2 in. diameter.

PERFORMANCE: High Speed: 426 mph (max. speeds vary by 3%(+/-).
Combat Radius of Action (2x 165-gal. drop tanks):
1221 miles, includes 15 min. Mil. Power combat at 25,000' plus 5 min. combat at WEP + landing reserve.

GENERAL: Normal Fuel: 370 gallons (internal). 270 gallons in main tank, 100 gallons in auxiliary tank.

PHOTO: *A camouflaged (in theater) P-47D-25-RE in the full markings of an 82FS, 78FG fighter in VIII FC as seen in flight with partial invasion markings (upper surface stripes deleted). The high-visibility cowling markings contrasted nicely with camouflage paint.*

P-47D-26-RA

DESCRIPTION: The Evansville plant completed the last of their Jugs in razorback form and rolled out their first bubble canopied P-47D-26-RA fighter just two and a half months after the first P-47D-25-RE was completed at Farmingdale on 3 February 1944. The only reportable differences in the two types involved installation of a 13-foot diameter Curtiss Electric propeller with the early style narrow paddle blades. That also made the RA versions 4 inches longer.

PHOTO: *Because of the impending launch of Operation OVERLORD, heavy deliveries of D-25 and D-26 airplanes went to the ETO. This 56FG Thunderbolt received its unique camouflage and hot color markings in theater; its code letters were HV★S, tying it to the 61FS. Col. Dave Schilling seemed partial to trying his hand with the triple-packaged M10/M14/M15 Bazooka 4.5-in. rocket launchers. Preparations were being made for loading one of the 108-gallon drop tanks on the centerline shackles. (The Bazooka was named in honor of comedian Bob Burns' homemade musical instrument.)*

P-47D-27-RE

DESCRIPTION: P-47D-27-RE airplanes incorporated minor changes which could not be included when the D-25 was scheduled into production. Some additional changes resulted from a Long Range Fighter Escort Capabilities test (ST No.4-44-20) conducted by the AAF Proving Ground Command, Eglin AFB, Florida. The Proof Division conducted the tests on a P-47D-25-RE. (These tests provided convincing Combat Radius of Action data.) The only significant difference from the D-26 series was continued use of the Hamilton-Standard propeller on Farmingdale production airplanes. (Although it does not show up on the P&W Characteristics chart, water injection horsepower supposedly rose by 130 hp to 2430 hp. An earlier 2300 hp figure is also a "no show" on that chart. They were quite obviously intended to be WEP characteristics authorized by the AAF.

PHOTO: *A flight of P-47D-27-RE fighters, each carrying two 110-gal. wing tanks and a 75-gal. belly drop tank are shown heading north over the Italian Alps while escorting B-25 bombers on a combat mission. Capt. Arthur Scramm, 345FS, 350FG, took the picture from his own P-47 with an aerial camera. The Twelfth Air Force mission was being flown in March 1945.*

P-47D-28-RA

DESCRIPTION: Evansville produced more than 1000 copies of the P-47D-28-RA fighter, all equipped as before with Curtiss Electric propellers. However, the majority of those airplanes had latest wide paddle blades in place of the narrower paddle blades used on the D-26 airplanes. If dorsal fins were introduced on this series, it had to be very late in the production run. However, some retrofit installation was accomplished in the field or in depots.

PHOTO: *Displaying the markings adopted for use in the Philippine Islands invasion (very wide fuselage and wing bands), this P-47D-28-RA appears to have been with the 35FG then assigned to the V FC.*

P-47D-28-RE

DESCRIPTION: By the Fourth of July 1944, the Farmingdale plant was producing near duplicates of the Evansville airplanes, even by returning to Curtiss Electric propellers. However, the paddle blade design used on the N.Y.-built airplanes was of the narrower type, similar to the shape of H-S paddle blades.

PHOTO: *While pilots of new T-bolts at the 56FG were, in a manner of speaking, applying random camouflage colors and patterns coupled with bright colors, all such 78FG P-47Ds were painted in a uniform pattern. This new P-47D-28-RE at AAF Station F-357, Duxford, England, on 26 August was a prime example. At the same time, unfortunately, conditions for Ninth AF personnel and airplanes on the Continent and Twelfth AF men and equipment in the MTO were crude at best.*

P-47D-30-RE

DESCRIPTION: In most respects, these fighters were coequals of the P-47D-26-RA. Propeller blades seemed to vary from the narrow paddle blade style to the broad blade type. Blunt leading edge ailerons were introduced, along with the first dive flaps seen on Jugs. Those were located at the 30-percent chord line. Several additional equipment and detail changes appeared.

PHOTO: *Zebra stripe markings applied to First Air Commando Group aircraft, such as this P-47D-30-RE were very distinctive. The Tenth Air Force fighter's metal control surfaces show to advantage in this picture. In the CBI, an Air Ministry directive led to deletion of the wide paint stripe on movable control surfaces when somebody was convinced control balance would be upset by the paint weight. It was an unwarranted belief.*

P-47D-30-RA

DESCRIPTION: In nearly every respect, Evansville-built P-47D-30-RAs duplicated D-30s built at Farmingdale.

GENERAL: All employees at the Evansville plant in Indiana had a great right to be proud of their achievements in building T-bolts of good quality and in massive quantities. They produced 1800 of the D-30s in just six months in a factory barely completed almost exactly two years earlier. In contrast, total Thunderbolt production by Curtiss Airplane Division in an existing plant was a pitiful 354 non-combat airplanes in approximately the same two years. Total Evansville P-47 production amounted to 6242 airplanes.

***PHOTO**: Photographed by the author at the Cleveland National Air Races on Labor Day weekend in 1948, this F-47D-30-RA was assigned to the District of Columbia National Guard. The wide-style paddle blades of the propeller and the patented cuffs show up well.*

P-47D-40-RA

DESCRIPTION: The only significant changes incorporated in the D-40 series Thunderbolts (665 built) were installation of the K-14 gunsight and installation of the dorsal fin on all aircraft. Ten zero-length rocket launcher mounts for 5-in. HVARs were included, as was tail-warning radar.

***PHOTO**: When Col. Robert Baseler returned to the Z.I. and his first stateside command, he had a camouflaged razorback P-47D bearing his "Big Stud" markings. While commanding the Brownsville, Texas, base he received a new P-47D-40-RA, promptly had it marked as the airplane he flew in Italy. (All "Big Stud" photos used in this book are from original negatives given to the author by Col. Bob Baseler.)*

XP-47E

DESCRIPTION: The last production P-47B airplane was converted at the Long Island factory to provide an experimental installation for a pressurized cabin fighter. Project initiation date was 1 Oct. 1941, well before the airplane was built. In virtually every respect it was a P-47B, except for 400 additional pounds associated with the pressurization equipment and structural changes. It was one of only four Thunderbolts to utilize a clamshell canopy and door in place of the more conventional sliding canopy. It retained the P-47B radio antenna although the slanting clearance for a sliding canopy was unneeded. At Wright Field, the R-2800-21 engine was eventually replaced by a -59 engine with a Hamilton-Standard paddle blade prop in lieu of the Curtiss propeller. Standard P-47B camouflage was applied to cover the chromate-yellow paint scheme.

***PHOTO**: The XP-47E revealed its ultimate configuration after delivery to Wright Field. With the -59 engine came the D-type cowling associated with the more powerful Double Wasp. The original turbine shroud was replaced by later production-style stainless steel parts. Pressurization modification cost was $160,000.*

XP-47F

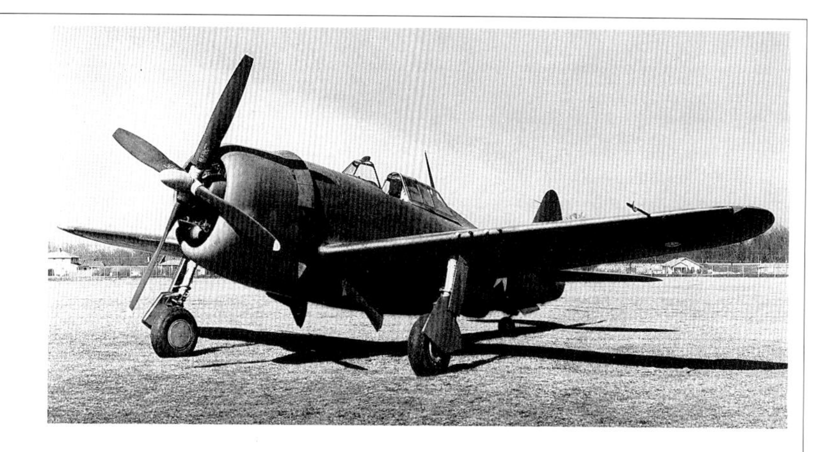

DESCRIPTION: This was a modification project to determine relative performance of a laminar flow wing compared to a standard wing on the same type aircraft. The standard section in this situation was the Republic S-3 airfoil pitted against an NACA low-drag airfoil. The Farmingdale Experimental Dept. modified one standard P-47B fighter on Project MX-116, with the airplane being accepted in August 1942. The airplane modified was the final unit in the initial production lot of 38 aircraft, s/n AC41-5938. Test work was performed at Wright Field and then at the NACA Lab at Langley Field. Cost of the modification phase was $183,147.00.

DIMENSIONS: Wing Span/Area: 42 ft./322 sq. ft.

PHOTO*: Pictured at Wright Field in September 1942, the XP-47F was equipped with dummy machine guns. The aircraft was lost in a fatal accident at Hot Springs, VA, 11 October 1943. (Aircraft spotters must have had their skills tested to the ultimate end when the XP-47F was flying. The only fighter with a similar wing planform was the Vultee P-66, but it had an entirely different tail shape.)*

P-47G-1-CU

DESCRIPTION: The initial batch of twenty Curtiss P-47G-CU airplanes was based on the early Republic P-47C-RE production model, lacking the 8-inch nose extension. Dimensions, weights and performance should have been about equal to its Republic contemporary. It did not have provisions for any external fuel tanks. A second batch of forty P-47G-I-CU fighter-trainers was based on the Republic P-47C-1-RE which was the first Thunderbolt model to feature the 8-in. forward fuselage extension.

PHOTO: *This side view of the Curtiss P-47G-1-CU was taken on 8 June 1943 in Buffalo, New York. The exhaust waste gate is located forward of the wing leading edge, proof that it had the extended nose section. The flexible VHF radio antenna may be seen atop the fuselage spine.*

P-47G-15-CU

DESCRIPTION: The largest block of Curtiss-built Thunderbolts was the P-47G-15-CU group of 154 airplanes. They approximated Republic P-47Ds in the -10-RE block.

PHOTO*: It must be clearly evident this one was well into the -15 block and it still does not have the bulged keel line. At least it had the generally finalized cowling with four additional cowl flaps, and it may have had an R-2800-63 engine with higher power ratings. In appearance, it could easily have been a P-47C-I-RE. Odd markings indicate the airplane was to be used for taxiing instruction only. The War Department finally cancelled all production, including 4220 still on order. (It is doubtful any Nazi aircraft factory filled with people from slave labor camps could have had a worse wartime production record, even with constant sabotage.)*

TP-47G-15-CU

DESCRIPTION: A Modification contract was initiated on 4 March 1944 to rework two Curtiss-built P-47G-15-CU airplanes into 2-seat gunnery trainers. Although Curtiss had produced some successful P-40N conversions, that company evidently declined to accept the project. The first of two airplanes made its initial flight in November at the Evansville, Indiana, Modification Center where all work on two airplanes was accomplished. Except for the lengthened canopy and a second cockpit which was fitted with full operational controls, the airplane was a typical P-47G-15-CU. It was necessary to remove the top half of the main gas tank, reducing total internal fuel capacity to 182 gallons. Estimated cost to accomplish the modification of two airplanes was $46,000.00 (came from Procurement Division funding).

***PHOTO**: The pilot (back seat) of the first TP-47G-16-CU is thought to be Capt. George Colchagoff. His female passenger was a Wasp (ferry pilot) according to legend. (Too bad they did not try to use one shortened bubble canopy and an elevated pilot's position with another bubble canopy.)*

XP-47H

DESCRIPTION: Like so many panic-inspired engine, airplane and weapon programs initiated during WWII, many of which can be properly described as "Pork Barrel" options, the Chrysler XV-2220 supercharged and inverted V-16 engine came under critical fire in mid-1943. Somewhat uncharacteristically, Col. Edwin Page – closely associated with such porkers as the Wright Tornado (non-engine), Lycoming O-1230-A and XH-2470-1, Continental XIV-1430-1, Menasco XIV-2040-1, P&W X-1800-SA2G (H-2600) and (maybe) the Miller L-510 – was exceptionally critical of the Chrysler program. Interestingly, he was superseded by Col. J.M. Gillespie toward the end of 1943. Surely some hero should have come to the rescue at an earlier date and engaged Packard to expand for production of the new Rolls-Royce Griffon and Chrysler to go into production on the latest Napier Sabre. All but one or two of the most promising Hyper engine projects should have been terminated. However, Gillespie managed to obtain approval for flight tests on the Chrysler XV-2220-11, an engine which had passed its 50-hour type test. Chrysler was given the contract to modify two P-47D airplanes, which they deftly sidestepped, and then managed to have the work subcontracted to the Evansville Modification Center. Two P-47D-15-RA airplanes were obtained from APGC at Orlando, Florida. In due time they were modified and tested at Evansville Airport late in the war.

DIMENSIONS: Wing Span/Area: 40 ft. 10 in./300 sq.ft.
Overall Length: 38 ft. 4 in.

WEIGHTS: Gross Weight (Actual): 15,138 lbs.
Gross Weight (Design): 14,010 lbs.
Gross Weight (Test): 13,427 lbs.
Empty Weight: 11,442 lbs.

POWERPLANT: Chrysler IV-2220-11/G.E. CH-5 supercharger rated at:
Takeoff: 2500 hp @ 3400 rpm., 71 in. Hg.
Normal: 2150 hp @ 3200 rpm., 62 in. Hg. @ S.L.
Normal: 2150 hp @ 3200 rpm., 62 in. Hg. @ 25,000 ft.
Military: 2500 hp @ 3400 rpm., 71 in. Hg. @ 25,000 ft.

PERFORMANCE: High Speed (30,000 ft.): 414 mph @ 116% rated power.
High Speed (20,000 ft.): 393 mph @ 2440 hp
Rate-of-Climb: 2800 fpm. @ S.L.; 2480 fpm. @ 20,000'
Service Ceiling: 36,000 ft.

GENERAL: Normal Fuel: 205 gallons (Internal)
Maximum Fuel: 295 gallons (Internal)

***PHOTO**: In this static 3/4 rear view of the second XP-47H, the absence of cuffs on the Curtiss Electric propeller is notable. When circular air intakes on the aft fuselage sides of the spine failed to provide adequate compressor cooling air, projecting scoops had to be added, reducing top speed.*

XP-47J

DESCRIPTION: This airplane was intended to be a very high-speed offensive fighter aircraft to be powered by a fan-cooled R-2800-61 engine driving an Aeroproducts contra-rotating propeller (six blades). It had a gross weight of 12,500 pounds. The -61 engine provided a guaranteed WEP rating of 2415 hp at altitudes between 30,000 and 35,000 feet. Two aircraft were ordered, but one was cancelled. The delivered airplane diverged from the plan to some extent, and the program was eventually dropped from production planning in favor of the same manufacturer's more powerful P-72A program. Manufacturer's Model Specification No. 505, rev. 6/12/44.

DIMENSIONS: Wing span and area: Standard
Overall Length: 36 ft. 3 in.

WEIGHTS: Gross Weight (Design): 12,400 lbs.
Empty Weight: 9662.5 lbs.

POWERPLANT: P&W R-2800-57 with G.E. CH-3* supercharger.
Rated at:
Takeoff: 2100 hp @ 2800 rpm. and 55 in. Hg.
Normal: 1700 hp @ 2600 rpm. @ S.L.
Military: 2100 hp @ 2800 rpm. @ 30K ft.; 54 in. Hg.
WEP: 2800 hp @ 2800 rpm. @ 36,000 ft.; 72 in. Hg.
* Changed to CH-5 later.

ARMAMENT: Six .50-cal. Browning M-2 machine guns.
Ammunition per gun: 267 rds.

PERFORMANCE: Maximum (WEP) Speed: 507 mph @ 34,300 ft. w/2800 hp (133% rated power).
High Speed (Military Power): 470 mph @ 2800 rpm. @ 34,300 ft. w/2100 hp (100% rated power).
Operating Speed: 435 mph @ 2600 rpm. @ 34,300 ft. w/1700 hp (81% rated power).
Cruising Speed: 400 mph @ 2300 rpm. @ 34,300 ft. w/1365 hp (65% rated power).
Range (Optimum Cruise at 25,000 ft.): 1070 miles.
Rate-of-Climb: 4900 fpm. at S.L.
Rate-of-Climb: 4400 fpm. at 20,000 ft.
Time to Climb to 20,000 ft.: 4.25 min.
Time to Climb to 30,000 ft.: 6.75 min.

GENERAL: Normal Fuel: 287 gallons, Grade 100/130.

PHOTO: *Viewing this photo, we must doff our caps to Chief Powerplant Engineer, Ray Higginbotham and his crew. The things still needed were even wider propeller blades like those on some of the Luftwaffe aircraft. The "Superman" emblem seems appropriate.*

XP-47K

DESCRIPTION: A standard P-47D-5-RE razorback fighter was taken directly from the production line for modification. All dimensions and weights standard for that model. Fuselage modified under CPFF contract to accept Hawker Typhoon Mk.IB type sliding, full-view Perspex canopy. This was authorized by Change Order 28 to Contract W535-ac-21080 on 17 June 1943. The aircraft, AC42-8702, was completed on 3 July and delivered in the same month. In contrast, it had taken Hawker Aircraft nearly eight months to modify a Typhoon Mk.IA because of the differences in structural problems. Tests on the Republic airplane, redesignated XP-47K, proved there was less than a 3 percent speed loss, while overall visibility from the cockpit was increased immensely. Sometime in 1944, the XP-47K, like an aircraft I.D.d as the "XP-47M" but in the C series, was used in experiments aimed at providing fuel tanks with additional internal fuel capacity of 65 gallons in each wing panel to create a long-range fighter. Conversion cost for installation of the free-blown full-vision canopy and flat-plate windshield: $64,464.24.

PHOTO: *It would take an expert in aircraft recognition or a very special crew chief to recognize this was not a P-47D-25-RE and even then it would be worth a bet he would fail. The XP-47K is shown at Wright Field in the summer of 1943. Just six months later, essentially identical, but improved and unpainted P-47D-25-REs were rolling out the doors of Bldg 17 daily at Farmingdale.*

YP-47M

DESCRIPTION: If it wasn't for the bright chromate-yellow cowling and tail, plus M2 emblazoned on the cowling sides *and* the fuselage sides, in early springtime 1944 (snow was still on the cold ground), it would have taken an "insider" to know it wasn't another P-47D-27-RE. The YP-47Ms (three of them) were supposedly created on a moments notice to deal with German V-1 Buzz Bombs which were falling around London with devastating and demoralizing effect. It just wasn't so. The rare "XP-47M" flown frequently by tester Carl Bellinger was a very much hot-rodded P-47C modified to test the new R-2800 C-series engines. (Those engines are fairly easy to detect because the reduction gear case is split circumferentially, mated with many bolts.) At another early date, when it seemed obvious that the XP-47J was unlikely to go into mass production, a brief study showed there was another route to a powerful and very fast Thunderbolt, and it was a "quick and not-so-dirty" approach of mating the XP-47J powerplant to airplanes currently in production. Without a defined need, it was not even proposed. An even more convincing argument against the V-1 theory: If 422-mph Hawker Typhoons and even faster Mk.V Tempests were chasing down the V-1s, what was to prevent 429-mph P-47Ds from doing the same thing? The AAF support effort was probably made at an early stage of the attacks when effectiveness of RAF aircraft and pilots was still an unknown. P-47M production was in high gear when the V-1 threat was essentially ended by October 1st. Supplies of R-2800C engines were already being stockpiled at Republic for the forthcoming P-47N production runs, and internal fuel capacity was up to 370 gallons in all bubble canopy Ds. When a need was expressed for that V-1 chaser, the tools and equipment were ready to produce. The YP-47M was born in a hurry. In nearly every respect except for the engine and performance, the airplanes were P-47D-27s. Initially powered by R-2800-14W engines which were not intended for use with turbosupercharging at seal level, the three service test/prototypes soon were flying with -57 engines. WEP ratings went up to 2800 hp, producing speeds in the 470-480 mph range with regularity. On the strength of YP-47M performance over Long Island, a production contract was created. The new CH-5 turbosupercharger, intended for the XP-47J type, was included in the power package. Factually, the YP-47M was an XP-47N with short-range wings.

DIMENSIONS: Same as P-47D-28-RE

WEIGHTS: Gross Wt. : 15,350 lbs.
Gross Wt. (Max.): 18,250 lbs.

POWERPLANT: P&W R-2800-14W/-57 w/G.E. CH-5 T-supercharger, rated:
Takeoff: 2100 hp @ 2800 rpm. (no S.L. turbo on - 14W)
Normal: 1700 @ 2600 rpm.
Military: 2100 hp @ 2800 rpm. @ 28,500 ft.
WEP: 2800 hp @ 2800 rpm.

PERFORMANCE: High Speed: 475 mph
Time to Climb: 4.9 min. to 15,000 ft.
Range (Combat): 530 miles

***PHOTO**: Bundled up against the biting cold and propeller blast, test pilot Mike Ritchie checked it out on the Farmingdale compass rose. The big embankment surrounding it was fairly effective in limiting excessive sound blasting people working outside on other T-bolts.*

P-47M-1-RE

DESCRIPTION: Derived directly from the mass-production P-47D-27-RE fighter – in fact the last four production airplanes were taken from the assembly schedule and three became YP-47Ms and one became the XP-47N. The P-47M-1-RE was in limited production at Farmingdale. Although enough airplanes were constructed to supply two fighter groups, virtually all were destined for the 56FG in the Eighth Air Force. All production units were nearly identical to the YP-47Ms, although there were detail differences. Several problems arose in the ETO when the aircraft were rushed into service. Most disturbing were high-altitude ignition problems and a plague of piston ring and piston failures. Vice-president Hart Miller flew to England to participate in problem solving. Col. Cass Hough's organization at Bovingdon was also involved. Large numbers of the P-47Ms were given the typical 56FG "camouflage" treatment. Most of the airplanes were, initially, flown without drop tank pylons, but at a slightly later date that condition was altered. Some great misconceptions about the part played by P-47Ms in combatting the V-1 NO BALL attacks on England from Continental launch sites have persisted for years. The first V-1 attacks came in June 1944. As sites were overrun by the Allied armies and as the RAF/RNZAF fighter squadrons and anti-aircraft guns destroyed hundreds of the flying bombs, those attacks dwindled to nothingness in September. The number of air-launched V-1s was far below the launch level attained by fixed coastal sites. As for P-47M participation, the first of the type rolled from Bldg. 17 on 8 October 1944. Snow was on the ground in Great Britain when the first Ms arrived from the U.S.A. for assignment to the 56FG, too late to chase Buzz Bombs.

DIMENSIONS & WEIGHTS: Refer to YP-47M.

PERFORMANCE: Refer to YP-47M.

***PHOTO**: In a winter setting, a new P-47M-1-RE received the attention of several men. No group or squadron markings had been applied and considering the attention being shown, it most certainly was a new arrival from BRD or Burtonwood Air Depot. No external stores pylons were fitted, but that was soon to be corrected.*

XP-47N

DESCRIPTION: Designed, tested and developed in record time to offset plans to cancel all P-47 Thunderbolt production, the XP-47N prototype and the production offshoots turned out to be far better than anticipated. This change was more extensive than the move from razorback Thunderbolts to the bubble canopy version. A newly developed R-2800C model engine was involved, a substantial rework of the wing was necessary, a much-improved CH-5 turbosupercharger was installed and operating systems and controls underwent significant change. Internal fuel capacity was increased by an amount essentially equal to the original P-47B main fuel tank capacity. The maximum gross weight increased amazingly, and yet the top speeds exceeded those of any P-47s except for the XP-47J and the companion P-47M series. Project 611-1 launched the program to develop and test a set of special wing panels "for application to the YP-47M airplane, all for the purpose of increasing the range of this type airplane." So wrote Capt. G. D Colchagoff on 19 May 1944.

Of great interest, preliminary tests were conducted using the P-47C previously referred to as the "XP-47M." Wings with 27-inch stubs (or plugs),not 18 or 20 inches, and squared wing tips made of wood were installed on the aircraft listed as the P-47C-707 airplane (AC41-6601). Something not previously revealed anywhere was a program conducted by Republic in November 1943 with stubs of various lengths installed on the same "mule" P-47C. At that time the tests were not too successful for several reasons, so the company dropped the project. No wingtip changes had been made.

Another point never mentioned by any media was the assignment of the third YP-47M (as in Mabel) for installation of the extended wings as of 26 June 1944. Defining the task to be accomplished, the assignment was to provide a P-47 fighter with a *combat radius of action* (and that does not mean ferry radius) of 1000 miles! In the meantime, the "Mule" ("XP-47M") had demonstrated a rate of roll "equal to or better than" any standard P-47. At that point, on 3 July, M/Gen. B.E. Meyers sent a teletype to B/Gen. F.O. Carroll, assigning the designation XP-47N to the YP-47M airplane being used for tests. At the moment, the airplane we have always seen listed as YP-47M, s/n AC42-27388, became *the XP-47N*. Republic Aviation has *always* listed that specific aircraft as the third YP-47M. The airplane made its first flight on 22 July 1944. On 31 July the XP-47N (388) flew nonstop from Farmingdale to Eglin Field, turned and headed back for N.Y., flying there non-refueled, but being forced to land at Millville, N.J. because of weather.

DIMENSIONS: Wing Span/Area: 42 ft. 6-1/16 in./322.0 sq.ft.
Overall Length: 36 ft. 1-3/4 in.

WEIGHTS: Gross Wt. (Normal): 16,300 lbs. (570 gals. Intern.)
Gross Wt. (Max. Normal): 18,545 lbs. (2x165-gal. ext.)
Gross Wt. (Max. Overload): 20,166 lbs. (2x300-gal.ext.)
Empty: 12,950 lbs.

POWERPLANT: P&W R-2800-57 w/G.E. CH-5 turbosupercharger, rated at:
Takeoff: 2100 hp @ 2700 rpm.
Normal: 1700 hp @ 2600 rpm. @ 25,000 ft.
Military: 2100 hp @ 2800 rpm. @ 28,500 ft.
WEP: 2800 hp @ 2800 rpm. (72 in. Hg., 5 min.)*
*This is an AAF authorized rating, not P&W.

ARMAMENT: 8x .50-in. M-2 Browning machine guns with 267-500 rpg.

PERFORMANCE: High Speed (Actual): 438.5 mph @ 2600 hp @ 26,000 ft.
High Speed (WEP Act.): 450 mph @ 2800 hp @ 26,000 ft.
Cruise Range (Combat): 2190 mi. @ 315 mph (1170 gal.)
Service Ceiling: 39,900 ft.
Time to Climb: 10 min. to 23,000 ft.
(All speeds are T.A.S.)

GENERAL: Rate of Roll: 98 deg./second @ 300 mph. (Comparable rate for P-47D-28-RE was 79 deg./second @ 300 mph.; at all speeds to 350 mph the XP-47N roll rate was *better* than P-47D-28.)

***PHOTO**: The XP-47N, s/n AC42-27387, was photographed at Eglin Field, Florida, as it was being prepared for a variety of tests with external fuel tanks, up to and including the 310-gallon Lockheed P-38 drop tanks and 330 gallon wooden versions.*

Suddenly, Capt. Colchagoff formally requested, and received, authorization to conduct tests on XP-47N AC42-27387(!) – the airplane pictured with these data – where the airplane left Wright Field with two 300-gallon drop tanks, climbed to 25,000 feet and flew at 315 mph TAS until the tanks were empty, returning to base with the tanks in place. (The 300/310/315-gallon drop tank nomenclature was the result of generalized reference and different tank suppliers. They were all basically 310-gallon tanks.) Immediately following, the airplane flew with two 165-gallon tanks, climbed to 25,000 ft. on external fuel, and dropped the tanks. At that point, the routine combat flight conditions were performed. Initial gross weight with 1170 gallons of fuel was 20,165 pounds. Combat radius of action was 1310 miles. Flight duration was...12 hours and 3 minutes...nonstop! The pilot survived.

Not having the entire history documentation relating to the XP-47N project, we do not know what became of YP-47M/XP-47N AC42-27388. Best bet: it was returned to its earlier configuration and the designation YP-47M was restored. (The most significant factors in devlopment of the long-range P-47N: Dire need, the P&W R-2800C engine at a good time, plus C. Hart Miller's attitude about getting things done. He was the sparkplug for Kartveli.) History also shows the XP-47K was equipped with one of the sets of experimental long-range wings. So the real history of the XP-47N is NOT embodied in one airplane. It involves the P-47C-5-RE, the true XP-47N, the YP-47M/XP-47N and the XP-47K. (Documentation is in the possession of the author.)

P-47N-1-RE

DESCRIPTION: The P-47N airplane was a long-range escort fighter, offensive fighter and with superior capabilities as an fighter-bomber. The wing span was increased over all other P-47 models by insertion of 18-inch stubs between the fuselage and the wing panels. Two internal fuel tanks in each stub wing carried a total of 100 gallons of gasoline, providing a total increase in internal fuel of 200 gallons. The airplane was capable of flying missions with 600 additional gallons of fuel in two 300-gallon wing drop tanks. Other arrangements included carrying two 165-gallon drop tanks on the wing pylons and a 110-gallon belly drop tank. The fighter has better rate of roll capabilities than the P-47D-28 model and could turn with that fighter.

DIMENSIONS: Wing Span/Area: 42 ft. 6-13/16 in./322 sq.ft. (Spec.)
Length: 36 ft. 1¾ in.
Tread: 18 ft. 5-13/16 in.

WEIGHTS: Gross Wt. (Normal): 13,823 lbs.
Gross Wt. (Max. w/2x165 gal. wing tanks, 1x110 gal C/L. tank): 19,149 lbs.
Gross Wt. (Max. w/2x300 gal. wing tanks): 20,160 lbs.
Empty Weight: 10,988 lbs.

POWERPLANT: P&W R-2800-57 w/G.E. CH-5 Turbosupercharger.
Rated at: See XP-47N.

PERFORMANCE: High Speed (Actual): 455 mph @ 38,750 ft.
High Speed (WEP-Crit.): 467 mph @ 32,000 ft.
High Speed (Crit. Alt.): 438 mph @ 42,000 ft.
High Speed (2100 hp. Mil.): 345 mph @ 5000 ft.
Time to Climb: 6.2 min. to 15,000 ft.
Time to Climb: 10.4 min. to 25,000 ft.

GENERAL: P-47Ns were frequently being flown in the FEAF at max. gross weights in the vicinity of 22,500 pounds. Although many hours of flight testing were accomplished carrying 600 gallons of fuel in wing drop tanks, APGC did not recommend this as a standard operational practice.

***PHOTO**: This high-quality photograph of the sixth Republic P-47N-1-RE, taken at Farmingdale, did not show any installation of the commonplace dorsal fin. (See P-47N-5-RA for typical dorsal fin.)*

P-47N-5-RE

DESCRIPTION: Probably produced in March 1945, this P-47N-5-RE was representative of one of the last series of P-47Ns dispatched to combat. The N-15 and later blocks were likely produced too late to reach the FEAF combat areas. Those aircraft were physically identical to aircraft through P-47N-25-RE, although equipment changes were ongoing. Some P-47N-5-REs were equipped with R-2800-73 engines, and some later blocks not involved in combat had -77 or -81 engines.

***PHOTO**: Bearing the aircraft identification markings almost universally applied to P-47 aircraft in the Far East Air Force (FEAF), this P-47N-5-RE was not yet marked with any group or squadron markings. The typical dorsal fin fitted to nearly all, if not all, airplanes in the N configuration is shown.*

P-47N-15-RE

***PHOTO**: Displaying ten zero-length Launcher Pads, this P-47N-15-RE was cold soaking at Patterson Field, Ohio, in the winter of 1944-45.*

P-47N-25-RE

DESCRIPTION: Physically unchanged from the P-47N-1-RE in all but details.

PHOTO: *Depicting a Georgia National Guard F-47N-25-RE in flight heading up to the North Georgia mountain area in the 1950s is really a summation of the production P/F-47 line. It shows the fighter-bomber configuration of the airplane with multiple gun, tank/bomb and rocket installation capabilities. And, yet, the P-47N was not done. Document TSEAL-2-4308-7-4 referred to "conduct of an investigation of an R-2800(-25 or similar) 2-stage, 2-speed supercharged engine (no turbo) and an I-40 or I-16 turbojet engine installed in a P-47N." C. Hart Miller, the power behind the P-47N idea, might well have pushed for this had the war continued for many months.*

XP-72

DESCRIPTION: It was expected, and logically so, the P-72 airplane would be the fastest, highest flying of all Thunderbolt fighters. Although it did not, in the experimental format, carry a P-47 designation, there was every possibility it might well have been allocated a P-47 designator, probably Q. (The P-51A was as far removed from the P-51H configuration as the XP-72 was from the P-47M, probably more so.) Even at an early stage of development, the XP-72 could boast of 3000 hp, with every reason to believe 3650 would be seen in the near future. A WEP rating would have been several hundred horsepower greater. It certainly would have been a 500+ mph fighter, and the probability remains that it would have soon seen a P-47N wing introduced. For greater detail, refer to Chapter 11.

DIMENSIONS: Wing Span/Area: 40 ft. 11-7/16 in./300 sq.ft.
Overall Length: 36 ft. 7-13/16 in. (w/contraprops)
Landing Gear Tread: 17 ft. 5½ in.

WEIGHTS: Empty Weight: 10,965 lbs.
Gross Weight (Normal): 13,859.8 lbs.
Gross Weight (Max. Fuel): 14,760 lbs.

POWERPLANT: P&W R-4360-13 Wasp Major 28-cylinder, aircooled radial engine. Single-stage integral blower and one large diameter remote variable-speed centrifugal supercharger. An engine-cooling fan was placed within the cowling behind the propeller. One 4-blade or two 3-blade dual-rotation propellers. Max. dia. of 13 ft. 6 inches. A large air-to-air intercooler was located aft of the extension-shaft-driven supercharger blower. Powerplant was rated:
Takeoff: 3000 hp @ 2700 rpm.
Military: 3000 hp @ 2700 rpm.

ARMAMENT: Six .50-cal. M-2 machine guns with 267-300 rpg.

GENERAL: Normal Fuel: 237 gals. Overload Fuel: 370 gals.

PHOTOS: *The elegant lines of the first XP-72 are revealed in this direct side view on rollout day. That cowling foretold the shape of the engine cowlings to be used on the XF-12 four-engine reconnaissance airplane. In the lower picture, the second XP-72 displayed its 6-blade dual-rotation propellers (identities deleted, but either Curtiss or Aeroproducts). Test pilot Ken Jernstedt made a hazardous dead-stick crash landing from high altitude.*

APPENDIX E

A CHRONOLOGY AND MODEL LIST

MODEL DESIG.	C/N	LICENSE/ CUST.NO.	1ST FLIGHT AND PILOT	TECH DATA/GENERAL NOTES
SEV-3	301	X2106	A. Seversky June 1933	Sport amphib.; Edo built. Design. by Seversky and M. Gregor, 1930-31. 420 hp. Wr. J-6-9E eng.; to WF as XP-944. (Orig. had 350hp R-975-ET eng.)
SEV-3L	301	X2106 XP-944	A. Seversky June 1934	SEV-3 with fixed land gr. as trainer demo. at WF, Dayton, Ohio, June 1934
SEV-3XAR	35	X2106	A. Seversky Sept. 1934	Modif. SEV-3L w/R-975E (Wright) engine and new cockpit arrangement. Won trainer contract (BT-8). Converted to SEV-3M.
SEV-2XP	2	X18Y	May 1935	2-seat pursuit; damaged prior competition. Wr. XR-1670, 775hp (?).
SEV-1XP (SEV-4)	(2) Mod.	X18Y	A. Seversky August 1935	1-seat pursuit modif. of SEV-2XP. Wright XR-1670. Retract. ldg. gear. Chg. to Wr. Cyclone SR-1820G4 at Wright Fld. Modified w/P&W R-1830B engine, new tail and cowling. To NY for modif.; Wr. Cyclone GR-1820G5 eng. instal.; by late April, P&W R-1830B is installed. Won P-35 order. Note: Modif. as racer/demo. Flown to 4th place in Bendix Race by Frank Sinclair 1937. Written off by J. Cochran, Miami, Fl., Dec.'37
SEV-3M-WW	1, 2, 3	X15391 X15689 X15928	August 1935	Three a/c plus one fuse. (to SEV-2XP) Kirkham blt. but incomplete. Seversky A/C completed 1936. Mfd. for Colombia. Amphibious
SEV-3M	301	NR2106	December 1936	Rebuilt with SR-1820-G2 Wright Cyclone and vert. tail from April 1936 SEV-1XP. Amphib. gear, newly painted. Sold Col. Fierro and in Mexico as XA-ABG. written off Banolas, Spain 2/39.
BT-8	36,	AC34-247 thru -27629	A. Seversky Jan. 1936	Air Corps BT-8 monoplane basic trainer. P&W 400 hp R-985-11 engine. #36 deliv. Feb. etc. 1936 to WF.
SEV-X-BT	6* (See Model 2PA Prototype)	X189M	August 1936	Advanced basic trainer, lost out to BT-9. Had SEV-1XP parts, P&W Wasp S3H-1, 550hp. Model may be *SEV-6 vs c/n 6. Had variable wing panels.
P-35		AC36-354 thru -429	A. Seversky 4 May 1937	Production version of SEV-1XP. Air Corps buy of 77 pursuit a/c. Last (AC36-430) to XP-41 for 1939 competition. First -354 rejected by AC[1].
AP-2	39	X1250 R1250	A. Seversky 1937	For 1937 Pursuit Comp., flush ldg. gr., fast but contract to Curtiss P-36. Floyd Bennett Fld. crash ldg. 9/1/37. Parted out.
AP-1	44	X1390 R1390 NR1390	A. Seversky 4 May 1937	Rejected AC36-354; flown by NACA Langley w/mods. Had diff. windshield and odd wheel fairings aft. Dest. 4/2/38, Miami, Fl. in hangar fire.
2PA Proto.		NR189M	Frank Sinclair 8 July 1937	"Convoy Fighter" Proto. Wr. Cyclone GR-1820-G3 875hp, guns; demo in S.A. Eng. failure; a/c left in Brazil unattended.
2PA-A		NX1307		Convertible amphibian, landplane, retract. gear. For USSR via Amtorg Corp. Wr. Cyclone R-1820G5 eng. Exported 10/23/37.
2PA-L	R		C. Hart Miller 2 Nov. 1937	No CAA license no. assigned. Flew armed, w/o any lic. markings. Built for USSR; bought via Amtorg Corp. Wr. Cyclone geared GR-1820-G7 engine.
SEV-S2	43	R70Y NR70Y	25 July 1937	Racing version of P-35; P&W R-1830-SCG eng. Won 1937 Bendix Trophy, 9/4/37 and 1939. Frank Fuller, Jr. owner-pilot. Exported to Ecuador AF, 1941 as C-2.

OF SEVERSKY AIRCRAFT

SEV-DS	42	NR1291	Sept. 1937	Doolittle Special for Shell Aviation Prod. Damage in ldg. acc., rebuilt. Wright Cyclone R-1820-G5. Exported to Ecuador as C-1.
NF1		X1254	C. Hart Miller A/C complete 6/1/37. Flown 6/3/37. At DC 9/24/37.	Quick opportunist entry for USN fighter competition. To Anacostia, DC, 9/24/37. Did not meet all requirements. Lost to Brewster XF2A-1. Wright R-1820-22 engine.
2PA-B3	126	X1321		Production test, first one as prototype. Wr. Cyclone R-1820-G3B engine. For Japan via Aircraft Trading Corp.. Fate unknown. Was not part of Project 64 shipments. Records conflict; were doctored.
2PA-B3	64-1 64-2 64-3	NX1391 NX1388 (?)	5 April 1938 19 April 1938	Two lic. for test and flite to Calif. for shipment from San Pedro. Eighteen shipped to Japan from NY. No Bill of Sale. NOTE: Bad weather, new model engine prevented much flight effort prior 9 April. No. 64-1 crated, shipped Hokkai Maru 4/12/38. Pilot F. Sinclair sailed 5/2/38. All twenty for Japanese Navy as A8V1 bomber convoy fighters. Nos. 64-2, -3 flew West Coast for shipment.
AP-7	145	NX1384	A. Seversky	Dev., Demo and record plane w/#13. J. Cochran records. Won 1938 Bendixrace. Modified wing + flush gear for 1939. Long fus.; to Ecuador 1941as C-3 ($40,000).
EP1-68	147	NX2587	G. Burrell Sept. 1938	To Europe for demo Nov.1938 w/2PA-202. Short and long wing panels. P&W R-1830S1C-G engine, 1200 hp. Sold to Ecuador AF as C-4, 1941.
AP-4	144	NX2597	F. Sinclair 22 Dec. 1938	G.E. turbochgr. in aft fus. Entry in 1939 Pursuit Comp. Better than XP-40, but lost anyway. AMD ordered 13 for service test as YP-43. Fire in flt., Sinclair jumped, a/c destroyed nr. Fairborn.
AP-9	148	NX2598	F. Sinclair	Interim Compet. entry for late XP-41 entry. Curtiss won huge contract for P-40s despite failure to meet the hi-alt. requirement. Dispo. not recorded. No lic. renewal. No purchase record.
2PA-202	146	NX2586	F. Sinclair 23 Oct. 1938	To Europe for demo Nov.1938 as 2PA-BX. 2-seat convoy ftr. version. Same length as EP1-68, same wings and R-1830 eng.; acc. Croydon Aerodrome 4/1/39. Written off, shipped NYC./ J.Hopla.
XP-41		AC36-430	March 1939 Deliv. 4 Mar.	Final (77th) on P-35 purch. contract. Almost 100% rev. Near duplicate of AP-4, but w/2-spd, 2-stg. mechanical super charger. To NACA test. Prob. disp.: Class 26.

Strangely, Seversky Aircraft Corp. was born in the Great Depression, rose to greatness in those daunting years, but failed just as war expansion came into view. Major Seversky never again designed or built aircraft. Republic Aviation Corp., the business successor, rose to towering heights basically on a design (the AP-4) only the Major had any confidence in. The competing corporate-sponsored AP-9 and Air Corps-sponsored XP-41 were both second rate.[2]

1 Became Seversky AP-1. Air Corps received another AC36-354!

2 For most of its decade-long history, Seversky Aircraft had a poor model-numbering system, and production control was often in chaos. The publisher would appreciate receiving any well documented information that may have duplicated files lost in the Fairchild takeover or during the demise of Republic as an entity. (See verso page in book front matter.)

RECOMMENDED

Bodie, Warren M.	THE LOCKHEED P-38 LIGHTNING - THE DEFINITIVE STORY
Boyne, Walter J.	SILVER WINGS, A HISTORY OF THE UNITED STATES AIR FORCE
Carter, Kit C. and Mueller, Robert	THE ARMY AIR FORCES IN WORLD WAR II -COMBAT CHRONOLOGY 1941-1945
Coffey, T.M.	DECISION OVER SCHWEINFURT
Copp, DeWitt S.	A FEW GREAT CAPTAINS
Dade, G.C. and Strnad, Frank	PICTURE HISTORY OF AVIATION ON LONG ISLAND 1908-1938
Davis, Larry	56TH FIGHTER GROUP
Davis II, A.H., Coffin, R.J. and Woodward, R.B. (Editors)	THE 56TH FIGHTER GROUP IN WORLD WAR II (Published 1948)
Ethell, Jeffrey L. and Sand, Robert T.	FIGHTER COMMAND
Fahey, James C.	U.S. ARMY AIRCRAFT - 1908-1946
Freeman, Roger A.	CAMOUFLAGE & MARKINGS - REPUBLIC P-47 THUNDERBOLT, USAAF, ETO & MTO, 1942-1945
Freeman, Roger A.	THE MIGHTY EIGHTH - A HISTORY OF THE U.S. 8TH ARMY AIR FORCE
Freeman, Roger	THE MIGHTY EIGHTH IN COLOR
Freeman, Roger	THUNDERBOLT, A DOCUMENTARY HISTORY OF THE REPUBLIC P-47
Freeman, Roger	REPUBLIC THUNDERBOLT
Goddard, B/Gen. G.W., w/Copp, DeW. S.	OVERVIEW
Green, William	WARPLANES OF THE THIRD REICH
Hagedorn, Dan	REPUBLIC P-47 THUNDERBOLT - THE FINAL CHAPTER
Hall, Jr., Grover C.	1000 DESTROYED
Hess, W.N. and Ivie, T.G.	FIGHTERS OF THE MIGHTY EIGHTH
Holley, Jr., I.B.	BUYING AIRCRAFT, MATERIEL PROCUREMENT FOR THE ARMY AIR FORCES
Johnson, Charles R.	THE HISTORY OF THE HELL HAWKS (THE 365FG)
Johnson, R.S. with Caidin, Martin	THUNDERBOLT
Kelsey, B/Gen. Benjamin S.	THE DRAGON'S TEETH?
Kenney, Gen. George C.	GENERAL KENNEY REPORTS
Ketchum, Richard M.	THE BORROWED YEARS 1938-1941 - AMERICA ON THE WAY TO WAR
Kupferer, A.J.	THE STORY OF THE 58TH FIGHTER GROUP IN WORLD WAR II
Lambert, J.W.	THE PINEAPPLE AIR FORCE
Lindbergh, Charles A.	THE WARTIME JOURNALS OF CHARLES A. LINDBERGH

READING

Mason, Francis K.	THE HAWKER TYPHOON AND TEMPEST
Maurer, Maurer	AVIATION IN THE U.S. ARMY, 1919-1939
Maurer, Maurer	COMBAT SQUADRONS OF THE AIR FORCE - WORLD WAR II
McDowell, E.R. and Hess, W.N.	CHECKERTAIL CLAN - THE 325TH FIGHTER GROUP IN NORTH AFRICA AND ITALY
McFarland, S.L. and Newton, W.P.	TO COMMAND THE SKY - THE BATTLE FOR AIR SUPERIORITY OVER GERMANY, 1942-1944
Powell, Jr., R.H. and Ivie, T.G.	THE BLUENOSE BASTARDS OF BODNEY (352FG)
Pratt & Whitney Div., UAC	THE PRATT & WHITNEY AIRCRAFT STORY (1950) TWENTY FIFTH ANNIVERSARY
Rust, Kenn C.	EIGHTH AIR FORCE STORY
Rust, Kenn C.	THE NINTH AIR FORCE IN WW II
Rust, K.C. and Hess, W.N.	THE SLYBIRD GROUP (353FG)
USAF Historical Div.	AIR FORCE COMBAT UNITS OF WORLD WAR II
Zemke, Hubert and Freeman, Roger A.	ZEMKE'S WOLF PACK

One set of books may be considered the most important and most reliable of all: Volumes One through Seven, THE ARMY AIR FORCES In World War II, by OFFICE OF AIR FORCE HISTORY, USAF; Edited by Craven, Wesley Frank, and by Cate, James Lea.

AAC	Army Air Corps
AAF	Army Air Forces
Abort	Terminate mission on command or malfunction
ACS	Army Chief of Staff. Also, Air Support Command
AEAF	Allied Expeditionary Air Force
AFM	Air Force Museum, W-PAFB, Ohio
AJAX	VIII Fighter Command, Bushey Hall, England
AEAF	Allied Expeditionary Air Force
AMC	Air Materiel Command (was Air Materiel Division)
ASC	Air Service Command
ATC	Air Transport Command
ATSC	Air Technical Service Command
ARGUMENT	Joint 8AF and 15AF Air Offensive against Germany
AVALANCHE	Invasion of Italy at Salerno, September 1943
BC	Bomber Command
BG	Bomb Group
BHP	Brake Horsepower
Big Friend	Nominally, USAAF or Allied multi-engine bomber
Bodenplatte Luftwaffe	Operation BASEPLATE; 1 Jan. 1945
BOLERO	Buildup of AAF in the UK (Invasion plan)
BOLERO Oper.	Transatlantic delivery of AAF a/c by air
CAF	Chinese Air Force
C.A.M.	Cradle of Aviation Museum, Mitchel Field, N.Y.
CAP	Combat air patrol (more often Navy term)
CBI	China-Burma-India Theater of Operations
CBO	Combined Bomber Offensive against Germany
CG	Commanding General
Circus	Decoy bomber mission with fighter escort
CO	Commanding Officer
D-Day	Invasion Day, June 6, 1944
DFC	Distinguished Flying Cross
DSC	Distinguished Service Cross
DUC	Distinguished Unit Citation
e/a	Enemy aircraft
ETO	European Theater of Operations
FC	Fighter Command
FEAC	Far East Air Command
FEAF	Far East Air Force
Festung Europa	Fortress Europe
FRANTIC	Operation for shuttle bombing of Axis Europe
GAF	German Air Force (*Luftwaffe*)
GHQ AF	General Headquarter Air Force
HUSKY	Invasion of Sicily
IAS	Indicated airspeed
Jabo	*Jagdbomber* (Fighter-bomber)
Jagd	Fighter/pursuit
JG	*Jagdgeschwader* (GAF fighter group)
Luftwaffe	German Air Force
MAAF	Mediterranean Allied Air Force
MAC	Mediterranean Air Command
Mach	Mach No. (relative speed of a/c to speed of sound)
MATAF	Mediterranean Allied Tactical Air Force
MoS	Ministry of Supply (British)
MTO	Mediterranean Theater of Operations
NACA	National Advisory Committee for Aeronautics
NASM	National Air & Space Museum (Smithsonian Instit.)
NRA	National Recovery Act
OPM	Office of Production Management
OVERLORD	Invasion of Occupied Western Europe
PG(F)	1942 transition period Pursuit Gp.(Fighter)
PINETREE	Eighth Bomber Command HQ, High Wycombe, England
POINTBLANK	Combined Bomber Offensive against German industry
PS(F)	1942 transition period Pursuit Sqdn.(Fighter)
RAC	Republic Aviation Corporation
RAE	Royal Aeronautical Establishment
RAF	Royal Air Force
Ramrod	Bomber escort mission
RCAF	Royal Canadian Air Force

Rhubarb	Ground strafing mission
Rodeo	Fighter sweep, several squadrons
SEAC	Southeast Asia Command
Sortie	Operational flight of one aircraft denoted
Strafe	Low-level "shoot up" of any ground/water object
S-2	Intelligence
TAC	Tactical Air Command
TAF	Tactical Air Force
TAS	True airspeed
T-bolt	Short name for Thunderbolt (also Jug)
TDY	Temporary duty
T/O	Target of opportunity.
TORCH	Invasion of North/Northwestern Africa, Nov. 1942
UAC	United Aircraft Corporation
USSAFE>USSTAF	United States Strategic Air Forces in Europe
WEP	War Emergency Power
WIDEWING	Eighth Air Force HQ at Bushy Park (pre-Doolittle)
WPA	Works Progress Administration
WPB	War Production Board
ZI	Zone of the Interior (United States)

Col. Robert Baseler, 325 FG commander in North Africa and Italy, peels off in his P-47D-40-RA near Brownsville, Texas, in early postwar days. Thunderbolts became the ultimate JABOs.

WHAT REPUBLIC FARMINGDALE LOOKS LIKE NOW

Perhaps I wasn't listening to the recent words spoken by my friends who had worked at Republic Aviation's Farmingdale, N.Y., plant during the war years and, in a few cases, well into the Viet Nam war years. Or, more likely, maybe I didn't want to know the truth.

As we drove near Conklin Street I was shocked to see the old Fairchild-Ranger hangar and parking lot had become a recycling center. Weeds – some waist high – were abundant everywhere. Living, breathing people were conspicuous in their absence; well, after all, it was Sunday. If the many broken windows in the 1941 factory Building 17 were eyeball pollution, the adjacent Administration building condition would absolutely spoil your day. Rotting curtains and dilapidated venetian blinds filled gaping wounds where window glass had been. Across Conklin Street, the dirty red brick factory building (No. 5) that had been constructed for Fairchild Aircraft in the 1920s and had been owned by Seversky/Republic since the mid-1930's was a bit better off. Somebody had cared enough to at least board up all windows with chip board. (Although almost 20 years newer, the administration building is open to the elements.)

Republic Airport is, in contrast, a viable commercial airport, accommodating some non-scheduled airline operations and several fixed-base operators. The latter operate from the three former Republic Aviation hangers. Nearly 1,000,000 square feet of factory area is slowly disintegrating, aided by unrestricted vandalism of course.

Over 50 years ago, that mostly modern factory began to produce the first of more than 9,000 P-47 Thunderbolt fighters. That those fighters played a major part in defeating the Axis powers is recorded in oral, written and pictorial history formats. The sad fact of life is millions of New Yorkers – and, of course, others – either know nothing of this or fail to realize the significance of what occurred in that factory complex, beginning in the worst of the Great Depression days before 1935. In many cases, they could probably care less.

For those of us who watched the aviation industry grow from the ashes of the stock market crash of 1929 into an Arsenal of Democracy – the likes of which will probably never be equalled in pure size and energy – and who participated (and had relatives involved) in World War II, the Farmingdale "disaster" is a gaping wound in our remembrances. It all seems to be a part of our "throw away" mentality in the second half of this century.

There are many strange contrasts in America. Out in the flat farmland country of Indiana, just to the north of Evansville, a large factory building and airport built in 1942 exclusively for production of P-47 Thunderbolts is nicely maintained and is obviously a viable manufacturing facility. Except for the paint colors, it might be 1942. The present occupant – Whirlpool – is a manufacturer of major home appliances.

Lockheed – its name structure having changed many times over the years to reflect a transformative business – has all but abandoned its original home in Burbank, California, and most recently moved corporate offices to the rolling hills of Calabasas (California) and transferred its famed "Skunk Works" to much more modern facilities in Palmdale, near Edwards AFB. Mass production facilities for airframe production have been centered at the huge WWII bomber plant in Marietta, Georgia. Lockheed Corporation has also purchased the entire General Dynamics facility at Fort Worth, Texas, as the aerospace industry begins to shrink. The Marietta factory has been well maintained for a half century. In recent years,the Burbank plant, largely constructed in the early 1940s, has been leveled. All of that real estate has become a valuable asset. And so it goes.

The best thing that could've happened back at Farmingdale is to have the structures leveled and materials recycled. This would present an active airport that could easily attract new business. Erection of a suitable P-47 monument could then preside as a permanent symbol by which we pay homage to those military and civilian personnel who contributed so much to restoring sanity to the world by defeating Germany, Italy and Japan.

Big
"STUD"